UNIFIED ANALYSIS
AND SOLUTIONS OF
HEAT AND MASS DIFFUSION

UNIFIED ANALYSIS AND SOLUTIONS OF HEAT AND MASS DIFFUSION

M. D. MIKHAILOV

Technical University
Sofia, Bulgaria

M. N. ÖZIŞIK

North Carolina State University
Raleigh, North Carolina

DOVER PUBLICATIONS, INC.

NEW YORK

Bibliographical Note

This Dover edition, first published in 1994, is an unabridged, corrected republication of the work first published by John Wiley & Sons, New York, 1984 (as "A Wiley-Interscience Publication").

Library of Congress Cataloging-in-Publication Data

Mikhaïlov, M. D. (Mikhail Dimitrov)
    Unified analysis and solutions of heat and mass diffusion / M. D. Mikhaïlov, M. N. Özişik.
        p.   cm.
    Includes bibliographical references and index.
    ISBN 0-486-67876-8 (pbk.)
    1. Thermal diffusivity.   2. Mass transfer.   I. Özişik, M. Necati.   II. Title.
QC321.7.T55M54   1993
530.4'75—dc20   93-34237
                                                                                      CIP

Manufactured in the United States of America
Dover Publications, Inc., 31 East 2nd Street, Mineola, N.Y. 11501

PREFACE

A vast amount of work at almost any level of mathematical and physical complexity and sophistication is available in the literature on the subject of heat and mass diffusion. In recent years, with the availability of large capacity high-speed digital computers and improved numerical schemes, significant advances have been made towards the solution of the more complicated types of these problems by purely numerical finite difference or finite element means. Despite such advances in numerical solution techniques, however, there still exists a vast class of problems of practical importance that can be studied by analytical approaches. Furthermore, analytical solutions, when available, are advantageous in that they provide a good insight into the significance of various system parameters affecting the transport phenomena, as well as accurate benchmarks for the numerical approach. Therefore, the literature on the analytic treatment of heat and mass diffusion problems is still growing, and the number and types of problems to be solved remains large because of the varied combination of physical parameters that governs the conservation equations, the boundary conditions, and the couplings between them.

A scrutiny of the analytical work available in the literature reveals that too often solutions are developed to provide answers only to specific problems. There is a need for generalized analytical solutions from which answers to a vast number of specific situations can be obtained as special cases. Therefore, the objective of this work is to attempt to fill this gap in the literature by providing some unification of the solutions of linear heat and mass diffusion problems. The field is so vast and the possibilities are so immense that we have decided it is best to approach this unification by taking, at first, a limited number of steps. Accordingly, we have chosen in this book to study only seven different classes of time dependent, linear heat, and mass diffusion problems. Such a choice, although somewhat arbitrary, covers a vast number of cases of practical interest as is apparent in the text.

The classes of problems considered in this book cover the following areas:

Class I problems include the steady and unsteady heat or mass diffusion (a single specie) in solids subject to generalized boundary and initial conditions.

Class II problems are the generalization of Class I problems to heat or mass diffusion in a finite, composite medium consisting of *n* subregions, subject to generalized boundary and initial conditions. Even though the Class I and II problems can be combined, we prefer for practical reasons, to treat them separately.

Class III problems are characterized by two nonhomogeneous time-dependent diffusion equations in which the coupling through the boundary conditions is more general than those for the Class II problems. The applications of *Class III* problems include, for example, heat transfer in concurrent-flow double-pipe heat exchangers, simultaneous heat and mass transfer in internal flow in a duct whose walls are coated with sublimated material, concurrent mass transfer between a fluid and a gas flowing separately in a parallel-plate channel, and numerous others.

Class IV problems are characterized by a set of diffusion equations in which the temperature or the mass concentration in every point in space is coupled through source-sink terms in the equations. Such problems occur, for example, in mass diffusion processes involving several chemically reacting components.

Class V problems are governed by two diffusion equations that are coupled through the source-sink terms in every point in space; in contrast to the Class IV problems, there is no symmetry between the coefficients that govern the coupling. Applications of this class of problems include, for example, heat transfer in fluid–solid mixtures flowing in the turbulent flow inside a pipe and diffusion in tubular reactors.

Class VI problems are also characterized by a set of diffusion equations, but their couplings through the boundary conditions are complicated by the fact that a gross mass balance is involved along the boundary surface. Applications include heating of bodies in a limited volume of well-stirred fluid, mass diffusion into a body from a limited volume of well-stirred fluid, flow development in the hydrodynamic entrance region of a conduit, and many others.

Class VII problems deal with a situation in which the diffusing substance is finite, while during the diffusion process some of the diffusing substance is absorbed by another with which it reacts chemically. For example, in diffusion into a textile fiber in which there are a number of active groups, the diffusing molecules can become attached to them, hence are immobilized. In diffusion within the pores of a solid body some of the diffusing substance may be absorbed in the pores, hence are considered immobilized. The diffusion of oxygen and carbon monoxide through the outer membrane of red blood corpuscles is accompanied by diffusion and chemical reaction in the case of concentrated hemoglobin.

We use the integral transform technique throughout the development of the general solutions in order to present a unified method of analysis for the solution of all seven different classes of problems. In order to separate the general theory from the applications, we devoted the first four chapters to the derivation of the general solutions. The remaining ten chapters deal with applications. The organization of the material in the book is as follows.

Chapter 1 deals with the definition and mathematical formulation of the different classes of problems. *Chapter 2* is devoted to the development of generalized time dependent three-dimensional formal solutions for each of these classes and *Chapter 3* gives the eigenvalue problems associated with them. The mathematical developments presented in Chapters 2 and 3 may be of more interest to the mathematically oriented reader, but the results can be easily followed by readers who are primarily interested only in the applications. *Chapter 4* summarizes, for ready reference, the one-dimensional time dependent form of the general solutions for each class of problem. The material in Chapters 5–14 constitutes a systematic illustration of the application of the general results to the solution of specific heat and mass diffusion problems in the following order.

Chapters 5–8 are devoted to the application of Class I results to the solution of various heat and mass diffusion problems. In particular, the one-dimensional and multidimensional steady-state diffusion in rectangular, cylindrical, and spherical coordinates is the subject of Chapter 5. The problem of heat flow through fins of varying cross-section, for a variety of fin geometries, is the main topic of Chapter 6. The solution of unsteady heat or mass diffusion problems in the rectangular, cylindrical, and spherical coordinate systems, subject to different types of boundary conditions is illustrated and transient temperature and heat flow charts are presented in Chapter 7. Thermal entry region heat transfer problems of forced convection for flow inside ducts having geometries such as a parallel plate or a circular cylinder are treated in Chapter 8.

Chapter 9 deals with the application of the Class II solution to transient diffusion in composite layers. *Chapter 10* is devoted to the application of Class III solutions to the analysis of heat and mass diffusion in a capillary, porous body governed by the Luikov system of equations, and to heat transfer in entrance concurrent flow. *Chapter 11* deals with the application of Class IV problems to the solution of one-dimensional transient diffusion in heterogeneous media. *Chapter 12* illustrates the application of Class V problems to the analysis of concentration and temperature distribution in chemically reacting flow inside conduits. *Chapter 13* gives the application of Class VI problems to the analysis of (1) heating of bodies in a limited volume of well-stirred fluid, (2) mass diffusion into a body from a limited volume of well-stirred fluid, and (3) flow development in the hydrodynamic entrance region of ducts. Finally, *Chapter 14* presents the application of the Class VII solutions to diffusion accompanied by a first order, reversible chemical reaction.

If the reader wishes to develop analytic solutions to a specific problem which is not covered in the illustrations presented in Chapters 5–14, the following procedure is suggested. First, the class of the problem is established and the correspondence between the general and specific problems is determined by comparing the governing differential equations and their boundary conditions. By utilizing the results of this correspondence regarding the values of various coefficients, the solution to the specific problem is obtained by proper simplification of the general solution. Thus the general solutions developed in this book for each of the seven different classes of heat and mass diffusion problems are expected to be of interest to scientists, researchers, practicing engineers, and graduate students in the fields of mechanical, chemical, and civil engineering, as well as biological and biomedical research. The book can also serve as a graduate level textbook on heat and mass diffusion theory in engineering schools.

As a closing note, we wish to state that this book is the outcome of an equally divided effort by the two authors. Therefore, instead of flipping a coin to determine the order of our names on the cover, we have chosen to list them alphabetically. Regarding the initiation of this project, we both are indebted to the late Professor A. V. Luikov. In 1974, Professor Luikov contacted us regarding the writing of a two-volume book, on the theory of heat diffusion (MNO) and on convective heat and mass transfer (MDM). Unfortunately, both projects terminated without even a start because of the untimely loss of Professor Luikov in the same year.

The present project was initiated between the two authors in 1976, and it required six years of continuous correspondence, enhanced from time to time by personal contacts, and several iterations on the manuscript before its completion. Therefore, we dedicate this work to the memory of Professor Luikov who contributed so much to the advancement of the art and science of heat and mass transfer.

M. D. Mikhailov   **M. N. Özışık**
Sofia, Bulgaria   *Raleigh, North Carolina*

June 1983

CONTENTS

CHAPTER ONE

Basic Relations

There is one basic cause for all effects.

Giordano Bruno
1548?–1600

To explain all effects of heat and mass transfer processes, one needs mathematical models. Such models are developed by utilizing the fundamental principles of physics—the conservation principles. In this chapter, the basic laws of heat and mass diffusion are introduced, the conservation equations are derived, the pertinent boundary conditions are discussed, and seven different classes of generalized boundary value problems of heat and mass diffusion, the subject of analysis in this book, are stated. For convenience and reference in the subsequent chapters, the transformation from the rectangular coordinate system to an orthogonal coordinate system is presented, and the reduction of three-dimensional problems to the two- and one-dimensional ones is discussed.

1.1 DERIVATION OF THE BALANCE EQUATION

We consider the transfer of some *substance* (i.e., energy, mass, etc.) in a *material volume* V moving in space with a velocity v and deforming while in motion. Let S be the surface of this material volume V and ϵ denote *the amount of this substance contained per unit volume*; ϵ is referred to as the *substance density*. We assume that the medium is physically continuous or is a continuum; that is, properties such as density and velocity vary in a continuous manner in space or they are smooth, continuous functions of the space variables.

We introduce a vector quantity $\mathbf{J}$, called the *substance-flux vector*, to denote substance flow at a spatial position $\mathbf{x}$, at any instant t. The magnitude of this substance-flux vector is equal to the quantity of substance crossing a unit area, normal to the direction of substance flow, at the position considered, per unit time.

The fundamental principle of conservation for the balance of the substance within the finite volume element V, at any given instant, can be stated as

$$\begin{pmatrix} \text{Rate of subs-} \\ \text{tance storage in} \\ V \end{pmatrix} = \begin{pmatrix} \text{Rate of substance en-} \\ \text{tering } V \text{ through its} \\ \text{bounding surfaces} \end{pmatrix} + \begin{pmatrix} \text{Rate of substance gen-} \\ \text{eration in } V \end{pmatrix}$$

(1.1)

Various terms in this relation are now evaluated.

The storage term is determined by taking the *substantial derivative* (or the *total derivative*) of the substance contained in the material volume V, namely,

$$(\text{Rate of substance storage in } V) = \frac{D}{Dt} \int_V \epsilon \, dV \qquad (1.2)$$

where D/Dt denotes the substantial derivative. By making use of the *Reynolds transport theorem* [1], we can transform the substantial derivative of a volume integral into the form

$$\frac{D}{Dt} \int_V \epsilon \, dV = \int_V \frac{\partial \epsilon}{\partial t} \, dV + \int_S \epsilon \mathbf{v} \cdot \mathbf{n} \, dS \qquad (1.3)$$

where $\mathbf{n}$ is the outward-drawn normal unit direction vector. Now using the *divergence theorem* [2] we write

$$\int_S \epsilon \mathbf{v} \cdot \mathbf{n} \, dS = \int_V \nabla \cdot (\epsilon \mathbf{v}) \, dV \qquad (1.4)$$

Introducing Equations (1.3) and (1.4) into Equation (1.2) we obtain

$$(\text{Rate of substance storage in } V) = \int_V \left[\frac{\partial \epsilon}{\partial t} + \nabla \cdot (\epsilon \mathbf{v}) \right] dV \qquad (1.5)$$

The substance entering V through an elemental area dS on the boundary surface is $-\mathbf{J} \cdot \mathbf{n}$, where $\mathbf{J}$ is the substance-flux vector at dS, $\mathbf{n}$ is the outward-drawn normal unit direction vector, and the minus sign is used to indicate that the substance flows into the volume V. Then the substance entering V through the entire boundary S is given by

$$\begin{pmatrix} \text{Rate of substance entering} \\ V \text{ through its bounding} \\ \text{surfaces} \end{pmatrix} = -\int_S \mathbf{J} \cdot \mathbf{n} \, dS = -\int_V \nabla \cdot \mathbf{J} \, dV \quad (1.6)$$

If I denotes the *production* (or *generation*) *rate of the substance per unit volume*, then

$$\text{(Rate of substance generation in } V) = \int_V I \, dV \qquad (1.7)$$

Substituting Equations (1.5), (1.6), and (1.7) into Equation (1.1) we obtain

$$\int_V \left[\frac{\partial \epsilon}{\partial t} + \nabla \cdot (\epsilon \mathbf{v}) + \nabla \cdot \mathbf{J} - I \right] dV = 0 \qquad (1.8)$$

Equation (1.8) is derived for an arbitrary finite volume V; this volume element may be chosen so small as to remove the integral. Then we obtain the *balance equation* as

$$\frac{\partial \epsilon}{\partial t} + \nabla \cdot (\epsilon \mathbf{v}) + \nabla \cdot \mathbf{J} = I \qquad (1.9)$$

Utilizing the relations

$$\nabla \cdot (\epsilon \mathbf{v}) = \epsilon \nabla \cdot \mathbf{v} + \mathbf{v} \cdot \nabla \epsilon \qquad (1.10)$$

and

$$\frac{D}{Dt} \equiv \frac{\partial}{\partial t} + \mathbf{v} \cdot \nabla \qquad (1.11)$$

the balance Equation (1.9) can be expressed in the alternative form as

$$\frac{D\epsilon}{Dt} + \epsilon \nabla \cdot \mathbf{v} + \nabla \cdot \mathbf{J} = I \qquad (1.12)$$

Here ϵ is the *substance density* (i.e., substance contained per unit volume of the material volume V), $\mathbf{J}$ is the *substance-flux vector*, and I is the *substance production rate per unit volume* of V of the substance considered.

If ρ is *the density of the material volume V*, we introduce a quantity ξ, called the *specific substance*, defined as

$$\xi = \frac{\epsilon}{\rho} \qquad (1.13)$$

Thus ξ represents *the substance per unit mass* of the material volume. Then in terms of ξ Equation (1.12) can be expressed in the alternative form as

$$\rho \frac{D\xi}{Dt} + \xi \left(\frac{D\rho}{Dt} + \rho \nabla \cdot \mathbf{v} \right) + \nabla \cdot \mathbf{J} = I \qquad (1.14)$$

Here the quantities I, $\mathbf{J}$, $\mathbf{v}$, ρ, ϵ, and ξ are in general functions of time and space coordinates.

1.2   EQUATION OF CONSERVATION OF MASS FOR A
MULTICOMPONENT SYSTEM

In the previous section we derived the balance equation for any given *substance* characterized by its *substance density* ϵ or its *specific substance* ξ. We now consider a system consisting of n components among which chemical reactions are possible. Let $\epsilon \equiv \rho_k$ ($k = 1, 2, \ldots, n$) denote the mass per unit volume of the kth component and ρ the *total density* of all components, namely,

$$\rho = \sum_{k=1}^{n} \rho_k \qquad (1.15)$$

If $\mathbf{v}_k$ is the velocity of the kth component with respect to stationary coordinate axes, then for the system consisting of n components; the local *center of mass* (or *barycentric* or *mass average*) *velocity* $\mathbf{v}$ is defined as [3]

$$\mathbf{v} = \frac{1}{\rho} \sum_{k=1}^{n} \rho_k \mathbf{v}_k \qquad (1.16)$$

Then the *diffusion mass flow* of the kth component relative to the barycentric motion is defined as

$$\mathbf{J}_k = \rho_k (\mathbf{v}_k - \mathbf{v}) \qquad (1.17)$$

Now summing up Equation (1.17) over all components ($k = 1, 2, \ldots, n$) and utilizing the results given by Equations (1.15) and (1.16), we find

$$\sum_{k=1}^{n} \mathbf{J}_k = 0 \qquad (1.18)$$

The general balance Equation (1.12) is now applied for the kth component in this multicomponent system; namely, by letting $\epsilon \equiv \rho_k$, $\mathbf{J} \equiv \mathbf{J}_k$, and $I \equiv I_k$, Equation (1.12) becomes

$$\frac{D\rho_k}{Dt} + \rho_k \nabla \cdot \mathbf{v} + \nabla \cdot \mathbf{J}_k = I_k \qquad (k = 1, 2, \ldots, n) \qquad (1.19)$$

where I_k is the production rate per unit volume of the kth component as a result of chemical reactions [3] and $\mathbf{J}_k$ is the mass-flux vector for the kth component. Since mass is conserved in each separate chemical reaction, we have

$$\sum_{k=1}^{n} I_k = 0 \qquad (1.20)$$

Finally, summing up Equations (1.19) over all components ($k = 1, 2, \ldots, n$)

utilizing Equations (1.15), (1.18), and (1.20), we obtain the *law of conservation of mass* for the entire system in the form

$$\frac{D\rho}{Dt} + \rho \nabla \cdot \mathbf{v} = 0 \qquad (1.21)$$

This law expresses the fact that the total mass in the material volume is conserved; namely, the total mass in the material volume can only change if the matter flows into (or out of) the volume element [3].

The law of conservation of mass given by Equation (1.21) can be utilized to simplify the balance Equation (1.14) to the form

$$\rho \frac{D\xi}{Dt} + \nabla \cdot \mathbf{J} = I \qquad (1.22)$$

If the velocity $\mathbf{v}$ of the material volume V is zero, the volume V is fixed in space. Then according to Equation (1.11), the substantial derivative reduces to the partial derivative with respect to time and Equation (1.22) simplifies to

$$\rho \frac{\partial \xi}{\partial t} + \nabla \cdot \mathbf{J} = I \qquad (1.23)$$

The physical significance of the *specific substance* ξ for the multicomponent mass transfer process considered here is envisioned better if we recall its definition $\xi = \epsilon / \rho$ given by Equation (1.13). For the kth component we set $\epsilon \equiv \rho_k$ and $\xi \equiv \xi_k$, and obtain $\xi_k = \rho_k / \rho$, which implies that ξ_k is the *mass fraction* of the kth component.

The foregoing Equations (1.22) and (1.23) contain the substance-flux vector $\mathbf{J}$ and they are not yet in useful form for obtaining the distribution of the substance unless the substance-flux vector $\mathbf{J}$ is replaced by the appropriate expression involving an appropriately chosen potential that controls the substance flow. This matter is discussed in the next section.

1.3   DIFFERENTIAL EQUATION OF LINEAR DIFFUSION

In every physical phenomenon the flow of the substance (i.e., energy, mass, etc.) can be related to an appropriately chosen *potential T* (i.e., temperature, concentration, etc.). The potential at any location in space is completely defined by its numerical value because it is a scalar quantity. When the potential distribution is not uniform throughout the space, experience has shown that a substance flow arises in the direction of decreasing potential. It is shown empirically that for a large class of irreversible phenomena and a wide range of experimental conditions, the substance flow is a linear function of the potential gradient. This observation forms the basis in establishing a linear relation between the substance flow and the potential gradient. For example,

Fourier's law of heat conduction gives a linear relationship between the heat flow and the temperature gradient; *Fick's law* for mass diffusion establishes a linear relationship between the mass flux and the concentration gradient [3].

Therefore, in the following analysis we transfer the balance Equation (1.22) into a useful form by replacing the *specific substance* ξ and the *substance-flux vector* **J** in terms of appropriately chosen potentials.

Let T be the appropriately chosen potential (i.e., temperature, concentration, etc.). Then the relationship between the substance-flux and the potential gradient, according to the *linear diffusion law*, is given as

$$\mathbf{J} = -k\nabla T \qquad (1.24)$$

where the proportionality factor k is a scalar quantity for *isotropic materials* (i.e., the property does not vary with direction). Inasmuch as the substance flow is in the direction of decreasing potential, the minus sign is included to make the substance flow a positive quantity when the substance-flux vector **J** points in the direction of decreasing potential.

The specific substance ξ can be expressed as a function of the potential T and the average pressure p

$$\xi = \xi(T, p) \qquad (1.25)$$

For constant average pressure, we utilize Equation (1.25) to obtain

$$\frac{D\xi}{Dt} = C_{\xi p}\frac{DT}{Dt} \qquad (1.26)$$

where $C_{\xi p}$ is the *constant pressure substance capacity per unit mass* defined as

$$C_{\xi p} \equiv \left(\frac{\partial \xi}{\partial T}\right)_p \qquad (1.27)$$

Substituting Equations (1.24) and (1.26) into Equation (1.22), we obtain the balance equation in the form

$$\rho C_{\xi p}\frac{DT}{Dt} - \nabla\cdot(k\nabla T) = I \qquad (1.28)$$

where $k \equiv k(\mathbf{x})$, $T \equiv T(\mathbf{x}, t)$.

If the material volume V is fixed in space (i.e., $\mathbf{v} = 0$), the specific substance ξ can be expressed as a function of the potential T and density ρ

$$\xi \equiv \xi(T, \rho) \qquad (1.29)$$

We assume that Equations (1.25) and (1.29) hold everywhere in time and space; that is, we impose the assumption of local equilibrium. Inasmuch as the density ρ is constant, which is consistent with the assumption $\mathbf{v} = 0$, we utilize Equation (1.29) to obtain

$$\frac{\partial \xi}{\partial t} = c_{\xi V} \frac{\partial T}{\partial t} \qquad (1.30)$$

where $c_{\xi V}$ is the *constant volume substance capacity per unit mass* defined as

$$c_{\xi V} \equiv \left(\frac{\partial \xi}{\partial T} \right)_{\rho} \qquad (1.31)$$

For solids we have $c_{\xi p} \simeq c_{\xi V}$.

Now, substituting Equations (1.24) and (1.30) into Equation (1.23), we obtain

$$\rho c_{\xi V} \frac{\partial T}{\partial t} - \nabla \cdot (k \nabla T) = I \qquad (1.32)$$

where $k \equiv k(\mathbf{x})$ and $T \equiv T(\mathbf{x}, t)$. The production term I, in general, depends on the potential. We assume a linear relation in the form

$$I = P(\mathbf{x}, t) - d(\mathbf{x}) T(\mathbf{x}, t) \qquad (1.33)$$

where $P(\mathbf{x}, t)$ is the *source or sink term*. Then Equation (1.32) becomes

$$w(\mathbf{x}) \frac{\partial T(\mathbf{x}, t)}{\partial t} + LT(\mathbf{x}, t) = P(\mathbf{x}, t) \qquad (1.34)$$

where we defined

$$w(\mathbf{x}) \equiv \rho c_{\xi V} \qquad (1.35)$$

$$L \equiv - \nabla \cdot [k(\mathbf{x}) \nabla] + d(\mathbf{x}) \qquad (1.36)$$

and $\mathbf{x}$ is a point in V (i.e., $\mathbf{x} \in V$).

1.4   THE LINEAR BOUNDARY CONDITIONS

The differential Equation (1.34) will have numerous solutions unless initial and boundary conditions are prescribed. Here we consider only the linear boundary conditions that cover most cases of practical interest [4].

We assume that the initial distribution of the potential $T(\mathbf{x}, t)$ is a prescribed function $f(\mathbf{x})$ in the region V; that is,

$$T(\mathbf{x}, 0) = f(\mathbf{x}), \qquad \mathbf{x} \in V \qquad (1.37)$$

For generality, we consider a boundary condition given in the form

$$BT(\mathbf{x}, t) = \phi(\mathbf{x}, t), \qquad \mathbf{x} \in S \qquad (1.38)$$

where B is the linear boundary condition operator defined as

$$B \equiv \alpha(\mathbf{x}) + \beta(\mathbf{x})k(\mathbf{x})\frac{\partial}{\partial \mathbf{n}} \qquad (1.39a)$$

and $\partial/\partial \mathbf{n}$ is the normal derivative at the boundary surface S in the outward direction, defined as

$$\frac{\partial}{\partial \mathbf{n}} \equiv \ell_x\frac{\partial}{\partial x} + \ell_y\frac{\partial}{\partial y} + \ell_z\frac{\partial}{\partial z} \qquad (1.39b)$$

with ℓ_x, ℓ_y, and ℓ_z being the direction cosines.

Here $\alpha(\mathbf{x})$, $\beta(\mathbf{x})$ are prescribed boundary coefficients defined on the boundary surface S; the function $\phi(\mathbf{x}, t)$ which is the nonhomogeneous part of the boundary condition (1.38) is a prescribed function which, in general, may be a function of both position and time. The coefficient $k(\mathbf{x})$ is associated with the definition of the linear diffusion law given by Equation (1.24).

The general boundary conditions defined by Equations (1.38) and (1.39) for the case of $\alpha(\mathbf{x}) \neq 0$ and $\beta(\mathbf{x}) \neq 0$ are referred to as the *boundary condition of the third kind*. If the nonhomogeneous term $\phi(\mathbf{x}, t)$ is zero, we have the *homogeneous boundary condition of the third kind*. In the case of heat transfer, the potential $T(\mathbf{x}, t)$ is the temperature; then the homogeneous boundary condition of the third kind represents heat transfer, according to Newton's law of cooling, from the boundary surface into an environment at zero temperature.

The special cases of the boundary conditions (1.38) and (1.39) include the following.

The case $\alpha(\mathbf{x}) = 0$, $\beta(\mathbf{x}) = 1$, $\mathbf{x} \in S$ is called the *boundary condition of the second kind*. This boundary condition is equivalent to that of prescribing the magnitude of the substance flux along the boundary. If the nonhomogeneous term $\phi(\mathbf{x}, t)$ is zero, we have the *homogeneous boundary condition of the second kind* which represents zero substance flux at the boundary surface.

The case $\alpha(\mathbf{x}) = 1$, $\beta(\mathbf{x}) = 0$, $\mathbf{x} \in S$ is called the *boundary condition of the first kind*, which represents the magnitude of the potential being prescribed at the boundary surface. If the nonhomogeneous term $\phi(\mathbf{x}, t)$ is zero, we have the *homogeneous boundary condition of the first kind*, which corresponds to zero potential at the boundary surface. A boundary surface kept at a constant potential T_0 also satisfies the homogeneous boundary condition of the first kind if the potential is measured in excess of T_0.

Coupled diffusion processes involve more complicated boundary conditions. The linear cases of such boundary conditions are discussed in the following section.

1.5   CLASSIFICATION OF LINEAR BOUNDARY VALUE PROBLEMS OF HEAT AND MASS DIFFUSION

In this book we are concerned with the analysis of a large variety of heat and mass diffusion problems that are encountered in numerous practical applications. In order to handle such problems with a systematic and unified approach, we classify the linear boundary value problems of heat and mass diffusion that are studied in this book into the following categories.

Class I   The problems in this class include the unsteady heat or mass diffusion in a finite region of arbitrary geometry characterized by Equation (1.34) subject to the initial and boundary conditions (1.37) and (1.38). The mathematical formulation of such problems is given as

$$w(\mathbf{x})\frac{\partial T(\mathbf{x}, t)}{\partial t} + LT(\mathbf{x}, t) = P(\mathbf{x}, t), \qquad \mathbf{x} \in V \qquad (1.40\text{a})$$

subject to the initial condition

$$T(\mathbf{x}, 0) = f(\mathbf{x}), \qquad \mathbf{x} \in V \qquad (1.40\text{b})$$

and the boundary conditions

$$BT(\mathbf{x}, t) = \phi(\mathbf{x}, t), \qquad \mathbf{x} \in S \qquad (1.40\text{c})$$

where the linear operators L and B are defined as

$$L \equiv -\nabla \cdot [k(\mathbf{x})\nabla] + d(\mathbf{x}) \qquad (1.40\text{d})$$

and

$$B \equiv \alpha(\mathbf{x}) + \beta(\mathbf{x})k(\mathbf{x})\frac{\partial}{\partial \mathbf{n}} \qquad (1.40\text{e})$$

Class II   The problems in this class include the unsteady heat or mass diffusion in a finite composite region V, bounded by a surface S, and subdivided into n finite subregions V_k ($k = 1, 2, \ldots, n$) [5]. The contact conductance $h_{kp}(\mathbf{x})$ between two adjacent subregions V_k and V_p is considered space dependent. For every subregion V_k the initial boundary value problem is obtainable from Equations (1.34), (1.37), and (1.38), but they are coupled through the boundary condition between two adjacent regions requiring special consideration. The mathematical formulation of the problems belonging to this class is given as

$$w_k(\mathbf{x})\frac{\partial T_k(\mathbf{x}, t)}{\partial t} + L_k T_k(\mathbf{x}, t) = P_k(\mathbf{x}, t), \qquad \mathbf{x} \in V_k \qquad (k = 1, 2, \ldots, n)$$

$$(1.41\text{a})$$

subject to the initial condition

$$T_k(\mathbf{x}, 0) = f_k(\mathbf{x}), \qquad \mathbf{x} \in V_k \qquad (k = 1, 2, \ldots, n) \qquad (1.41b)$$

and to the boundary conditions at the outer surface $\mathbf{x} \in S$ given as

$$B_k T_k(\mathbf{x}, t) = \phi_k(\mathbf{x}, t), \qquad \mathbf{x} \in S \qquad (k = 1, 2, \ldots, n) \qquad (1.41c)$$

where the linear operators L_k and B_k are defined as

$$L_k \equiv -\nabla \cdot \left[k_k(\mathbf{x}) \nabla \right] + d_k(\mathbf{x}) \qquad (1.41d)$$

and

$$B_k \equiv \alpha_k(\mathbf{x}) + \beta_k(\mathbf{x}) k_k(\mathbf{x}) \frac{\partial}{\partial \mathbf{n}} \qquad (1.41e)$$

Here $\partial/\partial \mathbf{n}$ represents the normal derivative at the outer surface S in region V_k in the outward direction.

The boundary condition at the interfaces $\mathbf{x} \in S_{kp}$ ($k \neq p$), is given as

$$k_k(\mathbf{x}) \frac{\partial T_k(\mathbf{x}, t)}{\partial \mathbf{n}_{kp}} = k_p(\mathbf{x}) \frac{\partial T_p(\mathbf{x}, t)}{\partial \mathbf{n}_{kp}} = h_{kp}(\mathbf{x}) \left[T_k(\mathbf{x}, t) - T_p(\mathbf{x}, t) \right]$$

$$(1.41f)$$

where $\partial/\partial \mathbf{n}_{kp}$ is the normal derivative at the interface S_{kp} between the regions V_k, V_p in the sense from V_k to V_p.

Class III The problems belonging to this class are characterized by two diffusion equations of the type given by Equation (1.34) coupled at the boundary, but the boundary conditions associated with this coupling are more general than those discussed previously, hence their solution requires special mathematical techniques [6]. There are numerous practical problems that can be characterized by such a system of equations: for example, the heat transfer in concurrent flow double pipe heat exchangers, simultaneous heat and mass transfer in internal flow in a duct whose walls are coated with a sublimable material, concurrent mass transfer between a fluid and a gas flowing separately in a parallel plate channel, and many others [7]. The mathematical formulation of the problems belonging to Class III is given as

$$w_k(\mathbf{x}) \frac{\partial T_k(\mathbf{x}, t)}{\partial t} + L_k T_k(\mathbf{x}, t) = P_k(\mathbf{x}, t), \qquad \mathbf{x} \in V \qquad (k = 1, 2)$$

$$(1.42a)$$

subject to the initial condition

$$T_k(\mathbf{x}, 0) = f_k(\mathbf{x}), \qquad \mathbf{x} \in V \qquad (k = 1, 2) \qquad (1.42\text{b})$$

and the boundary conditions

$$\alpha_{k1} T_1(\mathbf{x}, t) + \alpha_{k2} T_2(\mathbf{x}, t) + \beta_{k1} k_1(\mathbf{x}) \frac{\partial T_1(\mathbf{x}, t)}{\partial \mathbf{n}}$$

$$+ \beta_{k2} k_2(\mathbf{x}) \frac{\partial T_2(\mathbf{x}, t)}{\partial \mathbf{n}} = \phi_k(\mathbf{x}, t), \qquad \mathbf{x} \in S_1 \qquad (k = 1, 2) \quad (1.42\text{c})$$

$$B_k T_k(\mathbf{x}, t) = \phi_k(\mathbf{x}, t), \qquad \mathbf{x} \in S_2 \qquad (k = 1, 2) \qquad (1.42\text{d})$$

where the operators L_k and B_k are defined by Equations (1.41d) and (1.41e), respectively.

Class IV The problems in this class are characterized by a set of diffusion equations in which the potentials $T_k(\mathbf{x}, t)$, $k = 1, 2, 3, \ldots, n$, in every point in space $\mathbf{x} \in V$ are coupled through source-sink terms. Such a mathematical model is introduced in [8] to describe the components' temperatures of a heterogeneous medium (concrete, soil, water and/or oil-saturated sand layers, etc.). Some diffusion processes at the presence of chemical reactions are coupled through source-sink terms as well. The mathematical formulation of the problems in this class is given as [14]

$$w_k(\mathbf{x}) \frac{\partial T_k(\mathbf{x}, t)}{\partial t} + L_k T_k(\mathbf{x}, t)$$

$$= P_k(\mathbf{x}, t) + b(\mathbf{x}) \sum_{p=1}^{n} \alpha_{kp} \left[T_p(\mathbf{x}, t) - T_k(\mathbf{x}, t) \right],$$

$$\mathbf{x} \in V, \qquad \alpha_{kp} = \alpha_{pk} \qquad (k = 1, 2, \ldots, n) \quad (1.43\text{a})$$

with the initial condition

$$T_k(\mathbf{x}, 0) = f_k(\mathbf{x}), \qquad \mathbf{x} \in V \qquad (k = 1, 2, \ldots, n) \qquad (1.43\text{b})$$

and the boundary conditions

$$B_k T_k(\mathbf{x}, t) = \phi_k(\mathbf{x}, t), \qquad \mathbf{x} \in S \qquad (k = 1, 2, \ldots, n) \qquad (1.43\text{c})$$

where the operators L_k and B_k are defined by Equations (1.41d) and (1.41e), respectively.

Class V In the problems belonging to Class IV, we assumed symmetry relation between the conductances α_{kp} (i.e., $\alpha_{kp} = \alpha_{pk}$) as given by Equations (1.43a). However, in some diffusion problems such as the heat transfer in fluid–solid mixture flowing in turbulent flow inside a pipe [9] or diffusion in a tubular reactor [10], there is no symmetry between conductances α_{kp}, namely, $\alpha_{kp} \neq \alpha_{pk}$. Therefore, in Class V we consider the problems with $\alpha_{kp} \neq \alpha_{pk}$ for the case $n = 2$; the mathematical formulation of these problems is given as [11]

$$w_k(\mathbf{x}) \frac{\partial T_k(\mathbf{x}, t)}{\partial t} + L_k T_k(\mathbf{x}, t)$$

$$= P_k(\mathbf{x}, t) + (-1)^k b(\mathbf{x}) \left[\sigma_{k-1} T_1(\mathbf{x}, t) - \sigma_{k+1} T_2(\mathbf{x}, t) \right],$$

$$\mathbf{x} \in V \quad (k = 1, 2) \quad (1.44a)$$

with the initial and boundary conditions

$$T_k(\mathbf{x}, 0) = f_k(\mathbf{x}), \quad \mathbf{x} \in V \quad (k = 1, 2) \quad\quad\quad (1.44b)$$

$$B_k T_k(\mathbf{x}, t) = \phi_k(\mathbf{x}, t), \quad \mathbf{x} \in S \quad (k = 1, 2) \quad\quad\quad (1.44c)$$

where the operators L_k and B_k are defined by Equations (1.41d) and (1.41e), respectively.

Class VI The problems in this class are characterized by a set of diffusion equations of the type given by Equation (1.34) for $d(\mathbf{x}) = 0$, but their coupling through the boundary conditions is complicated by the fact that a gross substance balance is involved along the boundary surface. There are many applications of the problems belonging to this class. In the process of extracting substance from a porous solid medium, one can consider the extraction of sugar from sugar beet turnings, the extraction of plant oil from seeds, and many others. In the production of mineral salts in chemical industry, one can consider the extraction of the soluble components from the row mineral by means of water or water salts. An analogous process in extractive metallurgy involves the use of weak solutions of acids and salts for the dissolution purposes. All these processes have one common mechanism, that is, the mass transfer from the inside of the porous body by a diffusion process described by Equation (1.34) for $d(\mathbf{x}) = 0$. The boundary condition for this problem is complicated by the fact that the accumulation of the substance in the extracting liquid and the separation of the substance from the porous particles are coupled. The desired boundary condition for such a process can be developed from the boundary condition (1.38) by proper modification of the nonhomogeneous term $\phi(t)$. Namely, the nonhomogeneous term $\phi(t)$ now is an unknown function; it should be related to the variation of the concentration of the surrounding liquid through a mass balance. Mathematical formulation of the problem belonging to Class VI is given as [12]

$$w_k(\mathbf{x})\frac{\partial T_k(\mathbf{x},t)}{\partial t} + L_k T_k(\mathbf{x},t) = P_k(\mathbf{x},t), \qquad \mathbf{x} \in V_k \qquad (k = 1,2,\ldots,n)$$

$$(1.45a)$$

with the boundary and initial conditions

$$B_k T_k(\mathbf{x},t) = \phi(t), \qquad \mathbf{x} \in S_k \qquad (k = 1,2,\ldots,n) \qquad (1.45b)$$

$$\frac{d\phi(t)}{dt} + \sum_{k=1}^{n} \gamma_k \int_{S_k} k_k(\mathbf{x})\frac{\partial T_k(\mathbf{x},t)}{\partial \mathbf{n}} dS = Q(t) \qquad (1.45c)$$

$$T_k(\mathbf{x},0) = f_k(\mathbf{x}), \qquad \mathbf{x} \in V_k \qquad (k = 1,2,\ldots,n) \qquad (1.45d)$$

$$\phi(0) = \phi_0 \qquad (1.45e)$$

where the operators L_k and B_k are defined as

$$L_k \equiv -\nabla \cdot [k_k(\mathbf{x})\nabla] \qquad (1.45f)$$

$$B_k \equiv \alpha_k + \beta_k(\mathbf{x})k_k(\mathbf{x})\frac{\partial}{\partial \mathbf{n}} \qquad (1.45g)$$

Here $Q(t)$ is the substance source in surrounding liquid and α_k is the constant coefficient.

Class VII The general problem in this class can be conveniently stated in terms of a *solute* diffusion from a well-stirred limited volume of *solution* into a material region of arbitrary geometry. The solution being well stirred, the concentration of the solute in the solution depends only on time, and is determined by a mass balance at the boundaries of the material region; namely, the rate at which solute leaves the solution should be equal to that at which it enters the material region through its boundary surfaces. As solute diffuses through the material region, a first-order, reversible reaction takes place and, as a result, a nondiffusing product is formed at any point at a rate proportional to the concentration of the solute free to diffuse. The product formed then disappears at a rate proportional to its own concentration [13]. The mathematical formulation of this problem is given as [15]

$$w(\mathbf{x})\left[\frac{\partial T_1(\mathbf{x},t)}{\partial t} + \frac{\partial T_2(\mathbf{x},t)}{\partial t}\right] + LT_1(\mathbf{x},t) = P(\mathbf{x},t), \qquad \mathbf{x} \in V$$

$$(1.46a)$$

$$\frac{\partial T_2(\mathbf{x},t)}{\partial t} = \sigma_1 T_1(\mathbf{x},t) - \sigma_2 T_2(\mathbf{x},t), \qquad \mathbf{x} \in V \qquad (1.46b)$$

with the boundary and initial conditions

$$BT_1(\mathbf{x}, t) = \phi(t), \qquad \mathbf{x} \in S \tag{1.46c}$$

$$\frac{d\phi(t)}{dt} + \gamma \int_S k(\mathbf{x}) \frac{\partial T_1(\mathbf{x}, t)}{\partial \mathbf{n}} dS = 0 \tag{1.46d}$$

$$T_k(\mathbf{x}, 0) = f_k(\mathbf{x}), \qquad \mathbf{x} \in V \qquad (k = 1, 2) \tag{1.46e}$$

$$\phi(0) = \phi_0 \tag{1.46f}$$

where the operators L and B are defined as

$$L \equiv -\nabla \cdot [k(\mathbf{x})\nabla] \tag{1.46g}$$

$$B \equiv \alpha + \beta(\mathbf{x})k(\mathbf{x})\frac{\partial}{\partial \mathbf{n}} \tag{1.46h}$$

Here α is a constant coefficient.

1.6 TRANSFORMATION OF COORDINATES

In the mathematical formulation of the various boundary value problems previously discussed, the governing equations are presented without a reference to a particular coordinate system. When these equations are to be solved for a medium having a specific configuration, the coordinate system chosen should be such that it should fit to the boundary surfaces of the region under consideration. For example, rectangular bodies require a rectangular coordinate system, a circular cylinder requires a cylindrical coordinate system, and so on. Therefore, we present a brief discussion of the transformation from the rectangular coordinate system to an orthogonal coordinate system.

Consider a Cartesian coordinate system (x_1, x_2, x_3) and an orthogonal curvilinear coordinate system (u_1, u_2, u_3). A differential length ds in the Cartesian coordinate system is given as

$$(ds)^2 = (dx_1)^2 + (dx_2)^2 + (dx_3)^2 \tag{1.47}$$

Let the functional relationship between the two systems be

$$x_i = x_i(u_1, u_2, u_3) \qquad (i = 1, 2, 3) \tag{1.48}$$

A differential length dx_i along the x_i axis is related to the coordinates (u_1, u_2, u_3) by

$$dx_i = \sum_{k=1}^3 \frac{\partial x_i}{\partial u_k} du_k \tag{1.49}$$

Substituting Equation (1.49) into Equation (1.47), and noting that the cross products must vanish when u_1, u_2, and u_3 are mutually orthogonal, yields

$$(ds)^2 = a_1^2(du_1)^2 + a_2^2(du_2)^2 + a_3^2(du_3)^2 \qquad (1.50)$$

where

$$a_k = \sum_{i=1}^{3}\left(\frac{\partial x_i}{\partial u_k}\right)^2 \qquad (k = 1,2,3) \qquad (1.51)$$

The coefficients a_k are called *scale factors*; they may be constants or functions of the coordinates. When the functional relationship between the coordinates of the Cartesian and the orthogonal curvilinear coordinate system is given, the scale factors a_k are evaluated using Equation (1.51). Once the scale factors are known, transformation of linear operator $\nabla \cdot [k(\mathbf{x})\nabla]$ is given by the following relationship

$$\nabla \cdot [k(\mathbf{x})\nabla] = \frac{1}{a_1 a_2 a_3}\sum_{k=1}^{3}\frac{\partial}{\partial u_k}\left[\frac{a_1 a_2 a_3}{a_k^2}k(u_1, u_2, u_3)\frac{\partial}{\partial u_k}\right] \qquad (1.52)$$

If $k(\mathbf{x})$ is uniform (i.e., independent of position), Equation (1.52) gives the Laplacian

$$\nabla^2 = \frac{1}{a_1 a_2 a_3}\sum_{k=1}^{3}\frac{\partial}{\partial u_k}\left(\frac{a_1 a_2 a_3}{a_k^2}\frac{\partial}{\partial u_k}\right) \qquad (1.53)$$

Example 1.1 We consider the transformation of the Laplacian operator from the rectangular coordinate system (x, y, z) to the cylindrical coordinate system (r, ϕ, z) shown in Figure 1.1.

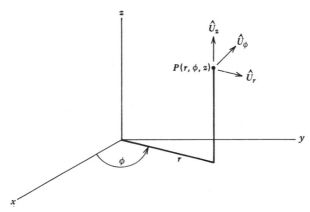

Figure 1.1 The cylindrical coordinate system.

The functional relationship between the coordinates is given by

$$x = r\cos\phi, \qquad y = r\sin\phi, \qquad z = z \qquad (1.54)$$

Let a_r, a_ϕ, a_z be the scale factors for the (r, ϕ, z) system. They are evaluated by using Equation (1.51) as follows:

$$a_r^2 = \left(\frac{\partial x}{\partial r}\right)^2 + \left(\frac{\partial y}{\partial r}\right)^2 + \left(\frac{\partial z}{\partial r}\right)^2 = \cos^2\phi + \sin^2\phi + 0 = 1$$

$$a_\phi^2 = \left(\frac{\partial x}{\partial \phi}\right)^2 + \left(\frac{\partial y}{\partial \phi}\right)^2 + \left(\frac{\partial z}{\partial \phi}\right)^2 = (-r\sin\phi)^2 + (r\cos\phi)^2 + 0 = r^2$$

$$a_z^2 = \left(\frac{\partial x}{\partial z}\right)^2 + \left(\frac{\partial y}{\partial z}\right)^2 + \left(\frac{\partial z}{\partial z}\right)^2 = 0 + 0 + 1 = 1$$

Hence the scale factors for the cylindrical coordinate system are

$$a_r = 1, \qquad a_\phi = r, \qquad a_z = 1 \qquad (1.55)$$

The Laplacian in the cylindrical coordinate system, from Equation (1.53), is given by

$$\nabla^2 = \frac{1}{r}\frac{\partial}{\partial r}\left(r\frac{\partial}{\partial r}\right) + \frac{1}{r^2}\frac{\partial^2}{\partial \phi^2} + \frac{\partial^2}{\partial z^2} \qquad (1.56)$$

Example 1.2 We now consider the transformation of the Laplacian operator from the rectangular coordinate system (x, y, z) to the spherical coordinate system (r, ϕ, θ) illustrated in Figure 1.2.

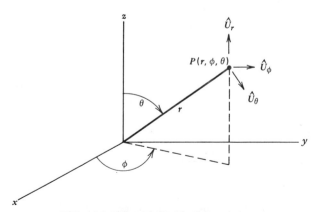

Figure 1.2 The spherical coordinate system.

The functional relationship between the coordinate systems is given as

$$x = r \sin \theta \cos \phi, \quad y = r \sin \theta \sin \phi, \qquad z = r \cos \theta \qquad (1.57)$$

Then the scale factors become

$$a_r = 1, \qquad a_\phi = r \sin \theta, \qquad a_\theta = r \qquad (1.58)$$

and the Laplacian in the spherical coordinate system takes the form

$$\nabla^2 = \frac{1}{r^2} \frac{\partial}{\partial r}\left(r^2 \frac{\partial}{\partial r} \right) + \frac{1}{r^2 \sin \theta} \frac{\partial}{\partial \theta}\left(\sin \theta \frac{\partial}{\partial \theta} \right) + \frac{1}{r^2 \sin^2 \theta} \frac{\partial^2}{\partial \phi^2} \qquad (1.59)$$

1.7   REDUCTION TO TWO- AND ONE-DIMENSIONAL FORMS

In Section 1.5 we presented the mathematical formulation of seven different classes of boundary value problems in the general three-dimensional domain. It is instructive to describe briefly the physical considerations leading to the reduction of the three-dimensional problem to the two- and one-dimensional ones.

Reduction to Two-Dimensional Form

Let us first consider a finite region bounded by a surface S_2 whose generating lines are parallel to the z axis, and by the surface S_0 at $z = z_0$ and S_1 at $z = z_1$ as illustrated in Figure 1.3.

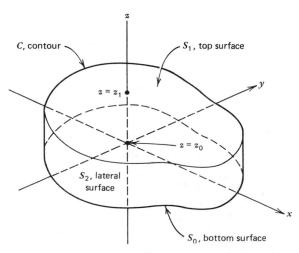

Figure 1.3   Nomenclature for reduction into two-dimensional form.

In order to reduce the general three-dimensional problem (1.40) to the two-dimensional form, we choose all parameters in this system to be independent of the coordinate z and set

$$\alpha(\mathbf{x}) = 0, \quad \beta(\mathbf{x}) = 1, \quad \phi(\mathbf{x}, t) = 0 \quad \text{on } \mathbf{x} \in S_0 \quad \text{and} \quad \mathbf{x} \in S_1 \quad (1.60)$$

The solution domain is the area $S_0 = S_1$ multiplied by $z_1 - z_0$ and the bounding surface is the bounding curve C multiplied by $z_1 - z_0$. The value of $z_1 - z_0$ can be arbitrary since the problem does not depend on z.

Then the problem (1.40) has the same form for the two-dimensional case but V *denotes the area* of the solution domain and S *denotes the bounding curve* of this domain.

Reduction to One-Dimensional Form

To illustrate the basic concept, we again consider the boundary value problem of Class I defined by Equations (1.40). In the three-dimensional rectangular coordinate system we have

$$\mathbf{x} = \mathbf{i}x + \mathbf{j}y + \mathbf{k}z, \qquad \mathbf{x} \in V \qquad (1.61a)$$

$$\nabla \equiv \mathbf{i}\frac{\partial}{\partial x} + \mathbf{j}\frac{\partial}{\partial y} + \mathbf{k}\frac{\partial}{\partial z}, \qquad \mathbf{x} \in V \qquad (1.61b)$$

Suppose we consider a finite region bounded by a surface $S = S_2$ whose generating lines are parallel to the x axis, and by the surface $S = S_0$ at $x = x_0$ and $S = S_1$ at $x = x_1$ as illustrated in Figure 1.4.

In order to reduce the general three-dimensional problem (1.40) to the one-dimensional form in the x variable, we choose the various parameters in this system as

$$
\begin{array}{llll}
k(\mathbf{x}) = k(x), & d(\mathbf{x}) = d(x), & P(\mathbf{x}, t) = P(x, t), & f(\mathbf{x}) = f(x) \\
\alpha(\mathbf{x}) = \alpha_0, & \beta(\mathbf{x}) = \beta_0, & \phi(\mathbf{x}, t) = \phi_0(t) & \text{on } \mathbf{x} \in S_0 \\
\alpha(\mathbf{x}) = \alpha_1, & \beta(\mathbf{x}) = \beta_1, & \phi(\mathbf{x}, t) = \phi_1(t) & \text{on } \mathbf{x} \in S_1 \\
\alpha(\mathbf{x}) = 0, & \beta(\mathbf{x}) = 1, & \phi(\mathbf{x}, t) = 0 & \text{on } \mathbf{x} \in S_2
\end{array}
$$

$$(1.62)$$

When the identities (1.62) are introduced into Equations (1.40), it follows that the variation of $T(\mathbf{x}, t)$ occurs only in the variables x and t; that is,

$$T(\mathbf{x}, t) = T(x, t) \qquad (1.63)$$

Then, $\partial/\partial y = \partial/\partial z = 0$ and the operator L defined by Equations (1.40d) becomes

$$L \equiv -\frac{\partial}{\partial x}\left[k(x)\frac{\partial}{\partial x}\right] + d(x) \quad \text{in } x_0 < x < x_1 \qquad (1.64a)$$

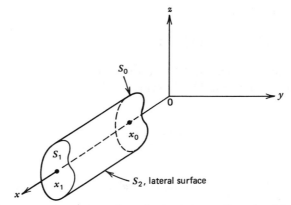

Figure 1.4 Nomenclature for reduction into one-dimensional form.

Similarly, the boundary condition operator B defined by Equation (1.40e) takes the form

$$B_0 \equiv \alpha_0 - \beta_0 k(x_0)\frac{\partial}{\partial x} \quad \text{at } x = x_0 \qquad (1.64b)$$

$$B_1 \equiv \alpha_1 + \beta_1 k(x_1)\frac{\partial}{\partial x} \quad \text{at } x = x_1 \qquad (1.64c)$$

The minus sign appearing in Equation (1.64b) is due to the fact that the outward-drawn normal at the boundary surface $x = x_0$ is in the negative x direction. Therefore, $\partial/\partial \mathbf{n} \equiv -\partial/\partial x$ at the boundary $x = x_0$.

The boundary conditions operators B_0 and B_1, given by Equations (1.64b, c) can be written more compactly as

$$B_k \equiv \alpha_k - (-1)^k \beta_k k(x_k)\frac{\partial}{\partial x} \quad \text{at } x = x_k, \qquad k = 0, 1 \quad (1.64d, e)$$

Thus the one-dimensional form of the Class I problem becomes

$$w(x)\frac{\partial T(x, t)}{\partial t} = \frac{\partial}{\partial x}\left[k(x)\frac{\partial T(x, t)}{\partial x}\right] - d(x)T(x, t) + P(x, t)$$

$$\text{in } x_0 < x < x_1, \qquad t > 0 \quad (1.65a)$$

subject to the boundary conditions

$$\alpha_k T(x_k, t) - (-1)^k \beta_k k(x_k)\frac{\partial T(x_k, t)}{\partial x} = \phi(x_k, t) \quad \text{at } x = x_k$$

$$(k = 0, 1) \qquad t > 0 \quad (1.65b, c)$$

and the initial condition

$$T(x,0) = f(x) \quad \text{in } x_0 \le x \le x_1, \qquad t = 0 \qquad (1.65d)$$

Other three-dimensional problems considered in this chapter can be reduced to the one-dimensional form by following similar procedures.

REFERENCES

1. S. Whitaker, *Elementary Heat Transfer Analysis*, Pergamon, New York, 1976.
2. G. A. Korn and T. M. Korn, *Mathematical Handbook for Scientists and Engineers*, McGraw-Hill, New York, 1961.
3. S. R. de Groot and P. Mazur, *Non-Equilibrium Thermodynamics*, North-Holland, Amsterdam, 1962.
4. M. N. Özişik, *Heat Conduction*, Wiley, New York, 1980.
5. Y. Yener and M. N. Özişik, On the Solution of Unsteady Heat Conduction in Multi-Region Media with Time Dependent Heat Transfer Coefficient, *Proc. 5th Int. Heat Transfer Conference*, Tokyo, Sept. 1974.
6. M. D. Mikhailov, General Solution of the Diffusion Equations Coupled at Boundary Conditions, *Int. J. Heat Mass Transf.*, 16, 2155–2164 (1973).
7. M. D. Mikhailov and B. K. Shishedjiev, Coupled at Boundary Mass or Heat Transfer in Entrance Concurrent Flow, *Int. J. Heat Mass Transf.*, 19, 553–557 (1976).
8. L. I. Rubinstein, *Temperature Fields in Oil Saturated Layers* (Russ.), Nedra Press, Moscow, 1972.
9. C. L. Tien, Heat Transfer by a Turbulently Flowing Fluid–Solids Mixture in a Pipe, *J. Heat Transf.*, 83C, 183–188 (1961).
10. P. Sibra, *Reversible Reaction and Diffusion in an Isothermal Tubular Reactor*, Master's Thesis, University of Texas, 1959.
11. M. D. Mikhailov, General Solutions of the Coupled Diffusion Equations, *Int. J. Eng. Sci.*, 11, 235–241 (1973).
12. M. D. Mikhailov, General Solution of Diffusion Processes in Solid–Liquid Extraction, *Int. J. Heat Mass Transf.*, 20, 1409–1415 (1977).
13. J. Crank, The Mathematics of Diffusion, 2nd ed., Clarendon Press, London, 1975.
14. M. D. Mikhailov, M. N. Özişik, and B. K. Shishedjiev, Diffusion in Heterogeneous Media, *J. Heat Transf.*, 104, 781–787 (1982).
15. M. D. Mikhailov and M. N. Özişik, A General Solution of Solute Diffusion with Reversible Reaction, *Int. J. Heat Mass Transf.*, 24, 81–87 (1981).

CHAPTER TWO

The General Solutions
of Seven Different Classes
of Heat and Mass Diffusion Problems

Knowledge is of two kinds. We know a subject ourselves,
or we know where we can find information upon it.

Samuel Johnson
1709–1784

A vast amount of published work exists in the field of heat and mass transfer and the literature is evergrowing. It is sometimes easier to solve and compute the results for a specific problem rather than going through a literature search to find out whether the problem has a published solution. Therefore, it is quite useful to have general solutions available for a number of sufficiently general classes of problems from which solutions to a variety of specific problems can be obtained as special cases. The objective of this chapter is to develop formal three-dimensional general solutions for each of the seven classes of problems described previously. To achieve this objective, the generalized integral transform technique is used to obtain formal analytic solutions for each of the seven different classes of boundary value problems of heat and mass diffusion defined earlier. The formal solutions developed in this chapter form the basis for the solution of specific problems considered in subsequent chapters.

The reader should consult References 1 and 2 for the fundamental theory of integral transform technique and its application to the solution of partial differential equations of physical problems in several other fields, References 4–6 for a summary of various integral transforms, and References 7–20 for the application of the integral transform technique in the solution of various heat and mass diffusion problems.

2.1   GENERAL SOLUTION OF PROBLEMS OF CLASS I

A general application of the finite integral transform technique to the solution of the problems of Class I is given in References 11 and 14. Here we present a discussion of this approach and further developments in the analysis to obtain a formal solution to the following boundary value problem of a finite region discussed in Chapter I.

$$w(\mathbf{x})\frac{\partial T(\mathbf{x}, t)}{\partial t} + LT(\mathbf{x}, t) = P(\mathbf{x}, t), \qquad \mathbf{x} \in V \qquad (2.1a)$$

subject to the boundary conditions

$$BT(\mathbf{x}, t) = \phi(\mathbf{x}, t), \qquad \mathbf{x} \in S \qquad (2.1b)$$

and the initial condition

$$T(\mathbf{x}, 0) = f(\mathbf{x}), \qquad \mathbf{x} \in V \qquad (2.1c)$$

where the linear operators are defined as

$$L \equiv -\nabla \cdot [k(\mathbf{x})\nabla] + d(\mathbf{x}) \qquad (2.1d)$$

$$B \equiv \alpha(\mathbf{x}) + \beta(\mathbf{x})k(\mathbf{x})\frac{\partial}{\partial \mathbf{n}} \qquad (2.1e)$$

Here $\alpha(\mathbf{x})$ and $\beta(\mathbf{x})$ are prescribed boundary coefficients defined on the boundary surface S, $k(\mathbf{x})$ is associated with the definition of the linear diffusion law given by Equation (1.24), and $\partial/\partial\mathbf{n}$ represents the normal derivative at the boundary surface S in the *outward* direction. Here the coefficient $k(\mathbf{x})$ is considered a continuous function of the space variable x. The case of discontinuous $k(\mathbf{x})$ can be treated as a Class II problem that is presented in the next section.

   In the following analysis we first develop the finite integral transform and the inversion formula needed for the solution of the preceding boundary value problem. The integral transform pair thus developed is then applied to solve the problem. In addition, the splitting up of the general problem into simpler problems is described, and the treatment of the special cases involving boundary condition of the second kind for all boundaries, as well as the periodically varying boundary conditions, is discussed.

Development of the Integral Transform Pair

The integral transform pair consisting of the finite integral transform and the inversion formula that is needed for the solution of the preceding problem can be developed by considering the following *auxiliary* homogeneous problems for the space variable function $\psi(\mathbf{x})$ in the same region V:

$$\mu^2 w(\mathbf{x})\psi(\mathbf{x}) = L\psi(\mathbf{x}), \qquad \mathbf{x} \in V \tag{2.2a}$$

subject to the homogeneous boundary conditions

$$B\psi(\mathbf{x}) = 0, \qquad \mathbf{x} \in S \tag{2.2b}$$

where the linear operators L and B are defined by Equations (2.1d) and (2.1e), respectively. The homogeneous problem of the type given by Equations (2.2) is called a *Sturm–Liouville problem (or system)* in honor of J. C. F. Sturm and J. Liouville who published the first extensive developments of the theory of such systems in 1836–1838.

System (2.2) has nontrivial solutions only for certain values of the parameter $\mu^2 \equiv \mu_i^2$ $(i = 1, 2, \ldots, \infty)$ called the *eigenvalues*, and the corresponding nontrivial solutions $\psi(\mu_i, \mathbf{x}) \equiv \psi_i(\mathbf{x})$ called the *eigenfunctions*.

We now consider the representation of the desired potential $T(\mathbf{x}, t)$ in terms of the eigenfunctions $\psi_i(\mathbf{x})$ of this auxiliary problem over the region V in the form

$$T(\mathbf{x}, t) = \sum_{i=1}^{\infty} A_i(t)\psi_i(\mathbf{x}), \qquad \mathbf{x} \in V \tag{2.3}$$

where the summation is taken over all discrete spectra of eigenvalues μ_i. To determine the expansion coefficients $A_i(t)$ we need the *orthogonality* property of these eigenfunctions which is developed as now described.

Equation (2.2a) is written for two different eigenfunctions $\psi_i(\mathbf{x})$ and $\psi_j(\mathbf{x})$, corresponding to two different eigenvalues μ_i and μ_j

$$\mu_i^2 w(\mathbf{x})\psi_i(\mathbf{x}) = L\psi_i(\mathbf{x}), \qquad \mathbf{x} \in V \tag{2.4a}$$

$$\mu_j^2 w(\mathbf{x})\psi_j(\mathbf{x}) = L\psi_j(\mathbf{x}), \qquad \mathbf{x} \in V \tag{2.4b}$$

The first equation is multiplied by $\psi_j(\mathbf{x})$, the second by $\psi_i(\mathbf{x})$, and the results are subtracted and then integrated over the region V

$$\left(\mu_i^2 - \mu_j^2\right)\int_V w(\mathbf{x})\psi_i(\mathbf{x})\psi_j(\mathbf{x})\, dv$$

$$= \int_V \left\{\psi_i(\mathbf{x})\nabla\cdot\left[k(\mathbf{x})\nabla\psi_j(\mathbf{x})\right] - \psi_j(\mathbf{x})\nabla\cdot\left[k(\mathbf{x})\nabla\psi_i(\mathbf{x})\right]\right\}\, dv \tag{2.5a}$$

The volume integral on the right-hand side is transformed to the surface integral by the Green theorem

$$\left(\mu_i^2 - \mu_j^2\right) \int_V w(\mathbf{x}) \psi_i(\mathbf{x}) \psi_j(\mathbf{x}) \, dv = \int_S k(\mathbf{x}) \left[\psi_i(\mathbf{x}) \frac{\partial}{\partial \mathbf{n}} \psi_j(\mathbf{x}) - \psi_j(\mathbf{x}) \frac{\partial}{\partial \mathbf{n}} \psi_i(\mathbf{x}) \right] ds$$

(2.5b)

This expression is now written more compactly in the form

$$(\psi_i, \psi_j) = \frac{1}{\mu_i^2 - \mu_j^2} \int_S k(\mathbf{x}) \begin{vmatrix} \psi_i(\mathbf{x}) & \dfrac{\partial}{\partial \mathbf{n}} \psi_i(\mathbf{x}) \\ \psi_j(\mathbf{x}) & \dfrac{\partial}{\partial \mathbf{n}} \psi_j(\mathbf{x}) \end{vmatrix} ds \quad \text{for } i \neq j \quad (2.6)$$

where we defined the *scalar product* of two functions as [21]

$$(f_1, f_2) \equiv \int_V w(\mathbf{x}) f_1(\mathbf{x}) f_2(\mathbf{x}) \, dv$$

(2.7)

For $\mu_j \rightarrow \mu_i$, the right-hand side of Equation (2.6) is evaluated by L'Hospital's rule and Equation (2.6) takes the form

$$N_i \equiv (\psi_i, \psi_i) = \frac{1}{2\mu_i} \int_S k(\mathbf{x}) \begin{vmatrix} \left(\dfrac{\partial \psi(\mathbf{x})}{\partial \mu}\right)_{\mu=\mu_i} & \left(\dfrac{\partial^2 \psi(\mathbf{x})}{\partial \mathbf{n}\, \partial \mu}\right)_{\mu=\mu_i} \\ \psi_i(\mathbf{x}) & \dfrac{\partial \psi_i(\mathbf{x})}{\partial \mathbf{n}} \end{vmatrix} ds \quad (2.8)$$

where $N_i^{1/2}$ is called the *norm* of the eigenfunction $\psi_i(\mathbf{x})$. In view of Equation (2.2b), the result in Equation (2.6) can be written more compactly in the form

$$(\psi_i, \psi_j) = \delta_{ij} N_i$$

(2.9)

where

$$\delta_{ij} \equiv \text{Kronecker delta} = \begin{cases} 0 & \text{for } i \neq j \\ 1 & \text{for } i = j \end{cases}$$

The expression given by Equation (2.9) proves that the eigenfunctions $\psi_i(\mathbf{x})$ of the Sturm–Liouville system (2.2) are orthogonal with respect to the weighting function $w(\mathbf{x})$ in the region V.

The foregoing orthogonality relation is now used to determine the expansion coefficients $A_i(t)$ in Equation (2.3) as now described. We operate on both sides of Equation (2.3) by the operator

$$\int_V w(\mathbf{x}) \psi_j(\mathbf{x}) \, dx$$

[i.e., we multiply both sides of Equation (2.3) by $w(\mathbf{x})\psi_j(\mathbf{x})$ and integrate over the region V] and utilize the orthogonality relation (2.9) to obtain

$$A_i(t) = \frac{1}{N_i} \int_V w(\mathbf{x}) \psi_i(\mathbf{x}) T(\mathbf{x}, t)\, dv \equiv \frac{(\psi_i, T)}{N_i} \qquad (2.10)$$

This expression for $A_i(t)$ is introduced into Equation (2.3) and the resulting expansion for $T(\mathbf{x}, t)$ is split into two parts to define the desired finite integral transform pair as

Inversion formula: $\quad T(\mathbf{x}, t) = \sum_{i=1}^{\infty} \frac{1}{N_i} \psi_i(\mathbf{x}) \tilde{T}_i(t) \qquad (2.11)$

Finite integral transform: $\quad \tilde{T}_i(t) = \int_V w(\mathbf{x}) \psi_i(\mathbf{x}) T(\mathbf{x}, t)\, dv \equiv (\psi_i, T)$

$$(2.12)$$

The given finite integral transform pair is now used to solve the boundary value problem of Class I defined by Equations (2.1) as described next.

Method of Solution

Equation (2.1a) is multiplied by $\psi_i(\mathbf{x})$, Equation (2.2a) by $T(\mathbf{x}, t)$, the results are added, integrated over the volume V, and the definition of the integral transform (2.12) is utilized. We find

$$\frac{d\tilde{T}_i(t)}{dt} + \mu_i^2 \tilde{T}_i(t) = \int_V [\psi_i \nabla \cdot (k\nabla T) - T\nabla \cdot (k\nabla \psi_i)]\, dv + \int_V \psi_i P\, dv$$

$$(2.13)$$

When the first volume integral on the right is changed to the surface integral as in Equations (2.5) and the resulting expression is written in the form given by Equation (2.6), Equation (2.13) becomes

$$\frac{d\tilde{T}_i(t)}{dt} + \mu_i^2 \tilde{T}_i(t) = g_i(t) \qquad (2.14a)$$

where

$$g_i(t) = \int_S k(\mathbf{x}) \begin{vmatrix} \psi_i(\mathbf{x}) & \dfrac{\partial}{\partial \mathbf{n}} \psi_i(\mathbf{x}) \\ T(\mathbf{x}, t) & \dfrac{\partial}{\partial \mathbf{n}} T(\mathbf{x}, t) \end{vmatrix} ds + \int_V \psi_i(\mathbf{x}) P(\mathbf{x}, t)\, dv \quad (2.14b)$$

The integrand of the preceding surface integral is now determined by solving Equations (2.1b) and (2.2b) for $\alpha(\mathbf{x})$ and $\beta(\mathbf{x})$, adding the resulting expressions for $\alpha(\mathbf{x})$ and $\beta(\mathbf{x})$. We find

$$k(\mathbf{x}) \begin{vmatrix} \psi_i(\mathbf{x}) & \dfrac{\partial}{\partial \mathbf{n}}\psi_i(\mathbf{x}) \\[2mm] T(\mathbf{x}, t) & \dfrac{\partial}{\partial \mathbf{n}}T(\mathbf{x}, t) \end{vmatrix} = \phi(\mathbf{x}, t)\,\frac{\psi_i(\mathbf{x}) - k(\mathbf{x})[\partial\psi_i(\mathbf{x})/\partial \mathbf{n}]}{\alpha(\mathbf{x}) + \beta(\mathbf{x})} \qquad (2.15)$$

Then Equations (2.14) take the form

$$\frac{d\tilde{T}_i(t)}{dt} + \mu_i^2\tilde{T}_i(t) = g_i(t) \qquad (2.16a)$$

where

$$g_i(t) = \int_S \phi(\mathbf{x}, t)\,\frac{\psi_i(\mathbf{x}) - k(\mathbf{x})[\partial\psi_i(\mathbf{x})/\partial \mathbf{n}]}{\alpha(\mathbf{x}) + \beta(\mathbf{x})}\,ds + \int_V \psi_i(\mathbf{x}) P(\mathbf{x}, t)\,dv$$

$$(2.16b)$$

Thus we transformed the partial differential Equation (2.1a) and its boundary condition (2.1b) into an ordinary differential equation for the transform $\tilde{T}_i(t)$ of the potential $T(\mathbf{x}, t)$ as given by Equations (2.16). The initial condition needed for the solution of Equations (2.16) is obtained by taking the integral transform of the initial condition (2.1c) according to the definition of the integral transform given by Equation (2.12). We obtain

$$\tilde{T}_i(0) = \int_V w(\mathbf{x})\psi_i(\mathbf{x})f(\mathbf{x})\,dv \equiv (\psi_i, f) \equiv \tilde{f}_i \qquad (2.17)$$

The solution of the first order linear differential Equation (2.16) subject to the initial condition (2.17) gives

$$\tilde{T}_i(t) = e^{-\mu_i^2 t}\left[\tilde{f}_i + \int_0^t g_i(t')e^{\mu_i^2 t'}\,dt'\right] \qquad (2.18)$$

Substituting the integral transform $\tilde{T}_i(t)$ given by Equation (2.18) into the inversion formula (2.11) the solution of the problems of Class I, defined by Equations (2.1), is obtained as

$$T(\mathbf{x}, t) = \sum_{i=1}^{\infty} \frac{1}{N_i}e^{-\mu_i^2 t}\psi_i(\mathbf{x})\left[\tilde{f}_i + \int_0^t g_i(t')e^{\mu_i^2 t'}\,dt'\right] \qquad (2.19a)$$

where

$$
N_i = \frac{1}{2\mu_i} \int_S k(\mathbf{x}) \left| \begin{array}{cc} \left(\dfrac{\partial \psi(\mathbf{x})}{\partial \mu} \right)_{\mu=\mu_i} & \left(\dfrac{\partial^2 \psi(\mathbf{x})}{\partial \mathbf{n}\, \partial \mu} \right)_{\mu=\mu_i} \\[1em] \psi_i(\mathbf{x}) & \dfrac{\partial \psi_i(\mathbf{x})}{\partial \mathbf{n}} \end{array} \right| ds \qquad (2.19b)
$$

$$
g_i(t) = \int_S \phi(\mathbf{x}, t) \frac{\psi_i(\mathbf{x}) - k(\mathbf{x})[\partial \psi_i(\mathbf{x})/\partial \mathbf{n}]}{\alpha(\mathbf{x}) + \beta(\mathbf{x})} \, ds + \int_V \psi_i(\mathbf{x}) P(\mathbf{x}, t) \, dv
$$

$$
(2.19c)
$$

$$
\tilde{f}_i = \int_V w(\mathbf{x}) \psi_i(\mathbf{x}) f(\mathbf{x}) \, dv \equiv (\psi_i, f) \qquad (2.19d)
$$

It is to be noted that the eigenfunctions $\psi_i(\mathbf{x})$ are arbitrary within multiplication constant $C(\mu_i)$. This arbitrariness is removed after substituting $\psi_i(\mathbf{x})$ in solution (2.19).

The mean potential $T_{av}(t)$ is defined as

$$
T_{av}(t) = \frac{\displaystyle\int_V w(\mathbf{x}) T(\mathbf{x}, t)\, dv}{\displaystyle\int_V w(\mathbf{x})\, dv} \equiv \frac{(1, T)}{(1, 1)} \qquad (2.20)
$$

Substituting the solution (2.19a) into Equation (2.20), the expression for the mean value of the potential is obtained as

$$
T_{av}(t) = \frac{1}{(1, 1)} \sum_{i=1}^{\infty} \frac{1}{N_i} e^{-\mu_i^2 t} (1, \psi_i) \left[\tilde{f}_i + \int_0^t g_i(t') e^{\mu_i^2 t'} \, dt' \right] \qquad (2.21)
$$

The expressions (2.19a) and (2.21), although representing the exact solutions for the problem, are not in a convenient form for practical purposes when the problem involves nonhomogeneous source functions. The reason for this is that the solution Equation (2.19), obtained by the eigenfunction expansion technique, converges fast at the points in the interior of the region but does not converge uniformly at the boundaries because the eigenconditions are satisfied at the boundaries. To alleviate this difficulty, we present the following splitting-up procedure.

For the cases in which the time-dependent source functions $\phi(\mathbf{x}, t)$ and $P(\mathbf{x}, t)$ are represented by exponentials in t and q-order polynomials in t, it is better for computational purposes to split up the previous solution of the general problem (2.1) into a system of quasisteady-state and transient parts as now described.

Splitting up the General Problem

We now separate the general problem given by Equations (2.1) into a set of simpler problems containing: a homogeneous transient problem and a set of steady-state problems for which separate solutions can be obtained with methods other than the eigenfunction expansion technique.

When the nonhomogeneous terms $P(\mathbf{x}, t)$ and $\phi(\mathbf{x}, t)$ are represented by exponentials and q-order polynomials of time in the form

$$P(\mathbf{x}, t) = P_e(\mathbf{x})e^{-d_P t} + \sum_{j=0}^{q} P_j(\mathbf{x})t^j, \qquad \mathbf{x} \in V \qquad (2.22a)$$

$$\phi(\mathbf{x}, t) = \phi_e(\mathbf{x})e^{-d_\phi t} + \sum_{j=0}^{q} \phi_j(\mathbf{x})t^j, \qquad \mathbf{x} \in S \qquad (2.22b)$$

where d_P and d_ϕ are constants, then the solution $T(\mathbf{x}, t)$ of the general boundary value problem (2.1) can be split up into the solution of simpler problems in the form [19]

$$T(\mathbf{x}, t) = T_\phi(\mathbf{x})e^{-d_\phi t} + T_P(\mathbf{x})e^{-d_P t} + \sum_{j=0}^{q} T_j(\mathbf{x})t^j + T_t(\mathbf{x}, t), \qquad \mathbf{x} \in V$$

$$(2.23)$$

The reasoning behind choosing the splitting-up process in the form given by Equation (2.23) is as follows.

The source terms $P(\mathbf{x}, t)$ and $\phi(\mathbf{x}, t)$ for the differential equation and the boundary conditions, respectively, being represented by exponentials and q-order polynomials of time, similar functions are tried for the separated solution, Equation (2.23). For simplicity in the analysis, we considered only one exponential term in Equations (2.22) for the $P(\mathbf{x}, t)$ and $\phi(\mathbf{x}, t)$ functions. However, if there were more exponential terms in Equations (2.22), the superposition principle permits us to use as many exponential terms in Equation (2.23) as necessary to correspond to those used in Equations (2.22).

When $T(\mathbf{x}, t)$ is represented as given by Equation (2.23), the functions $T_\phi(\mathbf{x})$, $T_P(\mathbf{x})$, and $T_j(\mathbf{x})$ are the solutions of the following steady-state problems.

$$LT_\phi(\mathbf{x}) = d_\phi w(\mathbf{x})T_\phi(\mathbf{x}), \qquad \mathbf{x} \in V \qquad (2.24a)$$

$$BT_\phi(\mathbf{x}) = \phi_e(\mathbf{x}), \qquad \mathbf{x} \in S \qquad (2.24b)$$

$$LT_P(\mathbf{x}) = d_P w(\mathbf{x})T_P(\mathbf{x}) + P_e(\mathbf{x}), \qquad \mathbf{x} \in V \qquad (2.25a)$$

$$BT_P(\mathbf{x}) = 0, \qquad \mathbf{x} \in S \qquad (2.25b)$$

and

$$LT_q(\mathbf{x}) = P_q(\mathbf{x}), \qquad \mathbf{x} \in V \tag{2.26a}$$

$$BT_q(\mathbf{x}) = \phi_q(\mathbf{x}), \qquad \mathbf{x} \in S \tag{2.26b}$$

$$LT_j(\mathbf{x}) + (j + 1)w(\mathbf{x})T_{j+1}(\mathbf{x}) = P_j(\mathbf{x}), \qquad \mathbf{x} \in V \tag{2.27a}$$

$$BT_j(\mathbf{x}) = \phi_j(\mathbf{x}), \qquad \mathbf{x} \in S \tag{2.27b}$$

where $j = q - 1, q - 2, \ldots, 1, 0$.

The function $T_t(\mathbf{x}, t)$ is the solution of the following transient homogeneous problem.

$$w(\mathbf{x}) \frac{\partial T_t(\mathbf{x}, t)}{\partial t} + LT_t(\mathbf{x}, t) = 0, \qquad \mathbf{x} \in V \tag{2.28a}$$

$$BT_t(\mathbf{x}, t) = 0, \qquad \mathbf{x} \in S \tag{2.28b}$$

$$T_t(\mathbf{x}, 0) = f(\mathbf{x}) - T_\phi(\mathbf{x}) - T_p(\mathbf{x}) - T_0(\mathbf{x}), \qquad \mathbf{x} \in V \tag{2.28c}$$

The validity of the foregoing splitting-up process defined by the systems (2.24)–(2.28) can be readily verified by substituting the solution given by Equation (2.23) into the boundary value problem defined by Equations (2.1).

The solutions of these simpler problems defined by Equations (2.24)–(2.28) are determined as now described.

The solution of the transient homogeneous problem (2.28) is immediately obtainable from the general solution (2.19) as a special case; we find

$$T_t(\mathbf{x}, t) = \sum_{i=1}^{\infty} \frac{1}{N_i} e^{-\mu_i^2 t} \psi_i(\mathbf{x}) \left(\tilde{f}_i - \tilde{T}_{\phi i} - \tilde{T}_{pi} - \tilde{T}_{0i} \right), \qquad \mathbf{x} \in V \tag{2.29}$$

where $\tilde{f}_i$, $\tilde{T}_{\phi i}$, $\tilde{T}_{pi}$, and $\tilde{T}_{0i}$ are the finite integral transforms of the functions $f(\mathbf{x})$, $T_\phi(\mathbf{x})$, $T_p(\mathbf{x})$, and $T_0(\mathbf{x})$, respectively, according to the transform (2.12). Here the transforms $\tilde{T}_{\phi i}$, $\tilde{T}_{pi}$, and $\tilde{T}_{0i}$ are obtainable by the application of the finite integral transform (2.12) to the problems (2.24)–(2.27) by the integral transform procedure described previously [i.e., to determine, say, $\tilde{T}_{\phi i}$, Equation (2.24a) is multiplied by $\psi_i(\mathbf{x})$, Equation (2.2a) by $T_\phi(\mathbf{x})$, the results are added, integrated over the region V, the definition of the integral transform (2.12) is utilized, the volume integral is transformed to the surface integral, and the integrand of the surface integral is determined according to Equation (2.15)].

We summarize these results.

$$\tilde{T}_{\phi i} = \frac{g_{\phi i}}{\mu_i^2 - d_\phi} \tag{2.30a}$$

$$\tilde{T}_{pi} = \frac{g_{pi}}{\mu_i^2 - d_P} \tag{2.30b}$$

$$\tilde{T}_{qi} = \frac{g_{qi}}{\mu_i^2} \tag{2.30c}$$

$$\mu_i^2 \tilde{T}_j + (j+1)\tilde{T}_{j+1} = g_{ji} \qquad (j = q-1, q-2, \ldots, 1, 0) \tag{2.30d}$$

where

$$g_{\phi i} = \int_S \phi_e(\mathbf{x}) \frac{\psi_i(\mathbf{x}) - k(\mathbf{x})[\partial\psi_i(\mathbf{x})/\partial n]}{\alpha(\mathbf{x}) + \beta(\mathbf{x})} \, ds \tag{2.30e}$$

$$g_{pi} = \int_V \psi_i(\mathbf{x}) P_e(\mathbf{x}) \, dv \tag{2.30f}$$

$$g_{ji} = \int_S \phi_j(\mathbf{x}) \frac{\psi_i(\mathbf{x}) - k(\mathbf{x})[\partial\psi_i(\mathbf{x})/\partial n]}{\alpha(\mathbf{x}) + \beta(\mathbf{x})} \, ds$$

$$+ \int_V \psi_i(\mathbf{x}) P_j(\mathbf{x}) \, dv \quad \text{for } j = 0, 1, 2, \ldots, q \tag{2.30g}$$

Then $\tilde{T}_{0i}$ is evaluated from Equations (2.30c) and (2.30d) as

$$\tilde{T}_{0i} = \sum_{j=0}^{q} \frac{(-1)^j j! g_{ji}}{\mu_i^{2(j+1)}} \tag{2.30h}$$

where $j! = 1.2.3.\ldots.(j-1)\cdot j$.

The substitution of Equations (2.30a, b, h) into Equation (2.29) completes the solution $T_t(\mathbf{x}, t)$ of the transient homogeneous problem (2.28).

The solutions of the steady-state problems (2.24)–(2.27) are readily obtained by introducing the integral transforms given by Equations (2.30a)–(2.30d) into the inversion formula (2.11).

In the case of one-dimensional problems, the solutions of the steady-state problems (2.24)–(2.27) are obtainable by direct integration.

The solution $T(\mathbf{x}, t)$ given by Equation (2.23) is more convenient for computational purposes than the one given by Equation (2.19) provided that explicit analytic solutions are obtained for the functions $T_\phi(\mathbf{x})$, $T_P(\mathbf{x})$, and $T_j(\mathbf{x})$ defined by Equations (2.24)–(2.27), respectively. Then the solution for $T(\mathbf{x}, t)$ converges fast because the $T_t(\mathbf{x}, t)$ function is the solution of an homogeneous problem that converges fast.

Boundary Conditions All of the Second Kind

If $\alpha(\mathbf{x}) = 0$ and $\beta(\mathbf{x}) = 1$ for all boundaries in the general problem (2.1), then all the boundary conditions are of the second kind. If we further assume that $d(\mathbf{x}) = 0$ in the L operator (2.1d), then $\mu_0 = 0$ is also an eigenvalue and $\psi_0 = $ constant is the corresponding eigenfunction in the eigenvalue problem (2.2).

If Equation (2.2a) is integrated over the region V for $d(\mathbf{x}) = 0$ and the volume integral involving the L operator is changed into the surface integral, we obtain

$$(1, \psi_i) = -\frac{1}{\mu_i^2} \int_S k(\mathbf{x}) \frac{\partial \psi_i(\mathbf{x})}{\partial \mathbf{n}} ds \qquad (2.31a)$$

For a boundary condition of the second kind for all boundaries, we have $\partial \psi_i(\mathbf{x})/\partial \mathbf{n} = 0$ over the entire boundary surface and Equation (2.31a) reduces to

$$(1, \psi_i) = 0 \qquad (2.31b)$$

Then by utilizing this result and noting that for such a case $\mu_0 = 0$ is also an eigenvalue, Equations (2.21) and (2.19a), respectively, take the form

$$T_{av}(t) = \frac{1}{(1,1)} \left\{ (1, f) + \int_0^t \left[\int_S \phi(\mathbf{x}, t') \, ds + \int_V P(\mathbf{x}, t') \, dv \right] dt' \right\} \qquad (2.32)$$

and

$$T(\mathbf{x}, t) = T_{av}(t) + \sum_{i=1}^{\infty} \frac{1}{N_i} e^{-\mu_i^2 t} \psi_i(\mathbf{x}) \left[\tilde{f}_i + \int_0^t g_i(t') e^{\mu_i^2 t'} \, dt' \right] \qquad (2.33)$$

Clearly, the additional term $T_{av}(t)$ appearing in Equation (2.33) is an average potential over the region.

The expression (2.33), although an exact solution, is not in a convenient form for practical purposes. When the nonhomogeneous terms $P(\mathbf{x}, t)$ and $\phi(\mathbf{x}, t)$ are represented by exponentials and q-order polynomials of time as given by Equations (2.22), then the general solution (2.33) of the problem (2.1) for $d(\mathbf{x}) = 0$, $\alpha(\mathbf{x}) = 0$ can be split in the form [19]

$$T(\mathbf{x}, t) = T_{av}(t) + T_\phi(\mathbf{x}) e^{-d_\phi t} + T_p(\mathbf{x}) e^{-d_p t} + \sum_{j=0}^{q} T_j(\mathbf{x}) t^j + T_t(\mathbf{x}, t),$$

$$\mathbf{x} \in V \quad (2.34)$$

where $T_{av}(t)$ is the average potential defined by Equation (2.32) and the functions $T_\phi(\mathbf{x})$, $T_p(\mathbf{x})$, and $T_j(\mathbf{x})$ are the solution of the following steady-state problems.

$$\frac{w(\mathbf{x})}{(1,1)} \int_S \phi_e(\mathbf{x})\, ds + LT_\phi(\mathbf{x}) = d_\phi w(\mathbf{x}) T_\phi(\mathbf{x}), \qquad \mathbf{x} \in V \qquad (2.35a)$$

$$BT_\phi(\mathbf{x}) = \phi_e(\mathbf{x}), \qquad \mathbf{x} \in S \quad \text{and} \quad (1, T_\phi) = 0 \qquad (2.35b,c)$$

$$\frac{w(\mathbf{x})}{(1,1)} \int_V P_e(\mathbf{x})\, dv + LT_p(\mathbf{x}) = d_p w(\mathbf{x}) T_p(\mathbf{x}) + P_e(\mathbf{x}), \qquad \mathbf{x} \in V$$

$$(2.36a)$$

$$BT_p(\mathbf{x}) = 0, \qquad \mathbf{x} \in S \quad \text{and} \quad (1, T_p) = 0 \qquad (2.36b,c)$$

$$\frac{w(\mathbf{x})}{(1,1)} \left[\int_S \phi_q(\mathbf{x})\, ds + \int_V P_q(\mathbf{x})\, dv \right] + LT_q(\mathbf{x}) = P_q(\mathbf{x}), \qquad \mathbf{x} \in V$$

$$(2.37a)$$

$$BT_q(\mathbf{x}) = \phi_q(\mathbf{x}), \qquad \mathbf{x} \in S \quad \text{and} \quad (1, T_q) = 0 \qquad (2.37b,c)$$

$$\frac{w(\mathbf{x})}{(1,1)} \left[\int_S \phi_j(\mathbf{x})\, ds + \int_V P_j(\mathbf{x})\, dv \right] + LT_j(\mathbf{x}) + (j+1)w(\mathbf{x})T_{j+1}(\mathbf{x}) = P_j(\mathbf{x}),$$

$$\mathbf{x} \in V \quad (2.38a)$$

$$BT_j(\mathbf{x}) = \phi_j(\mathbf{x}), \qquad \mathbf{x} \in S \quad \text{and} \quad (1, T_j) = 0, \qquad j = q-1, q-2, \ldots, 1, 0$$

$$(2.38b,c)$$

and the function $T_t(\mathbf{x}, t)$ is the solution of the following transient homogeneous problem.

$$w(\mathbf{x}) \frac{\partial T_t(\mathbf{x}, t)}{\partial t} + LT_t(\mathbf{x}, t) = 0, \qquad \mathbf{x} \in V \qquad (2.39a)$$

$$BT_t(\mathbf{x}, t) = 0, \qquad \mathbf{x} \in S \quad \text{and} \quad (1, T_t) = 0 \qquad (2.39b,c)$$

$$T_t(\mathbf{x}, 0) = f(\mathbf{x}) - \frac{(1, f)}{(1,1)} - T_\phi(\mathbf{x}) - T_p(\mathbf{x}) - T_0(\mathbf{x}), \qquad \mathbf{x} \in V \quad (2.39d)$$

The validity of the splitting-up procedure defined by the systems (2.35)–(2.39) can be readily verified by substituting Equation (2.34) into Equations (2.1) for $d(\mathbf{x}) = 0$ and $\alpha(\mathbf{x}) = 0$. The additional conditions given by Equations (2.35c),

(2.36c), (2.37c), (2.38c), and (2.39c) [i.e., $(1, T_\phi) = 0$, $(1, T_p) = 0$, etc.] result when Equation (2.34) is introduced into Equation (2.20) and the T_{av} term cancelled out.

The steady-state problems defined by Equations (2.35) to (2.38) can readily be solved by applying the finite integral transform (2.12) and taking into account the result given by Equation (2.31b); then we obtain a system of transforms $\tilde{T}_{\phi i}, \tilde{T}_{pi}, \tilde{T}_{qi}$ with $j = 0, 1, 2, \ldots, q$ identical to those given by Equations (2.30). When these transforms are inverted by the inversion formula (2.11), the solutions $T_\phi(\mathbf{x})$, $T_p(\mathbf{x})$, and $T_q(\mathbf{x})$, $j = 0, 1, \ldots, q$ of the preceding steady-state problems are obtained.

The solution of the transient homogeneous problem (2.39) coincides with the one given by Equation (2.29). Therefore, Equation (2.29) is the solution of both transient problems defined by Equations (2.28) and (2.39).

Periodically Varying Boundary Conditions

Here we consider two different types of periodically varying boundary conditions for the problems in Class I. That is, one type of periodically varying boundary condition is associated with the dependence of the source function $\phi(\mathbf{x}, t)$ on time in the form $\phi(\mathbf{x}, t) = \phi_e(\mathbf{x})\exp(i\omega t)$. Practical examples include the oscillatory behavior of heating and cooling loads in buildings exposed to periodic variations of outside temperature, the oscillatory behavior of temperature in the cylinder walls of an internal combustion engine, and many others. Usually such processes continue so long that the transient effects pass and the temperature variation within the body can be treated as quasi-stationary thermal waves. When the source function is periodic in time in the form given previously, the analysis of such problems is relatively straightforward because the transient part $T_t(\mathbf{x}, t)$ of the solution has very little influence on the temperature distribution. Then the solution for the quasi-stationary thermal waves can be obtained from Equations (2.24) or (2.35) by setting $d_\phi = -i\omega$.

The other type is associated with such engineering applications as those encountered in the theory of heat transfer in regenerators [22], and the determination of thermal contact resistance between periodically contacting surfaces. The thermal analysis of such problems is more complicated than that of the type discussed earlier. This is because in this type of problem the boundary surface is subjected to a certain type of boundary condition during one of the cycles, say, in the time interval $0 < t < t_1$, and to a different type of boundary condition during the following cycle in the time interval $t_1 < t < (t_1 + t_2)$. These two cycles are repeated continuously for sufficiently long times that the quasi-steady-state temperature distribution is achieved in the medium. The thermal analysis of problems of this type for the quasi-steady-state conditions has been performed approximately for one-dimensional bodies [22–25] and exactly for a three-dimensional finite region subject to general boundary conditions [26, 33].

In the following analysis we present the quasi-steady-state solution to the boundary value problems of Class I resulting from the periodic variation of the boundary condition in such a manner that the boundary condition and the source functions are of a certain type during one cycle and of another type during the following cycle. The two distinct cycles are repeated continuously for a long time until the quasi-steady-state is reached. Then the analysis is concerned with the determination of the quasi-steady-state potentials $T_k(\mathbf{x}, t)$, $k = 1, 2$ for each of these two cycles, $0 \le t \le t_k$, $k = 1, 2$, in the medium, for the boundary value problem given as

$$w_k(\mathbf{x})\frac{\partial T_k(\mathbf{x}, t)}{\partial t} + L_k T_k(\mathbf{x}, t) = P_k(\mathbf{x}, t),$$

$$\mathbf{x} \in V, \qquad 0 \le t \le t_k, \qquad k = 1 \text{ or } 2 \quad (2.40a)$$

$$B_k T_k(\mathbf{x}, t) = \phi_k(\mathbf{x}, t), \qquad \mathbf{x} \in S, \qquad 0 \le t \le t_k, \qquad k = 1 \text{ or } 2$$

$$(2.40b)$$

$$T_k(\mathbf{x}, 0) = f_k(\mathbf{x}), \qquad \mathbf{x} \in V, \qquad k = 1 \text{ or } 2 \qquad (2.40c)$$

where

$$L_k \equiv -\nabla \cdot [k_k(\mathbf{x})\nabla] + d_k(\mathbf{x}) \qquad (2.40d)$$

$$B_k \equiv \alpha_k(\mathbf{x}) + \beta_k(\mathbf{x})k_k(\mathbf{x})\frac{\partial}{\partial \mathbf{n}} \qquad (2.40e)$$

Inasmuch as the quasi-steady-state is assumed to be established, *the final distribution of the potential $T_k(\mathbf{x}, t_k)$ at the end of the interval t_k for one of the cycles should be equal to the initial distribution of the potential for the following cycle.* With this consideration, the following relation is written

$$T_k(\mathbf{x}, t_k) = f_{3-k}(\mathbf{x}), \qquad k = 1 \text{ or } 2 \qquad (2.41)$$

The preceding problem now involves the determination of the two potentials $T_k(\mathbf{x}, t)$ for each of the cycles and the two unknown distributions $f_k(\mathbf{x})$, $k = 1$ or 2.

The solution of the problem given by equations (2.40) is immediately obtainable from Equations (2.19) as

$$T_k(\mathbf{x}, t) = \sum_{i=1}^{\infty} \frac{1}{N_{ki}} e^{-\mu_{ki}^2 t} \psi_{ki}(\mathbf{x}) \left[\tilde{f}_{ki} + \int_0^t g_{ki}(t') e^{\mu_{ki}^2 t'} dt' \right], \qquad k = 1 \text{ or } 2$$

$$(2.42a)$$

where

$$N_{ki} = (\psi_{ki}, \psi_{ki}) \tag{2.42b}$$

$$\tilde{f}_{ki} = (\psi_{ki}, f_k) \tag{2.42c}$$

$$g_{ki}(t) = \int_S \phi_k(\mathbf{x}, t) \frac{\psi_{ki}(\mathbf{x}) - k_k(\mathbf{x})[\partial \psi_{ki}(\mathbf{x})/\partial \mathbf{n}]}{\alpha_k(\mathbf{x}) + \beta_k(\mathbf{x})} \, ds + \int_V \psi_{ki}(\mathbf{x}) P_k(\mathbf{x}, t) \, dv$$

$$\tag{2.42d}$$

In the expressions (2.42), μ_{ki} and $\psi_{ki}(\mathbf{x})$ are the eigenvalues and the eigenfunctions of the Sturm–Liouville problem (2.2), namely,

$$\mu_k^2 w_k(\mathbf{x}) \psi_k(\mathbf{x}) = L_k \psi_k(\mathbf{x}), \qquad \mathbf{x} \in V, \qquad k = 1 \text{ or } 2 \tag{2.43a}$$

$$B_k \psi_k(\mathbf{x}) = 0, \qquad \mathbf{x} \in S, \qquad k = 1 \text{ or } 2 \tag{2.43b}$$

where L_k and B_k are as defined by Equations (2.40d) and (2.40e). It is assumed that the eigenfunctions and eigenvalues of this eigenvalue problem are known. Then *to compute the temperatures* $T_k(\mathbf{x}, t)$, $k = 1$ *or* 2, *from Equations* (2.42) *one needs to determine the unknown functions* $\tilde{f}_k$, $k = 1$ *or* 2. The procedure for the determination of $\tilde{f}_k$ is now described.

The solution (2.42a) is introduced into Equation (2.41) to give

$$f_{3-k}(\mathbf{x}) = \sum_{i=1}^{\infty} \frac{1}{N_{ki}} e^{-\mu_{ki}^2 t_k} \psi_{ki}(\mathbf{x}) \left[\tilde{f}_{ki} + \int_0^{t_k} g_{ki}(t') e^{\mu_{ki}^2 t'} \, dt' \right], \qquad k = 1 \text{ or } 2$$

$$\tag{2.44}$$

Both sides of Equation (2.44) are operated on by the operator

$$\int_V w_{3-k}(\mathbf{x}) \psi_{3-k, j}(\mathbf{x}) \, dv$$

and the definition of the finite integral transform given by Equation (2.12) is utilized; we obtain

$$\tilde{f}_{3-k, j} = c_{kj} + \sum_{i=1}^{p} d_{kji} \tilde{f}_{ki}, \qquad j = 1, 2, \dots, p, \qquad p \to \infty \tag{2.45a}$$

where

$$d_{kji} \equiv \frac{1}{N_{ki}} e^{-\mu_{ki}^2 t_k} \int_V w_{3-k}(\mathbf{x}) \psi_{3-k, j}(\mathbf{x}) \psi_{ki}(\mathbf{x}) \, dv \tag{2.45b}$$

$$c_{kj} \equiv \sum_{i=1}^{p} d_{kji} \int_0^{t_k} g_{ki}(t') e^{-\mu_{ki}^2 t'} \, dt' \tag{2.45c}$$

The system (2.45) can be written in the matrix form as

$$F_{3-k} = C_k + D_k F_k, \qquad k = 1, 2 \tag{2.46}$$

where

$$F_{3-k} = \begin{bmatrix} \tilde{f}_{3-k,1} \\ \tilde{f}_{3-k,2} \\ \vdots \\ \tilde{f}_{3-k,p} \end{bmatrix}, \qquad C_k = \begin{bmatrix} c_{k1} \\ c_{k2} \\ \vdots \\ c_{kp} \end{bmatrix}, \qquad D_k = \begin{bmatrix} d_{k11} & d_{k12} & \cdots & d_{k1p} \\ d_{k21} & d_{k22} & & d_{k2p} \\ \vdots & & & \\ d_{kp1} & d_{kp2} & \cdots & d_{kpp} \end{bmatrix}$$

$$\tag{2.47}$$

Equations (2.46) can be written in the alternative form as

$$F_k = C_{3-k} + D_{3-k} F_{3-k}, \qquad k = 1, 2 \tag{2.48}$$

Substituting Equation (2.46) into (2.48), we find

$$F_k = (I - D_{3-k} D_k)^{-1} (C_{3-k} + D_{3-k} C_k) \tag{2.49}$$

where I is the unit matrix.

Equation (2.49) is the desired expression for the determination of the functions $\tilde{f}_{ki}$. Once the $\tilde{f}_{ki}$ are known, the distribution of the potentials $T_k(\mathbf{x}, t)$, $k = 1$ or 2, for both cycles is calculated from Equations (2.42) and the functions $f_{3-k}(\mathbf{x})$, $k = 1$ or 2 from Equation (2.44).

The foregoing analysis is also applicable for the case of boundary conditions all of the second kind, namely, for $\alpha(\mathbf{x}) = 0$; then if $d(\mathbf{x}) = 0$, the solution (2.33) is used to replace (2.42a).

It is also possible to use the alternative solutions given by Equations (2.23) or (2.34) depending on the type of the problem in order to improve the convergence of the series.

Alternative Solution of Problems of Class I for Steady-State

The number of infinite series contained in the steady-state solution of Class I problems can be reduced by following the alternative solution technique as now described.

Consider the steady-state form of the boundary value problem of Class I obtained by setting $[\partial T(\mathbf{x}, t)/\partial t] = 0$ in Equations (2.1).

$$LT(\mathbf{x}) = P(\mathbf{x}), \qquad \mathbf{x} \in V \tag{2.50a}$$

$$BT(\mathbf{x}) = \phi(\mathbf{x}), \qquad \mathbf{x} \in S \tag{2.50b}$$

where

$$L \equiv -\nabla \cdot [k(\mathbf{x})\nabla] + d(\mathbf{x}) \qquad (2.50c)$$

$$B \equiv \alpha(\mathbf{x}) + \beta(\mathbf{x})k(\mathbf{x})\frac{\partial}{\partial \mathbf{n}} \qquad (2.50d)$$

We note that in the preceding formulation the inhomogeneous terms $P(\mathbf{x})$ and $\phi(\mathbf{x})$ are independent of time and the problem involves no initial condition.

The solution of the steady-state problem (2.50) is obtainable from the general solution (2.19a) by noting that $g_i(t)$ is independent of time, performing the integration with respect to t, and in the resulting expression setting $t \rightarrow \infty$; then the steady-state solution becomes

$$T(\mathbf{x}) = \sum_{i=1}^{\infty} \frac{1}{N_i} \psi_i(\mathbf{x}) \frac{g_i}{\mu_i^2} \qquad (2.51)$$

where $\psi_i(\mathbf{x})$ and μ_i are the eigenfunctions and eigenvalues of the Sturm–Liouville problem (2.2), and N_i and g_i are defined, respectively, by Equations (2.19b) and (2.19c), with the inhomogeneous terms $\phi(\mathbf{x})$ and $P(\mathbf{x})$ being independent of time.

The solution (2.51) can also be obtained by applying the finite integral transform (2.12) to the steady-state problem (2.50) to remove from the differential Equation (2.50a) *all the partial derivatives with respect to the space variables*. The resulting expression is an *algebraic equation* for $\tilde{T}_i$. When $\tilde{T}_i$ is determined from this equation and inverted by the inversion formula (2.11), we again obtain the same solution as given by Equation (2.51).

The foregoing approaches for the solution of the steady-state problem (2.50) are equivalent to the removal from the system of all partial derivatives with respect to the space variables by the application of the finite integral transform and then inverting the resulting transform of the function by the inversion formula. As a result, the solution contains triple summation for a three-dimensional problem and double summation for a two-dimensional problem.

In an alternative approach, the steady-state system (2.50) is reduced to an ordinary differential equation in one of the space variables by integral transformation, the resulting ordinary differential equation is solved, and the transform is inverted. *The solution obtained with this approach contains one less infinite series than that given by solution (2.51), hence it is better for computational purposes* [6].

The steady-state solutions of problems of Class I have numerous important applications in various branches of science and engineering. A number of useful solutions of heat conduction and mass diffusion problems are given in References [27–29]. Here we present a generalized, systematic approach for the solution of two and three-dimensional steady-state problems of Class I by the *alternative approach* referred to previously by following the formalism developed in Reference [30].

In the following analysis of the steady-state problem, in order to develop the solution with a formalism similar to that used in the solution of the time-dependent problems of Class I, *we shall use the variable t to denote one of the space variables in the multidimensional problem*. Then we write the mathematical formulation of the preceding steady-state problem (2.50) in the alternative form as

$$[w(\mathbf{x})L_t + L]T(\mathbf{x}, t) = P(\mathbf{x}, t), \qquad \mathbf{x} \in V, \qquad t_0 \le t \le t_1 \quad (2.52a)$$

$$BT(\mathbf{x}, t) = 0, \qquad \mathbf{x} \in S \qquad (2.52b)$$

$$\left[\delta_k - (-1)^k \gamma_k \frac{\partial}{\partial t}\right] T(\mathbf{x}, t_k) = f_k(\mathbf{x}), \qquad k = 0, 1 \qquad (2.52c, d)$$

where

$$L \equiv -\nabla \cdot [k(\mathbf{x})\nabla] + d(\mathbf{x}) \qquad (2.52e)$$

$$B \equiv \alpha(\mathbf{x}) + \beta(\mathbf{x})k(\mathbf{x})\frac{\partial}{\partial \mathbf{n}} \qquad (2.52f)$$

$$L_t \equiv -a(t)\frac{\partial}{\partial t}\left[b(t)\frac{\partial}{\partial t}\right] \qquad (2.52g)$$

where $a(t)$ and $b(t)$ are prescribed functions.

In the above formulation *the variable t is not the time variable. We use the notation t to denote the space variable that is not to be transformed by the application of the integral transform*. That is, in the following analysis, we apply the integral transform to remove from the system (2.52) only the partial derivatives associated with the differential operator L. The system is then reduced to an ordinary differential equation in the space variable t as now described.

We assume further that the $\alpha(\mathbf{x})$ do not vanish simultaneously at all points $\mathbf{x} \in S$.

Appropriate eigenvalue problems needed for the removal of the differential operator L from the system (2.52) by the integral transform technique is exactly the same as given by Equations (2.2).

The finite integral transform pair needed for the solution of problem (2.52) is the same as that given by Equations (2.11) and (2.12).

By applying the transform (2.12) we take the integral transform of the system (2.52) to yield

$$L_t \tilde{T}_i(t) + \mu_i^2 \tilde{T}_i(t) = g_i(t) \qquad (2.53a)$$

$$\left[\delta_k - (-1)^k \gamma_k \frac{d}{dt}\right] \tilde{T}_i(t_k) = (\psi_i, f_k), \qquad k = 0, 1 \qquad (2.53b, c)$$

where

$$g_i(t) = \int_V \psi_i(\mathbf{x}) P(\mathbf{x}, t)\, dv \qquad (2.53d)$$

We note that Equation (2.53a) coincides with the Equation (2.14a) if the operator L_t is chosen as $L_t \equiv \partial/\partial t$.

Let $u(\mu_i, t)$ and $v(\mu_i, t)$ be two linearly independent solutions of the homogeneous part of the differential Equation (2.53a), that is,

$$\left(L_t + \mu_i^2\right) u(\mu_i, t) = 0 \qquad (2.54a)$$

$$\left(L_t + \mu_i^2\right) v(\mu_i, t) = 0 \qquad (2.54b)$$

The general solution for $\tilde{T}_i(t)$ is constructed by taking linear combinations of the solutions $u(\mu_i, t)$ and $v(\mu_i, t)$ in the form

$$\tilde{T}_i(t) = C(t) u(\mu_i, t) + D(t) v(\mu_i, t) \qquad (2.54c)$$

where the coefficients $C(t)$ and $D(t)$ should be so determined that the solution (2.54c) satisfies the nonhomogeneous differential Equation (2.53a). The procedure for the determination of $C(t)$ and $D(t)$ is as follows.

We introduce the solution (2.54c) into Equation (2.53a), perform the indicated operations, and, by following Lagrange, impose the condition

$$C'(t) u(\mu_i, t) + D'(t) v(\mu_i, t) = 0 \qquad (2.54d)$$

Then we obtain

$$C'(t) u'(\mu_i, t) + D'(t) v'(\mu_i, t) = -\frac{g_i(t)}{a(t) b(t)} \qquad (2.54e)$$

where primes denote differentiation with respect to the variable t.

Equations (2.54d) and (2.54e) provide two relations for the determination of $C'(t)$ and $D'(t)$. Once $C'(t)$ and $D'(t)$ are determined, the resulting expressions are integrated from the initial value t_0 to t to obtain $C(t)$ and $D(t)$. When the coefficients $C(t)$ and $D(t)$ are introduced into Equation (2.54c), the solution for $\tilde{T}_i(t)$ becomes

$$\tilde{T}_i(t) = C_0 u(\mu_i, t) + D_0 v(\mu_i, t) + W_i(t) \qquad (2.55a)$$

where

$$W_i(t) = \int_{t_0}^{t} \frac{g_i(\tau)}{a(\tau) b(\tau)} \frac{u(\mu_i, t) v(\mu_i, \tau) - v(\mu_i, t) u(\mu_i, \tau)}{u(\mu_i, \tau) v'(\mu_i, \tau) - u'(\mu_i, \tau) v(\mu_i, \tau)} d\tau$$

where C_0 and D_0 are the values of $C(t)$ and $D(t)$ for $t = t_0$, that is,

$$C_0 \equiv C(t_0) \quad \text{and} \quad D_0 \equiv D(t_0)$$

If the solution (2.55a) should satisfy the boundary conditions (2.53b, c), we have

$$C_0 U(\mu_i, t_k) + D_0 V(\mu_i, t_k) = (\psi_i, f_k) + \int_{t_0}^{t_1} Z(\mu_i, t_k, \tau) \, d\tau \quad (2.55b)$$

where

$$U(\mu_i, t_k) = \delta_k u(\mu_i, t_k) - (-1)^k \gamma_k u'(\mu_i, t_k) \qquad (2.55c)$$

$$V(\mu_i, t_k) = \delta_k v(\mu_i, t_k) - (-1)^k \gamma_k v'(\mu_i, t_k) \qquad (2.55d)$$

$$Z(\mu_i, t_k, \tau) = \frac{g_i(\tau)}{a(\tau)b(\tau)} \frac{u(\mu_i, \tau)V(\mu_i, t_k) - U(\mu_i, t_k)v(\mu_i, \tau)}{u(\mu_i, \tau)v'(\mu_i, \tau) - u'(\mu_i, \tau)v(\mu_i, \tau)}$$

$$(2.55e)$$

with $k = 0$ or 1.

The coefficients C_0 and D_0 are determined from the solution of Equations (2.55b) for $k = 0$ and 1. When these coefficients are introduced into Equation (2.55a) and the resulting expression for the integral transform $\tilde{T}_i(t)$ is inverted by the inversion formula (2.11), the general solution for $T(\mathbf{x}, t)$ is determined as

$$T(\mathbf{x}, t) = \sum_{i=1}^{\infty} \frac{\psi_i(\mathbf{x})}{N_i} \left\{ \frac{1}{\displaystyle\sum_{k=0}^{1} (-1)^k U(\mu_i, t_k)V(\mu_i, t_{1-k})} \right.$$

$$\times \sum_{k=0}^{1} (-1)^{k+1} \left[(\psi_i, f_{1-k}) + \int_{t_0}^{t_1} Z(\mu_i, t_{1-k}, \tau) \, d\tau \right]$$

$$\left. \times [V(\mu_i, t_k)u(\mu_i, t) - U(\mu_i, t_k)v(\mu_i, t)] + W(t) \right\} \quad (2.56a)$$

where $W(t)$ is defined in Equation (2.55a) and the normalization integral N_i is defined by Equation (2.8). Also, the terms involving the functions $V(\mu_i, t_k)$ and $U(\mu_i, t_k)$ can be written in the explicit form by utilizing the definitions of these functions given by Equations (2.55c, d). We obtain

$$V(\mu_i, t_k)u(\mu_i, t) - U(\mu_i, t_k)v(\mu_i, t)$$

$$= \delta_k[v(\mu_i, t_k)u(\mu_i, t) - u(\mu_i, t_k)v(\mu_i, t)]$$

$$- (-1)^k \gamma_k[v'(\mu_i, t_k)u(\mu_i, t) - u'(\mu_i, t_k)v(\mu_i, t)] \quad (2.56b)$$

$$\sum_{k=0}^{1} (-1)^k U(\mu_i, t_k)V(\mu_i, t_{1-k})$$

$$= \delta_0\delta_1[u(\mu_i, t_0)v(\mu_i, t_1) - u(\mu_i, t_1)v(\mu_i, t_0)]$$

$$- \gamma_0\gamma_1[u'(\mu_i, t_0)v'(\mu_i, t_1) - u'(\mu_i, t_1)v'(\mu_i, t_0)]$$

$$+ \delta_0\gamma_1[u(\mu_i, t_0)v'(\mu_i, t_1) - u'(\mu_i, t_1)v(\mu_i, t_0)]$$

$$- \delta_1\gamma_0[u'(\mu_i, t_0)v(\mu_i, t_1) - u(\mu_i, t_1)v'(\mu_i, t_0)] \quad (2.56c)$$

Special Cases

We now consider a special case of solution (2.56a) corresponding to

$$\delta_0 = 0, \qquad f_0(\mathbf{x}) = 0, \qquad P(\mathbf{x}, t) = 0, \quad \text{and} \quad t_0 = 0 \qquad (2.57a)$$

Then if the condition

$$\lim_{t \to 0} \frac{u'(\mu_i, t_0)}{v'(\mu_i, t_0)} = 0 \qquad (2.57b)$$

is satisfied, solution (2.56a) reduces to

$$T(\mathbf{x}, t) = \sum_{i=1}^{\infty} \frac{\psi_i(\mathbf{x})}{N_i}(\psi_i, f_1)\frac{u(\mu_i, t)}{\delta_1 u(\mu_i, t_1) + \gamma_1 u'(\mu_i, t_1)} \qquad (2.57c)$$

Clearly, several specific cases are obtainable from the general solutions (2.56a) and (2.57c) depending on how the operator L_t is defined. In the following we examine the cases in which the operator L_t is defined in different forms including $L_t \equiv -\partial^2/\partial t^2$, $L_t \equiv -(1/t)(\partial/\partial t)(t\partial/\partial t)$, $L_t \equiv -(t\partial/\partial t)(t\partial/\partial t)$, and $L_t \equiv -(\partial/\partial t)(t^2\partial/\partial t)$.

Case 1 The operator L_t is defined as

$$L_t \equiv -\frac{\partial^2}{\partial t^2} \qquad (2.58a)$$

Then Equation (2.53a) becomes

$$\frac{d^2\tilde{T}_i(t)}{dt^2} - \mu_i^2\tilde{T}_i(t) = 0 \qquad (2.58b)$$

and the solution of which is taken as

$$\tilde{T}_i(t) = C\cosh(\mu_i t) + D\sinh(\mu_i t) \qquad (2.58c)$$

By comparing Equations (2.54c) and (2.58c) we find

$$u(\mu_i, t) = \cosh(\mu_i t), \qquad v(\mu_i, t) = \sinh(\mu_i t) \qquad (2.59a, b)$$

Introducing Equations (2.59) into the general solution (2.56), we obtain

$$T(\mathbf{x}, t) = \sum_{i=1}^{\infty} \left| \frac{\psi_i(\mathbf{x})}{N_i} \right.$$

$$\times \frac{\sum_{k=0}^{1}(-1)^{1-k}(\psi_i, f_{1-k})\left\{\delta_k \sinh[\mu_i(t_k - t)] - (-1)^k \gamma_k \mu_i \cosh[\mu_i(t_k - t)]\right\}}{(\delta_0\delta_1 + \gamma_0\gamma_1\mu_i^2)\sinh[\mu_i(t_1 - t_0)] + (\delta_0\gamma_1 + \delta_1\gamma_0)\mu_i\cosh[\mu_i(t_1 - t_0)]} \left. \right)$$

$$(2.60a)$$

For the special case $\delta_0 = 0$, $f_0(\mathbf{x}) = 0$, and $t_0 = 0$, Equation (2.60a) can be simplified according to Equation (2.57c) since the solutions (2.59) satisfy the condition (2.57b). We find

$$T(\mathbf{x}, t) = \sum_{i=1}^{\infty} \frac{\psi_i(\mathbf{x})}{N_i}(\psi_i, f_1)\frac{\cosh(\mu_i t)}{\delta_1\cosh(\mu_i t_1) + \gamma_1\mu_i\sinh(\mu_i t_1)} \qquad (2.60b)$$

Case 2   The operator L_t is defined as

$$L_t \equiv -\frac{1}{t}\frac{\partial}{\partial t}\left(t\frac{\partial}{\partial t}\right) \qquad (2.61a)$$

Then Equation (2.53a) becomes

$$\frac{d^2\tilde{T}_i(t)}{dt^2} + \frac{1}{t}\frac{d\tilde{T}_i(t)}{dt} - \mu_i^2\tilde{T}_i(t) = 0 \qquad (2.61b)$$

and the solution of this equation is taken as

$$\tilde{T}_i(t) = CI_0(\mu_i t) + DK_0(\mu_i t) \tag{2.61c}$$

By comparing Equations (2.54c) and (2.61c) we find

$$u(\mu_i, t) = I_0(\mu_i t), \qquad v(\mu_i, t) = K_0(\mu_i t) \tag{2.62a, b}$$

and we also write

$$u'(\mu_i, t) = \mu_i I_1(\mu_i t), \qquad v'(\mu_i, t) = -\mu_i K_1(\mu_i t) \tag{2.62c, d}$$

where $I_\nu(z)$ and $K_\nu(z)$ are modified Bessel functions of order ν of the first and second kind, respectively.

Substituting Equations (2.62) into the general solution (2.56), we obtain the desired solution.

The solutions (2.62) satisfy the condition (2.57b); then for the special case $\delta_0 = 0$, $f_0(x) = 0$, and $t_0 = 0$, the resulting solution for the problem is simplified according to Equation (2.57c) as

$$T(x, t) = \sum_{i=1}^{\infty} \frac{\psi_i(x)}{N_i}(\psi_i, f_1)\frac{I_0(\mu_i t)}{\delta_1 I_0(\mu_i t_1) + \gamma_1 \mu_i K_0(\mu_i t_1)} \tag{2.63}$$

Case 3 The operator L_t is defined as

$$L_t \equiv -t\frac{\partial}{\partial t}\left(t\frac{\partial}{\partial t}\right) \tag{2.64a}$$

Then Equation (2.53a) becomes

$$t\frac{d}{dt}\left(t\frac{d\tilde{T}_i(t)}{dt}\right) - \mu_i^2 \tilde{T}_i(t) = 0 \tag{2.64b}$$

or

$$t^2\frac{d^2\tilde{T}_i(t)}{dt^2} + t\frac{d\tilde{T}_i(t)}{dt} - \mu_i^2 \tilde{T}_i(t) = 0 \tag{2.64c}$$

This is Euler's homogeneous (i.e., equidimensional) differential equation and its solution is taken as

$$\tilde{T}_i(t) = Ct^{\mu_i} + Dt^{-\mu_i} \quad \text{for } \mu_i \neq 0 \tag{2.64d}$$

By comparing Equations (2.54c) and (2.64d) we find

$$u(\mu_i, t) = t^{\mu_i}, \qquad v(\mu_i, t) = t^{-\mu_i} \tag{2.65a, b}$$

Substituting Equations (2.65) into the general solution (2.56), we obtain the desired solution as

$$T(\mathbf{x}, t) = \sum_{i=1}^{\infty} \frac{\psi_i(\mathbf{x})}{N_i} \frac{A_i}{B_i} \tag{2.66a}$$

where

$$A_i = \sum_{k=0}^{1} (-1)^{1-k} (\psi_i, f_{1-k}) \left\{ \delta_k \left[(t/t_k)^{\mu_i} - (t_k/t)^{\mu_i} \right] + (-1)^k (\gamma_k \mu_i/t_k) \right.$$

$$\left. \times \left[(t/t_k)^{\mu_i} + (t_k/t)^{\mu_i} \right] \right\}$$

$$B_i = \left[\delta_0 \delta_1 + (\gamma_0 \gamma_1/t_0 t_1) \mu_i^2 \right] \left[(t_0/t_1)^{\mu_i} - (t_1/t_0)^{\mu_i} \right]$$

$$- \mu_i \left[(\delta_0 \gamma_1/t_1) + (\delta_1 \gamma_0/t_0) \right] \left[(t_0/t_1)^{\mu_i} + (t_1/t_0)^{\mu_i} \right]$$

The solutions (2.65) satisfy the condition (2.57b); then for the special case of $\delta_0 = 0$, $f_0(\mathbf{x}) = 0$, and $t_0 = 0$, the solution for $T(\mathbf{x}, t)$ is simplified according to Equation (2.57c) as

$$T(\mathbf{x}, t) = \sum_{i=1}^{\infty} \frac{\psi_i(\mathbf{x})}{N_i} (\psi_i, f_1) \frac{(t/t_1)^{\mu_i}}{\delta_1 + (\gamma_1/t_1)\mu_i} \tag{2.66b}$$

Case 4 The operator L_t is defined as

$$L_t \equiv -\frac{\partial}{\partial t} \left(t^2 \frac{\partial}{\partial t} \right) \tag{2.67a}$$

Then Equation (2.53a) takes the form

$$\frac{d}{dt} \left(t^2 \frac{d\tilde{T}_i(t)}{dt} \right) - \mu_i^2 \tilde{T}_i(t) = 0 \tag{2.67b}$$

The solution of this equation is taken as

$$\tilde{T}_i(t) = C t^{a_i} + D t^{-(a_i+1)} \tag{2.67c}$$

where

$$a_i(a_i + 1) = \mu_i^2 \tag{2.67d}$$

The validity of this solution can be verified by direct substitution of Equation (2.67c) into Equation (2.67b). By comparing Equations (2.54c) and (2.67c) we find

$$u(\mu_i, t) = t^{a_i}, \qquad v(\mu_i, t) = t^{-(a_i+1)} \tag{2.68a, b}$$

and we also write

$$u'(\mu_i, t) = a_i t^{a_i - 1}, \qquad v'(\mu_i, t) = -(a_i + 1)t^{-(a_i+2)} \tag{2.68c, d}$$

Substituting Equations (2.68) into the general solution (2.56), we obtain the desired solution.

The solutions (2.68) satisfy the condition (2.57b); for the special case of $\delta_0 = 0$, $f_0(x) = 0$, and $t_0 = 0$ the solution for $T(x, t)$ is simplified according to Equation (2.57c) as

$$T(x, t) = \sum_{i=1}^{\infty} \frac{\psi_i(x)}{N_i}(\psi_i, f_1)\frac{(t/t_1)^{a_i}}{\delta_1 + (\gamma_1/t_1)a_i} \tag{2.69}$$

2.2   GENERAL SOLUTION OF PROBLEMS OF CLASS II

The problems in this class, as discussed in Chapter 1, include the unsteady heat or mass diffusion in a finite, composite region V, bounded by a surface S_0 and subdivided into n finite regions V_k, $k = 1, 2, 3, \ldots, n$. Each subregion has different physical properties $w_k(x)$ and $k_k(x)$ which may depend on the space variable x, contain distributed substance sources $P_k(x, t)$ which are prescribed functions of time and space, and has arbitrary prescribed initial potential distributions $f_k(x)$. The coefficients $h_{kp}(x)$ between two adjacent subregions V_k and V_p may be space dependent. Then the mathematical formulation of the general, three-dimensional, time dependent boundary value problems of Class II include the balance equations given as

$$w_k(x)\frac{\partial T_k(x, t)}{\partial t} + L_k T_k(x, t) = P_k(x, t), \qquad x \in V_k \qquad (k = 1, 2, \ldots, n)$$
$$\tag{2.70a}$$

subject to the boundary conditions

$$B_k T_k(x, t) = \phi_k(x, t), \qquad x \in S_{k0} \tag{2.70b}$$

at the outer surface S_{k0} of the subregion V_k, and

$$k_k(x)\frac{\partial T_k(x, t)}{\partial n_{kp}} = k_p(x)\frac{\partial T_p(x, t)}{\partial n_{kp}} = h_{kp}(x)\big[T_k(x, t) - T_p(x, t)\big],$$

$$x \in S_{kp} \qquad (k \neq p) \tag{2.70c}$$

at the interface S_{kp} of the subregions V_k and V_p, and the initial conditions

$$T_k(\mathbf{x}, 0) = f_k(\mathbf{x}), \qquad \mathbf{x} \in V_k \tag{2.70d}$$

where

$$L_k \equiv -\nabla \cdot [k_k(\mathbf{x})\nabla] + d_k(\mathbf{x}) \tag{2.70e}$$

$$B_k \equiv \alpha_k(\mathbf{x}) + \beta_k(\mathbf{x})k_k(\mathbf{x})\frac{\partial}{\partial \mathbf{n}_{k0}} \tag{2.70f}$$

Here $\partial/\partial \mathbf{n}_{kp}$ denotes the normal derivative at the interface S_{kp} between the subregions V_k and V_p from V_k to V_p, and $\partial/\partial \mathbf{n}_{k0}$ the normal derivative at the outer surface S_{k0} of the subregion V_k in the outward direction.

The analytical solution of the foregoing boundary value problem for the case of time dependent coefficients $h_{kp}(\mathbf{x}, t)$ is given in Reference 16. In the following analysis we present the application of this approach to the solution to the preceding boundary value problem (2.70).

Development of the Integral Transform Pair

The first step in the analysis is the development of the finite integral transform and the inversion formula needed in the solution of this problem by the integral transform technique. The desired integral transform pair can be constructed by considering the following *auxiliary* homogeneous problem for the space variable function $\psi_k(\mathbf{x})$, in the same subregions V_k as for the original problem:

$$\mu^2 w_L(\mathbf{x})\psi_k(\mathbf{x}) = L_k\psi_k(\mathbf{x}), \qquad \mathbf{x} \in V_k \tag{2.71a}$$

subject to the homogeneous boundary conditions

$$B_k\psi_k(\mathbf{x}) = 0, \qquad \mathbf{x} \in S_{k0} \tag{2.71b}$$

$$k_k(\mathbf{x})\frac{\partial \psi_k(\mathbf{x})}{\partial \mathbf{n}_{kp}} = k_p(\mathbf{x})\frac{\partial \psi_p(\mathbf{x})}{\partial \mathbf{n}_{kp}} = h_{kp}(\mathbf{x})\left[\psi_k(\mathbf{x}) - \psi_p(\mathbf{x})\right],$$

$$\mathbf{x} \in S_{kp} \qquad (k \neq p) \tag{2.71c}$$

where the linear operators L_k and B_k are defined by Equations (2.70e) and (2.70f), respectively.

The system (2.71) has nontrivial solutions for a discrete spectrum of eigenvalues μ_i, $i = 1, 2, \ldots$, corresponding to the eigenfunctions $\psi_k(\mu_i, \mathbf{x}) \equiv \psi_{ki}(\mathbf{x})$.

We now expand the desired potential $T(\mathbf{x}, t)$ in terms of the eigenfunctions $\psi_{ki}(\mathbf{x})$ of the preceding auxiliary problem over the subregion V_k in the form

$$T(\mathbf{x}, t) = \sum_{i=1}^{\infty} A_i(t)\psi_{ki}(\mathbf{x}), \qquad \mathbf{x} \in V_k \qquad (2.72)$$

where the summation is taken over all discrete spectra of eigenvalues μ_i.

To determine the coefficients $A_i(t)$, the orthogonality condition for these eigenfunctions is needed, but the orthogonality condition (2.9) developed in the previous section cannot be used for this purpose because the auxiliary problem (2.71) does not belong to the conventional Sturm–Liouville system. Therefore, the orthogonality condition appropriate for the system (2.71) is derived as described in the following.

Equation (2.71a) is written for two different eigenvalues μ_i and μ_j. The first of these equations is multiplied by $\psi_{kj}(\mathbf{x})$, the second by $\psi_{ki}(\mathbf{x})$, the results are subtracted, integrated over the volume V_k, the volume integral is changed to the surface integral as in Equations (2.5b), and the resulting expression is summed up for $k = 1, 2, \dots, n$ taking into consideration the homogeneous boundary conditions (2.71b) and (2.71c). Then the desired orthogonality condition for the eigenfunctions $\psi_{ki}(\mathbf{x})$ of the auxiliary problem (2.71) becomes

$$\sum_{k=1}^{n} \left(\psi_{ki}, \psi_{kj}\right) = \delta_{kj} N_i \qquad (2.73)$$

where

$$N_i = \sum_{k=1}^{n} \left(\psi_{ki}, \psi_{ki}\right) \qquad (2.74)$$

and δ_{ij} is the Kronecker delta. We note that for a problem involving a single region (i.e., $n = 1$), the preceding orthogonality condition reduces to that given by Equation (2.9) of the Sturm–Liouville problem.

The orthogonality condition given by Equation (2.73) is used to determine the coefficients $A_i(t)$ in Equation (2.72) as now described. We operate on both sides of Equation (2.72) by the operator

$$\int_{V_k} w_k(\mathbf{x})\psi_{kj}(\mathbf{x}) \, dv$$

sum up the results for $k = 1, 2, \dots, n$, and utilize the orthogonality property (2-73). We obtain

$$A_i(t) = \frac{1}{N_i} \sum_{k=1}^{n} \left(\psi_{ki}, T_k\right) \qquad (2.75)$$

This expression for $A_i(t)$ is introduced into Equation (2.72) and the resulting expansion for $T(\mathbf{x}, t)$ is split up into two parts to define the desired finite integral transform pair as

Inversion formula: $\quad T_k(\mathbf{x}, t) = \sum\limits_{i=1}^{\infty} \dfrac{1}{N_i} \psi_{ki}(\mathbf{x}) \tilde{T}_i(t) \qquad (k = 1, 2, \ldots, n)$

$$(2.76)$$

Finite integral transform: $\quad \tilde{T}_i(t) = \sum\limits_{k=1}^{n} (\psi_{ki}, T_k) \qquad\qquad (2.77)$

We note that this integral transform pair for the n region composite medium reduces to the integral transform pair for a single region given by Equations (2.11) and (2.12) when $n = 1$. The integral transform pair given by Equation (2.76) and (2.77) is now used to solve the boundary value problem of Class II defined by Equations (2.70) as described next.

Method of Solution

Equation (2.70a) is multiplied by $\psi_{ki}(\mathbf{x})$, Equation (2.71a) by $T_k(\mathbf{x}, t)$, the results are added, integrated over the volume V_k, the expression obtained in this manner is summed up over $k = 1, 2, \ldots, n$, the definition of the finite integral transform (2.77) is utilized, and the volume integral is transformed into the surface integral as in Equations (2.5). We obtain

$$\frac{d\tilde{T}_i(t)}{dt} + \mu_i^2 \tilde{T}_i(t) = g_i(t) \qquad\qquad (2.78a)$$

where

$$g_i(t) = \sum_{k=1}^{n} \left[\int_{S_{k0}} k_k(\mathbf{x}) \begin{vmatrix} \psi_{ki}(\mathbf{x}) & \dfrac{\partial}{\partial \mathbf{n}} \psi_{ki}(\mathbf{x}) \\[2mm] T_k(\mathbf{x}, t) & \dfrac{\partial}{\partial \mathbf{n}} T_k(\mathbf{x}, t) \end{vmatrix} ds + \int_{V_k} \psi_{ki}(\mathbf{x}) P_k(\mathbf{x}, t)\, dv \right]$$

$$(2.78b)$$

We note that these expressions are similar to those given by Equations (2.14) except for the summation over the regions $k = 1, 2, \ldots, n$. The integrand of the surface integral is evaluated by utilizing the boundary conditions (2.70b), (2.70c), (2.71b) and (2.71c) in a similar manner to that of obtaining the expression (2.15) for the single region problem. Then Equations (2.78) take the form

$$\frac{d\tilde{T}_i(t)}{dt} + \mu_i^2 \tilde{T}_i(t) = g_i(t) \qquad\qquad (2.79a)$$

where

$$g_i(t) = \sum_{k=1}^{n} \left\{ \int_{S_{k0}} \phi_k(\mathbf{x}, t) \frac{\psi_{ki}(\mathbf{x}) - k_k(\mathbf{x})[\partial \psi_{ki}(\mathbf{x})/\partial \mathbf{n}]}{\alpha_k(\mathbf{x}) + \beta_k(\mathbf{x})} ds \right.$$

$$\left. + \int_{V_k} \psi_{ki}(\mathbf{x}) P_k(\mathbf{x}, t) \, dv \right\} \qquad (2.79b)$$

which is similar to those given by Equations (2.16) for the single region problem, except for the n-region composite medium problem considered here, the term $g_i(t)$ involves a summation over all subregions.

The initial condition needed for the solution of Equation (2.79a) is obtained by taking the integral transform of the initial condition (2.70d) according to the transform (2.76)

$$\tilde{T}_i(0) = \sum_{k=1}^{n} (\psi_{ki}, f_k) \equiv \tilde{f}_i \qquad (2.80)$$

When the solution of Equation (2.79a) subject to the initial condition (2.80) is introduced into the inversion formula (2.76), the desired solution of problem (2.70) becomes

$$T_k(\mathbf{x}, t) = \sum_{i=1}^{\infty} \frac{1}{N_i} e^{-\mu_i^2 t} \psi_{ki}(\mathbf{x}) \left[\tilde{f}_i + \int_0^t g_i(t') e^{\mu_i^2 t'} dt' \right] \qquad (k = 1, 2, \ldots, n)$$

$$(2.81)$$

where N_i, $g_i(t')$, and $\tilde{f}_i$ are defined by Equations (2.74), (2.79b), and (2.80), respectively. It is interesting to compare solution (2.81) for the n-region composite medium with solution (2.19a) for the single region. They are exactly of the same form except for the definitions of the integral transforms.

Splitting up the General Problem

The expression (2.81), although representing the exact solution for the considered problem, is not in a convenient form for practical purposes when the problem involves nonhomogeneous source functions. For the cases in which the nonhomogeneous terms $P_k(\mathbf{x}, t)$ and $\phi_k(\mathbf{x}, t)$ are represented in the form

$$P_k(\mathbf{x}, t) = P_{ke}(\mathbf{x}) e^{-d_p t} + \sum_{j=0}^{q} P_{kj}(\mathbf{x}) t^j, \qquad \mathbf{x} \in V_k \qquad (2.82a)$$

$$\phi_k(\mathbf{x}, t) = \phi_{ke}(\mathbf{x}) e^{-d_\phi t} + \sum_{j=0}^{q} \phi_{kj}(\mathbf{x}) t^j, \qquad \mathbf{x} \in S_{k0} \qquad (2.82b)$$

where d_p and d_ϕ are constants, the solution $T_k(\mathbf{x}, t)$ of the general boundary value problem (2.70) can be split up into the solution of simpler problems in the form

$$T_k(\mathbf{x}, t) = T_{k\phi}(\mathbf{x})e^{-d_\phi t} + T_{kp}(\mathbf{x})e^{-d_p t} + \sum_{j=0}^{q} T_{kj}(\mathbf{x})t^j + T_{kt}(\mathbf{x}, t), \qquad \mathbf{x} \in V_k$$

$$(2.83)$$

It is to be noted that this representation is exactly of the same form as that given by Equation (2.23) for the single region medium. In this expression the functions $T_{k\phi}(\mathbf{x})$, $T_{kp}(\mathbf{x})$, and $T_{kj}(\mathbf{x})$ are the solutions of the following steady-state problems

$$L_k T_{k\phi}(\mathbf{x}) = d_\phi w_k(\mathbf{x}) T_{k\phi}(\mathbf{x}), \qquad \mathbf{x} \in V_k \qquad (k = 1, 2, \ldots, n) \quad (2.84a)$$

$$B_k T_{k\phi}(\mathbf{x}) = \phi_{ke}(\mathbf{x}), \qquad \mathbf{x} \in S_{k0} \qquad (2.84b)$$

$$k_k(\mathbf{x}) \frac{\partial T_{k\phi}(\mathbf{x})}{\partial \mathbf{n}_{kp}} = k_p(\mathbf{x}) \frac{\partial T_{p\phi}(\mathbf{x})}{\partial \mathbf{n}_{kp}} = h_{kp}(\mathbf{x}) \left[T_{k\phi}(\mathbf{x}) - T_{p\phi}(\mathbf{x}) \right], \qquad \mathbf{x} \in S_{kp}$$

$$(2.84c)$$

$$L_k T_{kp}(\mathbf{x}) = d_p w_k(\mathbf{x}) T_{kp}(\mathbf{x}) + P_{ke}(\mathbf{x}), \qquad \mathbf{x} \in V_k \qquad (2.85a)$$

$$B_k T_{kp}(\mathbf{x}) = 0, \qquad \mathbf{x} \in S_{k0} \qquad (2.85b)$$

$$k_k(\mathbf{x}) \frac{\partial T_{kp}(\mathbf{x})}{\partial \mathbf{n}_{kp}} = k_p(\mathbf{x}) \frac{\partial T_{pp}(\mathbf{x})}{\partial \mathbf{n}_{kp}} = h_{kp}(\mathbf{x}) \left[T_{kp}(\mathbf{x}) - T_{pp}(\mathbf{x}) \right], \qquad \mathbf{x} \in S_{kp}$$

$$(2.85c)$$

and

$$L_k T_{kq}(\mathbf{x}) = P_{kq}(\mathbf{x}), \qquad \mathbf{x} \in V_k \qquad (2.86a)$$

$$B_k T_{kq}(\mathbf{x}) = \phi_{kq}(\mathbf{x}), \qquad \mathbf{x} \in S_{k0} \qquad (2.86b)$$

$$k_k(\mathbf{x}) \frac{\partial T_{kq}(\mathbf{x})}{\partial \mathbf{n}_{kp}} = k_p(\mathbf{x}) \frac{\partial T_{pq}}{\partial \mathbf{n}_{kp}} = h_{kp}(\mathbf{x}) \left[T_{kq}(\mathbf{x}) - T_{pq}(\mathbf{x}) \right], \qquad x \in S_{kp}$$

$$(2.86c)$$

$$L_{kj} T_{kj}(\mathbf{x}) + (j + 1) w_k(\mathbf{x}) T_{k, j+1}(\mathbf{x}) = P_{kj}(\mathbf{x}), \qquad \mathbf{x} \in V_k \quad (2.87a)$$

$$B_k T_{kj}(\mathbf{x}) = \phi_{kj}(\mathbf{x}), \qquad \mathbf{x} \in S_{k0} \qquad (2.87b)$$

$$k_k(\mathbf{x}) \frac{\partial T_{kj}(\mathbf{x})}{\partial \mathbf{n}_{kp}} = k_p(\mathbf{x}) \frac{\partial T_{pj}(\mathbf{x})}{\partial \mathbf{n}_{kp}} = h_{kp}(\mathbf{x}) \left[T_{kj}(\mathbf{x}) - T_{pj}(\mathbf{x}) \right], \qquad \mathbf{x} \in S_{kp}$$

$$\text{where } j = q - 1, q - 2, \ldots, 1, 0 \quad (2.87c)$$

The function $T_{kt}(\mathbf{x}, t)$ is the solution of the following transient homogeneous problem

$$w_k(\mathbf{x}) \frac{\partial T_{kt}(\mathbf{x}, t)}{\partial t} + L_k T_{kt}(\mathbf{x}, t) = 0, \qquad \mathbf{x} \in V_k \qquad (2.88a)$$

$$B_k T_{kt}(\mathbf{x}, t) = 0, \qquad \mathbf{x} \in S_{k0} \qquad (2.88b)$$

$$k_k(\mathbf{x}) \frac{\partial T_{kt}(\mathbf{x}, t)}{\partial \mathbf{n}_{kp}} = k_p(\mathbf{x}) \frac{\partial T_{pt}(\mathbf{x}, t)}{\partial \mathbf{n}_{kp}} = h_{kp}(\mathbf{x}) \big[T_{kt}(\mathbf{x}, t) - T_{pt}(\mathbf{x}, t) \big],$$

$$\mathbf{x} \in S_{kp} \qquad (2.88c)$$

$$T_{kt}(\mathbf{x}, 0) = f_k(\mathbf{x}) - T_{k\phi}(\mathbf{x}) - T_{kp}(\mathbf{x}) - T_{k0}(\mathbf{x}), \qquad \mathbf{x} \in V_k \quad (2.88d)$$

The validity of this splitting-up process defined by system (2.84)–(2.88) can be verified by substituting the solution given by Equation (2.83) into the boundary value problem (2.70). A comparison of system (2.84)–(2.88) for the composite medium with the corresponding system (2.24)–(2.28) for the single region medium reveals that they are exactly of the same form.

The solution of the transient homogeneous problem (2.88) is obtained as a special case from the general solution (2.81) as

$$T_{kt}(\mathbf{x}, t) = \sum_{i=1}^{\infty} \frac{1}{N_i} e^{-\mu_i^2 t} \psi_{ki}(\mathbf{x}) \big(\tilde{f}_i - \tilde{T}_{\phi i} - \tilde{T}_{pi} - \tilde{T}_{0i} \big),$$

$$\mathbf{x} \in V_k \qquad (k = 1, 2, \ldots, n) \quad (2.89)$$

where $\tilde{f}_i$, $\tilde{T}_{\phi i}$, $\tilde{T}_{pi}$, and $\tilde{T}_{0i}$ are the finite integral transforms of the functions $f_k(\mathbf{x})$, $T_{k\phi}(\mathbf{x})$, $T_{kp}(\mathbf{x})$, and $T_{k0}(\mathbf{x})$, respectively. The integral transforms $\tilde{T}_{\phi i}$, $\tilde{T}_{pi}$, and $\tilde{T}_{0i}$ are obtainable by the application of the finite integral transform (2.77) to the steady-state problems (2.84)–(2.87) as described previously in the solution of similar problems for the single-region medium [i.e., problems (2.24)–(2.27)]. We summarize the results.

$$\tilde{T}_{\phi i} = \frac{g_{\phi i}}{\mu_i^2 - d_\phi} \qquad (2.90a)$$

$$\tilde{T}_{pi} = \frac{g_{pi}}{\mu_i^2 - d_p} \qquad (2.90b)$$

$$\tilde{T}_{0i} = \sum_{j=0}^{q} (-1)^j j! \frac{g_{ji}}{\mu_i^{2(j+1)}} \qquad (2.90c)$$

where

$$g_{\phi i} = \sum_{k=1}^{n} \int_{S_{k0}} \phi_{ke}(\mathbf{x}) \frac{\psi_{ki}(\mathbf{x}) - k_k(\mathbf{x})[\partial\psi_{ki}(\mathbf{x})/\partial\mathbf{n}_{k0}]}{\alpha_k(\mathbf{x}) + \beta_k(\mathbf{x})} \, ds \qquad (2.90\text{d})$$

$$g_{pi} = \sum_{k=1}^{n} \int_{V_k} \psi_{ki}(\mathbf{x}) P_{ke}(\mathbf{x}) \, dv \qquad (2.90\text{e})$$

$$g_{ji} = \sum_{k=1}^{n} \left\{ \int_{S_{k0}} \phi_{kj}(\mathbf{x}) \frac{\psi_{ki}(\mathbf{x}) - k_k(\mathbf{x})[\partial\psi_{ki}(\mathbf{x})/\partial\mathbf{n}_{k0}]}{\alpha_k(\mathbf{x}) + \beta_k(\mathbf{x})} \, ds \right.$$

$$\left. + \int_{V_k} \psi_{ki}(\mathbf{x}) P_{kj}(\mathbf{x}) \, dv \right\} \qquad \text{for } j = 0, 1, 2, \ldots, q \quad (2.90\text{f})$$

which are similar to those given by Equations (2.30) except for the summation over the number of regions $k = 1$ to n in the foregoing results.

The steady-state solutions $T_{k\phi}(\mathbf{x})$, $T_{kp}(\mathbf{x})$, and $T_{k0}(\mathbf{x})$ are obtained by introducing these transforms into the inversion formula (2.76).

Boundary Conditions All of the Second Kind

In the preceding composite region problem if $\alpha(\mathbf{x}) = 0$ for all boundaries, then all boundary conditions for the problem are of the second kind. We also assume that $d(\mathbf{x}) = 0$. Then in the eigenvalue problem (2.71), $\mu_0 = 0$ is also an eigenvalue and $\psi_{k0} = \psi_{p0} = \psi_0 = $ constant is the corresponding eigenfunction. For such cases the solution (2.81) includes an additional term corresponding to the eigenvalue $\mu_0 = 0$. That is, for boundary condition of the second kind for all boundaries, Equation (2.81) takes the form

$$T_k(\mathbf{x}, t) = \frac{1}{\sum\limits_{k=1}^{n} (1,1)_k}$$

$$\times \left(\sum_{k=1}^{n} \left\{ (1, f_k) + \int_0^t \left[\int_{S_{k0}} \phi_k(\mathbf{x}, t') \, ds + \int_{V_k} P_k(\mathbf{x}, t') \, dv \right] dt' \right\} \right)$$

$$+ \sum_{i=1}^{\infty} \frac{1}{N_i} e^{-\mu_i^2 t} \psi_{ki}(\mathbf{x}) \left[\tilde{f}_i + \int_0^t g_i(t') e^{\mu_i^2 t'} \, dt' \right] \qquad (2.91\text{a})$$

where

$$(1,1)_k \equiv \int_{V_k} w_k(\mathbf{x}) \, dv \qquad (2.91\text{b})$$

Here N_i, $g_i(t')$, and $\tilde{f}_i$ are defined by Equations (2.74), (2.79b), and (2.80), respectively. We now show that the additional term in Equation (2.91a) is the mean potential $T_{av}(\mathbf{x})$ over the region.

Equation (2.71a) is integrated over the region V_k for $d(\mathbf{x}) = 0$, the volume integral is changed to the surface integral, the resulting expression is summed up for $k = 1, 2, \ldots, n$, and the fact that $\partial\psi_{ki}/\partial\mathbf{n} = 0$ for all boundaries is taken into account; we obtain

$$\sum_{k=1}^{n} (1, \psi_{ki}) = 0 \qquad (2.92)$$

We note that for $k = 1$ this result reduces to that given by Equation (2.31b) for the single region medium. Now the mean potential $T_{av}(t)$ over the entire region, by definition, is given by

$$T_{av}(t) = \frac{\displaystyle\sum_{k=1}^{n} (1, T_k)}{\displaystyle\sum_{k=1}^{n} (1, 1)_k} \qquad (2.93a)$$

When solution (2.91) is substituted into Equation (2.93a) and the result (2.92) is utilized, it becomes apparent that the first term on the right-hand side of Equation (2.91a) is indeed the mean potential over the region; that is,

$$T_{av}(t) = \frac{1}{\displaystyle\sum_{k=1}^{n} (1, 1)_k}$$
$$\times \left(\sum_{k=1}^{n} \left\{ (1, f_k) + \int_0^t \left[\int_{S_{k0}} \phi_k(\mathbf{x}, t')\, ds + \int_{V_k} P_k(\mathbf{x}, t')\, dv \right] dt' \right\} \right)$$

$$(2.93b)$$

For the special case of $k = 1$, this result reduces to that given by Equation (2.32) for the single region medium with all boundary conditions of the second kind.

The expression (2.91), although an exact solution, is not convenient for computational purposes. When the nonhomogeneous terms $P(\mathbf{x}, t)$ and $\phi(\mathbf{x}, t)$ are represented by exponentials and q-order polynomials of time as given by Equations (2.82), the general solution $T_k(\mathbf{x}, t)$ of the boundary value problem of Class II for $\alpha_k(\mathbf{x}) = 0$, $d_k(\mathbf{x}) = 0$ can be split up in the form

$$T_k(\mathbf{x}, t) = T_{av}(t) + T_{k\phi}(\mathbf{x})e^{-d_\phi t} + T_{kp}(\mathbf{x})e^{-d_p t} + \sum_{j=0}^{q} T_{kj}(\mathbf{x})t^j + T_{kt}(\mathbf{x}, t)$$

$$(2.94)$$

where $T_{av}(t)$ is given by Equation (2.93b) and the functions $T_{k\phi}(\mathbf{x})$, $T_{kp}(\mathbf{x})$, and $T_{kj}(\mathbf{x})$ are the solutions of the following steady-state problems

$$\frac{w_k(\mathbf{x})}{\sum\limits_{k=1}^{n}(1,1)_k}\sum_{k=1}^{n}\int_{S_{k0}}\phi_e(\mathbf{x})\,ds + L_kT_{k\phi}(\mathbf{x}) = d_\phi w_k(\mathbf{x})T_{k\phi}(\mathbf{x}), \qquad \mathbf{x}\in V_k$$

$$(2.95a)$$

subject to the boundary conditions (2.84b) and (2.84c);

$$\frac{w_k(\mathbf{x})}{\sum\limits_{k=1}^{n}(1,1)_k}\sum_{k=1}^{n}\int_{V_k}P_{ke}(\mathbf{x})\,dv + L_kT_{kp}(\mathbf{x}) = d_p w_k(\mathbf{x})T_{kp}(\mathbf{x}) + P_{ke}(\mathbf{x}),$$

$$\mathbf{x}\in V_k \quad (2.95b)$$

subject to the boundary conditions (2.85b) and (2.85c);

$$\frac{w_k(\mathbf{x})}{\sum\limits_{k=1}^{n}(1,1)_k}\sum_{k=1}^{n}\left[\int_{S_{k0}}\phi_q(\mathbf{x})\,ds + \int_{V_k}P_{kq}(\mathbf{x})\,dv\right] + L_kT_{kq}(\mathbf{x}) = P_{kq}(\mathbf{x}),$$

$$\mathbf{x}\in V_k \quad (2.95c)$$

subject to the boundary conditions (2.86b) and (2.86c);

$$\frac{w_k(\mathbf{x})}{\sum\limits_{k=1}^{n}(1,1)_k}\sum_{k=1}^{n}\left[\int_{S_{k0}}\phi_{kj}(\mathbf{x})\,ds + \int_{V_k}P_{kj}(\mathbf{x})\,dv\right]$$

$$+L_kT_{kj}(\mathbf{x}) +(j+1)w_k(\mathbf{x})T_{k,j+1}(\mathbf{x}) = P_{kj}(\mathbf{x}),$$

$$j = q-1, q-2,\ldots,1,0, \qquad \mathbf{x}\in V_k \quad (2.95d)$$

subject to the boundary conditions (2.87b) and (2.87c).

The function $T_{kt}(\mathbf{x}, t)$ is the solution of the following transient homogeneous problem

$$w_k(\mathbf{x})\frac{\partial T_{kt}(\mathbf{x}, t)}{\partial t} + L_kT_{kt}(\mathbf{x}, t) = 0, \qquad \mathbf{x}\in V_k \qquad (2.95e)$$

subject to the boundary conditions (2.88b) and (2.88c) and the initial conditions

$$T_{kt}(\mathbf{x},0) = f_k(\mathbf{x}) - \frac{\displaystyle\sum_{k=1}^{n}(1,f_k)}{\displaystyle\sum_{k=1}^{n}(1,1)_k} - T_{k\phi}(\mathbf{x}) - T_{kp}(\mathbf{x}) - T_{k0}(\mathbf{x})   \quad (2.95\mathrm{f})$$

Finally, introducing Equation (2.94) into the definition (2.93a) we obtain the additional conditions for the problems defined by Equations (2.95) as

$$\sum_{k=1}^{n}(1,T_{k\phi}) = \sum_{k=1}^{n}(1,T_{kp}) = \sum_{k=1}^{n}(1,T_{kj}) = \sum_{k=1}^{n}(1,T_{kt}) = 0   \quad (2.95\mathrm{g})$$

We note that for the case of $k = 1$, the problems defined by Equations (2.95) reduce to those given by Equations (2.35)–(2.39) for the single region problem subject to boundary conditions of the second kind for all boundaries.

The validity of the preceding splitting-up process can be verified by introducing Equation (2.94) into the boundary value problem (2.70) for $\alpha_k(\mathbf{x}) = 0$ and $d_k(\mathbf{x}) = 0$.

2.3   GENERAL SOLUTION OF PROBLEMS OF CLASS III

The problems in this class, as discussed in Chapter 1, include heat or mass diffusion problems characterized by two diffusion type equations coupled at the boundary as given in the following.

$$w_k(\mathbf{x})\frac{\partial T_k(\mathbf{x},t)}{\partial t} + L_k T_k(\mathbf{x},t) = P_k(\mathbf{x},t), \quad k = 1,2, \quad \mathbf{x} \in V$$

$$(2.96\mathrm{a})$$

subject to the boundary conditions

$$B_{k1}T_1(\mathbf{x},t) + B_{k2}T_2(\mathbf{x},t) = \phi_k(\mathbf{x},t), \quad k = 1,2, \quad \mathbf{x} \in S_1 \quad (2.96\mathrm{b})$$

$$B_k T_k(\mathbf{x},t) = \phi_k(\mathbf{x},t), \quad k = 1,2, \quad \mathbf{x} \in S_2 \quad (2.96\mathrm{c})$$

and the initial condition

$$T_k(\mathbf{x},0) = f_k(\mathbf{x}), \quad k = 1,2, \quad \mathbf{x} \in V \quad (2.96\mathrm{d})$$

where

$$L_k \equiv -\nabla \cdot [k_k(\mathbf{x})\nabla] + d_k(\mathbf{x}) \tag{2.96e}$$

$$B_{k1} \equiv \alpha_{k1} + \beta_{k1}k_1(\mathbf{x})\frac{\partial}{\partial \mathbf{n}} \tag{2.96f}$$

$$B_{k2} \equiv \alpha_{k2} + \beta_{k2}k_2(\mathbf{x})\frac{\partial}{\partial \mathbf{n}} \tag{2.96g}$$

$$B_k \equiv \alpha_k(\mathbf{x}) + \beta_k(\mathbf{x})k_k(\mathbf{x})\frac{\partial}{\partial \mathbf{n}} \tag{2.96h}$$

The boundary surface S of the finite region V of arbitrary geometry is composed of two bounding surfaces S_1 and S_2 (i.e., $S = S_1 \cup S_2$).

In the following analysis, we first develop the appropriate finite integral transform pair and then apply this transform pair for the solution of the previous problem. In addition we discuss splitting up the general problem into simpler problems and consider the special case of all boundary conditions being of the second kind.

Development of the Integral Transform Pair

To develop the desired finite integral transform pair we consider the following homogeneous auxiliary problem [13, 14]

$$\mu^2 w_k(\mathbf{x})\psi_k(\mathbf{x}) = L_k\psi_k(\mathbf{x}), \qquad k = 1, 2, \qquad \mathbf{x} \in V \tag{2.97a}$$

subject to the boundary conditions

$$B_{k1}\psi_1(\mathbf{x}) + B_{k2}\psi_2(\mathbf{x}) = 0, \qquad k = 1, 2, \qquad \mathbf{x} \in S_1 \tag{2.97b}$$

$$B_k\psi_k(\mathbf{x}) = 0, \qquad k = 1, 2, \qquad \mathbf{x} \in S_2 \tag{2.97c}$$

where the linear operators L_k, B_{k1}, B_{k2}, and B_k are defined by Equations (2.96e)–(2.96h), respectively.

We now expand the desired potentials $T_k(\mathbf{x}, t)$, $k = 1, 2$ in terms of the eigenfunctions $\psi_{ki}(\mathbf{x})$ of the preceding eigenvalue problem in the form

$$T_k(\mathbf{x}, t) = \sum_{i=1}^{\infty} A_i(t)\psi_{ki}(\mathbf{x}), \qquad k = 1, 2, \qquad \mathbf{x} \in V \tag{2.98}$$

where the summation is taken over all discrete spectra of eigenvalues μ_i.

To determine the unknown coefficients $A_i(t)$, the orthogonality condition for these eigenfunctions is needed. The homogeneous system defined by Equations (2.97) is an eigenvalue problem having a common set of eigenvalues

μ_i, but different eigenfunctions; therefore, it does not belong to the conventional Sturm–Liouville system. Then the first step in the analysis is to establish the orthogonality condition appropriate to the system (2.97) as now described.

By the application of the conventional manipulations discussed in the previous sections in the derivation of Equation (2.6) or that of leading to Equation (2.73), we obtain

$$(\psi_{ki}, \psi_{kj}) = \frac{1}{\mu_i^2 - \mu_j^2} \int_S k_k(\mathbf{x}) \begin{vmatrix} \psi_{ki}(\mathbf{x}) & \dfrac{\partial \psi_{ki}(\mathbf{x})}{\partial \mathbf{n}} \\[3mm] \psi_{kj}(\mathbf{x}) & \dfrac{\partial \psi_{kj}(\mathbf{x})}{\partial \mathbf{n}} \end{vmatrix} ds, \qquad i \neq j \quad (2.99)$$

where $S = S_1 \cup S_2$ and $k = 1, 2$.

By utilizing the boundary conditions (2.97b), $\psi_1(\mathbf{x})$ and $k_1(\mathbf{x})[\partial\psi_1(\mathbf{x})/\partial\mathbf{n}]$ are expressed in terms of $\psi_2(\mathbf{x})$ and $k_2(\mathbf{x})[\partial\psi_2(\mathbf{x})/\partial\mathbf{n}]$; the resulting expressions are then utilized to compute the quantity

$$\begin{vmatrix} \psi_{1i} & k_1 \dfrac{\partial \psi_{1i}}{\partial \mathbf{n}} \\[3mm] \psi_{1j} & k_2 \dfrac{\partial \psi_{1j}}{\partial \mathbf{n}} \end{vmatrix}$$

After some manipulations the following expression is obtained

$$\sum_{k=1}^{2} \sigma_k k_k(\mathbf{x}) \begin{vmatrix} \psi_{ki}(\mathbf{x}) & \dfrac{\partial \psi_{ki}(\mathbf{x})}{\partial \mathbf{n}} \\[3mm] \psi_{kj}(\mathbf{x}) & \dfrac{\partial \psi_{kj}(\mathbf{x})}{\partial \mathbf{n}} \end{vmatrix} = 0, \qquad \mathbf{x} \in S_1 \quad (2.100a)$$

where

$$\sigma_1 = \begin{vmatrix} \beta_{11} & \alpha_{11} \\ \beta_{21} & \alpha_{21} \end{vmatrix} \quad \text{and} \quad \sigma_2 = \begin{vmatrix} \alpha_{12} & \beta_{12} \\ \alpha_{22} & \beta_{22} \end{vmatrix} \qquad (2.100b)$$

Also, from the boundary conditions (2.97c) for $i \neq j$ it follows that

$$\begin{vmatrix} \psi_{ki}(\mathbf{x}) & \dfrac{\partial \psi_{ki}(\mathbf{x})}{\partial \mathbf{n}} \\[3mm] \psi_{kj}(\mathbf{x}) & \dfrac{\partial \psi_{kj}(\mathbf{x})}{\partial \mathbf{n}} \end{vmatrix} = 0, \qquad \mathbf{x} \in S_2 \quad (2.100c)$$

Equation (2.99) is multiplied by σ_1 for $k = 1$, by σ_2 for $k = 2$, the results are added, and the expressions given by Equations (2.100) are taken into account; we obtain

$$\sum_{k=1}^{2} \sigma_k (\psi_{ki}, \psi_{kj}) = \delta_{ij} N_i \qquad (2.101a)$$

where

$$N_i \equiv \sum_{k=1}^{2} \sigma_k (\psi_{ki}, \psi_{ki})$$

$$= \frac{1}{2\mu_i} \sum_{k=1}^{2} \sigma_k \int_S k_k(\mathbf{x}) \begin{vmatrix} \left[\dfrac{\partial \psi_k(\mathbf{x})}{\partial \mu}\right]_{\mu=\mu_i} & \left[\dfrac{\partial^2 \psi_k(\mathbf{x})}{\partial \mathbf{n}\, \partial \mu}\right]_{\mu=\mu_i} \\ \psi_{ki}(\mathbf{x}) & \dfrac{\partial \psi_{ki}(\mathbf{x})}{\partial \mathbf{n}} \end{vmatrix} ds,$$

$$(2.101b)$$

This result is the desired orthogonality condition for the eigenvalue problem defined by Equations (2.97).

The above orthogonality condition is now used to determine the coefficients $A_i(t)$ in the expansion (2.98) as now described. We operate on both sides of Equation (2.98) by the operator

$$\sigma_k \int_V w_k(\mathbf{x}) \psi_{kj}(\mathbf{x})\, dv$$

sum up the results for $k = 1$ and 2, and utilize the orthogonality condition (2.101); we obtain

$$A_i(t) = \frac{1}{N_i} \sum_{k=1}^{2} \sigma_k (\psi_{ki}, T_k) \qquad (2.102)$$

This expression for $A_i(t)$ is introduced into the expansion (2.98) and the resulting expression is split up into two parts to define the desired finite integral transform pair as

Inversion formula: $T_k(\mathbf{x}, t) = \sum_{i=1}^{\infty} \frac{1}{N_i} \psi_{ki}(\mathbf{x}) \tilde{T}_i(t), \qquad k = 1, 2$

$$(2.103)$$

Finite integral transform: $\tilde{T}_i(t) = \sum_{k=1}^{2} \sigma_k (\psi_{ki}, T_k) \qquad (2.104)$

where N_i is as defined by Equation (2.101b).

Method of Solution

The boundary value problem (2.96) is now solved by the application of the integral transform pair given by Equations (2.103) and (2.104) in a manner similar to those described in connection with the solutions of problems of Class I and II. That is, Equations (2.96a) are multiplied by $\sigma_k \psi_{ki}(\mathbf{x})$, Equations (2.97a) by $\sigma_k T_k(\mathbf{x}, t)$, respectively, for $k = 1$ and 2, the resulting four expressions are added, integrated over the volume V, and the definition of the finite integral transform is utilized to obtain

$$\frac{d\tilde{T}_i(t)}{dt} + \mu_i^2 \tilde{T}_i(t)$$

$$= \sum_{k=1}^{2} \sigma_k \left\{ \int_V [\psi_{ki} \nabla \cdot (k_k \nabla T_k) - T_k \nabla \cdot (k_k \psi_{ki})] \, dv + \int_V \psi_{ki} P_k \, dv \right\}$$

$$(2.105a)$$

The first volume integral on the right-hand side is changed into a surface integral

$$\frac{d\tilde{T}_i(t)}{dt} + \mu_i^2 \tilde{T}_i(t) = g_i(t) \qquad (2.105b)$$

where

$$g_i(t) = \sum_{k=1}^{2} \sigma_k \int_{S_1} k_k(\mathbf{x}) \begin{vmatrix} \psi_{ki}(\mathbf{x}) & \dfrac{\partial \psi_{ki}(\mathbf{x})}{\partial \mathbf{n}} \\ T_k(\mathbf{x}, t) & \dfrac{\partial T_k(\mathbf{x}, t)}{\partial \mathbf{n}} \end{vmatrix} ds$$

$$+ \sum_{k=1}^{2} \sigma_k \left\{ \int_{S_2} k_k(\mathbf{x}) \begin{vmatrix} \psi_{ki}(\mathbf{x}) & \dfrac{\partial \psi_{ki}(\mathbf{x})}{\partial \mathbf{n}} \\ T_k(\mathbf{x}, t) & \dfrac{\partial T_k(\mathbf{x}, t)}{\partial \mathbf{n}} \end{vmatrix} ds + \int_V \psi_{ki}(\mathbf{x}) P_k(\mathbf{x}, t) \, dv \right\}$$

$$(2.105c)$$

The integrands of the surface integrals over S_1 and S_2 are evaluated as now described.

By utilizing the boundary conditions (2.96b), $T_1(\mathbf{x}, t)$, and $k_1(\mathbf{x})[\partial T_1(\mathbf{x})/\partial \mathbf{n}]$ are expressed in terms of $T_2(\mathbf{x}, t)$ and $k_2(\mathbf{x})[\partial T_2(\mathbf{x}, t)/\partial \mathbf{n}]$; similarly, by utilizing the boundary conditions (2.97b), $\psi_1(\mathbf{x})$ and $k_1(\mathbf{x})[\partial \psi_1(\mathbf{x})/\partial \mathbf{n}]$ are expressed in terms of $\psi_2(\mathbf{x})$ and $k_2(\mathbf{x})[\partial\psi_2(\mathbf{x})/\partial\mathbf{n}]$. The resulting expressions are then utilized to compute the quantity

$$
\begin{vmatrix}
\psi_{1i} & k_1\dfrac{\partial\psi_{1i}}{\partial\mathbf{n}} \\[2mm]
T_1 & k_1\dfrac{\partial T_1}{\partial\mathbf{n}}
\end{vmatrix}
$$

After some manipulations the following expression is obtained

$$
\sum_{k=1}^{2}\sigma_k k_k(\mathbf{x})
\begin{vmatrix}
\psi_{ki}(\mathbf{x}) & \dfrac{\partial\psi_{ki}(\mathbf{x})}{\partial\mathbf{n}} \\[3mm]
T_k(\mathbf{x},t) & \dfrac{\partial T_k(\mathbf{x},t)}{\partial\mathbf{n}}
\end{vmatrix}
=
\begin{vmatrix}
\phi_1(\mathbf{x},t) & B_{11}\psi_{1i}(\mathbf{x}) \\[2mm]
\phi_2(\mathbf{x},t) & B_{21}\psi_{1i}(\mathbf{x})
\end{vmatrix}, \qquad \mathbf{x}\in S_1
$$

$$(2.106\text{a})$$

The boundary conditions (2.96c) and (2.97c) are solved for $\alpha_k(\mathbf{x})$ and $\beta_k(\mathbf{x})$; the resulting expressions for $\alpha_k(\mathbf{x})$ and $\beta_k(\mathbf{x})$ are added to obtain

$$
k_k(\mathbf{x})
\begin{vmatrix}
\psi_{ki}(\mathbf{x}) & \dfrac{\partial\psi_{ki}(\mathbf{x})}{\partial\mathbf{n}} \\[3mm]
T_k(\mathbf{x},t) & \dfrac{\partial T_k(\mathbf{x},t)}{\partial\mathbf{n}}
\end{vmatrix}
= \phi_k(\mathbf{x},t)\frac{\psi_{ki}(\mathbf{x})-k_k(\mathbf{x})[\partial\psi_{ki}(\mathbf{x})/\partial\mathbf{n}]}{\alpha_k(\mathbf{x})+\beta_k(\mathbf{x})},
$$

$$
\mathbf{x}\in S_2 \quad (2.106\text{b})
$$

The expressions given by Equations (2.106) are introduced into the integrand of the surface integral in Equation (2.105c). Then Equations (2.105) become

$$
\frac{d\tilde{T}_i(t)}{dt}+\mu_i^2\tilde{T}_i(t)=g_i(t) \tag{2.107a}
$$

where

$$
g_i(t)=\int_{S_1}
\begin{vmatrix}
\phi_1(\mathbf{x},t) & B_{11}\psi_{1i}(\mathbf{x}) \\[2mm]
\phi_2(\mathbf{x},t) & B_{21}\psi_{1i}(\mathbf{x})
\end{vmatrix} ds
$$

$$
+\sum_{k=1}^{2}\sigma_k\Bigg\{\int_{S_2}\phi_k(\mathbf{x},t)\frac{\psi_{ki}(\mathbf{x})-k_k(\mathbf{x})[\partial\psi_{ki}(\mathbf{x})/\partial\mathbf{n}]}{\alpha_k(\mathbf{x})+\beta_k(\mathbf{x})}\,ds
$$

$$
+\int_V\psi_{ki}(\mathbf{x})P_k(\mathbf{x},t)\,dv\Bigg\} \tag{2.107b}
$$

The initial condition needed for the solution of Equation (2.107) is obtained by taking the integral transform of the initial condition (2.96d) according to the integral transform (2.104). We find

$$\tilde{T}_i(0) = \sum_{k=1}^{2} \sigma_k(\psi_{ki}, f_k) \equiv \tilde{f}_i \tag{2.108}$$

Finally, Equation (2.107) is solved subject to the initial condition (2.108) to obtain the finite integral transform $\tilde{T}_i(t)$. When this transform is inverted by the inversion formula (2.103), the desired solution $T_k(\mathbf{x}, t)$ of the boundary value problem of Class III defined by Equations (2.96) becomes

$$T_k(\mathbf{x}, t) = \sum_{i=1}^{\infty} \frac{1}{N_i} e^{-\mu_i^2 t} \psi_{ki}(\mathbf{x}) \left[\tilde{f}_i + \int_0^t g_i(t') e^{\mu_i^2 t'} dt' \right] \tag{2.109}$$

where N_i, $g_i(t')$, and $\tilde{f}_i$ are defined by Equations (2.101b), (2.107b), and (2.108), respectively.

Splitting up the General Problem

We now examine the splitting up of the Class III problem for the case when the nonhomogeneous terms $P_k(\mathbf{x}, t)$ and $\phi_k(\mathbf{x}, t)$ of the problem (2.96) are exponentials and q-order polynomials of time in the form

$$P_k(\mathbf{x}, t) = P_{ke}(\mathbf{x}) e^{-d_p t} + \sum_{j=0}^{q} P_{kj}(\mathbf{x}) t^j,$$

$$k = 1, 2, \qquad \mathbf{x} \in V \quad (2.110a)$$

$$\phi_k(\mathbf{x}, t) = \phi_{ke}(\mathbf{x}) e^{-d_\phi t} + \sum_{j=0}^{q} \phi_{kj}(\mathbf{x}) t^j, \qquad k = 1, 2, \qquad \mathbf{x} \in S$$

$$(2.110b)$$

where d_p and d_ϕ are constants. Then the general solution of the boundary value problem (2.96) can be written in the form

$$T_k(\mathbf{x}, t) = T_{k\phi}(\mathbf{x}) e^{-d_\phi t} + T_{kp}(\mathbf{x}) e^{-d_p t} + \sum_{j=0}^{q} T_{kj}(\mathbf{x}) t^j + T_{kt}(\mathbf{x}, t),$$

$$k = 1, 2, \qquad \mathbf{x} \in V \quad (2.111)$$

where the functions $T_{k\phi}(\mathbf{x})$, $T_{kp}(\mathbf{x})$, and $T_{kj}(\mathbf{x})$ are the solutions of the following steady-state problems

$$L_k T_{k\phi}(\mathbf{x}) = d_\phi w_k(\mathbf{x}) T_{k\phi}(\mathbf{x}), \qquad k = 1, 2, \qquad \mathbf{x} \in V \qquad (2.112a)$$

$$B_{k1} T_{1\phi}(\mathbf{x}) + B_{k2} T_{2\phi}(\mathbf{x}) = \phi_{ke}(\mathbf{x}), \qquad k = 1, 2, \qquad \mathbf{x} \in S_1 \qquad (2.112b)$$

$$B_k T_{k\phi}(\mathbf{x}) = \phi_{ke}(\mathbf{x}), \qquad k = 1, 2, \qquad \mathbf{x} \in S_2 \qquad (2.112c)$$

$$L_k T_{kp}(\mathbf{x}) = d_p w_k(\mathbf{x}) T_{kp}(\mathbf{x}) + P_{ke}(\mathbf{x}), \qquad k = 1, 2, \qquad \mathbf{x} \in V$$

$$(2.113a)$$

$$B_{k1} T_{1p}(\mathbf{x}) + B_{k2} T_{2p}(\mathbf{x}) = 0, \qquad k = 1, 2, \qquad \mathbf{x} \in S_1 \qquad (2.113b)$$

$$B_k T_{kp}(\mathbf{x}) = 0, \qquad k = 1, 2, \qquad \mathbf{x} \in S_2 \qquad (2.113c)$$

and

$$L_k T_{kq}(\mathbf{x}) = P_{kq}(\mathbf{x}), \qquad k = 1, 2, \qquad \mathbf{x} \in V \qquad (2.114a)$$

$$B_{k1} T_{1q}(\mathbf{x}) + B_{k2} T_{2q}(\mathbf{x}) = \phi_{kq}(\mathbf{x}), \qquad k = 1, 2, \qquad \mathbf{x} \in S_1 \qquad (2.114b)$$

$$B_k T_{kq}(\mathbf{x}) = \phi_{kq}(\mathbf{x}), \qquad k = 1, 2, \qquad \mathbf{x} \in S_2 \qquad (2.114c)$$

$$L_k T_{kj}(\mathbf{x}) + (j + 1) w_k(\mathbf{x}) T_{k, j+1}(\mathbf{x}) = P_{kj}(\mathbf{x}), \qquad k = 1, 2, \qquad \mathbf{x} \in V$$

$$(2.115a)$$

$$B_{k1} T_{1j}(\mathbf{x}) + B_{k2} T_{2j}(\mathbf{x}) = \phi_{kj}(\mathbf{x}), \qquad k = 1, 2, \qquad \mathbf{x} \in S_1 \qquad (2.115b)$$

$$B_k T_{kj}(\mathbf{x}) = \phi_{kj}(\mathbf{x}), \qquad k = 1, 2, \qquad \mathbf{x} \in S_2 \qquad (2.115c)$$

where $j = q - 1, q - 2, \ldots, 1, 0$.

The function $T_{kt}(\mathbf{x}, t)$ is the solution of the following transient homogeneous problem.

$$w_k(\mathbf{x}) \frac{\partial T_{kt}(\mathbf{x}, t)}{\partial t} + L_k T_{kt}(\mathbf{x}, t) = 0, \qquad k = 1, 2, \qquad \mathbf{x} \in V \qquad (2.116a)$$

$$B_{k1} T_{kt}(\mathbf{x}, t) + B_{k2} T_{kt}(\mathbf{x}, t) = 0, \qquad k = 1, 2, \qquad \mathbf{x} \in S_1 \qquad (2.116b)$$

$$B_k T_{kt}(\mathbf{x}, t) = 0, \qquad k = 1, 2, \qquad \mathbf{x} \in S_2 \qquad (2.116c)$$

$$T_{kt}(\mathbf{x}, 0) = f_k(\mathbf{x}) - T_{k\phi}(\mathbf{x}) - T_{kp}(\mathbf{x}) - T_{k0}(\mathbf{x}), \qquad k = 1, 2, \qquad \mathbf{x} \in V$$

$$(2.116d)$$

The validity of this splitting-up process can be verified by substituting Equation (2.111) into the boundary value problem (2.96).

The solution of the transient problem (2.116) is obtained as a special case from the general solution (2.109) as

$$T_{kt}(\mathbf{x}, t) = \sum_{i=1}^{\infty} \frac{1}{N_i} e^{-\mu_i^2 t} \psi_{ki}(\mathbf{x}) \big(\tilde{f}_i - \tilde{T}_{\phi i} - \tilde{T}_{pi} - \tilde{T}_{0i} \big), \qquad k = 1, 2, \qquad \mathbf{x} \in V$$

(2.117a)

where $\tilde{f}_i$, $\tilde{T}_{\phi i}$, $\tilde{T}_{pi}$, and $\tilde{T}_{0i}$ are the finite integral transforms of the functions $f_k(\mathbf{x})$, $T_{k\phi}(\mathbf{x})$, $T_{kp}(\mathbf{x})$, and $T_{k0}(\mathbf{x})$, respectively. The integral transforms $\tilde{T}_{\phi i}$, $\tilde{T}_{pi}$, and $\tilde{T}_{0i}$ are obtainable by the application of the finite integral transform (2.104) to the steady-state problems (2.112) to (2.115). We summarize the results

$$\tilde{T}_{\phi i} = \frac{g_{\phi i}}{\mu_i^2 - d_\phi}$$

(2.117b)

$$\tilde{T}_{pi} = \frac{g_{pi}}{\mu_i^2 - d_p}$$

(2.117c)

$$\tilde{T}_{0i} = \sum_{j=0}^{q} (-1)^j j! \frac{g_{ji}}{\mu_i^{2(j+1)}}$$

(2.117d)

where

$$g_{\phi i} = \int_{S_1} \begin{vmatrix} \phi_{1e}(\mathbf{x}) & B_{11}\psi_{1i}(\mathbf{x}) \\ \phi_{2e}(\mathbf{x}) & B_{21}\psi_{1i}(\mathbf{x}) \end{vmatrix} ds$$

$$+ \sum_{k=1}^{2} \sigma_k \int_{S_2} \phi_{ke}(\mathbf{x}) \frac{\psi_{ki}(\mathbf{x}) - k_k(\mathbf{x})[\partial\psi_{ki}(\mathbf{x})/\partial\mathbf{n}]}{\alpha_k(\mathbf{x}) + \beta_k(\mathbf{x})} ds \qquad (2.117e)$$

$$g_{pi} = \sum_{k=1}^{2} \sigma_k \int_V \psi_{ki}(\mathbf{x}) P_{ke}(\mathbf{x}) \, dv$$

(2.117f)

$$g_{ji} = \int_{S_1} \begin{vmatrix} \phi_{1j}(\mathbf{x}) & B_{11}\psi_{1i}(\mathbf{x}) \\ \phi_{2j}(\mathbf{x}) & B_{21}\psi_{1i}(\mathbf{x}) \end{vmatrix} ds$$

$$+ \sum_{k=1}^{2} \sigma_k \left\{ \int_{S_2} \phi_{kj}(\mathbf{x}) \frac{\psi_{ki}(\mathbf{x}) - k_k(\mathbf{x})[\partial\psi_{ki}(\mathbf{x})/\partial\mathbf{n}]}{\alpha_k(\mathbf{x}) + \beta_k(\mathbf{x})} ds \right.$$

$$\left. + \int_V \psi_{ki}(\mathbf{x}) P_{kj}(\mathbf{x}) \, dv \right\}$$

(2.117g)

we note that the g_{ji} function is similar in form to that defined previously by Equation (2.107b).

When the transforms $\tilde{T}_{\phi i}$, $\tilde{T}_{pi}$, and $\tilde{T}_{0i}$ are inserted into the inversion formula (2.103), the solutions $T_{k\phi}(\mathbf{x})$, $T_{kp}(\mathbf{x})$, and $T_{k0}(\mathbf{x})$ are obtained.

Boundary Conditions All of the Second Kind

If $\alpha_k(\mathbf{x}) = 0$, $d_k(\mathbf{x}) = 0$, and $\alpha_{11}\alpha_{22} - \alpha_{12}\alpha_{21} = 0$ [i.e., obtained from the boundary conditions (2.97b) for $k = 1$ and 2, and by setting $\partial\psi_k/\partial\mathbf{n} = 0$], then $\mu_0 = 0$ is also an eigenvalue of the problem (2.97). The corresponding constant eigenfunctions ψ_{10} and ψ_{20} must satisfy the relation

$$\alpha_{k1}\psi_{10} + \alpha_{k2}\psi_{20} = 0, \qquad k = 1, 2 \tag{2.118}$$

which is apparent from the boundary condition (2.97b). For such a case the solution (2.109) includes an additional term corresponding to the eigenvalue $\mu_0 = 0$; the resulting solution takes the form

$$
T_k(\mathbf{x}, t) = \frac{\psi_{k0}}{\displaystyle\sum_{k=1}^{2} \sigma_k(\psi_{k0}, \psi_{k0})} \left(\sum_{k=1}^{2} \sigma_k(\psi_{k0}, f_k) + \int_0^t \left\{ \psi_{10} \int_{S_1} \begin{vmatrix} \phi_1(\mathbf{x}, t') & \alpha_{11} \\ \phi_2(\mathbf{x}, t') & \alpha_{21} \end{vmatrix} ds \right. \right.
$$

$$
\left. \left. + \sum_{k=1}^{2} \sigma_k \psi_{k0} \left[\int_{S_2} \phi_k(\mathbf{x}, t') \, ds + \int_V P_k(\mathbf{x}, t') \, dv \right] \right\} dt' \right)
$$

$$
+ \sum_{i=1}^{\infty} \frac{1}{N_i} e^{-\mu_i^2 t} \psi_{ki}(\mathbf{x}) \left[\tilde{f}_i + \int_0^t g_i(t') e^{\mu_i^2 t'} \, dt' \right] \tag{2.119}
$$

2.4   GENERAL SOLUTION OF PROBLEMS OF CLASS IV

The boundary value problems in this class contain a set of diffusion equations in which the potentials $T_k(\mathbf{x}, t)$, $k = 1, 2, 3, \ldots, n$ in every point in the region V are coupled through the source-sink term. The solution of such a system of equations by the finite integral transform technique is discussed in Reference 14. The mathematical formulation of the problems in this class is given as [32]

$$
w_k(\mathbf{x}) \frac{\partial T_k(\mathbf{x}, t)}{\partial t} + L_k T_k(\mathbf{x}, t) = P_k(\mathbf{x}, t) + b(\mathbf{x}) \sum_{p=1}^{n} \alpha_{kp} \left[T_p(\mathbf{x}, t) - T_k(\mathbf{x}, t) \right],
$$

$$
\alpha_{kp} = \alpha_{pk}, \qquad k = 1, 2, \ldots, n, \qquad \mathbf{x} \in V \tag{2.120a}
$$

subject to the boundary and initial conditions

$$B_k T_k(\mathbf{x}, t) = \phi_k(\mathbf{x}, t), \qquad k = 1, 2, \ldots, n, \qquad \mathbf{x} \in S \qquad (2.120b)$$

$$T_k(\mathbf{x}, 0) = f_k(\mathbf{x}), \qquad k = 1, 2, \ldots, n, \qquad \mathbf{x} \in V \qquad (2.120c)$$

where

$$L_k \equiv -\nabla \cdot \left[k_k(\mathbf{x}) \nabla \right] + d_k(\mathbf{x}) \qquad (2.120d)$$

$$B_k \equiv \alpha_k(\mathbf{x}) + \beta_k(\mathbf{x}) k_k(\mathbf{x}) \frac{\partial}{\partial \mathbf{n}} \qquad (2.120e)$$

The finite integral transform pair that is needed for the solution of this boundary value problem is now developed.

Development of the Integral Transform Pair

To develop the desired integral transform pair we consider the following homogeneous auxiliary problem

$$\mu^2 w_k(\mathbf{x}) \psi_k(\mathbf{x}) + b(\mathbf{x}) \sum_{p=1}^{n} \alpha_{kp} \left[\psi_p(\mathbf{x}) - \psi_k(\mathbf{x}) \right] = L_k \psi_k(\mathbf{x}),$$

$$\alpha_{kp} = \alpha_{pk}, \qquad k = 1, 2, \ldots, n, \qquad \mathbf{x} \in V \quad (2.121a)$$

subject to the boundary conditions

$$B_k \psi_k(\mathbf{x}) = 0, \qquad k = 1, 2, \ldots, n, \qquad \mathbf{x} \in S \qquad (2.121b)$$

where L_k and B_k are defined by Equations (2.120d) and (2.120e), respectively.

We now expand the desired potential $T_k(\mathbf{x}, t)$ in terms of the eigenfunctions of the foregoing eigenvalue problem in the form

$$T_k(\mathbf{x}, t) = \sum_{i=1}^{\infty} A_i(t) \psi_{ki}(\mathbf{x}), \qquad k = 1, 2, \ldots, n, \qquad \mathbf{x} \in V \quad (2.122)$$

where the summation is taken over all discrete eigenvalues.

To determine the unknown coefficients $A_i(t)$, the orthogonality condition for these eigenfunctions is needed. The orthogonality property is determined by manipulations similar to those discussed in the previous section; the result is given as

$$\sum_{k=1}^{n} (\psi_{ki}, \psi_{kj}) = \delta_{ij} N_i \qquad (2.123a)$$

where

$$N_i = \sum_{k=1}^{n} (\psi_{ki}, \psi_{ki}) \tag{2.123b}$$

This orthogonality condition is used to determine the expansion coefficients $A_i(t)$ as described in the following.

Both sides of Equation (2.122) are operated on by the operator

$$\int_V w_k(\mathbf{x}) \psi_{kj}(\mathbf{x}) \, dv$$

the results are summed up for $k = 1, 2, \ldots, n$, and the orthogonality condition (2.123) is utilized to obtain the $A_i(t)$. This solution for $A_i(t)$ is introduced into the expansion (2.122) and the resulting expression is split up into two parts to define the finite integral transform pair as

Inversion formula: $\quad T_k(\mathbf{x}, t) = \sum_{i=1}^{\infty} \dfrac{1}{N_i} \psi_{ki}(\mathbf{x}) \tilde{T}_i(t) \quad$ (2.124)

Finite integral transform: $\quad \tilde{T}_i(t) = \sum_{k=1}^{n} (\psi_{ki}, T_k) \quad$ (2.125)

where N_i is defined by Equation (2.123b).

Method of Solution

The finite integral transform pair (2.124) and (2.125) is now applied to solve the boundary value problem (2.120) in a similar manner described in the previous sections for the solution of boundary value problems. That is, Equations (2.120a) are multiplied by $\psi_{ki}(\mathbf{x})$, Equations (2.121a) by $T_k(\mathbf{x}, t)$, the results are summed up for $k = 1, 2, \ldots, n$, integrated over the region V, the definition of the integral transform (2.125) is utilized, and the volume integral is transformed into the surface integral. We obtain

$$\frac{d\tilde{T}_i(t)}{dt} + \mu_i^2 \tilde{T}_i(t) = g_i(t) \tag{2.126a}$$

where

$$g_i(t) = \sum_{k=1}^{n} \left\{ \int_S k_k(\mathbf{x}) \begin{vmatrix} \psi_{ki}(\mathbf{x}) & \dfrac{\partial \psi_{ki}(\mathbf{x})}{\partial \mathbf{n}} \\ T_k(\mathbf{x}, t) & \dfrac{\partial T_k(\mathbf{x}, t)}{\partial \mathbf{n}} \end{vmatrix} ds + \int_V \psi_{ki}(\mathbf{x}) P_k(\mathbf{x}, t) \, dv \right\}$$

$$\tag{2.126b}$$

We note that these expressions are similar to those given by Equations (2.78). Here the integrand of the surface integral is evaluated by utilizing the boundary conditions (2.120b) and (2.121b); then $g_i(t)$ is expressed in the form

$$g_i(t) = \sum_{k=1}^{n} \left\{ \int_S \phi_k(\mathbf{x}, t) \frac{\psi_{ki}(\mathbf{x}) - k_k(\mathbf{x})[\partial\psi_{ki}(\mathbf{x})/\partial\mathbf{n}]}{\alpha_k(\mathbf{x}) + \beta_k(\mathbf{x})} ds \right.$$

$$\left. + \int_V \psi_{ki}(\mathbf{x}) P_k(\mathbf{x}, t)\, dv \right\} \qquad (2.126c)$$

The initial condition needed to solve Equation (2.126a) is determined by taking the finite integral transform of the initial condition (2.120c) according to the transform (2.125); that is,

$$\tilde{T}_i(0) = \sum_{k=1}^{n} (\psi_{ki}, f_k) \equiv \tilde{f}_i \qquad (2.127)$$

Equation (2.126a) is solved subject to the initial condition (2.127); the resulting transform is inverted by the inversion formula (2.124) to obtain the general solution of the boundary value problem of Class IV, defined by Equation (2.120), in the form

$$T_k(\mathbf{x}, t) = \sum_{i=1}^{\infty} \frac{1}{N_i} e^{-\mu_i^2 t} \psi_{ki}(\mathbf{x}) \left[\tilde{f}_i + \int_0^t g_i(t') e^{\mu_i^2 t'}\, dt' \right],$$

$$k = 1, 2, \ldots, n, \qquad \mathbf{x} \in V \quad (2.128)$$

where N_i, $g_i(t')$, and $\tilde{f}_i$ are defined by Equations (2.123b), (2.126c), and (2.127), respectively.

Boundary Conditions All of the Second Kind

In the foregoing boundary value problem, if $\alpha_k(\mathbf{x}) = 0$ for all boundaries and also $d_k(\mathbf{x}) = 0$, then $\mu_0 = 0$ is also an eigenvalue for the eigenvalue problem (2.121); the corresponding constant eigenfunctions ψ_{k0} and ψ_{p0} must satisfy the equations

$$\sum_{p=1}^{n} \alpha_{kp}(\psi_{p0} - \psi_{k0}) = 0 \qquad (k = 1, 2, \ldots, n) \qquad (2.129)$$

which is apparent from Equation (2.121a). Therefore, when the boundary conditions are all of the second kind and $d(\mathbf{x}) = 0$, $\mu_0 = 0$ is also an eigenvalue, then the solution (2.128) should include an additional term corresponding to the zero eigenvalue. Equation (2.128) takes the form

$$T_k(\mathbf{x}, t) = \frac{\psi_{k0}}{\displaystyle\sum_{k=1}^{n} (\psi_{k0}, \psi_{k0})}$$

$$\times \sum_{k=1}^{n} \psi_{k0}\left\{(1, f_k) + \int_0^t \left[\int_S \phi_k(\mathbf{x}, t')\, ds + \int_V P_k(\mathbf{x}, t')\, dv\right] dt'\right\}$$

$$+ \sum_{i=1}^{\infty} \frac{1}{N_i} e^{-\mu_i^2 t}\psi_{ki}(\mathbf{x})\left[\tilde{f}_i + \int_0^t g_i(t') e^{\mu_i^2 t'}\, dt'\right] \qquad (2.130)$$

If the source functions $P_k(\mathbf{x}, t)$ and $\phi_k(\mathbf{x}, t)$ are exponentials and q-order polynomials of time in the form given by Equations (2.110), then it is desirable to split up the general solution into a set of steady-state problems and a transient homogeneous problem in the same way as discussed in the previous sections in order to improve the convergence of the series.

2.5 GENERAL SOLUTION OF PROBLEMS OF CLASS V

The principal difference between the problems of Class IV and V is that in the latter the coefficients α_{kp} are not symmetric, namely, $\alpha_{kp} \neq \alpha_{pk}$. With this consideration we consider only the case $k = 1, 2$ (i.e., $n = 2$). The mathematical formulation of the boundary value problem in this class is given as

$$w_k(\mathbf{x})\frac{\partial T_k(\mathbf{x}, t)}{\partial t} + L_k T_k(\mathbf{x}, t)$$

$$= P_k(\mathbf{x}, t) + (-1)^k b(\mathbf{x})\left[\sigma_{k-1}T_1(\mathbf{x}, t) - \sigma_{k+1}T_2(\mathbf{x}, t)\right],$$

$$k = 1, 2, \qquad \mathbf{x} \in V \quad (2.131a)$$

subject to the boundary and initial conditions

$$B_k T_k(\mathbf{x}, t) = \phi_k(\mathbf{x}, t), \qquad k = 1, 2, \qquad \mathbf{x} \in S \qquad (2.131b)$$

$$T_k(\mathbf{x}, 0) = f_k(\mathbf{x}), \qquad k = 1, 2, \qquad \mathbf{x} \in V \qquad (2.131c)$$

where

$$L_k \equiv -\nabla \cdot \left[k_k(\mathbf{x})\nabla\right] + d_k(\mathbf{x}) \qquad (2.131d)$$

$$B_k \equiv \alpha_k(\mathbf{x}) + \beta_k(\mathbf{x})k_k(\mathbf{x})\frac{\partial}{\partial \mathbf{n}} \qquad (2.131e)$$

We now develop the finite integral transform pair that is needed for the solution of this boundary value problem.

Development of the Integral Transform Pair

To develop the desired integral transform pair we consider the following auxiliary homogeneous problem

$$\mu^2 w_k(\mathbf{x}) \psi_k(\mathbf{x}) + (-1)^k b(\mathbf{x})\left[\sigma_{k-1}\psi_1(\mathbf{x}) - \sigma_{k+1}\psi_2(\mathbf{x})\right] = L_k\psi_k(\mathbf{x}),$$

$$k = 1, 2, \qquad \mathbf{x} \in V \quad (2.132a)$$

subject to the boundary conditions

$$B_k\psi_k(\mathbf{x}) = 0, \qquad k = 1, 2, \qquad \mathbf{x} \in S \qquad\qquad (2.132b)$$

The desired potential $T_k(\mathbf{x}, t)$ is expanded in terms of the eigenfunctions of the previous problem in the form

$$T_k(\mathbf{x}, t) = \sum_{i=1}^{\infty} A_i(t)\psi_{ki}(\mathbf{x}), \qquad k = 1, 2, \qquad \mathbf{x} \in V \qquad (2.133)$$

To determine the unknown coefficients $A_i(t)$, the orthogonality condition for these eigenfunctions is needed. To establish the orthogonality condition, Equation (2.132a) is written for two distinct eigenvalues μ_i and μ_j; the first is multiplied by $\sigma_k\psi_{kj}(\mathbf{x})$, the second by $\sigma_k\psi_{ki}(\mathbf{x})$, the results are subtracted, integrated over the region V, and the volume integral is transformed into the surface integral according to Equations (2.5b). The resulting expression is summed up for $k = 1, 2$ to yield

$$\left(\mu_i^2 - \mu_j^2\right) \sum_{k=1}^{2} \sigma_k\left(\psi_{ki}, \psi_{kj}\right)$$

$$+ \sum_{k=1}^{2} (-1)^k \sigma_k\left[\sigma_{k-1}\int_V b(\mathbf{x}) \begin{vmatrix} \psi_{1i} & \psi_{ki} \\ \psi_{1j} & \psi_{kj} \end{vmatrix} dv - \sigma_{k+1}\int_V b(\mathbf{x}) \begin{vmatrix} \psi_{2i} & \psi_{ki} \\ \psi_{2j} & \psi_{kj} \end{vmatrix} dv\right]$$

$$= \sum_{k=1}^{2} \sigma_k \int_S k_k(\mathbf{x}) \begin{vmatrix} \psi_{ki} & \dfrac{\partial\psi_{ki}}{\partial\mathbf{n}} \\[2ex] \psi_{kj} & \dfrac{\partial\psi_{kj}}{\partial\mathbf{n}} \end{vmatrix} ds \quad (2.134)$$

It can readily be shown that the term inside the large bracket on the left-hand side vanishes; the integrand of the surface integral on the right-hand side is evaluated by utilizing boundary conditions (2.131b) and (2.132b) as has been done in obtaining Equation (2.9). Finally, the orthogonality condition is obtained as

$$\sum_{k=1}^{2} \sigma_k \left(\psi_{ki}, \psi_{kj} \right) = \delta_{kj} N_i \qquad (2.135a)$$

where

$$N_i \equiv \sum_{k=1}^{2} \sigma_k \left(\psi_{ki}, \psi_{ki} \right) = \frac{1}{2\mu_i} \sum_{k=1}^{2} \sigma_k \int_S k_k(\mathbf{x}) \left| \begin{array}{cc} \left[\dfrac{\partial \psi_k(\mathbf{x})}{\partial n} \right]_{\mu=\mu_i} & \left[\dfrac{\partial^2 \psi_k(\mathbf{x})}{\partial n\, \partial \mu} \right]_{\mu=\mu_i} \\ \psi_{ki}(\mathbf{x}) & \dfrac{\partial \psi_{ki}(\mathbf{x})}{\partial n} \end{array} \right| ds$$

$$(2.135b)$$

This orthogonality condition is applied to determine the expansion coefficients $A_i(t)$ as now described. Equation (2.133) is written first for $k = 1$ and multiplied by $\sigma_1 w_1(\mathbf{x})\psi_{1j}(\mathbf{x})$, then written for $k = 2$ and multiplied by $\sigma_2 w_2(\mathbf{x})\psi_{2j}(\mathbf{x})$, the results are added, integrated over the volume V, and the orthogonality conditions (2.135a) are utilized; one immediately obtains $A_i(t)$. Then $A_i(t)$ is introduced into Equation (2.133) and the resulting expansion is split up into two parts to define the desired integral transform pair as

Inversion formula: $\quad T_k(\mathbf{x}, t) = \sum_{i=1}^{\infty} \dfrac{1}{N_i} \psi_{ki}(\mathbf{x}) \tilde{T}_i(t), \qquad k = 1, 2$

$$(2.136)$$

Finite integral transform: $\quad \tilde{T}_i(t) = \sum_{k=1}^{2} \sigma_k \left(\psi_{ki}, T_k \right) \qquad (2.137)$

where N_i is defined by Equation (2.135b).

Method of Solution

The preceding finite integral transform pair is now applied to solve the boundary value problem (2.131). Equation (2.131a) is multiplied by $\sigma_k \psi_{ki}(\mathbf{x})$, Equation (2.132a) by $\sigma_k T_k(\mathbf{x}, t)$, each of the resulting expressions are written for $k = 1$ and 2, these four expressions are added, integrated over the region V, the volume integral is transformed to the surface integral in the usual manner, and the definition of the integral transform given by Equation (2.137) is utilized. One obtains

$$\frac{d\tilde{T}_i(t)}{dt} + \mu_i^2 \tilde{T}_i(t)$$

$$= \sum_{k=1}^{2} \sigma_k \left\{ \int_S k_k(\mathbf{x}) \begin{vmatrix} \psi_{ki}(\mathbf{x}) & \dfrac{\partial \psi_{ki}(\mathbf{x})}{\partial \mathbf{n}} \\ T_k(\mathbf{x}, t) & \dfrac{\partial T_k(\mathbf{x}, t)}{\partial \mathbf{n}} \end{vmatrix} ds + \int_V \psi_{ki}(\mathbf{x}) P_k(\mathbf{x}, t) \, dv \right\}$$

$$+ \sum_{k=1}^{2} (-1)^k \sigma_k \int_V b(\mathbf{x}) \left[\sigma_{k-1} (T_1 \psi_{ki} - T_k \psi_{1i}) - \sigma_{k+1} (T_2 \psi_{ki} - \psi_{2i} T_k) \right] dv$$

$$(2.138)$$

By expanding the last term on the right for $k = 1$ and 2, it can be shown that it vanishes. The integrand of the surface integral is evaluated by using the boundary conditions (2.131b) and (2.132b); then Equation (2.138) reduces to

$$\frac{d\tilde{T}_i(t)}{dt} + \mu_i^2 \tilde{T}_i(t) = g_i(t) \qquad (2.139a)$$

where

$$g_i(t) = \sum_{k=1}^{2} \sigma_k \left\{ \int_S \phi_k(\mathbf{x}, t) \frac{\psi_{ki}(\mathbf{x}) - k_k(\mathbf{x}) [\partial \psi_{ki}(\mathbf{x})/\partial \mathbf{n}]}{\alpha_k(\mathbf{x}) + \beta_k(\mathbf{x})} ds \right.$$

$$\left. + \int_V \psi_{kl}(\mathbf{x}) P_k(\mathbf{x}, t) \, dv \right\} \qquad (2.139b)$$

The initial condition for Equation (2.139a) is obtained by taking the integral transform of the initial condition (2.131c) according to the integral transform (2.137), that is,

$$\tilde{T}_i(0) = \sum_{k=1}^{2} \sigma_k (\psi_{ki}, f_k) \equiv \tilde{f}_i \qquad (2.139c)$$

Equation (2.139a) is solved subject to the initial condition (2.139c); the resulting solution for the transform is inverted by the inversion formula (2.136). Then the general solution of the boundary value problem of Class V, defined by Equations (2.131), is obtained as

$$T_k(\mathbf{x}, t) = \sum_{i=1}^{\infty} \frac{1}{N_i} e^{-\mu_i^2 t} \psi_{ki}(\mathbf{x}) \left[\tilde{f}_i + \int_0^t g_i(t') e^{\mu_i^2 t'} \, dt' \right], \qquad k = 1, 2,$$

$$(2.140)$$

where N_i, $g_i(t')$, and $\tilde{f}_i$ are defined by Equations (2.135b), (2.139b), and (2.139c), respectively.

Boundary Conditions All of the Second Kind

If $\alpha_k(\mathbf{x}) = 0$, $d_k(\mathbf{x}) = 0$, and $\sigma_0\sigma_3 - \sigma_1\sigma_2 = 0$ for $k = 1, 2$, then $\mu_0 = 0$ is also an eigenvalue of the eigenvalue problem (2.132) and the corresponding constant eigenfunctions ψ_{10} and ψ_{20} must satisfy the equations

$$\sigma_{k-1}\psi_{10} - \sigma_{k+1}\psi_{20} = 0, \qquad k = 1, 2 \qquad (2.141)$$

which is apparent from Equation (2.132a). In this case solution (2.140) should include an additional term corresponding to the eigenvalue $\mu_0 = 0$ and Equation (2.140) takes the form

$$T_k(\mathbf{x}, t) = \frac{\psi_{k0}}{\sum\limits_{k=1}^{2} \sigma_k(\psi_{k0}, \psi_{k0})}$$

$$\times \sum_{k=1}^{2} \sigma_k\psi_{k0}\left\{ (1, f_k) + \int_0^t \left[\int_S \frac{\phi(\mathbf{x}, t')}{\beta_k(\mathbf{x})}\, ds + \int_V P_k(\mathbf{x}, t')\, dv \right] dt' \right\}$$

$$+ \sum_{i=1}^{\infty} \frac{1}{N_i} e^{-\mu_i^2 t}\psi_{ki}(\mathbf{x})\left[\tilde{f}_i + \int_0^t g_i(t')e^{\mu_i^2 t'}\, dt' \right], \qquad k = 1, 2 \quad (2.142)$$

If the source functions $P_k(\mathbf{x}, t)$ and $\phi_k(\mathbf{x}, t)$ are exponentials and q-order polynomials of time in the form given by Equation (2.110), then the general solution should be split up into a set of steady-state problems and a transient homogeneous problem in the same way as discussed in Sections 2.1–2.3 in order to improve the convergence of the solution.

2.6 GENERAL SOLUTION OF PROBLEMS OF CLASS VI

The boundary value problems in this class, solved in [20] and discussed in Chapter 1, are defined by the following system of partial differential equations.

$$w_k(\mathbf{x})\frac{\partial T_k(\mathbf{x}, t)}{\partial t} + L_k T_k(\mathbf{x}, t) = P_k(\mathbf{x}, t), \qquad k = 1, 2, \ldots, n, \qquad \mathbf{x} \in V_k$$

$$(2.143a)$$

with the boundary and initial conditions

$$B_k T_k(\mathbf{x}, t) = \phi(t), \qquad k = 1, 2, \ldots, n, \qquad \mathbf{x} \in S_k \qquad (2.143\text{b})$$

$$\frac{d\phi(t)}{dt} + \sum_{k=1}^{n} \gamma_k \int_{S_k} k_k(\mathbf{x}) \frac{\partial T_k(\mathbf{x}, t)}{\partial \mathbf{n}} \, ds = Q(t) \qquad (2.143\text{c})$$

$$T_k(\mathbf{x}, 0) = f_k(\mathbf{x}), \qquad k = 1, 2, \ldots, n, \qquad \mathbf{x} \in V_k \qquad (2.143\text{d})$$

$$\phi(0) = \phi_0 \qquad (2.143\text{e})$$

where

$$L_k \equiv -\nabla \cdot \left[k_k(\mathbf{x}) \nabla \right] \qquad (2.143\text{f})$$

$$B_k \equiv \alpha_k + \beta_k(\mathbf{x}) k_k(\mathbf{x}) \frac{\partial}{\partial \mathbf{n}} \qquad (2.143\text{g})$$

Here α_k is a constant coefficient and $Q(t)$ is the substance source in the surrounding fluid.

To solve this problem it is convenient to apply the Laplace transform defined as [5]

$$\hat{T}_k(\mathbf{x}, p) = \int_0^{\infty} e^{-pt} T_k(\mathbf{x}, t) \, dt \qquad (2.144)$$

where p is the Laplace transform variable.

Taking the Laplace transform of Equations (2.143) we obtain

$$pw_k(\mathbf{x}) \hat{T}_k(\mathbf{x}, p) - w_k(\mathbf{x}) f_k(\mathbf{x}) + L_k \hat{T}_k(\mathbf{x}, p) = \hat{P}_k(\mathbf{x}, p),$$

$$k = 1, 2, \ldots, n, \qquad \mathbf{x} \in V_k \quad (2.145\text{a})$$

$$B_k \hat{T}_k(\mathbf{x}, p) = \hat{\phi}(p), \qquad k = 1, 2, \ldots, n, \qquad \mathbf{x} \in S \qquad (2.145\text{b})$$

$$p\hat{\phi}(p) + \sum_{k=1}^{n} \gamma_k \int_{S_k} k_k(\mathbf{x}) \frac{\partial \hat{T}_k(\mathbf{x}, p)}{\partial \mathbf{n}} \, ds = \phi_0 + \hat{Q}(p) \qquad (2.145\text{c})$$

Equation (2.145a) is integrated over the region V; the volume integral involving the L_k operator is changed into the surface integral and then solved for the term

$$\int_{S_k} k_k(\mathbf{x}) \frac{\partial \hat{T}_k(\mathbf{x}, p)}{\partial \mathbf{n}} \, ds$$

This result is introduced into Equation (2.145c), which is solved for $\hat{\phi}(p)$ and substituted into Equation (2.145b) to obtain

$$B_k \hat{T}_k(\mathbf{x}, p) + \sum_{k=1}^{n} \gamma_k(1, \hat{T}_k) = \hat{g}(p), \qquad k = 1, 2, \ldots, n, \qquad \mathbf{x} \in S$$

(2.146)

where

$$\hat{g}(p) = \frac{1}{p} \left\{ \phi_0 + \hat{Q}(p) + \sum_{k=1}^{n} \gamma_k \left[(1, f_k) + \int_{V_k} \hat{P}_k(\mathbf{x}, p) \, dv \right] \right\}$$ (2.147)

Now the problem is reduced to the solution of the system of partial differential Equations (2.145a) subject to the boundary conditions (2.146) for the Laplace transform $\hat{T}_k(\mathbf{x}, p)$ of the potential $T_k(\mathbf{x}, t)$, for $k = 1, 2, \ldots, n$. The finite integral transform technique is used to solve this problem as described in the following.

Development of the Integral Transform Pair

To develop the desired integral transform pair for the solution of the preceding boundary value problem we consider the following homogeneous auxiliary problem

$$\mu^2 w_k(\mathbf{x}) \psi_k(\mathbf{x}) = L_k \psi_k(\mathbf{x}), \qquad k = 1, 2, \ldots, n, \qquad \mathbf{x} \in V_k \quad (2.148a)$$

subject to the boundary conditions

$$B_k \psi_k(\mathbf{x}) + \sum_{k=1}^{n} \gamma_k(1, \psi_k) = 0, \qquad k = 1, 2, \ldots, n, \qquad \mathbf{x} \in S_k$$

(2.148b)

where the linear operators L_k and B_k are defined by Equations (2.143f) and (2.143g), respectively. We note that this system has a common set of eigenvalues but different eigenfunctions.

Let the Laplace transform $\hat{T}_k(\mathbf{x}, p)$ of the potential $T(\mathbf{x}, t)$ be expanded in terms of the eigenfunctions of the previous eigenvalue problem in the form

$$\hat{T}_k(\mathbf{x}, p) = \sum_{i=1}^{\infty} A_i(p) \psi_{ki}(\mathbf{x})$$ (2.149)

where the summation is over all discrete eigenvalues. To determine the expansion coefficients $A_i(p)$, the orthogonality condition for these eigenfunctions is needed. By the application of the conventional manipulations to Equation (2.148a) we obtain, analogously to Equation (2.99), the following expression

$$\left(\mu_i^2 - \mu_j^2\right)\left(\psi_{ki}, \psi_{kj}\right) = \int_{S_k} k_k(\mathbf{x}) \begin{vmatrix} \psi_{ki}(\mathbf{x}) & \dfrac{\partial \psi_{ki}(\mathbf{x})}{\partial \mathbf{n}} \\[2mm] \psi_{kj}(\mathbf{x}) & \dfrac{\partial \psi_{kj}(\mathbf{x})}{\partial \mathbf{n}} \end{vmatrix} ds \qquad (2.150)$$

To evaluate the integrand on the right-hand side, the terms $\psi_{ki}(\mathbf{x})$ and $\psi_{kj}(\mathbf{x})$ are obtained from the boundary condition (2.148b) and introduced into Equation (2.150); after some manipulation we find

$$\left(\mu_i^2 - \mu_j^2\right)\left(\psi_{ki}, \psi_{kj}\right) = \frac{1}{\alpha_k} \begin{vmatrix} -\displaystyle\sum_{k=1}^{n} \gamma_k(1, \psi_{ki}) & \displaystyle\int_{S_k} k_k(\mathbf{x}) \dfrac{\partial \psi_{ki}}{\partial \mathbf{n}} ds \\[4mm] -\displaystyle\sum_{k=1}^{n} \gamma_k(1, \psi_{kj}) & \displaystyle\int_{S_k} k_k(\mathbf{x}) \dfrac{\partial \psi_{kj}}{\partial \mathbf{n}} ds \end{vmatrix}$$

$$(2.151)$$

The surface integrals appearing in this expression are evaluated by integrating Equation (2.148a) over the region V_k to obtain

$$\int_{S_k} k_k(\mathbf{x}) \frac{\partial \psi_{ki}}{\partial \mathbf{n}} ds = -\mu_i^2(1, \psi_{ki}) \qquad (2.152)$$

Introducing this expression into the right-hand side of Equation (2.151), multiplying both sides of the resulting expression by $\alpha_k \gamma_k$, summing up for $k = 1, 2, \ldots, n$, and after some manipulating we obtain the following orthogonality relation

$$\sum_{k=1}^{n} \alpha_k \gamma_k\left(\psi_{ki}, \psi_{kj}\right) + \sum_{k=1}^{n} \gamma_k(1, \psi_{ki}) \sum_{k=1}^{n} \gamma_k(1, \psi_{kj}) = \delta_{ij} N_i \qquad (2.153a)$$

where

$$N_i = \sum_{k=1}^{n} \alpha_k \gamma_k\left(\psi_{ki}, \psi_{ki}\right) + \left[\sum_{k=1}^{n} \gamma_k(1, \psi_{ki})\right]^2 \qquad (2.153b)$$

This orthogonality condition is now used to determine the coefficients $A_i(p)$ in the expansion (2.149). That is, both sides of Equation (2.149) are multiplied by

$$\gamma_k w_k(\mathbf{x}) \left[\alpha_k \psi_{kj}(\mathbf{x}) + \sum_{k=1}^{n} \gamma_k(1, \psi_{kj})\right]$$

the result is added for $k = 1$ to n, integrated over the region V_k, and the orthogonality condition (2.153) is utilized; one immediately obtains the expression for $A_i(p)$. Then $A_i(p)$ is introduced into Equation (2.149) and the resulting expansion is split into two parts to define the desired integral transform pair as

Inversion formula $\qquad \hat{T}_k(\mathbf{x}, p) = \sum_{i=1}^{\infty} \frac{1}{N_i} \psi_{ki}(\mathbf{x}) \tilde{\hat{T}}_i(p) \quad (k = 1, 2, \ldots, n) \quad (2.154)$

Finite integral transform: $\qquad \tilde{\hat{T}}_i(p) = \sum_{k=1}^{n} \alpha_k \gamma_k (\psi_{ki}, \hat{T}_k)$

$$+ \sum_{k=1}^{n} \gamma_k(1, \psi_{ki}) \sum_{k=1}^{n} \gamma_k(1, \hat{T}_k) \quad (2.155)$$

where N_i is defined by Equation (2.153b).

For the case of $\hat{T}_k(\mathbf{x}, p) = 1/\alpha_k$, from Equations (2.154) and (2.155) one obtains the following identity

$$\sum_{i=1}^{\infty} \frac{1}{N_i} \psi_{ki}(\mathbf{x}) \sum_{k=1}^{n} \gamma_k(1, \psi_{ki}) = \left[\alpha_k + \alpha_k \sum_{k=1}^{n} \frac{\gamma_k}{\alpha_k}(1, 1)_k \right]^{-1} \quad (2.156a)$$

where

$$(1, 1)_k = \int_{V_k} w_k(\mathbf{x}) \, dv \quad (2.156b)$$

Method of Solution

The foregoing finite integral transform pair is now applied to solve Equations (2.145) subject to the boundary conditions (2.146). Equation (2.145a) is multiplied by $\psi_{ki}(\mathbf{x})$, Equation (2.148a) by $\hat{T}_k(\mathbf{x}, p)$, the results are added and integrated over the region V_k, and the volume integral involving the L_k operator is changed into the surface integral. We obtain

$$(\psi_{ki}, \hat{T}_k) = \frac{1}{p + \mu_i^2} \left[(\psi_{ki}, f_k) + \int_{V_k} \psi_{ki}(\mathbf{x}) \hat{P}_k(\mathbf{x}, p) \, dv \right.$$

$$\left. + \int_{S_k} k_k(\mathbf{x}) \begin{vmatrix} \psi_{ki}(\mathbf{x}) & \dfrac{\partial \psi_{ki}(\mathbf{x})}{\partial \mathbf{n}} \\[2mm] \hat{T}_k(\mathbf{x}, p) & \dfrac{\partial \hat{T}_k(\mathbf{x}, p)}{\partial \mathbf{n}} \end{vmatrix} ds \right] \quad (2.157)$$

This result is introduced into Equation (2.155) to obtain

$$\tilde{T}_i(p) = \frac{1}{p + \mu_i^2} \sum_{k=1}^{n} \alpha_k \gamma_k \left[(\psi_{ki}, f_k) + \int_{V_k} \psi_{ki}(\mathbf{x}) \hat{P}_k(\mathbf{x}, p)\, dv \right]$$

$$+ \frac{1}{p + \mu_i^2} \sum_{k=1}^{n} \alpha_k \gamma_k \int_{S_k} k_k(\mathbf{x}) \begin{vmatrix} \psi_{ki}(\mathbf{x}) & \dfrac{\partial \psi_{ki}(\mathbf{x})}{\partial \mathbf{n}} \\[2ex] \hat{T}_k(\mathbf{x}, p) & \dfrac{\partial \hat{T}_k(\mathbf{x}, p)}{\partial \mathbf{n}} \end{vmatrix} ds$$

$$+ \sum_{k=1}^{n} \gamma_k(1, \psi_{ki}) \sum_{k=1}^{n} \gamma_k(1, \hat{T}_k) \tag{2.158}$$

We note that in this expression the integrand of the surface integral looks much like the one appearing on the right-hand side of Equation (2.150). Therefore, to evaluate the integrand we follow a procedure similar to the one discussed previously in connection with Equation (2.150). That is, the quantities $\psi_{ki}(\mathbf{x})$ and $\hat{T}_k(\mathbf{x}, p)$ are obtained from the boundary conditions (2.148b) and (2.146), respectively, and introduced into Equation (2.158). After some manipulations the quantities

$$\int_{S_k} k_k(\mathbf{x}) \frac{\partial \psi_{ki}}{\partial \mathbf{n}}\, ds \quad \text{and} \quad \int_{S_k} k_k(\mathbf{x}) \frac{\partial \hat{T}_k}{\partial \mathbf{n}}\, ds$$

appear inside the integrand in a similar manner to those in Equation (2.151). These surface integrals are determined by integrating Equations (2.148a) and (2.145a) over the region V_k, respectively. Finally, Equation (2.158) takes the form

$$\tilde{T}_i(p) = \frac{1}{p} \left\{ \phi_0 + \hat{Q}(p) + \sum_{k=1}^{n} \gamma_k \left[(1, f_k) + \int_{V_k} \hat{P}_k(\mathbf{x}, p)\, dv \right] \right\} \sum_{k=1}^{n} \gamma_k(1, \psi_{ki})$$

$$+ \frac{1}{p + \mu_i^2} \left\{ \sum_{k=1}^{n} \gamma_k \alpha_k \left[(\psi_{ki}, f_k) + \int_{V_k} \psi_{ki}(\mathbf{x}) \hat{P}_k(\mathbf{x}, p)\, dv \right] \right.$$

$$\left. - [\phi_0 + \hat{Q}(p)] \sum_{k=1}^{n} \gamma_k(1, \psi_{ki}) \right\} \tag{2.159}$$

This result is introduced into the inversion formula (2.154), the identity given by Equation (2.156) is utilized, and the inversion of the Laplace transform is performed. Then the desired solution of the boundary value problem of Class

VI becomes

$$T_k(\mathbf{x}, t) = \left[\alpha_k + \alpha_k \sum_{k=1}^{n} \frac{\gamma_k}{\alpha_k} (1,1)_k \right]^{-1}$$

$$\times \left\{ \phi_0 + \int_0^t Q(t')\, dt' + \sum_{k=1}^{n} \gamma_k \left[(1, f_k) + \int_0^t \int_{V_k} P_k(\mathbf{x}, t')\, dv\, dt' \right] \right\}$$

$$+ \sum_{i=1}^{\infty} \frac{1}{N_i} e^{-\mu_i^2 t} \psi_{ki}(\mathbf{x}) \left\{ \sum_{k=1}^{n} \alpha_k \gamma_k \left[(\psi_{ki}, f_k) + \int_0^t e^{\mu_i^2 t'} \int_{V_k} \psi_{ki}(\mathbf{x}) P_k(\mathbf{x}, t')\, dv\, dt' \right] \right.$$

$$\left. - \left[\phi_0 + \int_0^t e^{\mu_i^2 t'} Q(t')\, dt' \right] \left[\sum_{k=1}^{n} \gamma_k (1, \psi_{ki}) \right] \right\},$$

$$k = 1, 2, \ldots, n, \qquad \mathbf{x} \in V_k \quad (2.160)$$

It is to be noted that, if part of the surfaces S_{k0} are insulated, that is, if $\partial T_k(\mathbf{x}, t)/\partial \mathbf{n} = 0$ for $\mathbf{x} \in S_{k0}$, then the solution (2.160) is still valid, but when determining the eigenvalues and eigenfunctions from the eigenvalue problem (2.148), one has to take into account the condition that $\partial \psi_{ki}(\mathbf{x})/\partial \mathbf{n} = 0$ for $\mathbf{x} \in S_{k0}$.

2.7 GENERAL SOLUTION OF PROBLEMS OF CLASS VII

The problems in this class involve diffusion from a solution into a material region of arbitrary geometry in which some of the diffusing substance is immobilized as a result of a reversible first order reaction. In the following analysis we are concerned with the determination of the distributions of the solute and the immobilized component in the region. The mathematical formulation of this problem is given as [31]

$$w(\mathbf{x}) \left[\frac{\partial T_1(\mathbf{x}, t)}{\partial t} + \frac{\partial T_2(\mathbf{x}, t)}{\partial t} \right] + L T_1(\mathbf{x}, t) = P(\mathbf{x}, t), \qquad \mathbf{x} \in V \quad (2.161a)$$

$$\frac{\partial T_2(\mathbf{x}, t)}{\partial t} = \sigma_1 T_1(\mathbf{x}, t) - \sigma_2 T_2(\mathbf{x}, t), \qquad \mathbf{x} \in V \quad (2.161b)$$

subject to the boundary conditions

$$B T_1(\mathbf{x}, t) = \phi(t), \qquad \mathbf{x} \in S \quad (2.161c)$$

$$\frac{d\phi(t)}{dt} + \gamma \int_S k(\mathbf{x}) \frac{\partial T_1(\mathbf{x}, t)}{\partial \mathbf{n}}\, ds = 0, \qquad \mathbf{x} \in S \quad (2.161d)$$

and the initial conditions

$$T_k(\mathbf{x}, 0) = f_k(\mathbf{x}), \qquad k = 1, 2, \qquad \mathbf{x} \in V \qquad\qquad (2.161\text{e})$$

$$\phi(0) = \phi_0 \qquad\qquad (2.161\text{f})$$

where

$$L \equiv -\nabla \cdot [k(\mathbf{x})\nabla] \qquad\qquad (2.161\text{g})$$

$$B \equiv \alpha + \beta(\mathbf{x})k(\mathbf{x})\frac{\partial}{\partial \mathbf{n}} \qquad\qquad (2.161\text{h})$$

and α is constant. This problem is now solved by the combined application of the Laplace transform and the finite integral transform as now described.

The Laplace transform (2.144) of Equations (2.161) yields

$$pw(\mathbf{x})\big[\hat{T}_1(\mathbf{x}, p) + \hat{T}_2(\mathbf{x}, p)\big] - w(\mathbf{x})\big[f_1(\mathbf{x}) + f_2(\mathbf{x})\big] + L\hat{T}_1(\mathbf{x}, p) = \hat{P}(\mathbf{x}, p),$$

$$\mathbf{x} \in V \quad (2.162\text{a})$$

$$p\hat{T}_2(\mathbf{x}, p) - f_2(\mathbf{x}) = \sigma_1 \hat{T}_1(\mathbf{x}, p) - \sigma_2 \hat{T}_2(\mathbf{x}, p), \qquad \mathbf{x} \in V \quad (2.162\text{b})$$

$$B\hat{T}_1(\mathbf{x}, p) = \hat{\phi}(p), \qquad \mathbf{x} \in S \qquad\qquad (2.162\text{c})$$

$$\hat{\phi}(p) = \frac{1}{p}\left[\phi_0 - \gamma \int_S k(\mathbf{x})\frac{\partial \hat{T}_1(\mathbf{x}, p)}{\partial \mathbf{n}}\,ds\right], \qquad \mathbf{x} \in S \quad (2.162\text{d})$$

Equation (2.162a) is integrated over the region V, the volume integral involving the L operator changed to the surface integral; we find

$$\int_S k(\mathbf{x})\frac{\partial \hat{T}_1(\mathbf{x}, p)}{\partial \mathbf{n}}\,ds = p\big(1, \hat{T}_1 + \hat{T}_2\big) - (1, f_1 + f_2) - \int_V \hat{P}(\mathbf{x}, p)\,dv$$

$$(2.163)$$

This result is introduced into Equation (2.162d) and the resulting expression for $\hat{\phi}(p)$ is substituted into Equation (2.162c); we obtain

$$B\hat{T}_1(\mathbf{x}, p) + \gamma\big(1, \hat{T}_1 + \hat{T}_2\big) = g(p) \qquad\qquad (2.164\text{a})$$

where

$$g(p) = \frac{1}{p}\left[\phi_0 + \gamma(1, f_1 + f_2) + \gamma \int_V \hat{P}(\mathbf{x}, p)\,dv\right] \qquad (2.164\text{b})$$

Now the problem is reduced to the solution of Equations (2.162a) and (2.162b) subject to the boundary condition (2.164). This problem is now solved by the application of the finite integral transform technique as described in the following.

Development of the Integral Transform Pair

To develop the integral transform pair appropriate for the solution of the preceding problem we consider the auxiliary problem

$$\mu^2 w(\mathbf{x})[\psi_1(\mathbf{x}) + \psi_2(\mathbf{x})] = L\psi_1(\mathbf{x}), \qquad \mathbf{x} \in V \qquad (2.165a)$$

$$\mu^2 \psi_2(\mathbf{x}) = \sigma_2 \psi_2(\mathbf{x}) - \sigma_1 \psi_1(\mathbf{x}), \qquad \mathbf{x} \in V \qquad (2.165b)$$

$$B\psi_1(\mathbf{x}) + \gamma(1, \psi_1 + \psi_2) = 0, \qquad \mathbf{x} \in S \qquad (2.165c)$$

where the linear operators L and B are defined by Equations (2.161g) and (2.161h), respectively.

We now consider the expansion of the function $\hat{T}_k(\mathbf{x}, p)$, $k = 1, 2$ in terms of the eigenfunctions of the foregoing eigenvalue problem in the form

$$\hat{T}_k(\mathbf{x}, p) = \sum_{i=1}^{\infty} A_i(p)\psi_{ki}(\mathbf{x}), \qquad k = 1, 2, \qquad \mathbf{x} \in V \qquad (2.166)$$

where the summation is taken over all discrete eigenvalues. To determine the coefficients $A_i(p)$, the orthogonality condition for these eigenfunctions is needed. By the manipulation of Equations (2.165a) and (2.165b), as discussed in the previous sections, we obtain

$$(\mu_i^2 - \mu_j^2)(\psi_{1i}, \psi_{1j}) = \int_S k(\mathbf{x}) \begin{vmatrix} \psi_{1i}(\mathbf{x}) & \dfrac{\partial \psi_{1i}(\mathbf{x})}{\partial \mathbf{n}} \\ \psi_{1j}(\mathbf{x}) & \dfrac{\partial \psi_{1j}(\mathbf{x})}{\partial \mathbf{n}} \end{vmatrix} ds$$

$$+ \sigma_2 \int_V w(\mathbf{x}) \begin{vmatrix} \psi_{1i}(\mathbf{x}) & \psi_{2i}(\mathbf{x}) \\ \psi_{1j}(\mathbf{x}) & \psi_{2j}(\mathbf{x}) \end{vmatrix} dv \qquad (2.167)$$

and

$$(\mu_i^2 - \mu_j^2)(\psi_{2i}, \psi_{2j}) = -\sigma_1 \int_V w(\mathbf{x}) \begin{vmatrix} \psi_{1i}(\mathbf{x}) & \psi_{2i}(\mathbf{x}) \\ \psi_{1j}(\mathbf{x}) & \psi_{2j}(\mathbf{x}) \end{vmatrix} dv \qquad (2.168)$$

Equation (2.167) is multiplied by σ_1, Equation (2.168) by σ_2, and the results are added

$$\left(\mu_i^2 - \mu_j^2\right) \sum_{k=1}^{2} \sigma_k\left(\psi_{ki}, \psi_{kj}\right) = \sigma_1 \int_S k(\mathbf{x}) \begin{vmatrix} \psi_{1i}(\mathbf{x}) & \dfrac{\partial \psi_{1i}(\mathbf{x})}{\partial \mathbf{n}} \\ \psi_{1j}(\mathbf{x}) & \dfrac{\partial \psi_{1j}(\mathbf{x})}{\partial \mathbf{n}} \end{vmatrix} ds \quad (2.169)$$

In this expression the integrand of the surface integral is similar to that on the right-hand side of Equation (2.150). Therefore, to evaluate the integrand we follow a procedure similar to that discussed previously. That is, $\psi_{1i}(\mathbf{x})$ and $\psi_{1j}(\mathbf{x})$ are obtained from the boundary conditions (2.165c) and introduced into the right-hand side of Equation (2.169). After some manipulation we obtain

$$\left(\mu_i^2 - \mu_j^2\right) \sum_{k=1}^{2} \sigma_k\left(\psi_{ki}, \psi_{kj}\right) = \frac{\sigma_1}{\alpha} \begin{vmatrix} -\gamma(1, \psi_{1i} + \psi_{2i}) & \displaystyle\int_S k\dfrac{\partial \psi_{1i}}{\partial \mathbf{n}} ds \\ -\gamma(1, \psi_{1j} + \psi_{2j}) & \displaystyle\int_S k\dfrac{\partial \psi_{1j}}{\partial \mathbf{n}} ds \end{vmatrix}$$

$$(2.170)$$

The integration of Equation (2.165a) over V gives

$$\int_S k(\mathbf{x}) \frac{\partial \psi_1}{\partial \mathbf{n}} ds = -\mu^2(1, \psi_1 + \psi_2) \quad (2.171)$$

Introducing this expression on the right-hand side of Equation (2.170) and after some manipulation, we find the following orthogonality relation

$$\alpha \sum_{k=1}^{2} \sigma_k\left(\psi_{ki}, \psi_{kj}\right) + \gamma\sigma_1(1, \psi_{1i} + \psi_{2i})(1, \psi_{1j} + \psi_{2j}) = \delta_{ij} N_i \quad (2.172a)$$

where

$$N_i \equiv \alpha \sum_{k=1}^{2} \sigma_k\left(\psi_{ki}, \psi_{ki}\right) + \gamma\sigma_1(1, \psi_{1i} + \psi_{2i})^2 \quad (2.172b)$$

This orthogonality relation is now used to determine the coefficients $A_i(p)$ in the expansion (2.166). That is, both sides of Equation (2.166) are multiplied by

$$w(\mathbf{x})\left[\alpha\sigma_k\psi_{kj}(\mathbf{x}) + \gamma\sigma_1(1, \psi_{1j} + \psi_{2j})\right]$$

and the result is added for $k = 1$ and 2, integrated over the region V, and the orthogonality condition (2.172) is utilized; one immediately obtains the expression for $A_i(p)$. When this expression is introduced into the expansion (2.166), we obtain

$$\hat{T}_k(\mathbf{x}, p) = \sum_{i=1}^{\infty} \frac{1}{N_i} \psi_{ki}(\mathbf{x}) \left[\alpha \sum_{k=1}^{2} \sigma_k(\psi_{ki}, \hat{T}_k) + \gamma \sigma_1(1, \psi_{1i} + \psi_{2i})(1, \hat{T}_1 + \hat{T}_2) \right]$$

(2.173a)

For the special case of $\hat{T}_k(\mathbf{x}, p) = 1/(\alpha \sigma_k)$ we obtain from this expansion the following identity

$$\sum_{i=1}^{\infty} \frac{1}{N_i} \psi_{ki}(\mathbf{x})(1, \psi_{1i} + \psi_{2i}) = \left[\alpha \sigma_k + \frac{\sigma_k}{\sigma_2} \gamma(1, \sigma_1 + \sigma_2) \right]^{-1}$$ (2.173b)

The expansion (2.173a) can be split into two parts to define the desired integral transform pair as

Inversion formula: $$\hat{T}_k(\mathbf{x}, p) = \sum_{i=1}^{\infty} \frac{1}{N_i} \psi_{ki}(\mathbf{x}) \tilde{\hat{T}}_i(p), \qquad k = 1, 2$$

(2.174)

Finite integral transform: $$\tilde{\hat{T}}_i(p) = \alpha \sum_{k=1}^{2} \sigma_k(\psi_{ki}, \hat{T}_k)$$

$$+ \gamma \sigma_1(1, \psi_{1i} + \psi_{2i})(1, \hat{T}_1 + \hat{T}_2)$$

(2.175)

where N_i is defined by Equation (2.172b).

Method of Solution

The preceding finite integral transform pair is applied to solve the problem defined by Equations (2.162a, b) subject to the boundary condition (2.164) as now described.

Equations (2.162a), (2.162b), (2.165a), and (2.165b) are multiplied by $\sigma_1 \psi_{1i}(\mathbf{x})$, $\sigma_2 w(\mathbf{x}) \psi_{2i}(\mathbf{x})$, $\sigma_1 \hat{T}_1(\mathbf{x}, p)$, and $\sigma_2 w(\mathbf{x}) \hat{T}_2(\mathbf{x}, p)$, respectively, integrated over the region V; the results are added, the volume integral involving the L operator is changed into the surface integral, and after some manipulations we obtain

$$\left(p + \mu_i^2 \right) \sum_{k=1}^{2} \sigma_k \left(\psi_{ki}, \hat{T}_k \right)$$

$$= \sum_{k=1}^{2} \sigma_k \left(\psi_{ki}, f_k \right) + \sigma_1 \int_S k(\mathbf{x}) \begin{vmatrix} \psi_{1i}(\mathbf{x}) & \dfrac{\partial \psi_{1i}(\mathbf{x})}{\partial \mathbf{n}} \\[2ex] \hat{T}_1(\mathbf{x}, p) & \dfrac{\partial \hat{T}_1(\mathbf{x}, p)}{\partial \mathbf{n}} \end{vmatrix} ds$$

$$+ \sigma_1 \int_V \psi_{1i}(\mathbf{x}) \hat{P}(\mathbf{x}, p)\, dv \qquad (2.176a)$$

To evaluate the integrand of the surface integral in this expression, $\psi_{1i}(\mathbf{x})$ and $\hat{T}_1(\mathbf{x}, p)$ are determined from Equations (2.165c) and (2.164a), respectively, and introduced into the integrand; the Equation (2.176a) takes the form

$$\left(p + \mu_i^2 \right) \sum_{k=1}^{2} \sigma_k \left(\psi_{ki}, \hat{T}_k \right) = \sum_{k=1}^{2} \sigma_k \left(\psi_{ki}, f_k \right)$$

$$+ \sigma_1 \left\{ \frac{1}{\alpha} \begin{vmatrix} -\gamma(1, \psi_{1i} + \psi_{2i}) & \int_S k(\mathbf{x}) \dfrac{\partial \psi_{1i}}{\partial \mathbf{n}} ds \\[2ex] g(p) - \gamma(1, \hat{T}_1 + \hat{T}_2) & \int_S k(\mathbf{x}) \dfrac{\partial \hat{T}_1}{\partial \mathbf{n}} ds \end{vmatrix} + \int_V \psi_{1i}(\mathbf{x}) \hat{P}(\mathbf{x}, p)\, dv \right\}$$

$$(2.176b)$$

By integrating Equations (2.165a) and (2.162a) over the region V and changing the volume integral to the surface integral we obtain, respectively,

$$\int_S k(\mathbf{x}) \frac{\partial \psi_1(\mathbf{x})}{\partial \mathbf{n}} ds = -\mu^2 (1, \psi_1 + \psi_2) \qquad (2.177a)$$

$$\int_S k(\mathbf{x}) \frac{\partial \hat{T}_1(\mathbf{x}, p)}{\partial \mathbf{n}} ds = p(1, \hat{T}_1 + \hat{T}_2) - (1, f_1 + f_2) - \left(\frac{1}{w}, \hat{P} \right) \qquad (2.177b)$$

These results are introduced into Equation (2.176b), the definition of the integral transform (2.175) is utilized, and after some manipulation we obtain

$$\tilde{\tilde{T}}_i(p) = \frac{1}{p + \mu_i^2}\tilde{f}_i + \frac{\alpha\sigma_1}{p + \mu_i^2}\int_V \psi_{1i}(\mathbf{x})\hat{P}(\mathbf{x}, p)\, dv + \sigma_1(1, \psi_{1i} + \psi_{2i})$$

$$\times \left\{ \left(\frac{1}{p} - \frac{1}{p + \mu_i^2}\right)[\phi_0 + \gamma(1, f_1 + f_2)] + \frac{\gamma}{p}\int_V \hat{P}(\mathbf{x}, p)\, dv \right\}$$

$$(2.178a)$$

where

$$\tilde{f}_i = \alpha \sum_{k=1}^{2} \sigma_k(\psi_{ki}, f_k) + \gamma\sigma_1(1, \psi_{1i} + \psi_{2i})(1, f_1 + f_2) \qquad (2.178b)$$

The integral transform (2.178a) is substituted into the inversion formula (2.174), the identity (2.173b) is utilized, and the inverse Laplace transform is taken. The solution of the problem of Class VII defined by Equations (2.161) becomes

$$T_k(\mathbf{x}, t) = \sigma_1 \left[\phi_0 + \gamma(1, f_1 + f_2) + \gamma\int_0^t \int_V P(\mathbf{x}, t')\, dv\, dt'\right]$$

$$\times \left[\alpha\sigma_k + \gamma(1, \sigma_1 + \sigma_2)\frac{\sigma_k}{\sigma_2}\right]^{-1} + \sum_{i=1}^{\infty} \frac{1}{N_i}e^{-\mu_i^2 t}\psi_{ki}(\mathbf{x})$$

$$\times \left\{\alpha \sum_{k=1}^{2}\sigma_k(\psi_{ki}, f_k) - \phi_0\sigma_1(1, \psi_{1i} + \psi_{2i})\right.$$

$$\left. + \alpha\sigma_1\int_0^t \int_V \psi_{1i}(\mathbf{x})P(\mathbf{x}, t')e^{\mu_i^2 t'}\, dv\, dt'\right\} \qquad (2.179)$$

REFERENCES

1. Ian N. Sneddon, *Fourier Transforms*, McGraw-Hill, New York, 1951.

2. Ian N. Sneddon, *The Use of Integral Transforms*, McGraw-Hill, New York, 1972.

3. A. Erdelyi, W. Magnus, F. Oberhettinger, and F. G. Tricomi, *Tables of Integral Transforms*, 2 vols., McGraw-Hill, New York 1954.

4. C. J. Tranter, *Integral Transforms in Mathematical Physics*, Wiley, New York, 1962.

5. V. A. Ditkin and A. P. Prudnikov, *Integral Transforms and Operational Calculus*, Pergamon, New York, 1965.

6. M. N. Özisik, *Boundary Value Problems of Heat Conduction*, International Textbook, Scranton, PA, 1968.

7. N. Y. Ölçer, On the Theory of Conductive Heat Transfer in Finite Regions, *Int. J. Heat Mass Transf.*, **7**, 307–314 (1964).

8. N. Y. Ölçer, On the Theory of Conductive Heat Transfer in Finite Regions with Boundary Conditions of the Second Kind, *Int. J. Heat Mass Transf.*, **8**, 529–556 (1965).

9. G. Cinelli, An Extension of the Finite Hankel Transform and Applications, *Int. J. Eng. Sci.*, **3**, 539–559 (1965).

10. N. Y. Ölçer, General Solutions to a Class of Unsteady Heat Conduction Problems in a Rectangular Parallelepiped, *Int. J. Heat Mass Transf.*, **12**, 393–411 (1969).

11. M. D. Mikhailov, General Solutions of Heat Equation in Finite Regions, *Int. J. Eng. Sci.*, **10**, 577–591 (1972).

12. M. D. Mikhailov, General Solutions of the Coupled Diffusion Equations, *Int. J. Eng. Sci.*, **11**, 235–241 (1973).

13. M. D. Mikhailov, General Solutions of the Diffusion Equations Coupled at the Boundary Conditions, *Int. J. Heat Mass Transf.*, **16**, 2155–2164 (1973).

14. M. D. Mikhailov, General Solutions of Heat and Mass Transfer Problems, pp. 135–165, and Mass and Heat Transfer Problems in Capillary Porous Body in Drying Processes, pp. 166–188, in *Mathematical and Physical Problems of Heat Mass Transfer*, The Institute for Heat and Mass Transfer, AS BSSR, Minsk, USSR, 1973.

15. M. N. Özişik and R. L. Murray, On the Solution of Linear Diffusion Problems with Variable Boundary Condition Parameters, *J. Heat Transf.*, **96c**, 48–51 (1974).

16. Y. Yener and M. N. Özişik, On the Solution of Unsteady Heat Conduction in Multi-Region Media with Time Dependent Heat Transfer Coefficient, *Proc. 5th. Int. Heat Trans. Conference*, Cu 2.5, pp. 188–192, Tokyo, Sept. 1974.

17. K. Kobayashi, N. Ohtani, and J. Jung, Solution of Two-Dimensional Diffusion Equation by the Finite Fourier Transformation, *Nuc. Sci. and Eng.*, **55**, 320–328 (1974).

18. M. D. Mikhailov, On the Solution of Heat Equation with Time Dependent Coefficient, *Int. J. Heat Mass Transf.*, **18**, 344–345 (1975).

19. M. D. Mikhailov, Splitting Up of Heat-Conduction Problems, *Letters Heat Mass Transf.*, **4**, 163–166 (1977).

20. M. D. Mikhailov, General Solution of Diffusion Processes in Solid–Liquid Extraction, *Int. J. Heat Mass Transf.*, **20**, 1409–1415 (1977).

21. A. Korn and T. M. Korn, *Mathematical Handbook for Scientists and Engineers*, McGraw-Hill, New York, 1961.

22. Max Jacob, *Heat Transfer*, Vol. II, Wiley, New York, 1957.

23. W. Heiligenstädt, *Arch. Eigenhütterw.*, **2**, 217 (1928).

24. J. R. Howard and A. E. Sutton, An Analogue Study of Heat Transfer Through Periodically Contacting Surfaces, *Int. J. Heat Mass Transf.*, **13**, 173–183 (1970).

25. J. R. Reed and G. Mullineux, Quasi-Steady Solution of Periodically Varying Phenomena, *Int. J. Heat Mass Transf.*, **16**, 2007–2012 (1973).

26. M. D. Mikhailov, Quasi-Steady State Temperature Distribution in Finite Regions with Periodically-Varying Boundary Conditions, *Int. J. Heat Mass Transf.*, **17**, 1475–1478 (1974).

27. M. N. Özişik, *Heat Conduction*, Wiley, New York, 1980.

28. H. S. Carslaw and J. C. Jaeger, *Conduction of Heat in Solids*, Clarendon Press, London, 1959.

29. J. Crank, *The Mathematics of Diffusion*, 2nd ed., Clarendon Press, London, 1975.

30. M. D. Mikhailov and M. N. Özişik, An Alternative General Solution of the Steady-State Heat Diffusion Equation, *Int. J. Heat and Mass Transf.*, **23**, 609–612 (1980).

31. M. D. Mikhailov and M. N. Özişik, A General Solution of Solute Diffusion with Reversible Reaction, *Int. J. Heat and Mass Transf.*, **24**, 81–87 (1981).

32. M. D. Mikhailov, M. N. Özişik, and B. K. Shishedjiev, Diffusion in Heterogeneous Media, *J. Heat Transf.*, **104**, 781–787 (1982).

33. B. Vick and M. N. Özişik, Quasi-Steady State Temperature Distribution in a Periodically Contacting Finite Regions, *J. Heat Transf.*, **103C**, 436–440 (1981).

═══ CHAPTER THREE ═══

Eigenvalue Problems

Give me but one firm spot on which to stand, and I will move the earth.

Archimedes
287–212 BC

The eigenvalue problems are the firm spot on which we stand in Chapter 2 to solve the seven different classes of problems defined previously for finite regions. The solutions for such eigenvalue problems for arbitrary form are not known and the exact analytical solutions are available for only special cases. Therefore, the objective of this chapter is to present a comprehensive discussion of the methods of solving one-dimensional eigenvalue problems relevant to the solutions of the seven different classes of boundary value problems described in the previous chapters. The recently advanced, very efficient *sign-count* method is presented for the solution of eigenvalue problems associated with the Class I and II problems. The analytical methods of solving the eigenvalue problems are presented, various important properties of the resulting solutions are examined, and the algorithms are described for numerically computing the eigenvalues and the eigenfunctions.

3.1 EIGENVALUE PROBLEMS OF CLASS I

One-Dimensional Case

The eigenvalue problem associated with the problems of Class I is given by the Sturm–Liouville system (2.2). The one-dimensional form of the L operator appearing in this system is given by

$$L \equiv -\frac{d}{dx}\left[k(x)\frac{d}{dx}\right] + d(x), \quad \text{in } x_0 < x < x_1 \qquad (3.1a)$$

Letting $\alpha(x_0) = \alpha_0$, $\beta(x_0) = \beta_0$, $\alpha(x_1) = \alpha_1$, and $\beta(x_1) = \beta_1$ in the operator B for the boundary conditions we write

$$B_0 \equiv \alpha_0 - \beta_0 k(x_0)\frac{d}{dx} \quad \text{at } x = x_0 \tag{3.1b}$$

$$B_1 \equiv \alpha_1 + \beta_1 k(x_1)\frac{d}{dx} \quad \text{at } x = x_1 \tag{3.1c}$$

Here the minus sign is included in Equation (3.1b), because at $x = x_0$ the outward-drawn normal is in the negative x direction. Then the one-dimensional form of the eigenvalue problem of Class I becomes

$$\frac{d}{dx}\left[k(x)\frac{d\psi(\mu, x)}{dx}\right] + [\mu^2 w(x) - d(x)]\psi(\mu, x) = 0 \quad \text{in } x_0 < x < x_1 \tag{3.2a}$$

$$\alpha_k \psi(\mu, x) - (-1)^k \beta_k k(x)\frac{d\psi(\mu, x)}{dx} = 0 \quad \text{at } x = x_k, \quad k = 0, 1 \tag{3.2b, c}$$

where the functions $k(x)$, $k'(x)$, $w(x)$, and $d(x)$ are real and continuous, and $k(x) > 0$, $w(x) > 0$ over the interval (x_0, x_1). The constants α_0, β_0, α_1, and β_1 are independent of the parameter μ.

Let $u(\mu, x)$ and $v(\mu, x)$ be two linearly independent solutions of Equation (3.2a). The general solution is constructed by taking a linear combination of these two solutions in the form

$$\psi(\mu, x) = Cu(\mu, x) + Dv(\mu, x) \tag{3.3}$$

where C and D are two arbitrary constants.

If the general solution (3.3) should satisfy the two boundary conditions (3.2b, c), we have

$$CU(\mu, x_k) + DV(\mu, x_k) = 0, \quad k = 0, 1 \tag{3.4a, b}$$

where

$$U(\mu, x_k) = \alpha_k u(\mu, x_k) - (-1)^k \beta_k k(x_k)u'(\mu, x_k) \tag{3.4c, d}$$

$$V(\mu, x_k) = \alpha_k v(\mu, x_k) - (-1)^k \beta_k k(x_k)v'(\mu, x_k) \tag{3.4e, f}$$

here we defined $u'(\mu, x) \equiv du(\mu, x)/dx$, $v'(\mu, x) \equiv dv(\mu, x)/dx$. Clearly, the problem has a trivial solution, that is, $\psi(\mu, x) = 0$, if the coefficients C and D

are both zero. These two coefficients are satisfied by the two homogeneous Equations (3.4a, b); C and D will not be both zero if the determinant of the coefficients of these two homogeneous equations vanishes. This requirement leads to

$$U(\mu, x_0)V(\mu, x_1) - U(\mu, x_1)V(\mu, x_0) = 0 \qquad (3.5a)$$

This equation is called the *eigencondition* since it establishes the permissible values of the eigenvalues μ_i so that the trivial solution $C = D = 0$ is excluded.

An alternative way of arriving at Equation (3.5a) is to solve Equations (3.4a, b) for the ratio D/C. We obtain

$$-\frac{D}{C} = \frac{U(\mu, x_0)}{V(\mu, x_0)} = \frac{U(\mu, x_1)}{V(\mu, x_1)} \qquad (3.5b)$$

Clearly, the right-hand side of this equality is identical to Equation (3.5a).

Equation (3.5b) implies that the ratio D/C is a function of the parameter μ. That is, $C \equiv C(\mu)$ and $D \equiv D(\mu)$ are not independent; they are related to each other.

In view of Equation (3.5b), the solution (3.3) is now written as

$$\psi(\mu_i, x) = \left[u(\mu_i, x) - \frac{U(\mu_i, x_0)}{V(\mu_i, x_0)} v(\mu_i, x) \right] C(\mu_i) \qquad (3.6)$$

We note that the eigenfunctions $\psi(\mu_i, x)$ are arbitrary within multiplication constants $C(\mu_i)$ since the boundary value problem (3.2) is homogeneous.

Thus the eigenvalue problem (3.2) has nontrivial solutions $\psi_i(x) \equiv \psi(\mu_i, x)$ only for certain values of the parameter $\mu = \mu_i$ ($i = 1, 2, \ldots, \infty$), and for other values of μ it has trivial solutions; that is, $\psi(\mu, x) = 0$. The nontrivial solutions $\psi_i(x)$ are the eigenfunctions and the corresponding parameters μ_i are the eigenvalues.

Having established formally the condition for the determination of the eigenvalues and the general form of the eigenfunctions of the eigenvalue problem (3.2), we now proceed to obtain a formal expression for the evaluation of the normalization integral N_i associated with this eigenvalue problem.

The one-dimensional form of the normalization integral (2.8) becomes

$$N_i = \frac{1}{2\mu_i} \sum_{k=0}^{1} (-1)^{k+1} k(x_k) \left| \begin{array}{cc} \left(\dfrac{\partial \psi(\mu, x_k)}{\partial \mu} \right)_{\mu=\mu_i} & \left(\dfrac{\partial^2 \psi(\mu, x_k)}{\partial x\, \partial \mu} \right)_{\mu=\mu_i} \\ \psi(\mu_i, x_k) & \psi'(\mu_i, x_k) \end{array} \right|$$

$$(3.7a)$$

The solution (3.3) is substituted into Equation (3.7a) to give

$$N_i = \frac{1}{2\mu_i} \left\{ \sum_{k=0}^{1} (-1)^{k+1} k(x_k) \begin{vmatrix} A_1 & A_2 \\ B_1 & B_2 \end{vmatrix} \right.$$

$$\left. + \left(D\frac{dC}{d\mu} - C\frac{dD}{d\mu} \right) \sum_{k=0}^{1} (-1)^k k(x_k) \begin{vmatrix} u(\mu, x_k) & u'(\mu, x_k) \\ v(\mu, x_k) & v'(\mu, x_k) \end{vmatrix} \right\}$$

$$(3.7b)$$

where

$$A_1 = C\left(\frac{\partial u(\mu, x_k)}{\partial \mu} \right)_{\mu=\mu_i} + D\left(\frac{\partial v(\mu, x_k)}{\partial \mu} \right)_{\mu=\mu_i}$$

$$A_2 = C\left(\frac{\partial^2 u(\mu, x_k)}{\partial x\, \partial \mu} \right)_{\mu=\mu_i} + D\left(\frac{\partial^2 v(\mu, x_k)}{\partial x\, \partial \mu} \right)_{\mu=\mu_i}$$

$$B_1 = Cu(\mu_i, x_k) + Dv(\mu_i, x_k)$$

$$B_2 = Cu'(\mu_i, x_k) + Dv'(\mu_i, x_k)$$

The second term on the right-hand side vanishes and Equation (3.7b) reduces to

$$N_i = \frac{1}{2\mu_i} \sum_{k=0}^{1} (-1)^{k+1} k(x_k) \begin{vmatrix} A_1 & A_2 \\ B_1 & B_2 \end{vmatrix} \qquad (3.7c)$$

where A_1, A_2, B_1, and B_2 are defined above. To prove that the second term on the right-hand side of Equation (3.7b) vanishes, Equation (3.2a) is written for the two independent solutions $u(\mu_i, x)$ and $v(\mu_i, x)$ as

$$\frac{d}{dx}\left[k(x)\frac{du(\mu_i, x)}{dx} \right] + \left[\mu_i^2 w(x) - d(x) \right] u(\mu_i, x) = 0 \qquad (3.8a)$$

$$\frac{d}{dx}\left[k(x)\frac{dv(\mu_i, x)}{dx} \right] + \left[\mu_i^2 w(x) - d(x) \right] v(\mu_i, x) = 0 \qquad (3.8b)$$

The first equation is multiplied by $v(\mu_i, x)$, the second by $u(\mu_i, x)$, and the results are subtracted and integrated over (x_0, x_1). We obtain

$$\sum_{k=0}^{1} (-1)^{k+1} k(x_k) \begin{vmatrix} u(\mu, x_k) & u'(\mu, x_k) \\ v(\mu, x_k) & v'(\mu, x_k) \end{vmatrix} = 0 \qquad (3.8c)$$

Consequently, the last term in Equation (3.7b) vanishes.

Application to Generalized Bessel Equation

To illustrate the application of this procedure for the determination of eigenfunctions, eigenvalues, and the normalization integral of the one-dimensional
Sturm–Liouville system (3.2), we consider the following special case

$$k(x) = x^{1-2m} e^{-2ax} \tag{3.9a}$$

$$w(x) = c^2 x^{2c-2} k(x) \tag{3.9b}$$

$$d(x) = -\left(a^2 + a\frac{2m-1}{x} + \frac{m^2 - c^2\nu^2}{x^2}\right) k(x) \tag{3.9c}$$

Substituting Equations (3.9) into Equation (3.2a) we obtain the generalized
Bessel equation developed by Douglas [1]

$$\psi''(\mu, x) + \left(\frac{1-2m}{x} - 2a\right)\psi'(\mu, x)$$

$$+ \left(\mu^2 c^2 x^{2c-2} + a^2 + a\frac{2m-1}{x} + \frac{m^2 - c^2\nu^2}{x^2}\right)\psi(\mu, x) = 0 \tag{3.10}$$

The eigencondition, eigenfunctions, and the normalization integral of Equation (3.10) are determined as now described.

The general solution of this equation is

$$\psi(\mu, x) = x^m e^{ax} Z_\nu(\mu x^c) \tag{3.11a}$$

where

$$Z_\nu(\mu x^c) = C J_\nu(\mu x^c) + D Y_\nu(\mu x^c) \tag{3.11b}$$

The functions $J_\nu(\mu x^c)$ and $Y_\nu(\mu x^c)$ are *Bessel functions* of order ν of the first
and second kind, respectively. For detailed treatment of Bessel functions, the
reader should consult standard texts [2, 3].

By comparing the solutions (3.3) and (3.11) we find

$$u(\mu, x) = x^m e^{ax} J_\nu(\mu x^c) \tag{3.12a}$$

$$v(\mu, x) = x^m e^{ax} Y_\nu(\mu x^c) \tag{3.12b}$$

The eigencondition for this problem is obtained by substituting Equations
(3.12) into Equation (3.5a)

$$\sum_{k=0}^{1} (-1)^k \frac{A_k J_\nu(\mu x_k^c) - (-1)^k B_k \mu J_{\nu-1}(\mu x_k^c)}{A_k Y_\nu(\mu x_k^c) - (-1)^k B_k \mu Y_{\nu-1}(\mu x_k^c)} = 0 \tag{3.13}$$

and the eigenfunctions $\psi_i(x)$ are obtained by introducing Equations (3.12) into Equation (3.6)

$$\psi_i(x) = x^m e^{ax} \left\{ J_\nu(\mu_i x^c) - \frac{A_0 J_\nu(\mu_i x_0^c) - B_0 \mu_i J_{\nu-1}(\mu_i x_0^c)}{A_0 Y_\nu(\mu_i x_0^c) - B_0 \mu_i Y_{\nu-1}(\mu_i x_0^c)} Y_\nu(\mu_i x^c) \right\} C(\mu_i)$$

(3.14)

where

$$A_k = \alpha_k - (-1)^k \beta_k x_k^{1-2m} e^{-2ax_k} \left(\frac{m - c\nu}{x_k} + a \right), \qquad k = 0, 1$$

(3.15a, b)

$$B_k = \beta_k c x_k^{c-2m} e^{-2ax_k}, \qquad k = 0, 1$$

(3.15c, d)

The normalization integral N_i is determined by substituting Equations (3.12) into Equation (3.7c)

$$N_i = \frac{c}{2} \sum_{k=0}^{1} (-1)^{k+1} x_k^{2c} \left[Z_\nu^2(\mu_i x_k^c) - Z_{\nu-1}(\mu_i x_k^c) Z_{\nu+1}(\mu_i x_k^c) \right] \quad (3.16a)$$

In view of Equations (3.11b) and (3.5b), Equation (3.16a) becomes

$$N_i = \frac{2c}{(\pi\mu_i)^2} \sum_{k=0}^{1} (-1)^{k+1} \frac{(\mu_i B_k)^2 + A_k \left[A_k - (-1)^k B_k 2\nu/x_k^c \right]}{\left[A_k Y_\nu(\mu_i x_k^c) - (-1)^k B_k \mu_i Y_{\nu-1}(\mu_i x_k^c) \right]^2} C^2(\mu_i)$$

(3.16b)

In the following examples we examine several specific cases in order to illustrate further applications.

Example 3.1 Consider the following eigenvalue problem for a finite region $x_0 \le x \le x_1$

$$\psi''(\mu, x) + \mu^2 \psi(\mu, x) = 0 \qquad (3.17a)$$

$$\alpha_k \psi(\mu, x_k) - (-1)^k \beta_k \psi'(\mu, x_k) = 0, \qquad k = 0, 1 \qquad (3.17b)$$

Determine the eigencondition, eigenfunction, and nomalization integral.

Solution By comparing the present eigenvalue problem (3.17) with that given by Equations (3.2), we write

$$k(x) = 1, \qquad w(x) = 1, \qquad \text{and} \quad d(x) = 0 \qquad (3.18a)$$

Also, by comparing Equations (3.18a) and (3.9) we find

$$m = \tfrac{1}{2}, \qquad a = 0, \qquad c = 1, \quad \text{and} \quad \nu = \tfrac{1}{2} \tag{3.18b}$$

Then from Equations (3.15) we have

$$A_k = \alpha_k, \qquad B_k = \beta_k, \qquad k = 0, 1 \tag{3.18c}$$

By making use of the following relations

$$J_{1/2}(x) = Y_{-1/2}(x) = \sqrt{\frac{2}{\pi x}} \sin x \tag{3.19a}$$

$$J_{-1/2}(x) = -Y_{1/2}(x) = \sqrt{\frac{2}{\pi x}} \cos x \tag{3.19b}$$

the eigencondition is immediately obtained from Equation (3.13) as

$$(\alpha_0\beta_1 + \alpha_1\beta_0)\cos[\mu(x_1 - x_0)] + (\alpha_0\alpha_1 - \beta_0\beta_1\mu^2)\frac{\sin[\mu(x_1 - x_0)]}{\mu} = 0 \tag{3.20a}$$

Now if we choose $C(\mu_i) = \sqrt{\pi/2\mu_i}\,[\alpha_0\cos(\mu_i x_0) + \beta_0\mu_i\sin(\mu_i x_0)]$, then the eigenfunctions are obtained from Equation (3.14) as

$$\psi_i(x) = \beta_0\cos[\mu_i(x - x_0)] + \frac{\alpha_0}{\mu_i}\sin[\mu_i(x - x_0)] \tag{3.20b}$$

and the normalization integral from Equation (3.16b) as

$$N_i = \frac{1}{2\mu_i^2}\left\{[\alpha_0^2 + (\beta_0\mu_i)^2]\left[\frac{\alpha_1\beta_1}{\alpha_1^2 + (\beta_1\mu_i)^2} + x_1 - x_0\right] + \alpha_0\beta_0\right\} \tag{3.20c}$$

We now consider a special case of this problem for $\alpha_0 = 0$, $\beta_0 = 1$, $x_0 = 0$, and $x_1 = 1$. Then Equations (3.20a, b, c), respectively, reduce to

$$\cot \mu = \frac{\beta_1}{\alpha_1}\mu \tag{3.21a}$$

$$\psi_i(x) = \cos(\mu_i x) \tag{3.21b}$$

$$N_i = \frac{1}{2}\left[1 + \frac{\beta_1}{\alpha_1} + \left(\frac{\beta_1}{\alpha_1}\mu_i\right)^2\right]\sin^2\mu_i \tag{3.21c}$$

Example 3.2 Determine the eigencondition, eigenfunctions, and normalization integral of the following eigenvalue problem for a finite region $x_0 \leq x \leq x_1$

$$\psi''(\mu, x) + \frac{1}{x}\psi'(\mu, x) + \left(\mu^2 - \frac{\nu^2}{x^2}\right)\psi(\mu, x) = 0 \qquad (3.22a)$$

$$\alpha_k \psi(\mu, x_k) - (-1)^k \gamma_k \psi'(\mu, x_k) = 0, \qquad k = 0,1 \qquad (3.22b, c)$$

Solution Equation (3.22a) is written in an alternative form in order to compare it readily with Equation (3.2a)

$$\frac{d}{dx}\left[x\frac{d\psi(\mu, x)}{dx}\right] + x\left(\mu^2 - \frac{\nu^2}{x^2}\right)\psi(\mu, x) = 0 \qquad (3.22d)$$

A comparison of the present system with that given by Equations (3.2) reveals that

$$k(x) = x, \quad w(x) = x, \quad d(x) = \frac{\nu^2}{x}, \quad \beta_k k(x_k) = \gamma_k, \quad \text{or} \quad \beta_k = \frac{\gamma_k}{x_k}$$

$$(3.23a)$$

Furthermore, by comparing Equations (3.23a) and (3.9) we find

$$m = 0, \quad a = 0, \quad c = 1 \qquad (3.23b)$$

Then from Equations (3.15) we obtain

$$A_k = \alpha_k + (-1)^k \frac{\gamma_k}{x_k}\nu, \qquad B_k = \gamma_k \qquad (3.23c)$$

Thus the eigencondition is determined from Equation (3.13) as

$$\sum_{k=0}^{1}(-1)^k \frac{\left[\alpha_k + (-1)^k \gamma_k \nu/x_k\right]J_\nu(\mu x_k) - (-1)^k \gamma_k \mu J_{\nu-1}(\mu x_k)}{\left[\alpha_k + (-1)^k \gamma_k \nu/x_k\right]Y_\nu(\mu x_k) - (-1)^k \gamma_k \mu Y_{\nu-1}(\mu x_k)} = 0$$

$$(3.24a)$$

With the choice of $C(\mu_i) = 1$, the eigenfunctions $\psi_i(x)$ are determined from Equation (3.14) as

$$\psi_i(x) = J_\nu(\mu_i x) - \frac{[\alpha_0 + \gamma_0\nu/x_0]J_\nu(\mu_i x_0) - \gamma_0\mu_i J_{\nu-1}(\mu_i x_0)}{[\alpha_0 + \gamma_0\nu/x_0]Y_\nu(\mu_i x_0) - \gamma_0\mu_i Y_{\nu-1}(\mu_i x_0)}Y_\nu(\mu_i x)$$

$$(3.24b)$$

Finally, the normalization integral N_i is obtained from Equation (3.16b) by setting $C(\mu_i) = 1$ as

$$N_i = \frac{2}{(\pi\mu_i)^2} \sum_{k=0}^{1} (-1)^{k+1}$$

$$\times \frac{\alpha_k^2 + (\mu_i\gamma_k)^2 - (\gamma_k \nu/x_k)^2}{\left\{\left[\alpha_k + (-1)^k \gamma_k\nu/x_k\right] Y_\nu(\mu_i x_k) - (-1)^k \gamma_k\mu_i Y_{\nu-1}(\mu_i x_k)\right\}^2}$$

$$(3.24c)$$

We now consider some special cases of this problem:

1. For the special case of $\nu = 0$, Equations (3.24a, b, c), respectively, become

$$\sum_{k=0}^{1} (-1)^k \frac{\alpha_k J_0(\mu x_k) + (-1)^k \gamma_k\mu J_1(\mu x_k)}{\alpha_k Y_0(\mu x_k) + (-1)^k \gamma_k\mu Y_1(\mu x_k)} = 0 \qquad (3.25a)$$

$$\psi_i(x) = J_0(\mu_i x) - \frac{\alpha_0 J_0(\mu_i x_0) + \gamma_0\mu_i J_1(\mu_i x_0)}{\alpha_0 Y_0(\mu_i x_0) + \gamma_0\mu_i Y_1(\mu_i x_0)} Y_0(\mu_i x) \qquad (3.25b)$$

$$N_i = \frac{2}{(\pi\mu_i)^2} \sum_{k=0}^{1} (-1)^{k+1} \frac{\alpha_k^2 + (\mu_i\gamma_k)^2}{\left[\alpha_k Y_0(\mu_i x_k) + (-1)^k \gamma_k\mu_i Y_1(\mu_i x_k)\right]^2}$$

$$(3.25c)$$

2. We now restrict special case 1 for a finite region $0 \le x \le 1$. Then by setting $x_0 = 0$, $x_1 = 1$, and letting $\alpha_0 = 0$, and noting that $Y_1(0) = \infty$, Equations (3.25a, b, c), respectively, reduce to

$$\frac{J_0(\mu)}{J_1(\mu)} = \frac{\beta_1}{\alpha_1} \mu \qquad (3.26a)$$

$$\psi_i(x) = J_0(\mu_i x) \qquad (3.26b)$$

$$N_i = \frac{1}{2} J_1^2(\mu_i)\left[1 + \left(\frac{\beta_1}{\alpha_1}\mu_i\right)^2\right] \qquad (3.26c)$$

To obtain Equation (3.26c) we utilized the Wronskian relationship for the Bessel functions together with the result (3.26a).

Example 3.3 Determine the eigencondition, eigenfunctions, and normalization integral of the following eigenvalue problem for a finite region $x_0 \leq x \leq x_1$.

$$\psi''(\mu, x) + \frac{2}{x}\psi'(\mu, x) + \left[\mu^2 - \frac{n(n+1)}{x^2}\right]\psi(\mu, x) = 0 \qquad (3.27a)$$

$$\alpha_k \psi(\mu, x_k) - (-1)^k \gamma_k \psi'(\mu, x_k) = 0, \qquad k = 0, 1 \qquad (3.27b, c)$$

where $n = 0, 1, 2, \ldots$.

Solution Equation (3.27a) is put into a form comparable to that given by Equation (3.2a); that is,

$$\frac{d}{dx}\left[x^2 \frac{d\psi(\mu, x)}{dx}\right] + x^2 \left[\mu^2 - \frac{n(n+1)}{x^2}\right]\psi(\mu, x) = 0 \qquad (3.27d)$$

A comparison of this system with that given by Equations (3.2) yields

$$k(x) = x^2, \qquad w(x) = x^2, \qquad d(x) = n(n+1),$$

$$\beta_k k(x_k) = \gamma_k, \quad \text{or} \quad \beta_k = \frac{\gamma_k}{x_k^2} \qquad (3.28a)$$

Also, by comparing Equations (3.28a) and (3.9) we find

$$m = -\tfrac{1}{2}, \qquad a = 0, \qquad c = 1, \quad \text{and} \quad \nu = n + \tfrac{1}{2} \qquad (3.28b)$$

Then from Equations (3.15) we obtain

$$A_k = \alpha_k + (-1)^k \frac{n+1}{x_k}\gamma_k, \qquad B_k = \gamma_k \qquad (3.28c)$$

By making use of the following relations

$$J_{n+(1/2)}(x) = \sqrt{\frac{2x}{\pi}}\, j_n(x), \qquad Y_{n+(1/2)}(x) = \sqrt{\frac{2x}{\pi}}\, y_n(x) \qquad (3.29a, b)$$

the eigencondition is obtained from Equation (3.13) as

$$\sum_{k=0}^{1}(-1)^k \frac{\left[\alpha_k x_k + (-1)^k(n+1)\gamma_k\right]j_n(\mu x_k) - (-1)^k \gamma_k \mu x_k j_{n-1}(\mu x_k)}{\left[\alpha_k x_k + (-1)^k(n+1)\gamma_k\right]y_n(\mu x_k) - (-1)^k \gamma_k \mu x_k y_{n-1}(\mu x_k)} = 0$$

$$(3.30a)$$

where $j_n(x)$ and $y_n(x)$ are called the *spherical Bessel functions* of the first and second kind, respectively [4].

Choosing $C(\mu_i) = \sqrt{\pi/(2\mu_i)}$, the eigenfunctions $\psi_i(x)$ and the normalization integral N_i are obtained from Equations (3.14) and (3.16b), respectively, as

$$\psi_i(x) = j_n(\mu_i x) - \frac{[\alpha_0 x_0 + (n+1)\gamma_0]j_n(\mu_i x_0) - \gamma_0 \mu_i x_0 j_{n-1}(\mu_i x_0)}{[\alpha_0 x_0 + (n+1)\gamma_0]y_n(\mu_i x_0) - \gamma_0 \mu_i x_0 y_{n-1}(\mu_i x_0)} y_n(\mu_i x)$$

(3.30b)

$$N_i = \frac{1}{2\mu_i^4} \sum_{k=0}^{1} (-1)^{k+1} \frac{D_k}{E_k}$$

(3.30c)

where

$$D_k = (\gamma_k \mu_i x_k)^2 + [\alpha_k x_k + (-1)^k (n+1)\gamma_k][\alpha_k x_k - (-1)^k n\gamma_k]$$

$$E_k = \left\{ [\alpha_k x_k + (-1)^k (n+1)\gamma_k] y_n(\mu_i x_k) - (-1)^k \gamma_k \mu_i x_k y_{n-1}(\mu_i x_k) \right\}^2$$

Example 3.4 Determine the eigencondition, eigenfunctions, and normalization integral of the following eigenvalue problem for a finite region $x_0 \le x \le x_1$.

$$\psi''(\mu, x) + \frac{2}{x}\psi'(\mu, x) + \mu^2\psi(\mu, x) = 0$$

(3.31a)

$$\delta_k\psi(\mu, x_k) - (-1)^k\gamma_k\psi'(\mu, x_k) = 0, \qquad k = 0,1$$

(3.31b, c)

Solution   The substitution of $\Omega(\mu, x)$ defined as

$$\psi(\mu, x) = x^{-1}\Omega(\mu, x)$$

(3.32)

transforms problem (3.31) to

$$\Omega''(\mu, x) + \mu^2\Omega(\mu, x) = 0$$

(3.33a)

$$[\delta_k x_k + (-1)^k\gamma_k]\Omega(\mu, x_k) - (-1)^k\gamma_k x_k\Omega'(\mu, x_k) = 0, \qquad k = 0,1$$

(3.33b, c)

A comparison of system (3.33) with system (3.17) yields

$$\alpha_k = \delta_k x_k + (-1)^k\gamma_k, \qquad \beta_k = \gamma_k x_k$$

(3.34)

Then the eigencondition, eigenfunctions, and normalization integral of problem (3.31) are immediately obtained from Equations (3.20a, b, c), respectively, as

$$[(\delta_0 x_0 + \gamma_0)\gamma_1 x_1 + (\delta_1 x_1 - \gamma_1)\gamma_0 x_0]\cos[\mu(x_1 - x_0)]$$

$$+ [(\delta_0 x_0 + \gamma_0)(\delta_1 x_1 - \gamma_1) - \gamma_0 x_0 \gamma_1 x_1 \mu^2]\frac{\sin[\mu(x_1 - x_0)]}{\mu} = 0 \quad (3.35a)$$

$$\psi_i(x) = \frac{\gamma_0 x_0}{x}\cos[\mu_i(x - x_0)] + \frac{\delta_0 x_0 + \gamma_0}{\mu_i x}\sin[\mu_i(x - x_0)] \quad (3.35b)$$

$$N_i = \frac{1}{2\mu_i^2}\left\{[(\delta_0 x_0 + \gamma_0)^2 + (\gamma_0 x_0 \mu_i)^2]\left[\frac{(\delta_1 x_1 - \gamma_1)\gamma_1 x_1}{(\delta_1 x_1 - \gamma_1)^2 + (\gamma_1 x_1 \mu_i)^2} + x_1 - x_0\right]\right.$$

$$\left. + (\delta_0 x_0 + \gamma_0)\gamma_0 x_0\right\} \quad (3.35c)$$

For the special case of $\delta_0 = 0$, $\gamma_0 = 1$, $x_0 = 0$, and $x_1 = 1$, Equations (3.35a, b, c), respectively, reduce to

$$\frac{j_0(\mu)}{j_1(\mu)} = \frac{\gamma_1}{\delta_1}\mu \quad (3.36a)$$

$$\psi_i(x) = j_0(\mu_i x) \quad (3.36b)$$

$$N_i = \frac{1}{2}\left[1 + \left(\frac{\gamma_1}{\delta_1}\mu_i\right)^2 - \frac{\gamma_1}{\delta_1}\right]j_1^2(\mu_i) \quad (3.36c)$$

where the spherical Bessel functions of the first kind are

$$j_0(\mu) = \frac{\sin\mu}{\mu} \quad \text{and} \quad j_1(\mu) = \frac{\sin\mu - \mu\cos\mu}{\mu^2} \quad (3.36d, e)$$

Example 3.5   Determine the eigencondition, eigenfunctions, and normalization integral of the following eigenvalue problem for a finite region $x_0 \le x \le x_1$.

$$\psi''(\mu, x) - g\psi'(\mu, x) + \mu^2\psi(\mu, x) = 0 \quad (3.37a)$$

$$\delta_k\psi(\mu, x_k) - (-1)^k\gamma_k\psi'(\mu, x_k) = 0, \quad k = 0,1 \quad (3.37b, c)$$

Solution   The substitution of $\Omega(\mu, x)$ defined as

$$\psi(\mu, x) = \Omega(\mu, x)e^{gx/2} \quad (3.38)$$

transforms problem (3.37) to

$$\Omega''(\lambda, x) + \lambda^2 \Omega(\lambda, x) = 0 \qquad (3.39a)$$

$$\left[\delta_k - (-1)^k \gamma_k \frac{g}{2} \right] \Omega(\lambda, x_k) - (-1)^k \gamma_k \Omega'(\lambda, x_k) = 0, \qquad k = 0, 1$$

$$(3.39b, c)$$

where

$$\lambda^2 = \mu^2 - \left(\frac{g}{2} \right)^2 \qquad (3.39d)$$

A comparison of system (3.39) with system (3.17) yields

$$\alpha_k = \delta_k - (-1)^k \gamma_k \frac{g}{2}, \qquad \beta_k = \gamma_k \qquad (3.40)$$

Then the eigencondition, eigenfunctions, and normalization integral of problem (3.37) are obtained from Equations (3.20a, b, c), respectively, as

$$\left[\left(\delta_0 - \gamma_0 \frac{g}{2} \right) \gamma_1 + \left(\delta_1 + \gamma_1 \frac{g}{2} \right) \gamma_0 \right] \cos[\lambda(x_1 - x_0)]$$

$$+ \left[\left(\delta_0 - \gamma_0 \frac{g}{2} \right) \left(\delta_1 + \gamma_1 \frac{g}{2} \right) - \gamma_0 \gamma_1 \lambda^2 \right] \frac{\sin[\lambda(x_1 - x_0)]}{\lambda} = 0 \quad (3.41a)$$

$$\psi_i(x) = \left\{ \gamma_0 \cos[\lambda_i(x - x_0)] + \frac{\delta_0 - \gamma_0(g/2)}{\lambda_i} \sin[\lambda_i(x - x_0)] \right\} e^{(g/2)x}$$

$$(3.41b)$$

$$N_i = \frac{1}{2\lambda_i^2} \left\{ \left[\left(\delta_0 - \gamma_0 \frac{g}{2} \right)^2 + (\gamma_0 \lambda_i)^2 \right] \left[\frac{(\delta_1 + \gamma_1 g/2) \gamma_1}{(\delta_1 + \gamma_1 g/2)^2 + (\gamma_1 \lambda_i)^2} + x_1 - x_0 \right] \right.$$

$$\left. + \left(\delta_0 - \gamma_0 \frac{g}{2} \right) \gamma_0 \right\} \qquad (3.41c)$$

Example 3.6 Solve the following eigenvalue problem for a finite region $0 \le x \le 1$.

$$\psi''(\mu, x) + \frac{1 - 2m}{x} \psi'(\mu, x) + \mu^2 \psi(\mu, x) = 0 \qquad (3.42a)$$

$$\psi'(\mu, 0) = 0 \qquad (3.42b)$$

$$\alpha_1 \psi(\mu, 1) + \beta_1 \psi'(\mu, 1) = 0 \qquad (3.42c)$$

Solution The solution of Equation (3.42a) that satisfies the boundary condition (3.42b) at $x = 0$ is

$$\psi_i(x) = C(\mu_i x)^m J_{-m}(\mu_i x) \qquad (3.43a)$$

The boundary condition (3.42c) at $x = 1$ is satisfied if

$$\frac{J_{-m}(\mu)}{J_{1-m}(\mu)} = \frac{\beta_1}{\alpha_1}\mu \qquad (3.43b)$$

The normalization integral is determined by introducing the solution (3.43a) into Equation (3.7a) and utilizing the expression (3.43b). We find

$$N_i = \frac{1}{2}\left[\mu_i^m J_{1-m}(\mu_i)\right]^2\left[1 + 2m\frac{\beta_1}{\alpha_1} + \left(\frac{\beta_1}{\alpha_1}\mu_i\right)^2\right]C^2 \qquad (3.43c)$$

The functions $x^m J_{-m}(x)$ and $x^m J_{1-m}(x)$ are determined for the special cases of $m = \pm\frac{1}{2}$ by making use of Equations (3.19) and (3.29); the results are given in Table 3.1.

Now if we choose $C = 1$ for the case of $m = 0$, $C = \sqrt{\pi/2}$ for the cases of $m = \pm\frac{1}{2}$, and apply the relations given by Table 3.1, we find that the results (3.21), (3.26), and (3.36) coincide with that given by Equations (3.43) for the special cases $m = \frac{1}{2}$, $m = 0$, and $m = -\frac{1}{2}$, respectively.

Table 3.1

m	$x^m J_{-m}(x)$	$x^m J_{1-m}(x)$
$\frac{1}{2}$	$\sqrt{\dfrac{2}{\pi}}\cos x$	$\sqrt{\dfrac{2}{\pi}}\sin x$
0	$J_0(x)$	$J_1(x)$
$-\frac{1}{2}$	$\sqrt{\dfrac{2}{\pi}}j_0(x)$	$\sqrt{\dfrac{2}{\pi}}j_1(x)$

Example 3.7 Solve the following eigenvalue problem for a finite region $0 \le x \le 1$.

$$\psi''(\mu, x) + \frac{n}{x}\psi'(\mu, x) + \mu^2(1 - x^2)\psi(\mu, x) = 0, \qquad n = 0 \text{ or } 1$$

$$(3.44a)$$

$$\psi'(\mu, 0) = 0 \qquad (3.44b)$$

$$\alpha_1\psi(\mu, 1) + \beta_1\psi'(\mu, 1) = 0 \qquad (3.44c)$$

Solution   The substitution of

$$\psi(\mu, x) = y(z)e^{-z/2}, \qquad z = \mu x^2 \tag{3.45a, b}$$

transforms Equation (3.44a) to *Kummer's differential equation* [4, p. 504]

$$zy''(z) + (b - z)y'(z) - ay(z) = 0 \tag{3.46a}$$

where

$$a = \frac{1 + n - \mu}{4}, \qquad b = \frac{1 + n}{2}, \qquad z = \mu x^2 \tag{3.46b}$$

The particular solution of Equation (3.46) is the *confluent hypergeometric function*

$$M(a, b, z) = 1 + \frac{a}{b}z + \frac{a(a + 1)}{b(b + 1)} \frac{z^2}{2!} + \cdots$$

$$+ \frac{a(a + 1)\ldots(a + k - 1)}{b(b + 1)\ldots(b + k - 1)} \frac{z^k}{k!} + \cdots \tag{3.46c}$$

The properties of this function have been extensively studied. The reader should consult Reference 4 for extensive tabulations of this function and for some of its properties such as the derivatives, recurrence relations, and asymptotic expansions.

For computational purposes the series (3.46c) can be represented in an alternative form by repetitively factoring as

$$M(a, b, z)$$

$$= 1 + \frac{a}{b}z\left(1 + \frac{a + 1}{b + 1}\frac{z}{2}\left\{1 + \frac{a + 2}{b + 2}\frac{z}{3}\left[1 + \cdots \left(1 + \frac{a + k - 1}{b + k - 1}\frac{z}{k}\right)\right]\right\}\right)$$

$$\tag{3.46d}$$

The solution of Equation (3.44a) that satisfies the boundary condition (3.44b) at $x = 0$ is

$$\psi_i(x) = M(a_i, b, z_i)e^{-z_i/2} \tag{3.47a}$$

This solution satisfies the boundary condition (3.44c) at $x = 1$, if

$$(\alpha_1 - \beta_1\mu)M(a, b, \mu) + 2\frac{a}{b}\beta_1\mu M(a + 1, b + 1, \mu) = 0 \tag{3.47b}$$

which is the eigencondition for the determination of the eigenvalues μ_i.

Multidimensional Case

We consider the three-dimensional eigenvalue system (2.2) for the case $k(\mathbf{x}) = 1$, $w(\mathbf{x}) = 1$, and $d(\mathbf{x}) = 0$. Then

$$\nabla^2 \psi(\mu, \mathbf{x}) + \mu^2 \psi(\mu, \mathbf{x}) = 0, \qquad \mathbf{x} \in V \qquad (3.48a)$$

$$\alpha(\mathbf{x}) \psi(\mu, \mathbf{x}) + \beta(\mathbf{x}) \frac{\partial \psi(\mu, \mathbf{x})}{\partial \mathbf{n}} = 0, \qquad \mathbf{x} \in S \qquad (3.48b)$$

Equation (3.48a) is the *Helmholtz* equation subject to the homogeneous boundary conditions (3.48b). The *separation of variables* is an effective method for solving the eigenvalue problem (3.48) provided that the coordinate system used permits the separation of variables. It is shown that there are 11 orthogonal coordinate systems in which the Helmholtz equation separates into ordinary differential equations [5, 6, 7]. Table 3-2 shows these 11 coordinate systems and the type of functions associated with the solutions.

Therefore, the method of separation is important for solving the Helmholtz equation, inasmuch as one of these 11 coordinate systems may be suitable for many practical problems. In the following we examine the separation of the Helmholtz equation in the rectangular, cylindrical, and spherical coordinate systems.

Separation in the Rectangular Coordinate System

We consider a three-dimensional finite region in the rectangular coordinate system (x, y, z) enclosed by the planes $x = x_k$, $y = y_k$, and $z = z_k$ where

Table 3.2. Orthogonal Coordinate Systems in Which Separation of Helmholtz Equation Is Possible

Coordinate System	Functions that Appear in Solution
1. Rectangular	Exponential, circular, hyperbolic
2. Circular-cylinder	Bessel, exponential, circular
3. Elliptic-cylinder	Mathieu, circular
4. Parabolic-cylinder	Weber, circular
5. Spherical	Legendre, power, circular
6. Prolate spheroidal	Legendre, circular
7. Oblate spheroidal	Legendre, circular
8. Parabolic	Bessel, circular
9. Conical	Lamé, power
10. Ellipsoidal	Lamé
11. Paraboloidal	Baer

$k = 0, 1$. Then Equation (3.48a) becomes

$$\left(\frac{\partial^2}{\partial x^2} + \frac{\partial^2}{\partial y^2} + \frac{\partial^2}{\partial z^2} + \mu^2 \right) \psi(\mu, x, y, z) = 0,$$

$$\text{in } x_0 < x < x_1, \quad y_0 < y < y_1 \quad \text{and} \quad z_0 < z < z_1 \qquad (3.49a)$$

and the boundary conditions (3.48b) take the form

$$\left[\alpha_{kx} - (-1)^k \beta_{kx} \frac{\partial}{\partial x} \right] \psi(\mu, x_k, y, z) = 0, \qquad k = 0, 1 \quad (3.49b, c)$$

$$\left[\alpha_{ky} - (-1)^k \beta_{ky} \frac{\partial}{\partial y} \right] \psi(\mu, x, y_k, z) = 0, \qquad k = 0, 1 \quad (3.49d, e)$$

$$\left[\alpha_{kz} - (-1)^k \beta_{kz} \frac{\partial}{\partial z} \right] \psi(\mu, x, y, z_k) = 0, \qquad k = 0, 1 \quad (3.49f, g)$$

where $\alpha_{k\ell}, \beta_{k\ell}$ ($\ell = x, y, z$ and $k = 0, 1$) are prescribed boundary surface coefficients.

We assume a separation of variables in the form

$$\psi(x, y, z) = X(x) Y(y) Z(z) \qquad (3.50a)$$

Then the eigenvalue problem (3.49) becomes

$$\frac{X''(x)}{X(x)} + \frac{Y''(y)}{Y(y)} + \frac{Z''(z)}{Z(z)} + \mu^2 = 0 \qquad (3.50b)$$

$$\left[\alpha_{kx} - (-1)^k \beta_{kx} \frac{d}{dx} \right] X(x_k) = 0, \qquad k = 0, 1 \qquad (3.50c, d)$$

$$\left[\alpha_{ky} - (-1)^k \beta_{ky} \frac{d}{dy} \right] Y(y_k) = 0, \qquad k = 0, 1 \qquad (3.50e, f)$$

$$\left[\alpha_{kz} - (-1)^k \beta_{kz} \frac{d}{dz} \right] Z(z_k) = 0, \qquad k = 0, 1 \qquad (3.50g, h)$$

Since the functions $X(x)$, $Y(y)$, and $Z(z)$ are independent of each other, the only way Equation (3.50b) can be satisfied is if each of the group equals to a constant; that is,

$$\frac{X''(x)}{X(x)} = -\lambda^2, \qquad \frac{Y''(y)}{Y(y)} = -\nu^2, \qquad \frac{Z''(z)}{Z(z)} = -\eta^2 \qquad (3.50i, j, k)$$

where λ, ν, and η are the separation parameters that are constants.

Summarizing, we split up the preceding three-dimensional eigenvalue problem (3.49) into the following three, separated, one-dimensional eigenvalue problems

$$X''(\lambda, x) + \lambda^2 X(\lambda, x) = 0 \quad \text{in } x_0 < x < x_1 \qquad (3.51a)$$

$$\left[\alpha_{kx} - (-1)^k \beta_{kx} \frac{d}{dx}\right] X(\lambda, x_k) = 0, \qquad k = 0, 1 \qquad (3.51b, c)$$

$$Y''(\nu, y) + \nu^2 Y(\nu, y) = 0 \quad \text{in } y_0 < y < y_1 \qquad (3.52a)$$

$$\left[\alpha_{ky} - (-1)^k \beta_{ky} \frac{d}{dy}\right] Y(\nu, y_k) = 0, \qquad k = 0, 1 \qquad (3.52b, c)$$

and

$$Z''(\eta, z) + \eta^2 Z(\eta, z) = 0 \quad \text{in } z_0 < z < z_1 \qquad (3.53a)$$

$$\left[\alpha_{kz} - (-1)^k \beta_{kz} \frac{d}{dz}\right] Z(\eta, z_k) = 0, \qquad k = 0, 1 \qquad (3.53b, c)$$

Every one of these problems (3.51), (3.52), and (3.53) coincides with the problem (3.17) considered in Example 3.1. Clearly, the eigenvalues $\lambda_\ell, \nu_m, \eta_n$ ($\ell, m, n = 1, 2, \ldots, \infty$), the eigenfunctions $X(\lambda_\ell, x), Y(\nu_m, y), Z(\eta_n, z)$, and the normalization integrals N_ℓ, N_m, N_n are immediately determined according to the expressions (3.20a, b, c), respectively. Then μ^2 is obtained as a triple index set as

$$\mu^2 \equiv \mu_{\ell mn}^2 = \lambda_\ell^2 + \nu_m^2 + \eta_n^2 \qquad (3.54a)$$

the three-dimensional eigenfunction $\psi(\mu, x, y, z)$ is given as a triple product in the form

$$\psi(\mu, x, y, z) \equiv \psi_{\ell mn} = X(\lambda_\ell, x) Y(\nu_m, y) Z(\eta_n, z) \qquad (3.54b)$$

and the normalization integral $N_{\ell mn}$, which is the scalar product $(\psi_{\ell mn}, \psi_{\ell mn})$, becomes

$$N_{\ell mn} = \int_{x_0}^{x_1} \int_{y_0}^{y_1} \int_{z_0}^{z_1} \psi_{\ell mn}^2 \, dx \, dy \, dz \equiv N_\ell N_m N_n \qquad (3.54c)$$

where

$$N_\ell = \int_{x_0}^{x_1} X^2(\lambda_\ell, x) \, dx, \qquad N_m = \int_{y_0}^{y_1} Y^2(\nu_m, y) \, dy, \qquad N_n = \int_{z_0}^{z_1} Z^2(\eta_n, z) \, dz$$

$$(3.54d, e, f)$$

are determined from Equation (3.20c).

Separation in the Cylindrical Coordinate System

We consider a three-dimensional finite cylindrical region, specified by the coordinates $r_0 \leq r \leq r_1$, $\phi_0 \leq \phi \leq \phi_1$, and $z_0 \leq z \leq z_1$ in the cylindrical coordinate system (r, ϕ, z). Then Equation (3.48a), with the Laplacian obtained from Equation (1.56), becomes

$$\left[\frac{1}{r} \frac{\partial}{\partial r}\left(r \frac{\partial}{\partial r} \right) + \frac{1}{r^2} \frac{\partial^2}{\partial \phi^2} + \frac{\partial^2}{\partial z^2} + \mu^2 \right] \psi(\mu, r, \phi, z) = 0$$

$$\text{in } r_0 < r < r_1, \qquad \phi_0 < \phi < \phi_1, \quad \text{and} \quad z_0 < z < z_1 \quad (3.55a)$$

For generality in the analysis we assume that $r_0 \neq 0$, characterizing a *hollow cylinder*, and the range of the ϕ variable is less than 2π corresponding to *a portion of a cylinder* (i.e., circle). then the boundary conditions (3.48b) take the form

$$\left[\alpha_{kr} - (-1)^k \beta_{kr} \frac{\partial}{\partial r} \right] \psi(\mu, r_k, \phi, z) = 0, \qquad k = 0, 1 \quad (3.55b, c)$$

$$\left[\alpha_{k\phi} - (-1)^k \beta_{k\phi} \frac{\partial}{\partial \phi} \right] \psi(\mu, r, \phi_k, z) = 0, \qquad k = 0, 1 \quad (3.55d, e)$$

$$\left[\alpha_{kz} - (-1)^k \beta_{kz} \frac{\partial}{\partial z} \right] \psi(\mu, r, \phi, z_k) = 0, \qquad k = 0, 1 \quad (3.55f, g)$$

where $\alpha_{k\ell}$, $\beta_{k\ell}$ ($\ell \equiv r, \phi, z$, and $k = 0, 1$) are the prescribed constant coefficients.

We assume a separation of variables in the form

$$\psi(r, \phi, z) = R(r) \Phi(\phi) Z(z) \qquad (3.56a)$$

Then the eigenvalue problem (3.55) becomes

$$\frac{1}{R(r)} \left[R''(r) + \frac{1}{r} R'(r) \right] + \frac{1}{r^2} \frac{1}{\Phi(\phi)} \Phi''(\phi) + \frac{1}{Z(z)} Z''(z) + \mu^2 = 0$$

$$(3.56b)$$

$$\left[\alpha_{kr} - (-1)^k \beta_{kr} \frac{d}{dr} \right] R(r_k) = 0, \qquad k = 0, 1 \qquad (3.56c, d)$$

$$\left[\alpha_{k\phi} - (-1)^k \beta_{k\phi} \frac{d}{d\phi} \right] \Phi(\phi_k) = 0, \qquad k = 0, 1 \qquad (3.56e, f)$$

$$\left[\alpha_{kz} - (-1)^k \beta_{kz} \frac{d}{dz} \right] Z(z_k) = 0, \qquad k = 0, 1 \qquad (3.56g, h)$$

Equation (3.56b) is satisfied only if each group of separated variables is set equal to a constant; that is,

$$\frac{\Phi''(\phi)}{\Phi(\phi)} = -\nu^2, \qquad \frac{Z''(z)}{Z(z)} = -\eta^2,$$

$$\frac{1}{R(r)}\left[R''(r) + \frac{1}{r}R'(r)\right] + \beta^2 - \frac{\nu^2}{r^2} = 0 \qquad (3.56i,j,k)$$

where $\beta^2 = \mu^2 - \eta^2$.

Summarizing, we split up the preceding three-dimensional eigenvalue problem (3.55) into the following three, separated, one-dimensional eigenvalue problems

$$\Phi''(\nu, \phi) + \nu^2\Phi(\nu, \phi) = 0 \quad \text{in } \phi_0 < \phi < \phi_1 \qquad (3.57a)$$

$$\left[\alpha_{k\phi} - (-1)^k\beta_{k\phi}\frac{d}{d\phi}\right]\Phi(\nu, \phi_k) = 0, \qquad k = 0,1 \qquad (3.57b,c)$$

$$Z''(\eta, z) + \eta^2 Z(\eta, z) = 0 \quad \text{in } z_0 < z < z_1 \qquad (3.58a)$$

$$\left[\alpha_{kz} - (-1)^k\beta_{kz}\frac{d}{dz}\right]Z(\eta, z_k) = 0, \qquad k = 0,1 \qquad (3.58b,c)$$

and

$$R''(\beta, r) + \frac{1}{r}R'(\beta, r) + \left(\beta^2 - \frac{\nu^2}{r^2}\right)R(\beta, r) = 0 \quad \text{in } r_0 < r < r_1$$

$$(3.59a)$$

$$\left[\alpha_{kr} - (-1)^k\beta_{kr}\frac{d}{dr}\right]R(\beta, r_k) = 0, \qquad k = 0,1 \qquad (3.59b,c)$$

Problems (3.57) and (3.58) are equivalent to problem (3.17) considered in Example 3.1; problem (3.59) coincides with problem (3.22) treated in Example 3.2. Therefore, the eigenvalues, eigenfunctions, and normalization integrals are readily determined by utilizing the appropriate expression given in problems (3.17) and (3.22). Then μ^2 is obtained as

$$\mu^2 \equiv \mu^2_{\ell mn} = \beta^2_{\ell m} + \eta^2_n \qquad (3.60a)$$

the function $\psi(\mu, r, \phi, z)$ is given by

$$\psi(\mu, r, \phi, z) \equiv \psi_{\ell mn} = R(\beta_{\ell m}, r)\Phi(\nu_m, \phi)Z(\eta_n, z) \qquad (3.60b)$$

and the normalization integral $N_{\ell mn}$, which is the scalar product $(\psi_{\ell mn}, \psi_{\ell mn})$, is defined as

$$N_{\ell mn} = \int_{r_0}^{r_1} \int_{\phi_0}^{\phi_1} \int_{z_0}^{z_1} \psi_{\ell mn}^2 r \, dr \, d\phi \, dz \equiv N_{\ell m} N_m N_n \qquad (3.60c)$$

where

$$N_{\ell m} = \int_{r_0}^{r_1} r R_\nu^2(\beta_{\ell m}, r) \, dr, \quad N_m = \int_{\phi_0}^{\phi_1} \Phi^2(\nu_m, \phi) \, d\phi, \quad N_n = \int_{z_0}^{z_1} Z^2(\eta_n, z) \, dz$$

$$(3.60d, e, f)$$

are determined from Equations (3.20c) and (3.24c).

The boundary conditions (3.57b, c) are applicable when the range of ϕ is less than 2π as in the case of a portion of a circle. For the problems of a *full cylinder* we have $0 \le \phi \le 2\pi$, or $\phi_0 = 0$ and $\phi_1 = 2\pi$. Then the boundary conditions (3.57b, c) should be replaced by the requirement that the function $\Phi(\phi)$ must be cyclic with a period 2π, namely,

$$\Phi(\phi) = \Phi(\phi + 2\pi) \qquad (3.61a)$$

or we must have

$$\cos(\nu\phi) = \cos[\nu(\phi + 2\pi)], \qquad \sin(\nu\phi) = \sin[\nu(\phi + 2\pi)]$$

$$(3.61b, c)$$

These conditions are satisfied if the eigenvalues ν_m are chosen as

$$\nu_m = m, \qquad m = 0, 1, 2, \dots \qquad (3.61d)$$

We note that every eigenvalue m corresponds to two independent eigenfunctions $\cos(m\phi)$ and $\sin(m\phi)$. Then for the problems of a full cylinder, the eigenfunctions (3.60b) are given by

$$\psi_{\ell mn} = R(\beta_{\ell m}, r) \begin{Bmatrix} \cos(m\phi) \\ \sin(m\phi) \end{Bmatrix} Z(\eta_n, z) \qquad (3.61e)$$

and the normalization integral N_m becomes

$$N_m = \int_0^{2\pi} \cos^2(m\phi) \, d\phi = \begin{cases} \pi & \text{for } m = 1, 2, 3, \dots \\ 2\pi & \text{for } m = 0 \end{cases} \qquad (3.61f)$$

$$N_m = \int_0^{2\pi} \sin^2(m\phi) \, d\phi = \pi \quad \text{for } m = 1, 2, 3, \dots \qquad (3.61g)$$

Separation in the Spherical Coordinate System

We now consider a three-dimensional finite spherical region, specified by the coordinates $r_0 \leq r \leq r_1$, $0 \leq \phi \leq 2\pi$, and $0 \leq \theta \leq \pi$ in the spherical coordinate system (r, θ, ϕ). Then Equation (3.48a), with the Laplacian obtained from (1.59), becomes

$$\left[\frac{1}{r^2} \frac{\partial}{\partial r} \left(r^2 \frac{\partial}{\partial r} \right) + \frac{1}{r^2 \sin \theta} \frac{\partial}{\partial \theta} \left(\sin \theta \frac{\partial}{\partial \theta} \right) + \frac{1}{r^2 \sin^2 \theta} \frac{\partial^2}{\partial \phi^2} + \mu^2 \right] \psi(\mu, r, \theta, \phi) = 0$$

$$\text{in } r_0 < r < r_1, \qquad 0 \leq \phi \leq 2\pi, \quad \text{and} \quad 0 \leq \theta \leq \pi \qquad (3.62a)$$

The region being a *hollow sphere*, the boundary conditions associated with Equation (3.62a) are specified only for $r = r_0$ and $r = r_1$; then the boundary conditions (3.48b) take the form

$$\left[\alpha_{kr} - (-1)^k \beta_{kr} \frac{\partial}{\partial r} \right] \psi(\mu, r_k, \phi, \theta) = 0, \qquad k = 0, 1 \quad (3.62\text{b, c})$$

Equation (3.62a) is put into a more convenient form by defining a new independent variable

$$z = \cos \theta \qquad (3.62\text{d})$$

Then Equations (3.62a, b, c) become

$$\left\{ \frac{\partial^2}{\partial r^2} + \frac{2}{r} \frac{\partial}{\partial r} + \frac{1}{r^2} \frac{\partial}{\partial z} \left[(1 - z^2) \frac{\partial}{\partial z} \right] \right.$$

$$\left. + \frac{1}{r^2(1 - z^2)} \frac{\partial^2}{\partial \phi^2} + \mu^2 \right\} \psi(\mu, r, \phi, z) = 0$$

$$\text{in } r_0 < r < r_1, \qquad 0 \leq \phi \leq 2\pi, \quad \text{and} \quad -1 < z < 1 \qquad (3.62\text{e})$$

$$\left[\alpha_{kr} - (-1)^k \beta_{kr} \frac{\partial}{\partial r} \right] \psi(\mu, r_k, \phi, z) = 0, \qquad k = 0, 1 \quad (3.62\text{f, g})$$

Here the range of z, $-1 \leq z \leq 1$ covers all the range in the θ domain inasmuch as the maximum range of θ is from 0 to π.

The problem (3.62) is separated by assuming a separation in the form

$$\psi(r, \phi, z) = R(r)\Phi(\phi)Z(z) \qquad (3.63a)$$

Then Equations (3.62e) and (3.62b, c), respectively, become

$$\frac{1}{R(r)}\left[R''(r) + \frac{2}{r}R'(r)\right] + \frac{1}{r^2}\frac{1}{Z(z)}\left[(1 - z^2)Z'(z)\right]'$$

$$+ \frac{1}{r^2(1 - z^2)}\frac{\Phi''(\phi)}{\Phi(\phi)} + \mu^2 = 0 \quad (3.63b)$$

$$\left[\alpha_{kr} - (-1)^k\beta_{kr}\frac{d}{dr}\right]R(r_k) = 0, \qquad k = 0,1 \qquad (3.63c, d)$$

Equation (3.63b) is satisfied only if each group of separated variables is set equal to a constant, and one way of achieving this separation is in the form

$$\Phi''(\phi) + m^2\Phi(\phi) = 0, \qquad 0 \le \phi \le 2\pi \qquad (3.64)$$

$$\left[(1 - z^2)Z'(z)\right]' + \left[n(n + 1) - \frac{m^2}{1 - z^2}\right]Z(z) = 0, \qquad -1 < z < 1$$

$$(3.65)$$

and

$$R''(r) + \frac{2}{r}R'(r) + \left[\mu^2 - \frac{n(n + 1)}{r^2}\right]R(r) = 0, \qquad r_0 < r < r_1 \quad (3.66a)$$

$$\left[\alpha_{kr} - (-1)^k\beta_{kr}\frac{d}{dr}\right]R(r_k) = 0, \qquad k = 0,1 \qquad (3.66b, c)$$

where m, n, and μ are the separation parameters. Permissible values of m and n are determined with the following considerations.

Equation (3.64) involve no boundary condition except the requirement that the function $\Phi(\phi)$ should be cyclic with a period of 2π, since the range of ϕ is $0 \le \phi \le 2\pi$. The two independent solutions $\{\cos m\phi; \sin m\phi\}$ of Equation(3.64) satisfy this requirement if the values of m are chosen as positive integers, including zero; that is,

$$m = 0, 1, 2, \ldots .$$

Equation (3.65) is called *Legendre's associated differential equation* and its two independent solutions include the functions $P_n^m(z)$ and $Q_n^m(z)$, which are called *Legendre's associated functions of degree n order m of the* first and second kind, respectively. The Legendre function of the second kind $Q_n^m(z)$ is convergent for $|z| < 1$ but becomes infinite at $z = \pm 1$. We note that Equation (3.65) does not involve any boundary conditions except the requirement that the region should include $z = \pm 1$. Since the functions $Q_n^m(z)$ become infinite at $z = \pm 1$, the function $Q_n^m(z)$ is excluded from the solution. The function $P_n^m(z)$ remains finite over the range $-1 \leq z \leq 1$ if the values of n are chosen as positive integers; furthermore, the function $P_n^m(z)$ vanishes for $m > n$. With these considerations, the function $P_n^m(z)$ becomes the only permissible solution of Equation (3.65) for the range of $-1 \leq z \leq 1$, the values of the separation constant n are chosen as positive integers including zero, and the values of m should be restricted to $m \leq n$; that is,

$$n = 0, 1, 2, \ldots \quad \text{and} \quad m \leq n.$$

The reader should consult the standard texts [6, 8, 9, 10, 11] for detailed treatment of Legendre functions.

Finally, the eigenvalue problem (3.66) coincides with problem (3.27) treated in Example 3.3, and the eigenvalues $\mu_{\ell m}$, eigenfunctions $R(\mu_{\ell m}, r)$, and normalization integral $N_{\ell n}$ are readily obtained from Equations (3.30a, b, c), respectively.

Then the solution of Equation (3.62e) subject to the boundary conditions (3.62f, g) is taken as

$$\psi_{\ell mn} = R(\mu_{\ell n}, r) \begin{Bmatrix} \cos(m\phi) \\ \sin(m\phi) \end{Bmatrix} P_n^m(z) \tag{3.67a}$$

where $m = 0, 1, 2, \ldots$; $n = 0, 1, 2, \ldots$; $m \leq n$. The normalization integral $N_{\ell mn}$ is the scalar product $(\psi_{\ell mn}, \psi_{\ell mn})$ and determined from

$$N_{\ell mn} = \int_{r_0}^{r_1} \int_0^{2\pi} \int_{-1}^1 \psi_{\ell mn}^2 r^2 \, dr \, d\phi \, dz = N_{\ell n} N_m N_{mn} \tag{3.67b}$$

where

$$N_{\ell m} = \int_{r_0}^{r_1} r^2 R(\mu_{\ell m}, r) \, dr, \tag{3.67c}$$

$$N_m = \int_0^{2\pi} \cos^2(m\phi) \, d\phi = \begin{cases} \pi & \text{for } m = 1, 2, 3, \ldots \\ 2\pi & \text{for } m = 0 \end{cases} \tag{3.67d}$$

$$N_m = \int_0^{2\pi} \sin^2(m\phi) \, d\phi = \pi \quad \text{for } m = 1, 2, 3, \ldots \tag{3.67e}$$

$$N_{mn} = \int_{-1}^1 [P_n^m(z)]^2 \, dz = \frac{2}{2n+1} \frac{(n+m)!}{(n-m)!} \tag{3.67f}$$

Sphere Cut Out by a Cone

We now consider a finite region consisting of the sphere cut out by the cone $\theta = \theta_0$. Let the region be specified by the coordinates $r_0 \leq r \leq r_1$, $0 \leq \phi \leq 2\pi$, and $0 \le \theta \le \theta_0 < \pi$ (or $z_0 \le z \le 1$, where $z_0 > -1$) as illustrated in Figure 3.1. The eigenvalue problem is defined by Equations (3.62e) and the boundary conditions (3.62f, g), plus a boundary condition for the surface $z = z_0$ (i.e., the surface of the cone). The solution of this eigenvalue problem can be expressed in the same form as that given by Equations (3.67), but the permissible values of n are chosen from the requirement that the boundary condition at $z = z_0$ is satisfied. This involves noninteger values of n, whereas the function $P_n^m(\mu)$ has singularity at $z = 1$ for the noninteger values of n. To alleviate this difficulty, the function $P_n^m(z)$ should be replaced by $P_n^{-m}(z)$ for the region $z_0 \le z \le 1$, $z_0 > -1$, because the function $P_n^{-m}(z)$ remains finite over this range of z for noninteger values of n. We consider the following cases.

1. The boundary condition at $z = z_0$ is of the first kind; that is, $\psi(\mu, r, \phi, z) = 0$ at $z = z_0$. Then the permissible values of n are determined from the requirement that

$$P_n^{-m}(z_0) = 0 \qquad (3.67g)$$

2. The boundary condition at $z = z_0$ is of the second kind; that is, $\partial\psi(\mu, r, \phi, z)/\partial z = 0$ at $z = z_0$. Then the permissible values of n are determined from the requirement that

$$\left.\frac{dP_n^{-m}(z)}{dz}\right|_{z=z_0} = 0 \qquad (3.67h)$$

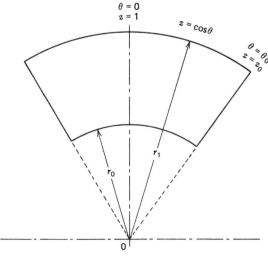

Figure 3.1 Sphere cut out of a cone $\theta = \theta_0$ in the spherical (r, ϕ, θ) or (r, ϕ, z), $z = \cos\theta$ coordinate system.

Asymptotic Formula

It is of interest to examine the solution of the eigenvalue problem (3.2) for the large values of μ_i^2. To perform such an analysis it is desirable to transform system (3.2) into a simpler form by a suitable change of variables. Therefore, new variables are defined as

$$z = \frac{\int_{x_0}^{x} \sqrt{[w(x)/k(x)]} \, dx}{\int_{x_0}^{x_1} \sqrt{[w(x)/k(x)]} \, dx} \tag{3.68a}$$

$$\Lambda = \mu \int_{x_0}^{x_1} \sqrt{\frac{w(x)}{k(x)}} \, dx \tag{3.68b}$$

$$\Omega(\Lambda, z) = [k(x)w(x)]^{1/4} \psi(\mu, x) \tag{3.68c}$$

Then the eigenvalue problem (3.2) is transformed to

$$\frac{d^2\Omega(\Lambda, z)}{dz^2} + [\Lambda^2 - \rho(z)]\Omega(\Lambda, z) = 0, \qquad 0 < z < 1 \tag{3.69a}$$

$$\left[A_k - (-1)^k B_k \frac{d}{dz} \right] \Omega(\Lambda, k) = 0, \qquad k = 0, 1 \tag{3.69b, c}$$

where

$$\rho(z) = \frac{d(x)}{w(x)} \left[\int_{x_0}^{x_1} \sqrt{\frac{w(x)}{k(x)}} \, dx \right]^2 + \frac{(d^2/dz^2)[k(x)w(x)]^{1/4}}{[k(x)w(x)]^{1/4}} \tag{3.69d}$$

$$A_k = \alpha_k + (-1)^k \beta_k \frac{(d/dz)[k(x)w(x)]^{1/2}_{x=x_k}}{2\int_{x_0}^{x_1} \sqrt{w(x)/k(x)} \, dx}, \qquad k = 0, 1$$

$$\tag{3.69e, f}$$

$$B_k = \beta_k \frac{[k(x)w(x)]^{1/2}_{x=x_k}}{\int_{x_0}^{x_1} \sqrt{w(x)/k(x)} \, dx}, \qquad k = 0, 1 \tag{3.69g, h}$$

The eigenvalue problem defined by Equations (3.69) is simpler in form than that given by Equations (3.2). For the higher eigenvalues, that is, for the large values of Λ^2, it is sufficiently accurate to take $1 - \rho(z)/\Lambda^2 \approx 1$; then the eigenvalue problem defined by Equations (3.69) coincides with the eigenvalue problem (3.17) considered in Example 3.1. Consequently, the eigencondition for the problem (3.69) is immediately written according to Equation (3.20a) as

$$\cot \Lambda = \frac{B_0 B_1 \Lambda^2 - A_0 A_1}{\Lambda (A_0 B_1 + A_1 B_0)} \tag{3.70a}$$

and the roots of this transcendental equation gives the values of Λ applicable for higher eigenvalues. The physical significance of this transcendental equation is envisioned better if it is written in the form

$$y = \cot \Lambda \tag{3.70b}$$

$$y = \frac{\Lambda}{C_1} - \frac{C_2}{\Lambda} \tag{3.70c}$$

where

$$C_1 = \frac{B_0 B_1}{A_0 B_1 + A_1 B_0}, \qquad C_2 = \frac{A_0 A_1}{A_0 B_1 + A_1 B_0}$$

    Equation (3.70b) is a cotangent curve and Equation (3.70c) is an hyperbola. Figure 3.2 illustrates a plot of these two curves; the intersections of these curves correspond to the eigenvalues Λ_i. We note that there are infinite numbers of these eigenvalues Λ_i ($i = 1, 2, \ldots, \infty$) inasmuch as there are infinite numbers of intersections. Only the positive roots are to be considered because

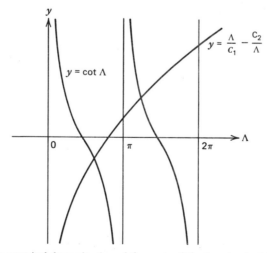

Figure 3.2   Geometrical determination of the roots of the transcendental Equation (3.70).

negative roots are equal in their absolute magnitude to positive roots. Reader should consult Reference 12 for the computation of the roots of the transcendental equations by numerical means. A tabulation of the roots of Equation (3.70) is given in Reference 14.

We note that as Λ becomes large, the eigenvalues approach πi. Following the so-called WKBJ method, in References 15–18 a more accurate asymptotic formula has been obtained for the eigenvalues in the form

$$\Lambda_i = \pi i + \epsilon \tag{3.71a}$$

where the constant ϵ depends on α_0, α_1, β_0, and β_1.

Finally, Equation (3.71a) is substituted into Equation (3.68b) to obtain the asymptotic formula in the form

$$\mu_i = \frac{\pi i + \epsilon}{\displaystyle\int_{x_0}^{x_1} \sqrt{w(x)/k(x)}\,dx} \tag{3.71b}$$

Algorithms for the Calculations of Eigenvalues and Eigenfunctions

When an eigenvalue problem cannot be handled analytically, it is necessary to calculate the eigenvalues and the eigenfunctions numerically. A number of schemes have been developed for calculating the eigenvalues and eigenfunctions of the Sturm–Liouville system. Here we first present a standard Runge-Kutta technique that utilizes Newton's iterative method for calculating the eigenvalues. The method is convenient for computing the lower eigenvalues, but difficulties may arise in calculating the higher order ones because of the rapid oscillation character of the higher eigenfunctions. When using such algorithms, extreme care must be exercised to ensure that no eigenvalues are missed. We then present a recently developed approach that provides a safe and automatic computation of the eigenvalues and the eigenfunctions.

The Standard Runge-Kutta Method

If the system (3.2) cannot be handled analytically, it is necessary to solve the problem by numerical means. To perform such calculations it is desirable to transform system (3.2) into a form more suitable for numerical integration. Therefore, new variables $y_1(\mu, x)$ and $y_2(\mu, x)$ are defined as

$$y_1(\mu, x) = k(x)\frac{d\psi(\mu, x)}{dx} \tag{3.72a}$$

$$y_2(\mu, x) = \psi(\mu, x) \tag{3.72b}$$

and the system (3.2) is transformed into

$$\frac{dy_1(\mu, x)}{dx} = \left[d(x) - \mu^2 w(x) \right] y_2(\mu, x) \qquad (3.73\text{a})$$

$$\frac{dy_2(\mu, x)}{dx} = \frac{1}{k(x)} y_1(\mu, x) \qquad (3.73\text{b})$$

$$\alpha_k y_2(\mu, x_k) - (-1)^k \beta_k y_1(\mu, x_k) = 0, \qquad k = 0,1 \quad (3.73\text{c, d})$$

Equations (3.73a, b) can be integrated numerically subject to the conditions

$$y_1(\mu, x_0) = \alpha_0, \qquad y_2(\mu, x_0) = \beta_0 \qquad (3.74\text{a, b})$$

Then Equation (3.73c) is identically satisfied and Equation (3.73d) becomes the eigencondition from which the eigenvalues μ_i are calculated.

To obtain an expression for the calculation of the normalization integral N_i, we define two more new variables as

$$y_3(\mu, x) = \frac{\partial y_1(\mu, x)}{\partial \mu}, \qquad y_4(\mu, x) = \frac{\partial y_2(\mu, x)}{\partial \mu} \qquad (3.75\text{a, b})$$

Equations (3.73a, b) are differentiated with respect to μ and Equations (3.75a, b) are introduced into the resulting expressions. We obtain

$$\frac{dy_3(\mu, x)}{dx} = \left[d(x) - \mu^2 w(x) \right] y_4(\mu, x) - 2\mu w(x) y_2(\mu, x) \quad (3.76\text{a})$$

$$\frac{dy_4(\mu, x)}{dx} = \frac{1}{k(x)} y_3(\mu, x) \qquad (3.76\text{b})$$

Equations (3.76a, b) can be integrated numerically subject to the conditions obtained from Equations (3.74) and (3.75) as

$$y_3(\mu, x_0) = 0, \qquad y_4(\mu, x_0) = 0 \qquad (3.76\text{c, d})$$

Then the normalization integral N_i, in view of the Equation (3.7a), is determined from

$$N_i = \frac{1}{2\mu_i} \begin{vmatrix} y_4(\mu_i, x_1) & y_3(\mu_i, x_1) \\ y_2(\mu_i, x_1) & y_1(\mu_i, x_1) \end{vmatrix} \qquad (3.77)$$

Based on the foregoing procedure, the following algorithm is recommended for solving the eigenvalue problem of Class I.

1.   Calculate the eigenvalue μ_i by the numerical integration (i.e., Runge-Kutta) method of Equations (3.73a, b) subject to the conditions (3.74a, b) and utilizing Newton's iterative method

$$\mu_i^{(n+1)} = \mu_i^{(n)} - \frac{\alpha_1 y_2\left[\mu_i^{(n)}, x_1\right] + \beta_1 y_1\left[\mu_i^{(n)}, x_1\right]}{\alpha_1 y_4\left[\mu_i^{(n)}, x_1\right] + \beta_1 y_3\left[\mu_i^{(n)}, x_1\right]}, \qquad n = 1, 2, 3, \ldots$$

(3.78)

so that Equation (3.73d) is also satisfied. The calculation is started with the initial approximation for $\mu_i^{(1)}$ ($i = 1, 2, 3, \ldots$) obtained from the asymptotic formula (3.71b).

2.   Once the eigenvalues μ_i are available, $y_1(\mu, x)$ and $y_2(\mu, x)$ are obtained from Equations (3.73); $y_3(\mu, x)$ and $y_4(\mu, x)$ are calculated by direct numerical integration of Equations (3.76a, b) subject to the conditions (3.76c, d).

3.   The normalization integral N_i is calculated by the formula (3.77). This commonly used approach for the determination of the eigenvalues suffers from the disadvantages discussed previously.

Sign-Count Method

The approaches of Datzeff [19, 20] and Wittrick and Williams [21, 22] have recently been applied [23, 24] for the solution of eigenvalue problems associated with the analytic solution of transient heat diffusion problems. The principal advantage of this method lies in the fact that as many eigenvalues as desired can be calculated without missing any one of them.

Here we consider the Sturm–Liouville problem given by Equations (3.2), where the constants α_0, β_0, α_1, and β_1 are real and independent of the parameter μ.

Let the coefficients $k(x)$, $w(x)$, and $d(x)$ be approximated by stepwise functions as:

$$k(x) = k_k, \qquad w(x) = w_k, \qquad d(x) = d_k \quad \text{for } x_{k-1} < x < x_k,$$

$$k = 1, 2, \ldots, n \quad (3.79)$$

The values of the constants k_k, w_k, and d_k, $k = 1, 2, \ldots, n$ can be chosen in various manners. For example, they can be taken as the values of the coefficients $k(x)$, $w(x)$, and $d(x)$ at the midpoints of the corresponding subintervals $\ell_k = x_k - x_{k-1}$, or they can be estimated as some weighted average of the values of the coefficients at the endpoints of subintervals. In the following analysis we assume the division of the original interval into subintervals in some suitable manner; however, there is a procedure for automatic generation of the subintervals [23].

In view of the assumptions, Equations (3.79), in any of the subintervals $\ell_k = x_k - x_{k-1}$, $k = 1, 2, \ldots, n$, Equation (3.2a) can be replaced by

$$\frac{d^2\psi_k(x, \mu)}{dx^2} + \omega_k^2\psi_k(x, \mu) = 0 \quad \text{in } x_{k-1} < x < x_k \qquad (3.80a)$$

where

$$\omega_k^2 = \frac{\mu^2 w_k - d_k}{k_k} \quad \text{and} \quad k = 1, 2, \ldots, n \qquad (3.80b)$$

Inasmuch as the original function $\psi(x, \mu)$ defined by Equation (3.2a) has a continuous first derivative in the interior of each subinterval, then the functions $\psi_k(x, \mu)$ and its first derivatives at the endpoints of the corresponding subintervals should satisfy the following conditions:

$$\psi_k(x, \mu) = \psi_{k+1}(x, \mu); \qquad x = x_k, \qquad k = 1, 2, \ldots, n - 1 \qquad (3.80c)$$

$$k_k\frac{d\psi_k(x, \mu)}{dx} = k_{k+1}\frac{d\psi_{k+1}(x, \mu)}{dx}; \qquad x = x_k, \qquad k = 1, 2, \ldots, n - 1$$

$$(3.80d)$$

whereas Equations (3.2b) and (3.2c), respectively, take the form:

$$\alpha_0\psi_1(x, \mu) - \beta_0 k_1\frac{d\psi_1(x, \mu)}{dx} = 0 \qquad x = x_0 \qquad (3.80e)$$

$$\alpha_n\psi_n(x, \mu) + \beta_n k_n\frac{d\psi_n(x, \mu)}{dx} = 0 \qquad x = x_n \qquad (3.80f)$$

In the last equation α_1 and β_1 are replaced, respectively, by α_n and β_n because this boundary condition now belongs to the endpoint of the last subregion.

Thus the original problem, Equations (3.2), has now been replaced by the problem defined by Equations (3.80). The computed eigenvalues of the latter problem represent the eigenvalues of the original problem, Equations (3.2), if sufficiently small subintervals are chosen.

In the subsequent analysis we assume that the parameter $\omega_k^2 > 0$. (The cases $\omega_k^2 = 0$ and $\omega_k^2 < 0$, encountered in other applications are not considered here.) For this particular case, the solution of Equation (3.80a) is taken as

$$\psi_k(x, \mu) = \psi_k(x_{k-1}, \mu)\frac{\sin[\omega_k(x_k - x)]}{\sin(\omega_k\ell_k)} + \psi_k(x_k, \mu)\frac{\sin[\omega_k(x - x_{k-1})]}{\sin(\omega_k\ell_k)}$$

$$(3.81)$$

We note that in Equation (3.81) the quantities $\psi_k(x_{k-1}, \mu)$ and $\psi_k(x_k, \mu)$ are actually the values of the eigenfunctions at the end points of the corresponding subinterval.

The solution, Equation (3.81), is now constrained to satisfy the boundary conditions (3.80c, d, e, f). We obtain

$$\left(\frac{\alpha_0}{\beta_0} + A_1\right)\psi_0^*(\mu) - B_1\psi_1^*(\mu) = 0 \tag{3.82a}$$

$$-B_k\psi_{k-1}^*(\mu) + (A_k + A_{k+1})\psi_k^*(\mu) - B_{k+1}\psi_{k+1}^*(\mu) = 0,$$
$$k = 1, 2, \ldots, n - 1 \tag{3.82b}$$

$$-B_n\psi_{n-1}^*(\mu) + \left(A_n + \frac{\alpha_n}{\beta_n}\right)\psi_n^*(\mu) = 0 \tag{3.82c}$$

where

$$B_k = \frac{k_k\omega_k}{\sin(\omega_k\ell_k)}; \qquad A_k = B_k\cos(\omega_k\ell_k) \tag{3.83a, b}$$

and the notation

$$\psi_k^*(\mu) \equiv \psi_k(x_k, \mu) = \psi_{k+1}(x_k, \mu) \tag{3.83c}$$

is introduced for simplicity. We note that the boundary conditions (3.80c) are automatically satisfied.

Equations (3.82) form a linear system of $(n + 1)$ homogeneous algebraic equations for the determination of the coefficients $\psi_k^*(\mu)$, $k = 0, 1, 2, \ldots, n$. These equations can be expressed in the matrix form as

$$[K(\mu)]\{\psi^*(\mu)\} = 0 \tag{3.84a}$$

where

$$[K(\mu)] = \begin{bmatrix} a_1 & -b_1 & 0 & 0 \cdots\cdots\cdots 0 & 0 & 0 \\ -b_1 & a_2 & -b_2 & 0 \cdots\cdots\cdots 0 & 0 & 0 \\ 0 & -b_2 & a_3 & & & 0 & 0 \\ \vdots & & & \cdots\cdots\cdots a_k \cdots\cdots & & \\ \vdots & & & & & -b_{n-1} & 0 \\ 0 & \cdots\cdots\cdots\cdots 0 & & -b_{n-1} & a_n & -b_n \\ 0 & 0 & 0 & 0 \cdots\cdots\cdots 0 & -b_n & a_{n+1} \end{bmatrix}$$

$$\tag{3.84b}$$

$$a_k = A_{k-1} + A_k, \qquad k = 1, 2, \ldots, (n + 1) \tag{3.84c}$$

$$b_k = B_k, \qquad k = 1, 2, \ldots, n \tag{3.84d}$$

$$A_0 = \frac{\alpha_0}{\beta_0} \qquad A_{n+1} = \frac{\alpha_n}{\beta_n} \tag{3.84e) (3.84f}$$

with A_k and B_k defined by Equations (3.83a, b) and

$$\{\psi^*(\mu)\}^T = \{\psi_0^*(\mu), \psi_1^*(\mu), \ldots, \psi_n^*(\mu)\} \tag{3.84g}$$

is the transpose of $\{\psi^*(\mu)\}$.

If the preceding system of equations has a nontrivial solution, the determinant of the coefficients should vanish, namely,

$$\det[K(\mu)] = 0 \tag{3.85}$$

The infinite number of real roots of this transcendental equation gives the eigenvalues of the eigenvalue problem, Equations (3.80). Customarily, this procedure is used for determining the eigenvalues of the system. However, in such a procedure there is always the risk of missing some of the eigenvalues and it is not possible to compute the higher order eigenvalues. Therefore, instead of following this standard approach, we present the *sign-count* method for the determination of the eigenvalues.

The Sign-Count Method for Determining the Eigenvalues

We now describe the procedure, advanced in References 21–24, for the determination of the eigenvalues of the eigenvalue problem given by Equations (3.80).

Wittrick and Williams [21, 22] show that the number of positive eigenvalues $N(\bar{\mu})$ lying between zero and some prescribed value of $\mu = \bar{\mu}$ is equal to:

$$N(\bar{\mu}) = N_0(\bar{\mu}) + s\{[K(\bar{\mu})]\} \tag{3.86}$$

where

$N_0(\bar{\mu}) =$ number of positive eigenvalues not exceeding $\bar{\mu}$, when all components of the vector $\{\psi^*(\mu)\}$ corresponding to $[K(\bar{\mu})]$ are zero (i.e., decoupled system).

$S([K(\bar{\mu})]) =$ sign-count of $[K(\bar{\mu})]$.

To find $N_0(\bar{\mu})$, one takes into account the fact that components of the vector $\{\psi^*(\bar{\mu})\}$ are zero; then the system of equations, Equation (3.80), degenerates into a decoupled set of equations:

$$\frac{d^2\psi_k(x, \bar{\mu})}{dx^2} + \bar{\omega}_k^2 \psi_k(x, \bar{\mu}) = 0; \quad x_{k-1} < x < x_k, \quad k = 1, 2, \ldots, n$$

$$\tag{3.87a}$$

subject to the boundary conditions:

$$\psi_k(x_{k-1}, \tilde{\mu}) = 0; \qquad \psi_k(x_k, \tilde{\mu}) = 0 \qquad (3.87b)$$

and ω_k is as defined by Equation (3.80b).

The eigencondition of the problem, Equation (3.87) is:

$$\sin(\tilde{\omega}_k \ell_k) = 0 \qquad (3.88a)$$

where $\ell_k = x_k - x_{k-1}$.

The transcendental Equation (3.88a) has an explicit solution only if $\tilde{\omega}_k^2 > 0$, namely,

$$\tilde{\omega}_k \ell_k = j\pi \qquad (j = 1, 2, 3, \ldots) \qquad (3.88b)$$

The number of eigenvalues not exceeding $\tilde{\mu}$, for the considered interval $x_{k-1} < x < x_k$, is given by "int($\tilde{\omega}_k \ell_k / \pi$)". Then the total number of eigenvalues for all the intervals in the region is given by

$$N_0(\tilde{\mu}) = \sum_{k=1}^{n} \text{int}\left(\frac{\tilde{\omega}_k \ell_k}{\pi}\right) \qquad (3.89)$$

where the symbol "int(z)" denotes the largest integer not exceeding the value of the argument z of the function.

The *sign-count*, $s\{[K(\tilde{\mu})]\}$, is shown [22] to be equal to the number of negative elements along the main diagonal of the matrix $[K^\Delta(\tilde{\mu})]$, which is the triangulated form of the matrix $[K(\tilde{\mu})]$. Using the Gauss elimination process we obtain the diagonal elements d_k of the triangulated matrix as:

$$d_k = a_k - \frac{b_{k-1}^2}{d_{k-1}}, \qquad k = 1, 2, \ldots, (n + 1) \qquad (3.90a)$$

with $b_0 = 0$. In this equation a_k and b_k are now replaced by their equivalents given by Equations (3.84c) and (3.84d), respectively. We obtain

$$d_k = A_{k-1} + A_k - \frac{B_{k-1}^2}{d_{k-1}}, \qquad k = 1, 2, \ldots, (n + 1) \qquad (3.90b)$$

where A_ks and B_ks are defined previously by Equations (3.83a, b).

In the case of boundary condition of the first kind at $x = x_0$ or at $x = x_n$ or at both endpoints, we have $\beta_0 = 0$, $\beta_n = 0$, or $\beta_0 = \beta_n = 0$ in the matrix (3.84b) and, as a result, $A_0 = \infty$, $A_{n+1} = \infty$, or $A_0 = A_{n+1} = \infty$, respectively. For any one of these three different cases, one simply neglects the corresponding row and column of $[K(\tilde{\mu})]$, as they do not influence the elimination process described by Equations (3.90). Such a situation implies that $\psi_1(x_0, \tilde{\mu}) = 0$, $\psi_n(x_n, \tilde{\mu}) = 0$, or $\psi_1(x_0, \tilde{\mu}) = \psi_n(x_n, \tilde{\mu}) = 0$.

Another way to impose boundary conditions of the first kind is to replace the corresponding zero coefficients β_0 and/or β_n by a small number, say 1×10^{-16}.

A simple but effective method for converging on a specified eigenvalue (say, the ith) based on the preceding result is as follows.

Try a value of $\tilde{\mu}$ and see if the corresponding $N(\tilde{\mu})$ defined by Equation (3.86) is $\geq i$. If it is, $\tilde{\mu}$ is an upper bound μ_u on the required eigenvalue. If not, double $\tilde{\mu}$ repetitively until an upper bound is obtained. An initial lower bound $\tilde{\mu}_\ell$ is zero. Every time a new value of

$$\tilde{\mu} = \tfrac{1}{2}(\tilde{\mu}_\ell + \tilde{\mu}_u)$$

is tried it will always give a new value for either $\tilde{\mu}_1$ or $\tilde{\mu}_u$, depending on whether N is $< i$, and the method can be made to converge to any specified accuracy.

Computation of Eigenfunctions

Once the eigenvalues μ_i are available, the eigenfunctions $\psi_k(x, \mu_i)$, at any location in the medium, can be computed from Equation (3.81) if the quantities $\psi_k(x_{k-1}, \mu) \equiv \psi_{k-1}^*(\mu)$ and $\psi_k(x_k, \mu) \equiv \psi_k^*(\mu)$ are known. To determine $\psi_k^*(\mu)$, $k = 0, 1, \ldots, n$, we utilize Equations (3.82) as now described. The choice of

$$\psi_1(x_0, \mu_i) = \beta_0 \quad \text{and} \quad k_1 \frac{d\psi_1(x_0, \mu)}{dx} = \alpha_0$$

satisfies Equation (3.80e) identically. Now utilizing the notation defined by Equation (3.83c), we write

$$\psi_0^*(\mu_i) = \beta_0 \tag{3.91a}$$

Introducing this result into Equation (3.82a), we obtain

$$\psi_1^*(\mu_i) = \frac{\alpha_0 + \beta_0 A_1}{B_1} \tag{3.91b}$$

Equation (3.82b) is rewritten in the form

$$\psi_{k+1}^*(\mu_i) = \frac{(A_k + A_{k+1})\psi_k^*(\mu_i) - B_k \psi_{k-1}^*(\mu_i)}{B_{k+1}}, \qquad k = 1, 2, \ldots, n - 1$$

$$\tag{3.91c}$$

where A_k and B_k are defined by Equations (3.83a, b).

Inasmuch as $\psi_0^*(\mu_i)$ and $\psi_1^*(\mu_i)$ are now available, Equations (3.91c) provide a recurrence relation for the computation of all the components of the vector $\{\psi^*(\mu_i)\}$ defined by Equation (3.84g), which are the values of the eigenfunctions $\psi(x, \mu_i)$ corresponding to the eigenvalue μ_i evaluated at the endpoints of the subintervals ℓ_k, $k = 1, 2, \ldots, n$.

Finally, Equation (3.82c) is used to estimate the magnitude of the global error involved in the computation of ψ_k^*. Namely, if the values of the ψ_k^* computed previously are exact, Equation (3.82c) is satisfied exactly. Any deviation of the right-hand side of this equation from zero is an estimate of the magnitude of the global error involved.

For the special case of a boundary condition of the first kind at $x = x_n$, Equation (3.82c) is not considered for the reason stated previously; for such a case, the accuracy estimate can be made from Equation (3.82b) and written for $k = n - 1$, noting that $\psi_n^*(\mu_i) = 0$.

Choosing Step Sizes

In the foregoing discussion it was assumed that the region was divided into subintervals $\ell_k = x_k - x_{k-1}$ ($k = 1, 2, \ldots, n$) in some suitable manner. We now describe a procedure for constructing these subdivisions in such a manner that the local errors in the values of the computed eigenfunctions could be held at a prescribed level and be of comparable magnitude for any of the subintervals necessary to approximate the original problem.

To adjust the step size $\ell_k = x_{k+1} - x_k$, an approximation for ψ_{k+1}^*/ψ_k^*, called $r^{(1)}$, is computed using Equation (3.91c) and step size ℓ. Then an approximation $r^{(2)}$ is computed using the step size $\ell/2$ twice. The accuracy is then tested in the following manner.

$$ \delta = \left| 1 - \frac{r^{(1)}}{r^{(2)}} \right| $$

Thus the value of δ is an approximate measure of the local truncation error at the point x_{k+1}. If δ is greater than a given tolerance ϵ, the step size ℓ is halved and the procedure is repeated at the point x_k. If δ is less than ϵ, the results $r^{(1)}$ and $r^{(2)}$ are considered to be correct.

Computation of the Norm

The norm N_i of the eigenfunction $\psi(x, \mu_i)$ is defined as

$$ N_i = \sum_{k=1}^{n} w_k \int_{x_{k-1}}^{x_k} \psi_k^2(x, \mu_i) \, dx \qquad (3.92) $$

The function $\psi_k(x, \mu_i)$ defined by Equation (3.81) is introduced into Equation

(3.92); after performing the integrations, the integral term becomes

$$\int_{x_{k-1}}^{x_k} \psi_k^2(x, \mu_i)\, dx = \{\psi_{k-1}^{*2} + \psi_k^{*2}\} \frac{\omega_k \ell_k - \sin(\omega_k \ell_k)\cos(\omega_k \ell_k)}{2\omega_k \sin^2(\omega_k \ell_k)}$$

$$+ \psi_{k-1}^* \psi_k^* \frac{1}{\omega_k} \left\{ \sin(\omega_k \ell_k) - \cos(\omega_k \ell_k) \right.$$

$$\left. \times \frac{\omega_k \ell_k - \sin(\omega_k \ell_k)\cos(\omega_k \ell_k)}{\sin^2(\omega_k \ell_k)} \right\} \qquad (3.93)$$

The relation Equations (3.83a, b) are utilized to eliminate the trigonometric terms appearing in Equation (3.93). In this procedure we utilized the expression $B_k^2 = A_k^2 + \omega_k^2 k_k^2$, which also follows from Equations (3.83a, b). After some manipulation, the integral term, Equation (3.93), is simplified and the resulting expression is introduced into Equation (3.92) to obtain the norm in the form

$$N_i = \sum_{k=1}^n w_k \frac{(\psi_{k-1}^{*2} + \psi_k^{*2})(B_k^2 \ell_k/k_k - A_k) + 2\psi_{k-1}^* \psi_k^* B_k (1 - A_k \ell_k/k_k)}{2\omega_k^2 k_k}$$

$$(3.94)$$

where the constants ψ_{k-1}^*, ψ_k^*, ω_k^2, A_k, and B_k are evaluated for $\mu = \mu_i$.

3.2   EIGENVALUE PROBLEM OF CLASS II

The eigenvalue problems of Class II are given by Equations (2.71). For the one-dimensional case the operator L_k becomes

$$L_k \equiv -\frac{d}{dx}\left[k_k(x)\frac{d}{dx}\right] + d_k(x), \qquad x_{k-1} < x < x_k, \qquad k = 1, 2, \ldots, n$$

$$(3.95a)$$

Letting $\alpha(x_0) = \alpha_0$, $\beta(x_0) = -\beta_0$, $\alpha(x_n) = \alpha_n$, and $\beta(x_n) = \beta_n$, the boundary condition operators B_0 and B_n applicable at the outer boundaries $x = x_0$ and $x = x_n$ become

$$B_0 \equiv \alpha_0 - \beta_0 k_1(x_0)\frac{d}{dx} \qquad \text{at } x = x_0 \qquad (3.95b)$$

$$B_n \equiv \alpha_n + \beta_n k_n(x_n)\frac{d}{dx} \qquad \text{at } x = x_n \qquad (3.95c)$$

Then the one-dimensional eigenvalue problem of Class II becomes

$$\frac{d}{dx}\left[k_k(x)\frac{d\psi_k(\mu, x)}{dx}\right] + \left[\mu^2 w_k(x) - d_k(x)\right]\psi_k(\mu, x) = 0$$

$$\text{in } x_{k-1} < x < x_k, \qquad k = 1, 2, \ldots, n \quad (3.96a)$$

subject to the boundary conditions

$$\alpha_0\psi_1(\mu, x_0) - \beta_0 k_1(x_0)\frac{d\psi_1(\mu, x_0)}{dx} = 0 \qquad (3.96b)$$

$$k_k(x_k)\frac{d\psi_k(\mu, x_k)}{dx} = h_{k, k+1}\left[\psi_{k+1}(\mu, x_k) - \psi_k(\mu, x_k)\right],$$

$$k = 1, 2, \ldots, (n-1) \quad (3.96c)$$

$$k_k(x_k)\frac{d\psi_k(\mu, x_k)}{dx} = k_{k+1}(x_k)\frac{d\psi_{k+1}(\mu, x_k)}{dx},$$

$$k = 1, 2, \ldots, (n-1) \quad (3.96d)$$

$$\alpha_n\psi_n(\mu, x_n) + \beta_n k_n(x_n)\frac{d\psi_n(\mu, x_n)}{dx} = 0 \qquad (3.96e)$$

where the functions $k_k(x)$, $k_k'(x)$, $w_k(x)$, and $d_k(x)$ are assumed to be real and continuous, and $k_k(x) > 0$, $w_k(x) > 0$ over the interval (x_{k-1}, x_k). The constants α_0, β_0, $h_{k, k+1}$, α_n, and β_n are real and independent of parameter μ, and $h_{k, k+1} = h_{k+1, k}$.

For the special case of perfect contact at the interfaces (i.e., $h_{k, k+1} \to \infty$), the boundary conditions, Equations (3.96c), reduce to

$$\psi_k(\mu, x_k) = \psi_{k+1}(\mu, x_k), \qquad k = 1, 2, \ldots, (n-1) \qquad (3.96f)$$

The safe and fast computation of the eigenvalues and eigenfunctions of the preceding eigenvalue problem, without missing any of the eigenvalues, is a very difficult matter and a discussion of various approaches for solving the eigenvalue problem is given in Reference 24. The *sign-count* method, advanced in Reference 24, appears to be the best approach known for solving the eigenvalue problems associated with the analysis of Class II problems. Here we present this method first for the case of perfect contact between the layers, and then discuss the case of contact resistance at the interfaces.

Perfect Contact at the Interfaces

Let $u_k(\mu, x)$ and $v_k(\mu, x)$ be two independent solutions of Equation (3.96a). The eigenfunctions $\psi_k(\mu, x)$ are generally constructed as a linear sum of these elementary solutions $u_k(\mu, x)$ and $v_k(\mu, x)$ in the form

$$\psi_k(\mu, x) = C_k u_k(\mu, x) + D_k v_k(\mu, x) \tag{3.97a}$$

However, in the computational procedure used here, it is desirable to replace the constants C_k and D_k with the value of the eigenfunctions evaluated at the end points $x = x_{k-1}$ and $x = x_k$, namely,

$$\psi_k(\mu, x_{k-1}) \equiv \psi_{k-1}^* \quad \text{and} \quad \psi_k(\mu, x_k) \equiv \psi_k^*, \qquad k = 1, 2, \ldots, n$$

To achieve this objective, we evaluate Equation (3.97a) for $x = x_{k-1}$ and $x = x_k$, and solve the resulting system for C_k and D_k. When the resulting constants C_k and D_k are introduced into Equation (3.97a), we obtain

$$\psi_k(\mu, x) = \psi_{k-1}^* U_k(\mu, x) + \psi_k^* V_k(\mu, x) \tag{3.97b}$$

where

$$U_k(\mu, x) = \frac{u_k(\mu, x)v_k(\mu, x_k) - u_k(\mu, x_k)v_k(\mu, x)}{u_k(\mu, x_{k-1})v_k(\mu, x_k) - u_k(\mu, x_k)v_k(\mu, x_{k-1})} \tag{3.98a}$$

$$V_k(\mu, x) = \frac{u_k(\mu, x_{k-1})v_k(\mu, x) - u_k(\mu, x)v_k(\mu, x_{k-1})}{u_k(\mu, x_{k-1})v_k(\mu, x_k) - u_k(\mu, x_k)v_k(\mu, x_{k-1})} \tag{3.98b}$$

We note that the functions $U_k(\mu, x)$, $V_k(\mu, x)$ as defined by Equations (3.98) satisfy the following conditions

$$U_k(\mu, x_{k-1}) = 1, \qquad U_k(\mu, x_k) = 0 \tag{3.99a}$$

$$V_k(\mu, x_{k-1}) = 0, \qquad V_k(\mu, x_k) = 1 \tag{3.99b}$$

Since we are considering here the perfect contact at the interfaces, we have $h_k \rightarrow \infty$, for $k = 1, 2, \ldots, n - 1$ at the interfaces. If the solution, Equation (3.97b), should satisfy the boundary conditions (3.96b, d, e), we have, respectively,

$$\left[\frac{\alpha_0}{\beta_0} - P_1(\mu, x_0) \right] \psi_0^* + P_1(\mu, x_1)\psi_1^* = 0 \tag{3.100a}$$

$$P_k(\mu, x_k)\psi_{k-1}^* + \{ Q_k(\mu, x_k) - P_{k+1}(\mu, x_k) \} \psi_k^* + P_{k+1}(\mu, x_{k+1})\psi_{k+1}^* = 0$$

$$k = 1, 2, \ldots, (n - 1) \tag{3.100b}$$

$$P_n(\mu, x_n)\psi_{n-1}^* + \left\{ Q_n(\mu, x_n) + \frac{\alpha_n}{\beta_n} \right\} \psi_n^* = 0 \tag{3.100c}$$

where

$$P_k(\mu, x) = k_k(x)U_k'(\mu, x) \tag{3.100d}$$

$$Q_k(\mu, x) = k_k(x)V_k'(\mu, x) \tag{3.100e}$$

$$k_k(x) = k_k x^{1-2m} \tag{3.100f}$$

and $m = \frac{1}{2}, 0, -\frac{1}{2}$ for slab, cylinder, and sphere, respectively.

In the derivation of Equations (3.100), the relation

$$P_k(\mu, x_k) + Q_k(\mu, x_{k-1}) = 0, \qquad k = 1, 2, \ldots, n \tag{3.101}$$

was substantially utilized.

Equation (3.101) is analogous to Equation (3.8c) and its validity can be established similar to the derivation of Equation (3.8c). Namely, one writes down Equation (3.96a) for the two linearly independent solutions $u_k(\mu, x)$ and $v_k(\mu, x)$:

$$\frac{d}{dx}\left[k_k(x)\frac{du_k(\mu, x)}{dx}\right] + \left[\mu^2 w_k(x) - d_k(x)\right]u_k(\mu, x) = 0 \tag{3.102a}$$

$$\frac{d}{dx}\left[k_k(x)\frac{dv_k(\mu, x)}{dx}\right] + \left[\mu^2 w_k(x) - d_k(x)\right]v_k(\mu, x) = 0 \tag{3.102b}$$

Equation (3.102a) is then multiplied by $v_k(\mu, x)$, and Equation (3.102b) by $u_k(\mu, x)$; the two results are subtracted, and integrated over the interval (x_{k-1}, x_k). Thus one obtains

$$k_k(x_k)\left[u_k'(\mu, x_k)v_k(\mu, x_k) - u_k(\mu, x_k)v_k'(\mu, x_k)\right]$$

$$-k_k(x_{k-1})\left[u_k'(\mu, x_{k-1})v_k(\mu, x_{k-1}) - u_k(\mu, x_{k-1})v_k'(\mu, x_{k-1})\right] = 0$$

$$\tag{3.103}$$

From Equations (3.98a, b), (3.100d, e), and Equation (3.103), the validity of Equation (3.101) can be established.

The system of Equations (3.100) forms the basis of our analysis for computing the eigenvalues and the eigenfunctions of the eigenvalue problem given by Equations (3.96).

Equations (3.100) form a linear system of $(n + 1)$ homogeneous equations for the determination of the quantity ψ_k^*, $k = 0, 1, 2, \ldots, n$. These equations can be expressed in the matrix form as

$$[K(\mu)]\{\psi^*\} = 0 \tag{3.104a}$$

where

$$\{\psi^*\}^T = \{\psi_0^*, \psi_1^*, \psi_2^*, \ldots, \psi_{n-1}^*, \psi_n^*\} \tag{3.104b}$$

is the transpose of $\{\psi^*\}$ and

$$[K] = \begin{bmatrix} a_0 & b_1 & & & & \\ b_1 & a_1 & b_2 & & & \\ \cdots & \cdots & \cdots & \cdots & \cdots & \cdots \\ & & b_{n-1} & a_{n-1} & b_n & \\ & & & b_n & a_n \end{bmatrix} \tag{3.104c}$$

$$a_0 = \frac{\alpha_0}{\beta_0} - P_1(\mu, x_0) \tag{3.104d}$$

$$b_k = P_k(\mu, x_k), \quad k = 1, 2, \ldots, n \tag{3.104e}$$

$$a_k = Q_k(\mu, x_k) - P_{k+1}(\mu, x_k), \quad k = 1, 2, 3, \ldots, (n-1) \tag{3.104f}$$

$$a_n = Q_n(\mu, x_n) + \frac{\alpha_n}{\beta_n} \tag{3.104g}$$

and A_k, B_k are defined by Equation (3.83). If the system (3.104) has a nontrivial solution, the determinant of the coefficients should vanish.

$$\det[K(\mu)] = 0 \tag{3.105}$$

The infinite number of real roots of this transcendental equation gives the eigenvalues of the eigenvalue problem (3.96). Customarily, such a procedure is used to determine the eigenvalues for the system. However, in such a procedure, there is always the risk of missing some of the eigenvalues and also computational difficulties arise in the determination of the higher order eigenvalues. Therefore, we are not following the standard procedure; instead, we apply the *sign-count* approach for the determination of the eigenvalues as now described.

Computation of Eigenvalues with the Sign-Count Method

As discussed previously for the case of Class I problems, the number of positive eigenvalues $N(\bar{\mu})$, lying between zero and some prescribed positive value of $\mu = \bar{\mu}$ of the eigenvalue parameter μ, is equal to

$$N(\bar{\mu}) = N_0(\bar{\mu}) + s\{[K(\bar{\mu})]\} \tag{3.106}$$

where $N_0(\bar{\mu})$ is the total number of positive eigenvalues not exceeding $\bar{\mu}$ when all components of the vector $\{\psi^*\}$ corresponding to $\tilde{\mu}$ are zero, and $s\{[K(\tilde{\mu})]\}$ denotes the sign-count of $[K(\tilde{\mu})]$ as discussed previously for the case of Class I problems.

To determine $N_0(\tilde{\mu})$, one takes into account the fact that, when all components of the vector $\{\psi^*\}$ are zero, the system of Equations (3.96a) degenerates into a decoupled set, namely,

$$\frac{d}{dx}\left[k_k(x)\frac{d\psi_k(\mu, x)}{dx}\right] + [\mu^2 w_k(x) - d_k(x)]\psi_k(\mu, x) = 0$$

$$\text{in } x_{k-1} < x < x_k, \qquad k = 1, 2, \ldots, n \quad (3.107a)$$

subject to the boundary conditions

$$\psi_k(\mu, x_{k-1}) = 0 \qquad \psi_k(\mu, x_k) = 0 \qquad (3.107b, c)$$

The eigencondition of the problem, defined by Equations (3.107), is

$$u_k(\mu, x_{k-1})v_k(\mu, x_k) - u_k(\mu, x_k)v_k(\mu, x_{k-1}) = 0 \quad (3.108a)$$

This transcendental equation should first be solved for each layer k, ($k = 1, 2, \ldots, n$) by a standard method. Generally, analytic expressions are available for simple cases. If an analytic expression cannot be found, then the eigenvalues for each layer k should be determined numerically and these eigenvalues should be stored in the memory of the digital computer. Once this information is available for a given value of $\tilde{\mu}$, the number of eigenvalues $N_{0,k}(\tilde{\mu})$ not exceeding $\tilde{\mu}$ for each layer can be determined. Then the total number of positive eigenvalues $N_0(\tilde{\mu})$ for the entire multilayered composite becomes

$$N_0(\tilde{\mu}) = \sum_{k=1}^{n} N_{0,k}(\tilde{\mu}) \qquad (3.108b)$$

The sign-count $s\{[K(\tilde{\mu})]\}$ is shown to be [21, 22] equal to the number of negative elements along the main diagonal of the matrix $[K^\Delta(\tilde{\mu})]$, which is the triangulated form of the matrix $[K(\tilde{\mu})]$ defined by Equations (3.104). Using the Gauss elimination process, the diagonal elements d_k, ($k = 1, 2, \ldots, n$) of the triangulated matrix are determined as

$$d_k = a_k - \frac{b_k^2}{d_{k-1}}, \qquad k = 0, 1, 2, \ldots, n \qquad (3.109)$$

with $b_0 = 0$. In this equation, a_k and b_k are defined by Equations (3.104).

In the case of boundary conditions of the first kind at $x = x_0$ and/or $x = x_n$, we have $\beta_0 = 0$ and/or $\beta_n = 0$. For such a case, $a_0 = \infty$ and/or $a_n = \infty$. In any of these cases one simply neglects the corresponding row and column of $[K(\tilde{\mu})]$, as they do not influence the computational process described by Equations (3.109), and imply that $\psi_1(\tilde{\mu}, x_0) = 0$ and/or $\psi_n(\tilde{\mu}, x_n) = 0$.

When both terms on the right-hand side of Equation (3.106) are computed as described, the quantity $N(\mu)$ becomes available. Knowing $N(\mu)$, an algorithm similar to the one discussed in the previous section can be used to determine the eigenvalues μ_i ($i = 1, 2, 3, \dots$) for the system.

Contact Resistance at the Interfaces

The foregoing analysis and the computational scheme are applicable if all layers are in perfect contact. However, the foregoing procedure can readily accommodate the situations involving contact conductance between the layers as now described.

Suppose there is contact conductance at the interface $x = x_k$ between the layers k and $k + 1$. We assume that the presence of interface resistance can be replaced by a fictitious layer of thickness $\ell^*_{k, k+1}$ so small that its capacitance can be neglected. Then the eigenvalue Equation (3.96a) in the local coordinate x^* degenerates into the following equation

$$\psi''_{k, k+1}(x^*) = 0 \quad \text{in } 0 < x^* < \ell^*_{k, k+1} \tag{3.110a}$$

The integration of this equation, first from 0 to x^* and then from 0 to $\ell^*_{k, k+1}$, after transforming to the system coordinate, can be expressed in the form

$$\left(\frac{k^*}{\ell^*}\right)_{k, k+1} [\psi(\mu, x_{k+1}) - \psi(\mu, x_k)] = k_k \psi'_k(\mu, x_k) \tag{3.110b}$$

A comparison of this result with Equation (3.96c) reveals that the two results coincide if

$$h_{k, k+1} = \left(\frac{k^*}{\ell^*}\right)_{k, k+1} \tag{3.110c}$$

This result implies that the presence of contact conductance $h_{k, k+1}$ is equivalent to a fictitious layer with $\omega_k \to 0$. The functions $P^*(\mu, x)$ and $Q^*(\mu, x)$ for such a fictitious layer are determined by a limiting process, by setting $\omega_k \to 0$, and $x_k - x_{k+1} = \ell^*$. We obtain

$$P_k(\mu) = -h_{k, k+1} \quad \text{and} \quad Q_k(\mu) = h_{k, k+1} \quad \text{at } x = x_k. \tag{3.111a, b}$$

The computational procedure now becomes apparent. Namely, whenever there is a contact resistance at an interface x_k, we treat this interface as an additional fictitious layer for which the functions $P_k(\mu, x)$ and $Q_k(\mu, x)$ are directly obtained from Equations (3.111).

Computation of Eigenfunctions

Once the eigenvalues μ_i are available, the eigenfunctions $\psi_k(x, \mu_i)$, at any location x in the medium can be computed from Equation (3.97b) if the quantities $\psi_k(\mu_i, x_{k-1}) \equiv \psi^*_{k-1}(\mu_i)$ and $\psi_k(\mu_i, x_k) \equiv \psi^*_k(\mu_i)$ are known. They can be determined by utilizing Equations (3.100a, b, c) as now described.

From Equations (3.100a, b) we have

$$\psi^*_1 = \frac{[P_1(\mu, x_0) - \alpha_0/\beta_0]\psi^*_0}{P_1(\mu, x_1)} \tag{3.112a}$$

$$\psi^*_{k+1} = \frac{[P_{k+1}(\mu, x_k) - Q_k(\mu, x_k)]\psi^*_k - P_k(\mu, x_k)\psi^*_{k-1}}{P_{k+1}(\mu, x_{k+1})}$$

$$k = 1, 2, \ldots, (n-1) \quad (3.112b)$$

Since the eigenfunctions are arbitrary within a multiplication constant, one can set

$$\psi^*_0 = \beta_0$$

and knowing ψ^*_0, the quantity ψ^*_1 is determined from Equation (3.112a).

Now, ψ^*_0 and ψ^*_1 being available, Equation (3.112b) provides a recurrence relation for the determination of the remaining ψ^*_k, $k = 2, 3, \ldots, (n-1)$.

In the case of boundary conditions of the first kind at $x = x_0$, as stated previously, Equation (3.112a) is no longer applicable. In that case an arbitrary value is chosen for ψ^*_1. Knowing $\psi^*_0 = 0$ and choosing ψ^*_1 arbitrarily, say, $\psi^*_1 = 1$, Equation (3.112b) can again be used as a recurrence relation for the computation of the remaining ψ^*_k, $k = 2, 3, \ldots, (n-1)$. Finally, Equation (3.100c), that is,

$$P_n(\mu, x_n)\psi^*_{n-1} + \left[Q_n(\mu, x_n) + \frac{\alpha_n}{\beta_n}\right]\psi^*_n = 0 \tag{3.112c}$$

is used to estimate the magnitude of the global error involved in the computation of ψ^*_k. Namely, if the eigenfunctions computed in the foregoing were exact, Equation (3.112c) would have been satisfied exactly. Any deviation of the right-hand side of Equation (3.112c) from zero is an estimate of the magnitude of the global error involved.

For the special case of the boundary condition of the first kind at $x = x_n$, Equation (3.112c) is not applicable for the reason stated previously. For such a case, the accuracy estimate can be made from Equation (3.112b), written for $k = n - 1$, and noting that $\psi^*_n = 0$.

Once the ψ^*_n and μ_i are evaluated, the eigenfunctions $\psi_k(\mu, x)$ for each region are determined according to Equation (3.97b). Knowing the eigenfunctions, the norm is determined from

$$N_i = \sum_{k=1}^{n} \int_{x_{k-1}}^{x_k} w_k(x)\psi_k^2(\mu_i, x)\, dx \tag{3.113}$$

3.3   EIGENVALUE PROBLEM OF CLASS III

The eigenvalue problem of Class III is given by Equations (2.97); for the one-dimensional case the operators L, B_k, B_{k1}, and B_{k2} become

$$L \equiv -\frac{d}{dx}\left[k_k(x)\frac{d}{dx}\right] + d_k(x) \tag{3.114a}$$

$$B_k \equiv \alpha_k - \beta_k k_k(x_0)\frac{d}{dx} \quad \text{at } x = x_0 \tag{3.114b}$$

$$B_{k1} \equiv \alpha_{k1} + \beta_{k1}k_1(x_1)\frac{d}{dx} \quad \text{at } x = x_1 \tag{3.114c}$$

$$B_{k2} \equiv \alpha_{k2} + \beta_{k2}k_2(x_1)\frac{d}{dx} \quad \text{at } x = x_1 \tag{3.114d}$$

where $x_0 \leq x \leq x_1$ and $k = 1, 2$.

Then the one-dimensional form of the eigenvalue problem of Class III becomes

$$\frac{d}{dx}\left[k_k(x)\frac{d\psi_k(\mu, x)}{dx}\right] + \left[\mu^2 w_k(x) - d_k(x)\right]\psi_k(\mu, x) = 0,$$

$$\text{in } x_0 < x < x_1 \quad (3.115a)$$

$$\alpha_k\psi_k(\mu, x_0) - \beta_k k_k(x_0)\frac{d\psi_k(\mu, x_0)}{dx} = 0 \tag{3.115b}$$

$$\alpha_{k1}\psi_1(\mu, x_1) + \alpha_{k2}\psi_2(\mu, x_1) + \beta_{k1}k_1(x_1)\frac{d\psi_1(\mu, x_1)}{dx}$$

$$+ \beta_{k2}k_2(x_1)\frac{d\psi_2(\mu, x_1)}{dx} = 0 \tag{3.115c}$$

where $k = 1, 2$.

The functions $k_k(x)$, $k_k'(x)$, $w_k(x)$, and $d_k(x)$ are assumed to be real and continuous, and $k_k(x) > 0$, $w_k(x) > 0$ over the interval (x_0, x_1). The constants α_k, β_k, α_{k1}, β_{k1}, α_{k2}, and β_{k2} are real and independent of μ.

Let $u_k(\mu, x)$ and $v_k(\mu, x)$ be two independent particular solutions of Equation (3.115a); then the general solution is constructed as a linear combination of these two solutions in the form

$$\psi_k(\mu, x) = C_k u_k(\mu, x) + D_k v_k(\mu, x) \tag{3.116}$$

If this solution should satisfy the boundary conditions (3.115b, c), we have

$$C_k U_k(\mu) + D_k V_k(\mu) = 0, \quad k = 1, 2 \tag{3.117a}$$

$$C_1 U_{k1}(\mu) + D_1 V_{k1}(\mu) + C_2 U_{k2}(\mu) + D_2 V_{k2}(\mu) = 0, \quad k = 1, 2 \tag{3.117b}$$

where

$$U_k(\mu) = \alpha_k u_k(\mu, x_0) - \beta_k k_k(x_0) u_k'(\mu, x_0) \tag{3.117c}$$

$$V_k(\mu) = \alpha_k v_k(\mu, x_0) - \beta_k k_k(x_0) v_k'(\mu, x_0) \tag{3.117d}$$

$$U_{kn}(\mu) = \alpha_{kn} u_n(\mu, x_1) + \beta_{kn} k_k(x_1) u_n'(\mu, x_1) \tag{3.117e}$$

$$V_{kn}(\mu) = \alpha_{kn} v_n(\mu, x_1) + \beta_{kn} k_k(x_1) v_n'(\mu, x_1) \tag{3.117f}$$

Equations (3.117a) are solved for the ratio D_k/C_k

$$\frac{D_k}{C_k} = -\frac{U_k(\mu)}{V_k(\mu)}, \quad k = 1, 2 \tag{3.118a}$$

By utilizing this ratio, Equations (3.117b) are written in the form

$$C_1\left[U_{k1}(\mu) - \frac{U_1(\mu)}{V_1(\mu)} V_{k1}(\mu)\right] + C_2\left[U_{k2}(\mu) - \frac{U_2(\mu)}{V_2(\mu)} V_{k2}(\mu)\right] = 0,$$
$$k = 1, 2 \tag{3.118b}$$

Equations (3.118b) are now solved for the ratio C_2/C_1 to obtain the eigencondition as

$$\frac{U_{11}(\mu)V_1(\mu) - U_1(\mu)V_{11}(\mu)}{U_{12}(\mu)V_2(\mu) - U_2(\mu)V_{12}(\mu)} = \frac{U_{21}(\mu)V_1(\mu) - U_1(\mu)V_{21}(\mu)}{U_{22}(\mu)V_2(\mu) - U_2(\mu)V_{22}(\mu)} \tag{3.119a}$$

Introducing the ratio (3.118a) into the solution (3.116) we obtain

$$\psi_k(\mu, x) = C_k\left[u_k(\mu, x) - \frac{U_k(\mu)}{V_k(\mu)} v_k(\mu, x)\right], \quad k = 1, 2 \tag{3.119b}$$

Thus once the eigenvalues are determined from Equation (3.119a), the eigenfunctions are given by Equation (3.119b).

Example 3.8 Determine the eigenfunctions and the eigencondition of the eigenvalue problem (3.115) for the following special case

$$k_k(x) = x^{1-2m_k}, \qquad w_k(x) = \omega_k^2 x^{1-2m_k}$$

$$d_k(x) = 0, \qquad x_0 = 0, \qquad x_1 = 1, \qquad \alpha_k = 0 \quad \text{where } k = 1, 2$$

$$(3.120)$$

Solution Introducing Equations (3.120) into the system (3.115) we obtain

$$\psi_k''(\mu, x) + \frac{1 - 2m_k}{x} \psi_k'(\mu, x) + (\mu\omega_k)^2 \psi_k(\mu, x) = 0 \quad \text{in } 0 < x < 1$$

$$(3.121a)$$

$$\psi_k'(\mu, 0) = 0 \tag{3.121b}$$

$$\alpha_{k1}\psi_1(\mu, 1) + \alpha_{k2}\psi_2(\mu, 1) + \beta_{k1}\psi_1'(\mu, 1) + \beta_{k2}\psi_2'(\mu, 1) = 0 \quad (3.121c)$$

where $k = 1, 2$.

The particular solution of Equation (3.121a) that satisfies the boundary conditions (3.121b) at $x = 0$ is

$$u_k(\mu, x) = (\mu\omega_k x)^{m_k} J_{-m_k}(\mu\omega_k x) \tag{3.122a}$$

Therefore, $U_k(\mu) = 0$ and $D_k = 0$; then the eigencondition is obtained from Equation (3.117b) or (3.119a) as

$$U_{11}(\mu)U_{22}(\mu) - U_{12}(\mu)U_{21}(\mu) = 0 \tag{3.122b}$$

where

$$U_{kn}(\mu) = \alpha_{kn}J_{-m_n}(\mu\omega_n) - \beta_{kn}\mu\omega_n J_{1-m_n}(\mu\omega_n) \tag{3.122c}$$

Algorithm for Numerical Calculation

To calculate the eigenvalues μ_i and the eigenfunctions $\psi_k(\mu_i, x)$ of the system (3.115) numerically, it is desirable to transform the system into a more suitable form for numerical integration. Therefore, new variables $y_{4k-2}(\mu, x)$ and $y_{4k-3}(\mu, x)$ are defined as [25]

$$\psi_k(\mu, x) = C_k y_{4k-2}(\mu, x) \tag{3.123a}$$

$$k_k(x) \frac{d\psi_k(\mu, x)}{dx} = C_k y_{4k-3}(\mu, x) \tag{3.123b}$$

Then system (3.115) is transformed to

$$y'_{4k-3}(\mu, x) = \left[d_k(x) - \mu^2 w_k(x) \right] y_{4k-2}(\mu, x) \qquad (3.124a)$$

$$y'_{4k-2}(\mu, x) = \frac{1}{k_k(x)} y_{4k-3}(\mu, x) \qquad (3.124b)$$

subject to the boundary conditions

$$\alpha_k y_{4k-2}(\mu, x_0) - \beta_k y_{4k-3}(\mu, x_0) = 0 \qquad (3.124c)$$

$$C_1 U_{k1}(\mu) + C_2 U_{k2}(\mu) = 0 \qquad (3.124d)$$

where $k = 1, 2,$ and

$$U_{kn}(\mu) = \alpha_{kn} y_{4n-2}(\mu, x_1) + \beta_{kn} y_{4n-3}(\mu, x_1) \qquad (3.124e)$$

Equations (3.124a, b) can be integrated numerically subject to the conditions

$$y_{4k-3}(\mu, x_0) = \alpha_k, \qquad y_{4k-2}(\mu, x_0) = \beta_k \qquad (3.125a, b)$$

Then Equation (3.124c) is identically satisfied. The two homogeneous Equations (3.124d) give the eigencondition

$$U_{11}(\mu) U_{22}(\mu) - U_{12}(\mu) U_{21}(\mu) = 0 \qquad (3.126a)$$

from which the eigenvalues μ_i are calculated.

Algorithm for a Specific Case

The calculation of μ_i is considerably facilitated if the intervals where the eigenvalues are to be found are known. Let us consider the location of the roots of Equation (3.126a) for the following special case [25]:

$$\alpha_k = 0, \qquad \beta_k = 1, \qquad \alpha_{11} = \alpha_{12} = 0, \qquad x_0 = 0, \qquad x_1 = 1$$

Then the eigencondition (3.126a) with $U_{kn}(\mu)$ defined by Equation (3.124e) becomes

$$-\alpha_{21}\beta_{12}\frac{y_2(\mu, 1)}{y_1(\mu, 1)} + \alpha_{22}\beta_{11}\frac{y_6(\mu, 1)}{y_5(\mu, 1)} + \beta_{11}\beta_{22} - \beta_{12}\beta_{21} = 0 \quad (3.126b)$$

Now we consider the special case treated in Example 3.8

$$k_k(x) = x^{1-2m_k}, \qquad w_k(x) = \omega_k^2 x^{1-2m_k}, \qquad d_k(x) = 0$$

Then eigencondition (3.126b) reduces to the following equation

$$\alpha_{21}\beta_{12}\frac{J_{-m_1}(\mu\omega_1)}{\mu\omega_1 J_{1-m_1}(\mu\omega_1)} - \alpha_{22}\beta_{11}\frac{J_{-m_2}(\mu\omega_2)}{\mu\omega_2 J_{1-m_2}(\mu\omega_2)} + \beta_{11}\beta_{22} - \beta_{12}\beta_{21} = 0$$

$$(3.126c)$$

For the case

$$m_1 = m_2 = \tfrac{1}{2}, \quad \alpha_{21}\beta_{12} = \alpha_{22}\beta_{11} = 1, \quad \beta_{11}\beta_{22} - \beta_{12}\beta_{21} = 0.5, \quad \omega_1 = 1$$

Equation (3.126c) becomes

$$\cot \mu - \frac{1}{\omega}\cot(\mu\omega) + 0.5\mu = 0 \qquad (3.126d)$$

Figure 3.3 shows the functions $y_1 = \cot \mu$, $y_2 = -(1/\omega)\cot(\mu\omega)$ for $\omega = 1.6$ and $y_3 = 0.5\mu$. The roots of Equation (3.126d) are determined from the points of intersection of the functions $y_1 + y_3$ and y_2. When $\beta_{12}\alpha_{21}$ and $\beta_{11}\alpha_{22}$ are of the opposite sign (i.e., $\alpha_{21}\alpha_{22}\beta_{11}\beta_{12} < 0$), and even if their values are different from 1, the roots of Equation (3.126d) lie between the consecutive roots of the equations

$$\sin(\mu) = 0 \quad \text{and} \quad \sin(\mu\omega) = 0 \qquad (3.127a)$$

By analogy the roots of Equation (3.126c) lie between the consecutive roots of the equations

$$J_{1-m_1}(\mu\omega_1) = 0 \quad \text{and} \quad J_{1-m_2}(\mu\omega_2) = 0 \qquad (3.127b)$$

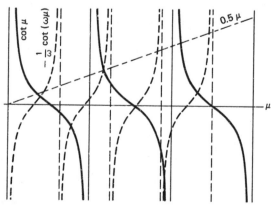

Figure 3.3 Functions $z_1 = \cot \mu$, $z_2 = -\cot(\mu\omega)/\omega$, and $z_3 = 0.5\,\mu$ for $\omega = 1.6$.

Having examined the behavior of the roots of Equation (3.126c, d), we now return to our problem on the determination of the behavior of the roots of Equation (3.126b). Equations (3.126b) and (3.126c) being analogical, we conclude that the behavior of the functions $\mu y_2(\mu, 1)/y_1(\mu, 1)$ and $\mu y_6(\mu, 1)/y_5(\mu, 1)$ should be similar to that of $\cot \mu$, hence the roots of Equation (3.126b) should lie between two consecutive roots of the equations

$$y_1(\mu, 1) = 0 \quad \text{and} \quad y_5(\mu, 1) = 0 \qquad (3.127c)$$

Here we note that if for some value of μ the roots of Equation (3.127) coincide, then this value of μ is a root of Equation (3.126).

With these considerations, the following algorithm has been successively applied [25] for the solution of a Class III problem for the special case of $\alpha_k = 0$, $\beta_k = 1$, $\alpha_{11} = \alpha_{12} = 0$, $x_0 = 0$, and $x_1 = 1$.

1. Calculate the eigenvalues μ_i^* by numerical integration (i.e., Runge-Kutta method) of Equations (3.124a, b) for $k = 1$ subject to the boundary conditions (3.125a, b), namely,

$$y_{4k-3}(\mu, 0) = 0, \qquad y_{4k-2}(\mu, 0) = 1 \qquad (3.127d)$$

by utilizing Newton's iterative method so that the boundary condition $y_1(\mu, 1) = 0$ is also satisfied.

2. Calculate the eigenvalues μ_i^{**} by numerical integration of Equations (3.124a, b) for $k = 2$ subject to the boundary conditions (3.127d) in the same way so that the boundary condition $y_5(\mu, 1) = 0$ is also satisfied. The calculations in steps 1 and 2 are started with the initial approximation given by Equation (3.71b).

3. The eigenvalues μ_i^* and μ_i^{**} so obtained are arranged according to their values in the increasing order. Then the eigenvalues μ_i of the eigencondition (3.126a) lie in the intervals defined by these eigenvalues.

4. Finally, the eigenvalues μ_i and the eigenfunctions $y_{4k-2}(\mu, x)$ and $y_{4k-3}(\mu, x)$ are determined through direct numerical integration of Equations (3.124) subject to the conditions (3.124c) using the method of successive bisection, so that Equation (3.126b) is identically satisfied.

3.4 EIGENVALUE PROBLEM OF CLASS IV

The eigenvalue problem of Class IV is given by Equations (2.121). Here we consider only the special case for

$$k_k(x) = \kappa_k k(x), \qquad w_k(x) = \omega_k b(x), \qquad d_k(x) = \kappa_k d(x) \quad (3.128)$$

Then the one-dimensional form of the eigenvalue problem of Class IV becomes

$$\kappa_k \frac{d}{dx}\left[k(x)\frac{d\psi_k(\mu, x)}{dx}\right] + \left[\mu^2\omega_k b(x) - \kappa_k d(x)\right]\psi_k(\mu, x)$$

$$+ b(x)\sum_{p=1}^{n}\alpha_{kp}\left[\psi_p(\mu, x) - \psi_k(\mu, x)\right] = 0$$

$$\text{in } x_0 < x < x_1, \qquad k = 1, 2, \ldots, n \quad (3.129\text{a})$$

$$\alpha_{0k}\psi_k(\mu, x_0) - \beta_{0k}k(x_0)\psi_k'(\mu, x_0) = 0 \qquad (3.129\text{b})$$

$$\alpha_{1k}\psi_k(\mu, x_1) + \beta_{1k}k(x_1)\psi_k'(\mu, x_1) = 0 \qquad (3.129\text{c})$$

A solution is assumed in the form

$$\psi_k(\mu, x) = E_k(\mu, \lambda)\psi(\lambda, x) \qquad (3.130\text{a})$$

where $\psi(\lambda, x)$ is taken as the solution of the equation

$$\frac{d}{dx}\left[k(x)\frac{d\psi(\lambda, x)}{dx}\right] + \left[\lambda^2 b(x) - d(x)\right]\psi(\lambda, x) = 0 \quad (3.130\text{b})$$

Then by substituting Equation (3.130a) into Equation (3.129a) and utilizing Equation (3.130b) it is shown that the functions $E_k(\mu, \lambda)$ satisfy the following n linear homogeneous equations

$$\left(\mu^2\omega_k - \kappa_k\lambda^2\right)E_k(\mu, \lambda) + \sum_{p=1}^{n}\alpha_{kp}\left[E_p(\mu, \lambda) - E_k(\mu, \lambda)\right] = 0$$

$$k = 1, 2, \ldots, n \quad (3.131)$$

The homogeneous system (3.131) possesses a nontrivial solution if the determinant of the coefficients vanishes, and this requirement is satisfied only for proper values of λ_m ($m = 1, 2, \ldots, n$). For each of these values of λ_m there are the corresponding values of $E_k(\mu, \lambda_m)$. Then the general solution $\psi_k(\mu, x)$ of Equation (3.129a) is constructed as a linear combination of these $E_k(\mu, \lambda_m)$ as

$$\psi_k(\mu, x) = \sum_{m=1}^{n} E_k(\mu, \lambda_m)\psi(\lambda_m, x) \qquad (3.132\text{a})$$

Let $u(\lambda, x)$ and $v(\lambda, x)$ be two independent particular solutions of Equation (3.130b). The solution $\psi(\lambda, x)$ is taken as

$$\psi(\lambda_m, x) = C_m u(\lambda_m, x) + D_m v(\lambda_m, x) \qquad (3.132\text{b})$$

Equation (3.132b) is introduced into Equation (3.132a)

$$\psi_k(\mu, x) = \sum_{m=1}^{n} E_k(\mu, \lambda_m)\left[C_m u(\lambda_m, x) + D_m v(\lambda_m, x)\right] \quad (3.132c)$$

This solution is substituted into the boundary conditions (3.129b, c); we find

$$\sum_{m=1}^{n} \left[C_m U_{km}(\mu, x_\ell) + D_m V_{km}(\mu, x_\ell)\right] = 0, \quad k = 1, 2, \ldots, n, \quad \ell = 0, 1$$

$$(3.133a)$$

where

$$U_{km}(\mu, x_\ell) = E_k(\mu, \lambda_m)\left[\alpha_{\ell k} u(\lambda_m, x_\ell) - (-1)^\ell \beta_{\ell k} k(x_\ell) u'(\lambda_m, x_\ell)\right]$$

$$(3.133b)$$

$$V_{km}(\mu, x_\ell) = E_k(\mu, \lambda_m)\left[\alpha_{\ell k} v(\lambda_m, x_\ell) - (-1)^\ell \beta_{\ell k} k(x_\ell) v'(\lambda_m, x_\ell)\right]$$

$$(3.133c)$$

which completes the formal solution of the problem. That is, the system (3.133a) provides $2n$ linear homogeneous equations for the determination of $2n$ arbitrary constants C_m and D_m, with $m = 1, 2, \ldots, n$; however, the system being homogeneous, these coefficients can be determined in terms of any one of them (i.e., a nonvanishing one) or within a multiple of an arbitrary constant. Finally, an additional relationship is needed for the determination of the unknown eigenvalues μ_i. This additional relationship is obtained from the requirement that the system of $2n$ homogeneous Equations (3.133a) has a nontrivial solution if the determinant of the coefficients vanishes.

3.5 EIGENVALUE PROBLEM OF CLASS V

The eigenvalue problem of Class V is given by Equations (2.132). Here we consider the one-dimensional case, simplified with the assumption (3.128), given in the form

$$\frac{d}{dx}\left[k(x)\frac{d\psi_k(\mu, x)}{dx}\right] + \left[\mu^2 \omega_k^2 b(x) - d(x)\right]\psi_k(\mu, x)$$

$$+ (-1)^k b(x)\left[\sigma_{k-1}\psi_1(\mu, x) - \sigma_{k+1}\psi_2(\mu, x)\right] = 0$$

$$\text{in } x_0 < x < x_1, \quad k = 1, 2 \quad (3.134a)$$

$$\alpha_{\ell k}\psi_k(\mu, x_\ell) - (-1)^\ell \beta_{\ell k} k(x_\ell)\psi_k'(\mu, x_\ell) = 0, \quad k = 1, 2, \quad \ell = 0, 1$$

$$(3.134b, c)$$

A solution is assumed in the form

$$\psi_k(\mu, x) = E_k(\mu, \lambda)\psi(\lambda, x) \qquad (3.135a)$$

where $\psi(\lambda, x)$ is taken as the solution of the equation

$$\frac{d}{dx}\left[k(x)\frac{d\psi(\lambda, x)}{dx}\right] + \left[\lambda^2 b(x) - d(x)\right]\psi(\lambda, x) = 0 \quad (3.135b)$$

Equation (3.135a) is substituted into Equation (3.134a) and Equation (3.135b) is utilized. It is found that the functions $E_1(\mu, \lambda)$ and $E_2(\mu, \lambda)$ satisfy the following two homogeneous equations

$$\left[(\mu\omega_1)^2 - \lambda^2 - \sigma_0\right]E_1(\mu, \lambda) + \sigma_2 E_2(\mu, \lambda) = 0 \qquad (3.136a)$$

$$\sigma_1 E_1(\mu, \lambda) + \left[(\mu\omega_2)^2 - \lambda^2 - \sigma_3\right]E_2(\mu, \lambda) = 0 \qquad (3.136b)$$

The homogeneous system (3.136) possesses a nontrivial solution if the determinant of the coefficients vanishes; this requirement leads to the following biquadratic equation for the determination of the permissible values of λ_m.

$$\lambda^4 + \left[\sigma_0 + \sigma_3 - \mu^2\left(\omega_1^2 + \omega_2^2\right)\right]\lambda^2 + \left(\sigma_0 - \mu^2\omega_1^2\right)\left(\sigma_3 - \mu^2\omega_2^2\right) - \sigma_1\sigma_2 = 0$$

$$(3.137a)$$

The solution for λ^2 is

$$\lambda_m^2 = \mu^2\frac{\omega_1^2 + \omega_2^2}{2}$$

$$-\frac{1}{2}\left\{\sigma_0 + \sigma_3 + (-1)^m\sqrt{\left[\sigma_0 - \sigma_3 + \mu^2\left(\omega_2^2 - \omega_1^2\right)\right]^2 + 4\sigma_1\sigma_2}\right\},$$

$$m = 1, 2 \quad (3.137b)$$

If we choose $E_1(\mu, \lambda_m) = 1$, Equation (3.136a) gives $E_2(\mu, \lambda_m) = [\sigma_0 + \lambda_m^2 - \mu^2\omega_1^2]/\sigma_2$. Thus the coefficients $E_k(\mu, \lambda_m)$ can be written as

$$E_k(\mu, \lambda_m) = \left[\frac{\sigma_0 + \lambda_m^2 - \mu^2\omega_1^2}{\sigma_2}\right]^{k-1}, \qquad k = 1, 2 \qquad (3.138a)$$

Then the general solution $\psi_k(\mu, x)$ of Equation (3.134a) is expressed in the form

$$\psi_k(\mu, x) = \sum_{m=1}^{2}\left[\frac{\sigma_0 + \lambda_m^2 - \mu^2\omega_1^2}{\sigma_2}\right]^{k-1}\psi(\lambda_m, x) \qquad (3.138b)$$

Let $u(\lambda, x)$ and $v(\lambda, x)$ be two independent particular solutions of Equation (3.135b). The solution $\psi(\lambda_m, x)$ is taken in the form

$$\psi(\lambda_m, x) = C_m u(\lambda_m, x) + D_m v(\lambda_m, x) \qquad (3.138c)$$

This solution is introduced into Equation (3.138b); we find

$$\psi_k(\mu, x) = \sum_{m=1}^{2} \left(\frac{\sigma_0 + \lambda_m^2 - \mu^2 \omega_1^2}{\sigma_2} \right)^{k-1} [C_m u(\lambda_m, x) + D_m v(\lambda_m, x)],$$

$$k = 1, 2 \quad (3.138d)$$

Finally, Equations (3.138d) are substituted into the boundary conditions (3.134b, c) to obtain

$$\sum_{m=1}^{2} [C_m U_{km}(\mu, x_\ell) + D_m V_{km}(\mu, x_\ell)] = 0, \qquad k = 1, 2, \qquad \ell = 0, 1$$

$$(3.139a)$$

where

$$U_{km}(\mu, x_\ell)$$

$$= \left[\frac{\sigma_0 + \lambda_m^2 - (\mu\omega_1)^2}{\sigma_2} \right]^{k-1} [\alpha_{\ell k} u(\lambda_m, x_\ell) - (-1)^\ell \beta_{\ell k} k(x_\ell) u'(\lambda_m, x_\ell)]$$

$$(3.139b)$$

$$V_{km}(\mu, x_\ell)$$

$$= \left[\frac{\sigma_0 + \lambda_m^2 - (\mu\omega_1)^2}{\sigma_2} \right]^{k-1} [\alpha_{\ell k} v(\lambda_m, x_\ell) - (-1)^\ell \beta_{\ell k} k(x_\ell) v'(\lambda_m, x_\ell)]$$

$$(3.139c)$$

The system (3.139a) provides four homogeneous equations for the determination of four unknown coefficients C_m, D_m with $m = 1, 2$; the system being homogeneous, these coefficients can be determined within a multiple of an arbitrary constant. Finally, if the homogeneous system (3.139a) should have a nontrivial solution, the determinant of the coefficients should vanish. This requirement leads to the following eigencondition or the determination of the eigenvalues μ_i.

$$\begin{vmatrix} U_{11}(\mu, x_0) & U_{12}(\mu, x_0) & V_{11}(\mu, x_0) & V_{12}(\mu, x_0) \\ U_{21}(\mu, x_0) & U_{22}(\mu, x_0) & V_{21}(\mu, x_0) & V_{22}(\mu, x_0) \\ U_{11}(\mu, x_1) & U_{12}(\mu, x_1) & V_{11}(\mu, x_1) & V_{12}(\mu, x_1) \\ U_{21}(\mu, x_1) & U_{22}(\mu, x_1) & V_{21}(\mu, x_1) & V_{22}(\mu, x_1) \end{vmatrix} = 0 \quad (3.140)$$

3.6   EIGENVALUE PROBLEM OF CLASS VI

The eigenvalue problem of Class VI is given by Equations (2.148). Here we consider the one-dimensional case for a finite region $x_0 \le x \le x_1$. The differential Equation (2.148a) takes the form

$$\frac{d}{dx}\left[k_k(x)\frac{d\psi_k(\mu, x)}{dx}\right] + \mu^2 w_k(x)\psi_k(\mu, x) = 0 \quad \text{in } x_0 < x < x_1$$

$$(3.141a)$$

and the boundary conditions at $x = x_0$ and $x = x_1$ are chosen in the form

$$\psi_k'(\mu, x_0) = 0 \quad (3.141b)$$

$$\alpha_k \psi_k(\mu, x_1) + \beta_k k_k(x_1)\psi_k'(\mu, x_1) + \sum_{k=1}^{n} \gamma_k \int_{x_0}^{x_1} w_k(x)\psi_k(\mu, x)\, dx = 0$$

$$(3.141c)$$

where $k = 1, 2, 3, \ldots, n$.

The boundary condition (3.141c) contains an integral. It is desirable to transform this integral into a derivative form since it is easier to handle the problem in the differential form than in the integral form. Therefore, we integrate Equation (3.141a) with respect to x from x_0 to x_1 and utilize the boundary condition (3.141b) to obtain

$$\int_{x_0}^{x_1} w_k(x)\psi_k(\mu, x)\, dx = -\frac{1}{\mu^2}k_k(x_1)\psi_k'(\mu, x_1) \quad (3.141d)$$

Then the boundary condition (3.141c) takes the form

$$\alpha_k \psi_k(\mu, x_1) + \beta_k k_k(x_1)\psi_k'(\mu, x_1) = \frac{1}{\mu^2}\sum_{k=1}^{n}\gamma_k k_k(x_1)\psi_k'(\mu, x_1)$$

$$(3.141e)$$

Let the particular solution of Equation (3.141a) that satisfies the boundary condition at $x = x_0$ be $u_k(\mu, x)$. The solution for $\psi_k(x, \mu)$ is written as

$$\psi_k(\mu, x) = C_k u_k(\mu, x) \tag{3.142}$$

If this solution should also satisfy the boundary condition (3.141e) at $x = x_1$, we should have

$$C_k U_k(\mu) = \frac{1}{\mu^2} \sum_{k=1}^{n} C_k \gamma_k k_k(x_1) u_k'(\mu, x_1), \qquad k = 1, 2, \ldots, n \tag{3.143a}$$

where

$$U_k(\mu) = \alpha_k u_k(\mu, x_1) + \beta_k k_k(x_1) u_k'(\mu, x_1) \tag{3.143b}$$

We note that the right-hand side of Equation (3.143a) does not depend on k; therefore, we write

$$C_1 U_1(\mu) = C_2 U_2(\mu) = \cdots = C_n U_n(\mu) \tag{3.144a}$$

or

$$C_1 U_1(\mu) = C_k U_k(\mu), \qquad k = 1, 2, \ldots, n \tag{3.144b}$$

or

$$\frac{C_k}{C_1} = \frac{U_1(\mu)}{U_k(\mu)}, \qquad k = 1, 2, \ldots, n \tag{3.144c}$$

Dividing both sides of Equation (3.143) with $C_k U_k(\mu)$ and utilizing the ratio (3.144c) we obtain the eigencondition for the problem as

$$\sum_{k=1}^{n} \gamma_k k_k(x_1) \frac{u_k'(\mu, x_1)}{U_k(\mu)} = \mu^2 \tag{3.145a}$$

Introducing the ratio (3.144c) into Equation (3.142), the eigenfunctions $\psi_k(\mu_i, x)$ become

$$\psi_k(\mu_i, x) = C_1 U_1(\mu_i) \frac{u_k(\mu_i, x)}{U_k(\mu_i)} \tag{3.145b}$$

Example 3.9 Determine the eigencondition and the eigenfunctions of the eigenvalue problem (3.141) for the following special case.

$$k_k(x) = x^{1-2m_k}, \qquad w_k(x) = \omega_k^2 x^{1-2m_k}, \qquad x_0 = 0 \quad \text{and} \quad x_1 = 1 \tag{3.146}$$

Solution   For this special case the eigenvalue problem (3.141) reduces to

$$\psi_k''(\mu, x) + \frac{1 - 2m_k}{x} \psi_k'(\mu, x) + (\mu\omega_k)^2 \psi_k(\mu, x) = 0 \quad \text{in } 0 \le x < 1$$

$$(3.147a)$$

$$\psi_k'(\mu, 0) = 0 \tag{3.147b}$$

$$\alpha_k \psi_k(\mu, 1) + \beta_k \psi_k'(\mu, 1) = \frac{1}{\mu^2} \sum_{k=1}^{n} \gamma_k \psi_k'(\mu, 1) \tag{3.147c}$$

We note that Equations (3.147a, b) are of similar form to Equations (3.42a, b). Therefore, a particular solution of (3.147a) that satisfies the boundary condition (3.147b) is immediately obtained from Equation (3.43a) as

$$u_k(\mu, x) = (\mu\omega_k x)^{m_k} J_{-m_k}(\mu\omega_k x) \tag{3.148a}$$

Introducing Equation (3.148a) into Equations (3.145), the eigencondition and the eigenfunctions are obtained, respectively, as

$$\mu^2 + \sum_{k=1}^{n} \frac{\mu\omega_k \gamma_k J_{1-m_k}(\mu\omega_k)}{\alpha_k J_{-m_k}(\mu\omega_k) - \beta_k \mu\omega_k J_{1-m_k}(\mu\omega_k)} = 0 \tag{3.148b}$$

$$\psi_k(\mu_i, x) = C_1 \frac{U_1(\mu_i)}{U_k(\mu_i)} (\mu_i \omega_k x)^{m_k} J_{-m_k}(\mu_i \omega_k x) \tag{3.148c}$$

where

$$U_k(\mu_i) = \left[\alpha_k J_{-m_k}(\mu_i \omega_k) - \beta_k \mu_i \omega_k J_{1-m_k}(\mu_i \omega_k) \right] (\mu_i \omega_k)^{m_k} \tag{3.148d}$$

3.7   EIGENVALUE PROBLEM OF CLASS VII

The eigenvalue problem of Class VII is given by Equation (2.165). For the one-dimensional case in a region $x_0 \le x \le x_1$, the Equations (2.165a, b), respectively, become

$$\frac{d}{dx}\left[k(x) \frac{d\psi_1(\mu, x)}{dx} \right] + \mu^2 w(x)[\psi_1(\mu, x) + \psi_2(\mu, x)] = 0$$

$$(3.149a)$$

$$\mu^2 \psi_2(\mu, x) = \sigma_2 \psi_2(\mu, x) - \sigma_1 \psi_1(\mu, x) \tag{3.149b}$$

We take the boundary conditions (2.165c) in the form

$$\psi_1'(\mu, x_0) = 0 \tag{3.149c}$$

$$\alpha\psi_1(\mu, x_1) + \beta k(x_1)\psi_1'(\mu, x_1) + \gamma\int_{x_0}^{x_1} w(x)[\psi_1(\mu, x) + \psi_2(\mu, x)]\, dx = 0$$

$$\tag{3.149d}$$

To replace the integral term in Equation (3.149d) by a derivative, Equation (3.149a) is integrated with respect to x from x_0 to x_1 and the boundary condition (3.149c) is utilized. We obtain

$$\int_{x_0}^{x_1} w(x)[\psi_1(\mu, x) + \psi_2(\mu, x)]\, dx = -\frac{1}{\mu^2}k(x_1)\psi_1'(\mu, x_1) \tag{3.149e}$$

Introducing Equation (3.149e) into Equation (3.149d), the boundary condition at $x = x_1$ is transformed to

$$\alpha\psi_1(\mu, x_1) + \left(\beta - \frac{\gamma}{\mu^2}\right)k(x_1)\psi_1'(\mu, x_1) = 0 \tag{3.149f}$$

Now Equation (3.149b) is written as

$$\psi_2(\mu, x) = \frac{\sigma_1}{\sigma_2 - \mu^2}\psi_1(\mu, x) \tag{3.150}$$

This expression is introduced into Equation (3.149a) and the function $\psi_2(\mu, x)$ is eliminated. We obtain

$$\frac{d}{dx}\left[k(x)\frac{d\psi_1(\lambda, x)}{dx}\right] + \lambda^2 w(x)\psi_1(\lambda, x) = 0 \tag{3.151a}$$

where

$$\lambda^2 = \mu^2\left(1 + \frac{\sigma_1}{\sigma_2 - \mu^2}\right) \tag{3.151b}$$

The solution for $\psi_1(\lambda, x)$ may be written in the form

$$\psi_1(\lambda, x) = C_1 u_1(\lambda, x) \tag{3.152a}$$

where $u_1(\lambda, x)$ is a particular solution of Equation (3.151a) that satisfies the boundary condition (3.149c) at $x = x_0$.

If the solution (3.152a) should satisfy the boundary condition (3.149f), we obtain the following eigencondition

$$\alpha u_1(\lambda, x_1) + \left(\beta - \frac{\gamma}{\mu^2}\right) k(x_1) u_1'(\lambda, x_1) = 0 \qquad (3.152b)$$

and the eigenvalues μ_i are the roots of this equation.

Example 3.10   Determine the eigenfunctions and eigencondition of the eigenvalue problem (3.149) for the following special case

$$k(x) = w(x) = x^{1-2m} \qquad (3.153a)$$

Solution   The alternative form of Equation (3.149a) is given by Equation (3.151a); a particular solution $u_1(\lambda, x)$ of Equation (3.151a) for this special case is

$$u_1(\lambda, x) = (\lambda x)^m J_m(\lambda x) \qquad (3.153b)$$

Introducing this solution into Equation (3.152b), the eigencondition is obtained as

$$\alpha J_{-m}(\lambda) + \left(\frac{\gamma}{\mu^2} - \beta\right) \lambda J_{1-m}(\lambda) = 0 \qquad (3.154a)$$

and the eigenvalues μ_i are the roots of this transcendental equation.

The eigenfunctions $\psi_1(\lambda, x)$ and $\psi_2(\lambda, x)$ are obtained from Equations (3.152a) and (3.150), respectively, by introducing Equation (3.153b) into these equations. We obtain

$$\psi_1(\mu_i, x) = C_1(\lambda_i x)^m J_{-m}(\lambda_i x) \qquad (3.154b)$$

$$\psi_2(\mu_i, x) = C_1 \frac{\sigma_1}{\sigma_2 - \mu_i^2} (\lambda_i x)^m J_{-m}(\lambda_i x) \qquad (3.154c)$$

where

$$\lambda_i = \mu_i \sqrt{1 + \frac{\sigma_1}{\sigma_2 - \mu_i^2}} \qquad (3.154d)$$

REFERENCES

1. T. K. Sherwood and C. E. Reed, *Applied Mathematics in Chemical Engineering*, McGraw-Hill, New York, 1939, p. 210.
2. G. N. Watson, *A Treatise on the Theory of Bessel Functions*, Cambridge University Press, London, 1966.

3. N. W. McLachlan, *Bessel Functions for Engineers*, 2nd ed., Clarendon Press, London, 1961.

4. M. Abramowitz and I. A. Stegun, *Handbook of Mathematical Functions*, National Bureau of Standards, Applied Mathematics Series 55, U.S. Government Printing Office, Washington, D.C., 1964.

5. Parry Moon and Domino Eberle Spencer, On the Classification of Ordinary Differential Equations of Field Theory, *Quart. Appl. Math.*, **16**, 1–10 (1950).

6. Parry Moon and Domino Eberle Spencer, *Field Theory for Engineers*, Van Nostrand, Princeton, N.J., 1961, p. 158.

7. M. Philip Morse and H. Feshbach, *Methods of Theoretical Physics*, Part I, McGraw-Hill, New York, 1953, p. 513.

8. T. M. MacRobert, *Spherical Harmonics*, 3rd ed., Pergamon, New York, 1967.

9. E. T. Whittaker and G. N. Watson, *A Course of Modern Analysis*, Cambridge University Press, London, 1965.

10. W. E. Byerly, *Fourier's Series and Spherical, Cylindrical and Ellipsoidal Harmonics*, Dover, New York, 1959.

11. E. W. Hobson, *The Theory of Spherical and Ellipsoidal Harmonics*, Cambridge University Press, London, 1932.

12. M. L. James, G. M. Smith, and J. C. Wolford, *Applied Numerical Methods for Digital Computations with Fortran and CSMP*, 2nd ed., IEP, New York, 1977.

13. M. N. Özişik, *Heat Conduction*, Wiley, New York, 1980.

14. M. D. Mikhailov, Nonstationary Temperature Fields in Skin, *Energiya*, Moscow, 1967.

15. J. R. Sellars, M. Tribus, and J. S. Klein, Heat Transfer to Laminar Flow in a Round Tube or Flat Conduit—The Graetz Problem Extended, *Trans. ASME*, **78**, 441–448 (1956).

16. R. E. Lundberg, P. A. McCuen, and W. C. Reynolds, Heat Transfer in Annular Passages Hydrodynamically Developed Laminar Flow with Arbitrary Prescribed Wall Temperatures and Heat Fluxes, *Int. J. Heat Mass Transf.*, **6**, 495–529 (1963).

17. B. M. Smol'sky, Z. P. Shulman, and V. M. Gorislavets, Rheodynamics and Heat Transfer of Non-linear Viscoplastic Materials, *Science and Technics*, Minsk, USSR, 1970.

18. C. A. Sleicher, R. H. Notter, and M. D. Crippen, A Solution for Turbulent Graetz Problem with Matched Asymptotic Expansion-I., *Chem. Eng. Sci.*, **25**, 845 (1970).

19. A. B. Datzeff, Über die Eigenwerte einiger Differentialgleichungen. I., *Comptes rendus de L'Académie Bulgare des Sciences*, **12**, 113–116 (1959).

20. A. B. Datzeff, Über die Eigenwerte einiger Differentialgleichungen. II., *Comptes rendus de L'Académie Bulgare des Sciences*, **12**, 285–288 (1959).

21. W. H. Wittrick and F. W. Williams, A General Algorithm for Computing Natural Frequencies of Elastic Structures; *Quart. J. Mech. Appl. Math.*, **24**, pt. 3, 263–284 (1971).

22. W. H. Wittrick and F. W. Williams, Buckling and Vibrations of Anisotropic or Isotropic Plate Assemblies under Combined Loading, *Int. J. Mech. Sci.*, **16**, 209–239 (1974).

23. M. D. Mikhailov and N. L. Vulchanov, A Computational Procedure for Sturm–Liouville Problems, *J. Comp. Phys.* **50**, 323–336 (1983).

24. M. D. Mikhailov, M. N. Özişik, and N. L. Vulchanov, Transient Heat Diffusion in One-Dimensional Composite Media and Automatic Solution of the Eigenvalue Problem, *Int. J. Heat Mass Transf.*, **26**, 1131–1141 (1983).

25. M. D. Mikhailov and B. V. Shishedjiev, Coupled at Boundary Mass or Heat Transfer in Entrance Countercurrent Flow, *Int. J. Heat Mass Transf.*, **19**, 553–557 (1976).

CHAPTER FOUR

One-Dimensional Problems

Science is organized knowledge.

Herbert Spencer
1820–1903

A proper organization of scientific knowledge is a key factor governing its usefulness in practice. The organization of the material in this chapter closely follows the pattern established in the previous chapters, and its contents form the basis on which the coming chapters stand.

In the first two chapters we presented the mathematical formulation and the general three-dimensional formal solution for each of the seven different classes of heat and mass diffusion problems. In the previous chapter, the methods of solving the one-dimensional eigenvalue problems associated with each of these seven classes of problems were described. The one-dimensional problems have numerous important applications in various branches of science and engineering. Therefore, in this chapter we present in explicit form the one-dimensional solutions for each of these classes of problems, systematically obtained from the general formal solutions. *The solutions presented in this chapter form sufficiently general expressions from which solutions to numerous specific one-dimensional, time dependent heat and mass diffusion problems encountered in a large variety of applications are obtainable as special cases. This matter is illustrated in the subsequent chapters with a large number of examples chosen from each of the seven different classes of problems.*

4.1 ONE-DIMENSIONAL PROBLEMS OF CLASS I

The general mathematical formulation of the boundary value problems of Class I is given by Equations (2.1). Here we focus our attention on a specific situation in which the potential $T(x, t)$ is a function of time t, and one of the space variables, say, of x, only. For such a case, the range of the x variable

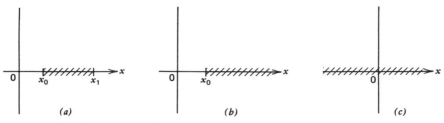

Figure 4.1   Coordinates for one-dimensional problems: (*a*) finite, (*b*) semi-infinite; (*c*) infinite regions.

may be finite, $x_0 \leq x \leq x_1$, semi-infinite, $x_0 \leq x < \infty$, or infinite, $-\infty < x < \infty$ as illustrated in Figure 4.1. For example, the region $x_0 \leq x \leq x_1$ represents a slab in the rectangular coordinate system, a hollow cylinder in the cyclindrical coordinate system, and a hollow sphere in the spherical coordinate system.

   In this section we present explicit solutions for the one-dimensional boundary value problems of Class I for finite, semi-infinite, and infinite regions.

Finite Region, $x_0 \leq x \leq x_1$

Consider the boundary value problem of Class I for the one-dimensional case given by Equations (1.65) in the form

$$w(x)\frac{\partial T(x,t)}{\partial t} = \frac{\partial}{\partial x}\left[k(x)\frac{\partial T(x,t)}{\partial x}\right] - d(x)T(x,t) + P(x,t)$$

$$\text{in } x_0 < x < x_1, \qquad t > 0 \quad (4.1a)$$

subject to the boundary conditions

$$\alpha_k T(x_k,t) - (-1)^k \beta_k k(x_k)\frac{\partial T(x_k,t)}{\partial x} = \phi(x_k,t)$$

$$\text{at } x = x_k, \qquad k = 0,1, \qquad t > 0 \quad (4.1b,c)$$

and the initial condition

$$T(x,0) = f(x) \quad \text{in } x_0 \leq x \leq x_1, \qquad t = 0 \qquad (4.1d)$$

   The solution of problem (4.1) is immediately obtainable from the general solution (2.19a) by restricting it to the one-dimensional case. We find

$$T(x,t) = \sum_{i=1}^{\infty} \frac{1}{N_i} e^{-\mu_i^2 t} \psi_i(x)\left[\tilde{f}_i + \int_0^t g_i(t')e^{\mu_i^2 t'}\,dt'\right] \quad \text{in } x_0 \leq x \leq x_1 \quad (4.2)$$

where N_i, $\tilde{f}_i$, and $g_i(t')$ are obtainable from Equations (2.19b, c, d), respectively, if the volume and surface integrals in Equations (2.19), for the one-dimensional case considered here, are evaluated according to the formulae

$$\int_V F(\mathbf{x}) \, dv = S_0 \int_{x_0}^{x_1} F(x) \, dx \qquad (4.3a)$$

$$\int_S \phi(\mathbf{x}, t) F(\mathbf{x}) \, ds = S_0 \sum_{k=0}^{1} \phi(x_k, t) F(x_k) \qquad (4.3b)$$

Then N_i, $\tilde{f}_i$, and $g_i(t)$ appear to be arbitrary within a multiplication constant equal to the cross-section area S_0; the solution (4.2) is not arbitrary because when N_i, $\tilde{f}_i$, and $g_i(t)$ are introduced into Equation (4.2), S_0 cancels out and the arbitrariness is removed. With this consideration we set $S_0 = 1$; then the expression defining N_i, $\tilde{f}_i$, and $g_i(t)$ for use in the one-dimensional solution (4.2) takes the form

$$N_i = \frac{1}{2\mu_i} \sum_{k=0}^{1} (-1)^{k+1} k(x_k) \left| \begin{array}{cc} \left[\dfrac{\partial \psi(\mu, x_k)}{\partial \mu} \right]_{\mu=\mu_i} & \left[\dfrac{\partial^2 \psi(\mu, x_k)}{\partial x \, \partial \mu} \right]_{\mu=\mu_i} \\[2mm] \psi_i(x_k) & \psi_i'(x_k) \end{array} \right|$$

$$(4.4a)$$

$$\tilde{f}_i = \int_{x_0}^{x_1} w(x) \psi_i(x) f(x) \, dx \qquad (4.4b)$$

$$g_i(t) = \sum_{k=0}^{1} \phi(x_k, t) \Omega_i(x_k) + \int_{x_0}^{x_1} \psi_i(x) P(x, t) \, dx \qquad (4.4c)$$

and $\Omega_i(x_k)$ is defined as

$$\Omega_i(x_k) = \frac{\psi_i(x_k) + (-1)^k k(x_k) \psi_i'(x_k)}{\alpha_k + \beta_k}$$

$$= \frac{\psi_i(x_k)}{\beta_k} = (-1)^k k(x_k) \frac{\psi_i'(x_k)}{\alpha_k}, \qquad k = 0, 1 \qquad (4.4d)$$

The eigenfunctions $\psi_i(x_k)$ and eigenvalues μ_i are the eigenfunctions and eigenvalues of the eigenvalue problem (3.2), that is,

$$\frac{d}{dx}\left[k(x)\frac{d\psi(\mu,x)}{dx}\right] + \left[\mu^2 w(x) - d(x)\right]\psi(\mu,x) = 0$$

$$\text{in } x_0 < x < x_1 \quad (4.5a)$$

$$\alpha_k \psi(\mu,x) - (-1)^k \beta_k k(x_k)\frac{d\psi(\mu,x)}{dx} = 0$$

$$\text{at } x = x_k, \quad k = 0,1 \quad (4.5b,c)$$

For the special case of $d(x) = 0$, the integral in Equation (4.4b) can be performed by utilizing Equation (4.5a). That is, we set in Equation (4.5a), $d(x) = 0$, multiply both sides by $f(x)$, integrate from $x = x_0$ to x_1, and perform the integration by parts. We find

$$\tilde{f}_i = \frac{1}{\mu_i^2}\sum_{k=0}^{1}(-1)^k k(x_k)\left[f(x_k)\psi_i'(x_k) - f'(x_k)\psi_i(x_k)\right]$$

$$- \frac{1}{\mu_i^2}\int_{x_0}^{x_1}\psi_i(x)\frac{d}{dx}\left[k(x)\frac{df(x)}{dx}\right]dx \quad (4.5d)$$

If $\alpha_0 = \alpha_1 = 0$, $\beta_0 = \beta_1 = 1$, that is, the boundary conditions at $x = x_0$ and $x = x_1$ are both of the second kind and, in addition, $d(x) = 0$, the solution of the problem (4.1) is obtainable from the one-dimensional form of the general solution (2.33). We find

$$T(x,t) = T_{av}(t) + \sum_{i=1}^{\infty}\frac{1}{N_i}e^{-\mu_i^2 t}\psi_i(x)\left[\tilde{f}_i + \int_0^t g_i(t')e^{\mu_i^2 t'}dt'\right] \quad (4.6a)$$

Here $T_{av}(t)$ is obtainable from the one-dimensional form of Equation (2.32) as

$$T_{av}(t) = \left\{\int_{x_0}^{x_1}w(x)f(x)\,dx + \int_0^t\left[\sum_{k=0}^{1}\phi(x_k,t') + \int_{x_0}^{x_1}P(x,t')\,dx\right]dt'\right\}$$

$$\times \left\{\int_{x_0}^{x_1}w(x)\,dx\right\}^{-1} \quad (4.6b)$$

If the one-dimensional eigenvalue problem (4.5) cannot be solved analytically, the two algorithms for numerical calculations discussed in Chapter 3 can be used for its solution.

Splitting Up the Solution

When the nonhomogeneous terms $\phi(x_0, t)$, $\phi(x_1, t)$, and $P(x, t)$ of the problem (4.1) are exponentials and q-order polynomials of time in the form

$$\phi(x_k, t) = \phi_e(x_k)e^{-d_\phi t} + \sum_{j=0}^{q} \phi_j(x_k)t^j, \qquad k = 0,1 \qquad (4.7a, b)$$

$$P(x, t) = P_e(x)e^{-d_p t} + \sum_{j=0}^{q} P_j(x)t^j \qquad (4.7c)$$

where d_ϕ and d_p are constants, then the solution of the problem (4.1) can be split up into the solution of simpler problems as given by Equation (2.23) in the form

$$T(x, t) = T_\phi(x)e^{-d_\phi t} + T_p(x)e^{-d_p t} + \sum_{j=0}^{q} T_j(x)t^j + T_t(x, t)$$

$$\text{in } x_0 \le x \le x_1 \quad (4.8a)$$

Here the functions $T_\phi(x)$, $T_p(x)$, and $T_j(x)$ are the solution of the one-dimensional forms of the steady-state problems given by Equations (2.24)–(2.27); the function $T_t(x, t)$ is the solution of the one-dimensional form of the transient homogeneous problem (2.28).

The solution of the problem (4.1) for $\alpha_0 = \alpha_1 = 0$, $\beta_0 = \beta_1 = 1$, and $d(x) = 0$, can be split up into the solutions of simpler problems as given by Equation (2.34) in the form

$$T(x, t) = T_{av}(t) + T_\phi(x)e^{-d_\phi t} + T_p(x)e^{-d_p t} + \sum_{j=0}^{q} T_j(x)t^j + T_t(x, t)$$

$$\text{in } x_0 \le x \le x_1 \quad (4.8b)$$

Here the functions $T_\phi(x)$, $T_p(x)$, and $T_j(x)$ are the solutions of the one-dimensional forms of the steady-state problems given by Equations (2.35)–(2.38); the function $T_t(x, t)$ for the transient part coincides with the solution given by Equation (2.28).

The solutions for the functions $T_\phi(x)$, $T_p(x)$, $T_j(x)$, and $T_t(x, t)$ appearing in the split up solutions (4.8) are determined as now described.

Solution for $T_\phi(x)$

The function $T_\phi(x)$ satisfies the one-dimensional form of the problem (2.24) given as

$$\frac{d}{dx}\left[k(x)\frac{dT_\phi(x)}{dx}\right] + \left[d_\phi w(x) - d(x)\right]T_\phi(x) = 0$$

$$\text{in } x_0 < x < x_1 \quad (4.9a)$$

$$\alpha_k T_\phi(x_k) - (-1)^k \beta_k k(x_k) T_\phi'(x_k) = \phi_e(x_k)$$

$$\text{at } x = x_k, \quad k = 0,1 \quad (4.9b,c)$$

We note that the homogeneous part of this problem (i.e., for $\phi_e = 0$) is similar to the eigenvalue problem defined by Equations (4.5); that is, for $\phi_e(x_k) = 0$ and $d_\phi \equiv \mu^2$, the problem (4.9) is identical to the eigenvalue problem (4.5). This eigenvalue problem was examined in Section 3.1 [see Equations (3.2)], and the eigencondition, eigenfunctions, and normalization integral were given by Equations (3.5)–(3.7), respectively. We now utilize these results to determine the solution $T_\phi(x)$ of the problem (4.9) as described in the following.

Let $u(\mu, x)$ and $v(\mu, x)$ be two linearly independent solutions of Equation (4.5a). Then $u(\sqrt{d_\phi}, x)$ and $v(\sqrt{d_\phi}, x)$ are the two linearly independent solutions of Equation (4.9a) and the solution for $T_\phi(x)$ is constructed by taking a linear combination of $u(\sqrt{d_\phi}, x)$ and $v(\sqrt{d_\phi}, x)$ as

$$T_\phi(x) = Cu\left(\sqrt{d_\phi}, x\right) + Dv\left(\sqrt{d_\phi}, x\right) \quad \text{in } x_0 \le x \le x_1 \quad (4.10)$$

where C and D are two arbitrary constants.

If the solution (4.10) should satisfy the two boundary conditions (4.9b, c), we have

$$CU\left(\sqrt{d_\phi}, x_k\right) + DV\left(\sqrt{d_\phi}, x_k\right) = \phi_e(x_k), \quad k = 0,1 \quad (4.11a,b)$$

where $U(\sqrt{d_\phi}, x_k)$ and $V(\sqrt{d_\phi}, x_k)$ are defined by Equations (3.4c, d) and (3.4e, f), respectively, in which μ is replaced by $\sqrt{d_\phi}$.

When Equations (4.11a, b) are solved for C and D and the results are introduced into Equation (4.10), the solution for $T_\phi(x)$ becomes

$$T_\phi(x)$$

$$= \frac{\displaystyle\sum_{k=0}^{1}(-1)^{k+1}\phi_e(x_{1-k})\left[V\left(\sqrt{d_\phi}, x_k\right)u\left(\sqrt{d_\phi}, x\right) - U\left(\sqrt{d_\phi}, x_k\right)v\left(\sqrt{d_\phi}, x\right)\right]}{\displaystyle\sum_{k=0}^{1}(-1)^k U\left(\sqrt{d_\phi}, x_k\right)V\left(\sqrt{d_\phi}, x_{1-k}\right)}$$

$$(4.12)$$

For the case $\alpha_0 = \alpha_1 = 0$, $\beta_0 = \beta_1 = 1$, and $d(x) = 0$, the function $T_\phi(x)$ satisfies the one-dimensional form of the steady-state problem (2.35) given as

$$\frac{d}{dx}\left[k(x)\frac{dT_\phi(x)}{dx}\right] + d_\phi w(x)T_\phi(x) = w(x)\frac{\sum\limits_{k=0}^{1}\phi_e(x_k)}{\int\limits_{x_0}^{x_1}w(x)\,dx} \quad \text{in } x_0 < x < x_1$$

$$(4.13a)$$

$$(-1)^{k+1}k(x_k)T_\phi'(x_k) = \phi_e(x_k)$$

$$\text{at } x = x_k, \quad k = 0,1 \quad (4.13b,c)$$

and
$$\int_{x_0}^{x_1}w(x)T_\phi(x)\,dx = 0 \qquad (4.13d)$$

Then the solution of the problem (4.13) for $T_\phi(x)$ for this special case becomes

$$T_\phi(x) = \frac{\sum\limits_{k=0}^{1}\phi_e(x_k)}{d_\phi\int\limits_{x_0}^{x_1}w(x)\,dx}$$

$$+ \frac{\sum\limits_{k=0}^{1}\phi_e(x_{1-k})k(x_k)\left[v'\left(\sqrt{d_\phi},x_k\right)u\left(\sqrt{d_\phi},x\right) - u'\left(\sqrt{d_\phi},x_k\right)v\left(\sqrt{d_\phi},x\right)\right]}{\sum\limits_{k=0}^{1}(-1)^{k+1}k(x_k)u'\left(\sqrt{d_\phi},x_k\right)k(x_{1-k})v'\left(\sqrt{d_\phi},x_{1-k}\right)}$$

$$(4.14)$$

Solution for $T_p(x)$

The function $T_p(x)$ satisfies the one-dimensional version of the problem (2.25) given as

$$\frac{d}{dx}\left[k(x)\frac{dT_p(x)}{dx}\right] + \left[d_p w(x) - d(x)\right]T_p(x) + P_e(x) = 0$$

$$\text{in } x_0 < x < x_1 \quad (4.15a)$$

$$\alpha_k T_p(x_k) - (-1)^k\beta_k k(x_k)T_p'(x_k) = 0$$

$$\text{at } x = x_k, \quad k = 0,1 \quad (4.15b,c)$$

We note that the homogeneous part of this problem (i.e., for $P_e = 0$) is similar to the eigenvalue problem defined by Equations (4.5) or to that defined by

Equations (3.2). Therefore, the eigenfunctions and the eigenvalues of the eigenvalue problem (3.2) can be used to construct the solution for $T_p(x)$ as now described.

We consider the solution of the problem (4.15) for *the special case of* $d(x) = 0$, $P_e(x) = w(x)P_w$, where P_w is a constant. If $u(\sqrt{d_p}, x)$ and $v(\sqrt{d_p}, x)$ are two linear independent solutions of the homogeneous part of Equation (4.15a) for $d(x) = 0$, the solution for $T_p(x)$ is constructed as

$$T_p(x) = Cu\left(\sqrt{d_p}, x\right) + Dv\left(\sqrt{d_p}, x\right) - \frac{P_w}{d_p} \quad \text{in } x_0 \le x \le x_1 \quad (4.16)$$

where C and D are two arbitrary constants and P_w/d_p is a particular solution corresponding to the nonhomogeneous term $P_e(x) = w(x)P_w$.

If the solution (4.16) should satisfy the two boundary conditions (4.15b, c), we have

$$CU\left(\sqrt{d_p}, x_k\right) + DV\left(\sqrt{d_p}, x_k\right) = \alpha_k \frac{P_w}{d_p}, \quad k = 0,1 \quad (4.17\text{a, b})$$

where $U(\sqrt{d_p}, x)$ and $V(\sqrt{d_p}, x)$ are defined by Equations (3.4c, d) and (3.4e, f), respectively, in which μ is replaced by $\sqrt{d_p}$.

When Equations (4.17a, b) are solved for C and D and the results are introduced into Equation (4.16), the solution for $T_p(x)$, for the special case of $d(x) = 0$ and $P_e(x) = w(x)P_w$, becomes

$$T_p(x) = \frac{P_w}{d_p}$$

$$\times \left\{ \frac{\displaystyle\sum_{k=0}^{1} (-1)^{k+1} \alpha_{1-k} \left[V\left(\sqrt{d_p}, x_k\right) u\left(\sqrt{d_p}, x\right) - U\left(\sqrt{d_p}, x_k\right) v\left(\sqrt{d_p}, x\right) \right]}{\displaystyle\sum_{k=0}^{1} (-1)^{k} U\left(\sqrt{d_p}, x_k\right) V\left(\sqrt{d_p}, x_{1-k}\right)} - 1 \right\}$$

$$(4.18)$$

For the special case of $\alpha_0 = \alpha_1 = 0$, $\beta_0 = \beta_1 = 1$, and $d(x) = 0$, the function $T_p(x)$ satisfies the one-dimensional form of the steady-state problem (2.36) given as

$$\frac{d}{dx}\left[k(x)\frac{dT_p(x)}{dx} \right] + d_p w(x)T_p(x) + P_e(x) = w(x)\frac{\displaystyle\int_{x_0}^{x_1} P_e(x)\,dx}{\displaystyle\int_{x_0}^{x_1} w(x)\,dx}$$

$$\text{in } x_0 < x < x_1 \quad (4.19\text{a})$$

$$T_p'(x_k) = 0 \quad \text{at } x = x_k, \quad k = 0,1 \quad (4.19\text{b, c})$$

and

$$\int_{x_0}^{x_1} w(x)T_p(x)\,dx = 0 \tag{4.19d}$$

If we assume further that $P_e(x) = w(x)P_w$, the solution (4.19) gives

$$T_p(x) = 0 \tag{4.20}$$

A Special Case of Functions $U(\sqrt{d}, x_k)$ and $V(\sqrt{d}, x_k)$

The functions $U(\sqrt{d}, x_k)$ and $V(\sqrt{d}, x_k)$, where $d \equiv d_\phi, d_p$, and $k = 0, 1$, appearing in the foregoing solutions for $T_\phi(x)$ and $T_p(x)$ are defined by Equations (3.4c, d) and (3.4e, f), respectively, in which μ is replaced by $\sqrt{d}$. The explicit form of these functions depends on the choice of the parameters $k(x)$, $w(x)$, and $d(x)$ in the eigenvalue problem (3.2). Here we consider a special case given by Equations (3.9), that is,

$$k(x) = x^{1-2m}e^{-2ax} \tag{4.21a}$$

$$w(x) = c^2 x^{2c-2}k(x) \tag{4.21b}$$

$$d(x) = -\left(a^2 + a\frac{2m-1}{x} + \frac{m^2 - c^2\nu^2}{x^2}\right)k(x) \tag{4.21c}$$

Then the functions $u(\sqrt{d}, x)$ and $v(\sqrt{d}, x)$ are given by Equations (3.12), that is,

$$u(\sqrt{d}, x) = x^m e^{ax}J_\nu(\sqrt{d}\,x^c) \tag{4.21d}$$

$$v(\sqrt{d}, x) = x^m e^{ax}Y_\nu(\sqrt{d}\,x^c) \tag{4.21e}$$

and the Equations (3.4c, d) and (3.4e, f), respectively, take the form

$$U(\sqrt{d}, x_k) = x_k^m e^{ax_k}\left[A_k J_\nu(\sqrt{d}\,x_k^c) - (-1)^k B_k\sqrt{d}\,J_{\nu-1}(\sqrt{d}\,x_k^c)\right], \qquad k = 0, 1$$

$$\tag{4.22a, b}$$

$$V(\sqrt{d}, x_k) = x_k^m e^{ax_k}\left[A_k Y_\nu(\sqrt{d}\,x_k^c) - (-1)^k B_k\sqrt{d}\,Y_{\nu-1}(\sqrt{d}\,x_k^c)\right], \qquad k = 0, 1$$

$$\tag{4.22c, d}$$

where the constants A_k and B_k are defined by Equations (3.15a, b) and (3.15c, d), respectively.

Solution for $T_j(x)$

The functions $T_j(x)$ satisfy the one-dimensional form of Equations (2.26) and (2.27), given as

$$\frac{d}{dx}\left[k(x)\frac{dT_j(x)}{dx}\right] - d(x)T_j(x) + P_j(x) = (j+1)w(x)T_{j+1}(x)$$

$$\text{in } x_0 \leq x < x_1 \quad (4.23a)$$

$$\alpha_k T_j(x_k) - (-1)^k \beta_k k(x_k) T_j'(x_k) = \phi_j(x_k)$$

$$\text{at } x = x_k, \quad k = 0,1 \quad (4.23b,c)$$

where $j = q, q-1, q-2, \ldots, 1, 0$ and

$$T_{q+1}(x) = 0 \quad (4.23d)$$

For $d(x) = 0$, the first and second integration of Equation (4.23a) from x_0 to x give

$$k(x)T_j'(x) = k(x_0)T_j'(x_0) - \int_{x_0}^{x}\left[P_j(x') - (j+1)w(x')T_{j+1}(x')\right]dx' \quad (4.24a)$$

and

$$T_j(x) = T_j(x_0) + k(x_0)T_j'(x_0)\int_{x_0}^{x}\frac{dx'}{k(x')}$$

$$- \int_{x_0}^{x}\frac{1}{k(x'')}\int_{x_0}^{x''}\left[P_j(x') - (j+1)w(x')T_{j+1}(x')\right]dx'\,dx'' \quad (4.24b)$$

The integration constants $T_j(x_0)$ and $k(x_0)T_j'(x_0)$ are determined by the application of the boundary conditions (4.23b, c). That is, for $k = 0$, Equation (4.23b) becomes

$$\alpha_0 T_j(x_0) - \beta_0 k(x_0)T_j'(x_0) = \phi_j(x_0) \quad (4.25a)$$

and the substitution of Equations (4.24a, b) into (4.23c) for $k = 1$ gives

$$\alpha_1 T_j(x_0) + \left[\beta_1 + \alpha_1\int_{x_0}^{x_1}\frac{dx}{k(x)}\right]k(x_0)T_j'(x_0)$$

$$= \phi_j(x_1) + \beta_1\int_{x_0}^{x_1}\left[P_j(x) - (j+1)w(x)T_{j+1}(x)\right]dx$$

$$+ \alpha_1\int_{x_0}^{x_1}\frac{1}{k(x)}\int_{x_0}^{x}\left[P_j(x') - (j+1)w(x')T_{j+1}(x')\right]dx'\,dx \quad (4.25b)$$

Equations (4.25) provide two relations for the determination of $T_j(x_0)$ and $k(x_0)T_j'(x_0)$; a simultaneous solution yields

$$T_j(x_0) = \frac{\Delta_1}{\Delta}, \qquad k(x_0)T_j'(x_0) = \frac{\Delta_2}{\Delta} \qquad (4.26a, b)$$

where

$$\Delta = \alpha_0\beta_1 + \alpha_1\beta_0 + \alpha_0\alpha_1 \int_{x_0}^{x_1} \frac{dx}{k(x)} \qquad (4.26c)$$

$$\Delta_1 = \phi_j(x_0)\left[\beta_1 + \alpha_1 \int_{x_0}^{x_1} \frac{dx}{k(x)}\right]$$

$$+ \beta_0\left\{\phi_j(x_1) + \beta_1 \int_{x_0}^{x_1}\left[P_j(x) - (j+1)w(x)T_{j+1}(x)\right]dx\right.$$

$$\left. + \alpha_1 \int_{x_0}^{x_1} \frac{1}{k(x)} \int_{x_0}^{x}\left[P_j(x') - (j+1)w(x')T_{j+1}(x')\right]dx'\,dx\right\} \qquad (4.26d)$$

and

$$\Delta_2 = -\phi_j(x_0)\alpha_1 + \alpha_0\left\{\phi_j(x_1) + \beta_1 \int_{x_0}^{x_1}\left[P_j(x) - (j+1)w(x)T_{j+1}(x)\right]dx\right.$$

$$\left. + \alpha_1 \int_{x_0}^{x_1} \frac{1}{k(x)} \int_{x_0}^{x}\left[P_j(x') - (j+1)w(x')T_{j+1}(x')\right]dx'\,dx\right\} \qquad (4.26e)$$

The substitution of Equations (4.26a, b) into Equation (4.24b) gives the solution for $T_j(x)$ as

$$T_j(x) = \left\{\left[\beta_1 + \alpha_1 \int_{x}^{x_1} \frac{dx'}{k(x')}\right]\phi_j(x_0)\right.$$

$$+ \left[\beta_0 + \alpha_0 \int_{x_0}^{x} \frac{dx'}{k(x')}\right]$$

$$\times \left[\phi_j(x_1) + \beta_1 \int_{x_0}^{x_1}\langle P_j(x) - (j+1)w(x)T_{j+1}(x)\rangle\,dx\right.$$

$$\left.\left. + \alpha_1 \int_{x_0}^{x_1} \frac{1}{k(x)} \int_{x_0}^{x}\langle P_j(x') - (j+1)w(x')T_{j+1}(x')\rangle\,dx'\,dx\right]\right\}$$

$$\times \left\{\alpha_0\beta_1 + \alpha_1\beta_0 + \alpha_0\alpha_1 \int_{x_0}^{x_1} \frac{dx}{k(x)}\right\}^{-1}$$

$$- \int_{x_0}^{x} \frac{1}{k(x'')} \int_{x_0}^{x''}\left[P_j(x') - (j+1)w(x')T_{j+1}(x')\right]dx'\,dx''$$

$$(4.27a)$$

where $j = q, q - 1, q - 2, \ldots, 1, 0$ and

$$T_{q+1}(x) = 0 \tag{4.27b}$$

For the case $\alpha_0 = \alpha_1 = 0$, $\beta_0 = \beta_1 = 1$ (i.e., a boundary condition of the second kind at both boundaries), and $d(x) = 0$, the function $T_j(x)$ satisfies the one-dimensional form of Equation (2.38) given as

$$\frac{d}{dx}\left[k(x)\frac{dT_j(x)}{dx}\right] + P_j(x) = (j + 1)w(x)T_{j+1}(x) + \frac{w(x)}{\int_{x_0}^{x_1} w(x)\,dx}$$

$$\times\left[\phi_j(x_0) + \phi_j(x_1) + \int_{x_0}^{x_1} P_j(x)\,dx\right]$$

$$\text{in } x_0 < x < x_1 \tag{4.28a}$$

$$-(-1)^k k(x_k)T_j'(x_k) = \phi_j(x_k) \quad \text{at } x = x_k, \quad k = 0,1 \tag{4.28b, c}$$

and

$$\int_{x_0}^{x_1} w(x)T_j(x)\,dx = 0 \tag{4.28d}$$

The integration of Equation (4.28a) from x_0 to x yields

$$k(x)T_j'(x) = k(x_0)T_j'(x_0)$$

$$+\left[\phi_j(x_0) + \phi_j(x_1) + \int_{x_0}^{x_1} P_j(x)\,dx\right]\frac{\int_{x_0}^{x} w(x)\,dx}{\int_{x_0}^{x_1} w(x)\,dx}$$

$$-\int_{x_0}^{x}\left[P_j(x') - (j + 1)w(x')T_{j+1}(x')\right]dx' \tag{4.29a}$$

The boundary condition (4.28b) immediately gives the integration constant $k(x_0)T_j'(x_0) = -\phi_j(x_0)$; then the equation (4.29a) becomes

$$k(x)T_j'(x) = -\phi_j(x_0) +\left[\phi_j(x_0) + \phi_j(x_1) + \int_{x_0}^{x_1} P_j(x)\,dx\right]h(x)$$

$$-\int_{x_0}^{x}\left[P_j(x') - (j + 1)w(x')T_{j+1}(x')\right]dx' \tag{4.29b}$$

where

$$h(x) = \frac{\int_{x_0}^{x} w(x)\, dx}{\int_{x_0}^{x_1} w(x)\, dx} \tag{4.29c}$$

Equation (4.29b) is integrated from x_0 to x as

$$T_j(x) = T_j(x_0) - \phi_j(x_0) \int_{x_0}^{x} \frac{dx'}{k(x')}$$

$$+ \left[\phi_j(x_0) + \phi_j(x_1) + \int_{x_0}^{x_1} P_j(x)\, dx \right] \int_{x_0}^{x} \frac{h(x')}{k(x')}\, dx'$$

$$- \int_{x_0}^{x} \frac{1}{k(x'')} \int_{x_0}^{x''} \left[P_j(x') - (j+1)w(x') T_{j+1}(x') \right]\, dx'\, dx''$$

$$\tag{4.30a}$$

and the condition (4.28d) is applied to give

$$T_j(x_0) = \left\{ \int_{x_0}^{x_1} w(x)\, dx \right\}^{-1}$$

$$\times \left\{ \phi_j(x_0) \int_{x_0}^{x_1} w(x) \int_{x_0}^{x} \frac{dx'}{k(x')}\, dx - \left[\phi_j(x_0) + \phi_j(x_1) + \int_{x_0}^{x_1} P_j(x)\, dx \right] \right.$$

$$\times \int_{x_0}^{x_1} w(x) \int_{x_0}^{x} \frac{h(x')}{k(x')}\, dx'\, dx + \int_{x_0}^{x_1} w(x) \int_{x_0}^{x} \frac{1}{k(x'')}$$

$$\left. \times \int_{x_0}^{x''} \left[P_j(x') - (j+1)w(x') T_{j+1}(x') \right]\, dx'\, dx''\, dx \right\} \tag{4.30b}$$

Introducing (4.30b) into (4.30a), the solution for the function $T_j(x)$ becomes

$$T_j(x) = \phi_j(x_0) \left\{ \left[\int_{x_0}^{x} \frac{h(x')}{k(x')}\, dx' - \int_{x_0}^{x} \frac{dx'}{k(x')} + \frac{1}{\int_{x_0}^{x_1} w(x)\, dx} \right. \right.$$

$$\times \left[\int_{x_0}^{x_1} w(x) \int_{x_0}^{x} \frac{dx'}{k(x')}\, dx - \int_{x_0}^{x_1} w(x) \int_{x_0}^{x} \frac{h(x')}{k(x')}\, dx'\, dx \right] \right\}$$

$$+ \left[\phi_j(x_1) + \int_{x_0}^{x_1} P_j(x)\, dx \right]$$

$$\times \left[\int_{x_0}^{x} \frac{h(x')}{k(x')} \, dx' - \frac{1}{\int_{x_0}^{x_1} w(x) \, dx} \int_{x_0}^{x_1} w(x) \int_{x_0}^{x} \frac{h(x')}{k(x')} \, dx' \, dx \right]$$

$$+ \frac{1}{\int_{x_0}^{x_1} w(x) \, dx} \int_{x_0}^{x_1} w(x) \int_{x_0}^{x} \frac{1}{k(x'')}$$

$$\times \int_{x_0}^{x''} \left[P_j(x') - (j+1)w(x')T_{j+1}(x') \right] dx' \, dx'' \, dx$$

$$- \int_{x_0}^{x} \frac{1}{k(x'')} \int_{x_0}^{x''} \left[P_j(x') - (j+1)w(x')T_{j+1}(x') \right] dx' \, dx'' \quad \text{(4.31a)}$$

Various integrals in Equation (4.31a) can be evaluated by integration by parts, according to the rule

$$\int_{x_0}^{x_1} f \, dg = [\, gf \,]_{x_0}^{x_1} - \int_{x_0}^{x_1} g \, df$$

For example, by letting $dg \equiv w(x) \, dx$, $g \equiv \int_{x_0}^{x} w(x') \, dx'$ and $df \equiv dx/k(x)$, $f \equiv \int_{x_0}^{x} dx'/k(x')$, we write

$$\int_{x_0}^{x_1} w(x) \left[\int_{x_0}^{x} \frac{dx'}{k(x')} \right] dx$$

$$= \int_{x_0}^{x_1} w(x) \, dx \int_{x_0}^{x_1} \frac{dx}{k(x)} - \int_{x_0}^{x_1} \frac{1}{k(x)} \left[\int_{x_0}^{x} w(x') \, dx' \right] dx \quad \text{(4.31b)}$$

similarly, we obtain

$$\int_{x_0}^{x_1} w(x) \left[\int_{x_0}^{x} \frac{h(x')}{k(x')} \, dx' \right] dx$$

$$= \int_{x_0}^{x_1} w(x) \, dx \int_{x_0}^{x_1} \frac{h(x)}{k(x)} \, dx - \int_{x_0}^{x_1} \frac{h(x)}{k(x)} \left[\int_{x_0}^{x} w(x') \, dx' \right] dx \quad \text{(4.31c)}$$

and

$$\int_{x_0}^{x_1} w(x) \left\{ \int_{x_0}^{x} \frac{1}{k(x'')} \int_{x_0}^{x''} \left[P_j(x') - (j+1)w(x')T_{j+1}(x') \right] dx' \, dx'' \right\} dx$$

$$= \int_{x_0}^{x_1} w(x) \, dx \int_{x_0}^{x_1} \frac{1}{k(x)} \int_{x_0}^{x} \left[P_j(x') - (j+1)w(x')T_{j+1}(x') \right] dx' \, dx$$

$$- \int_{x_0}^{x_1} \frac{1}{k(x)} \left\{ \int_{x_0}^{x} w(x') \, dx' \right\} \int_{x_0}^{x} \left[P_j(x') - (j+1)w(x')T_{j+1}(x') \right] dx' \, dx$$

$$\text{(4.31d)}$$

Various terms on the right-hand side of Equation (4.31a) are evaluated by utilizing equations (4.31b, c, d) and the definition of $h(x)$ given by Equation (4.29c) as

$$\int_{x_0}^{x} \frac{h(x')}{k(x')} dx' - \int_{x_0}^{x} \frac{dx'}{k(x')} + \frac{1}{\int_{x_0}^{x_1} w(x) \, dx}$$

$$\times \left[\int_{x_0}^{x_1} w(x) \int_{x_0}^{x} \frac{dx'}{k(x')} dx - \int_{x_0}^{x_1} w(x) \int_{x_0}^{x} \frac{h(x')}{k(x')} dx' \, dx \right]$$

$$= \int_{x_0}^{x_1} \frac{[h(x) - 1]^2}{k(x)} dx + \int_{x_0}^{x} \frac{h(x') - 1}{k(x')} dx' \qquad (4.32a)$$

and

$$\int_{x_0}^{x} \frac{h(x')}{k(x')} dx' - \frac{1}{\int_{x_0}^{x_1} w(x) \, dx} \int_{x_0}^{x_1} w(x) \int_{x_0}^{x} \frac{h(x')}{k(x')} dx' \, dx$$

$$= \int_{x_0}^{x_1} \frac{h^2(x)}{k(x)} dx - \int_{x}^{x_1} \frac{h(x')}{k(x')} dx' \qquad (4.32b)$$

Substituting Equations (4.32) and (4.31d) into Equation (4.31a), we obtain the solution for the function $T_j(x)$ (for the case $\alpha_0 = \alpha_1 = 0$, $\beta_0 = \beta_1 = 1$) as

$$T_j(x) = \phi_j(x_0) \left\{ \int_{x_0}^{x_1} \frac{[1 - h(x)]^2}{k(x)} dx - \int_{x_0}^{x} \frac{1 - h(x')}{k(x')} dx' \right\}$$

$$+ \left[\phi_j(x_1) + \int_{x_0}^{x_1} P_j(x') \, dx' \right] \left[\int_{x_0}^{x_1} \frac{h^2(x)}{k(x)} dx - \int_{x}^{x_1} \frac{h(x')}{k(x')} dx' \right]$$

$$+ \int_{x}^{x_1} \frac{1}{k(x'')} \int_{x_0}^{x''} \left[P_j(x') - (j + 1)w(x')T_{j+1}(x') \right] dx' \, dx''$$

$$- \int_{x_0}^{x_1} \frac{h(x)}{k(x)} \int_{x_0}^{x} \left[P_j(x') - (j + 1)w(x')T_{j+1}(x') \right] dx' \, dx \qquad (4.33a)$$

where $j = q, q - 1, \ldots, 1, 0$

$$T_{q+1}(x) = 0 \qquad (4.33b)$$

and $h(x)$ is defined by Equation (4.29c).

Solution for $T_t(x, t)$

The function $T_t(x, t)$ appearing in the solution (4.8a, b) satisfies the problem

$$w(x)\frac{\partial T_t(x, t)}{\partial t} = \frac{\partial}{\partial x}\left[k(x)\frac{\partial T_t(x, t)}{\partial x}\right] - d(x)T_t(x, t)$$

$$\text{in } x_0 < x < x_1, \qquad t > 0 \quad (4.34a)$$

$$\alpha_k T_t(x_k, t) - (-1)^k \beta_k k(x_k)\frac{\partial T_t(x_k, t)}{\partial x} = 0$$

$$\text{at } x = x_k, \qquad k = 0, 1, \qquad t > 0 \quad (4.34b)$$

$$T_t(x, t) = f(x) - T_\phi(x) - T_p(x) - T_0(x) \quad \text{in } x_0 \le x \le x_1, \qquad t = 0$$

$$(4.34c)$$

The solution of this homogeneous problem (4.34) is immediately obtainable from the one-dimensional form of the solutions given by Equations (2.29) and (2.30). We find

$$T_t(x, t) = \sum_{i=1}^{\infty} \frac{1}{N_i} e^{-\mu_i^2 t}\psi_i(x)\left\{\tilde{f}_i - \frac{1}{\mu_i^2 - d_\phi}\sum_{k=0}^{1}\phi_e(x_k)\Omega_i(x_k) - \frac{1}{\mu_i^2 - d_p}\right.$$

$$\times \int_{x_0}^{x_1}\psi_i(x')P_e(x')\,dx' - \sum_{j=0}^{q}(-1)^j\frac{j!}{\mu_i^{2(j+1)}}$$

$$\times \left.\left[\sum_{k=0}^{1}\phi_j(x_k)\Omega_i(x_k) + \int_{x_0}^{x_1}\psi_i(x)P_j(x)\,dx\right]\right\}$$

$$(4.35)$$

where the functions N_i, $\tilde{f}_i$, and $\Omega_i(x_k)$ are defined by Equations (4.4a), (4.4b), and (4.4d), respectively.

Circular Tube or a Parallel Plate with a Symmetry Boundary Condition at $x_0 = 0$

The solutions given previously are derived, for generality, for a finite region $x_0 \le x \le x_1$ which may represent an annular region or a parallel plate with nonsymmetrical boundary conditions. In the case of a circular cylinder or a parallel plate confined to a region $0 \le x \le x_1$ with a symmetry boundary condition at $x = 0$, the solution is obtainable from the foregoing results by introducing in those equations first the requirements of the symmetry boundary condition and then letting $x_0 \to 0$. We present the simplification of Equations (4.12) and (4.14) for the function $T_\phi(x)$ and of Equation (4.18) for the function $T_p(x)$, inasmuch as the results are not so readily apparent.

First we simplify Equations (4.21) by setting $a = 0$, $c = 1$, and $\nu = m$ to obtain

$$k(x) = x^{1-2m}, \qquad w(x) = x^{1-2m}, \qquad d(x) = 0 \quad (4.36\text{a, b, c})$$

$$u(\sqrt{d}, x) = x^m J_m(x\sqrt{d}) \qquad (4.36\text{d})$$

$$v(\sqrt{d}, x) = x^m Y_m(x\sqrt{d}) \qquad (4.36\text{e})$$

Equations (3.15) are simplified by the additional requirement of symmetry condition at the boundary $x = x_0$ (i.e., by setting $\alpha_0 = 0$ and $\beta_0 = 1$), to yield

$$A_0 = 0, \qquad A_1 = \alpha_1 \qquad (4.37\text{a, b})$$

$$B_0 = x_0^{1-2m}, \qquad B_1 = \beta_1 x^{1-2m} \qquad (4.37\text{c, d})$$

Then the functions $U(\sqrt{d}, x_0)$ and $V(\sqrt{d}, x_0)$ defined by Equations (4.22a) and (4.22c) become

$$U(\sqrt{d}, x_0) = -x_0^{1-m}\sqrt{d}\, J_{1-m}(x_0\sqrt{d}) \qquad (4.38\text{a})$$

$$V(\sqrt{d}, x_0) = -x_0^{1-m}\sqrt{d}\, Y_{1-m}(x_0\sqrt{d}) \qquad (4.38\text{b})$$

The symmetry boundary condition at $x = x_0$ implies that in Equation (4.9c) the function $\phi_e(x_0)$ should vanish, that is,

$$\phi_e(x_0) = 0$$

The preceding results are now utilized to simplify the functions $T_\phi(x)$ and $T_p(x)$ for the case of symmetry boundary condition.

Simplification of $T_\phi(x)$

We now introduce the foregoing results into Equation (4.12), let $x_0 \to 0$, and make use of the relation $\lim_{x_0 \to 0}[J_{m-1}(x_0\sqrt{d_\phi})/Y_{m-1}(x_0\sqrt{d_\phi})] = 0$. Then the function $T_\phi(x)$, valid in the region $0 \le x \le x_1$, with the symmetry boundary condition at $x = 0$ and a boundary condition of the first or third kind at $x = x_1$ is determined from Equation (4.12) as

$$T_\phi(x) = \phi_e(x_1)\frac{u(\sqrt{d_\phi}, x)}{U(\sqrt{d_\phi}, x_1)} \qquad (4.39\text{a})$$

where

$$u\left(\sqrt{d_\phi}, x\right) = x^m J_m\left(x\sqrt{d_\phi}\right) \tag{4.39b}$$

$$U\left(\sqrt{d_\phi}, x_1\right) = x_1^m\left[A_1 J_m\left(x_1\sqrt{d_\phi}\right) + B_1\sqrt{d_\phi}\, J_{m-1}\left(x_1\sqrt{d_\phi}\right)\right] \tag{4.39c}$$

$$A_1 = \alpha_1, \qquad B_1 = \beta_1 x_1^{1-2m} \tag{4.39d, e}$$

If the boundary condition at $x = x_1$ is of the second kind, we have $\alpha_0 = \alpha_1 = 0$ and $\beta_0 = \beta_1 = 1$. For this case we consider Equation (4.14) which now reduces to

$$T_\phi(x) = \frac{\phi_e(x_1)}{x_1^{2(1-m)}}\left[\frac{2(1-m)}{d_\phi} + \left(\frac{x}{x_1}\right)^m \frac{J_m\left(x\sqrt{d_\phi}\right)}{\sqrt{d_\phi}\, J_{m-1}\left(x_1\sqrt{d_\phi}\right)}\right] \tag{4.39f}$$

Simplification of $T_p(x)$

The solution $T_p(x)$, valid for the region $0 \le x \le x_1$, with the symmetry boundary condition at $x = 0$ and a boundary condition of the first or third kind at $x = x_1$ is obtained by the simplification of Equation (4.18) as discussed previously. We find

$$T_p(x) = \frac{p_w}{d_p}\left[\frac{u\left(\sqrt{d_p}, x\right)}{U\left(\sqrt{d_p}, x_1\right)} - 1\right] \tag{4.40a}$$

where

$$u\left(\sqrt{d_p}, x\right) = x^m J_m\left(x\sqrt{d_p}\right) \tag{4.40b}$$

$$U\left(\sqrt{d_p}, x_1\right) = x_1^m\left[A_1 J_m\left(x_1\sqrt{d_p}\right) + B_1\sqrt{d_p}\, J_{m-1}\left(x_1\sqrt{d_p}\right)\right] \tag{4.40c}$$

$$A_1 = \alpha_1, \qquad B_1 = \beta_1 x_1^{1-2m} \tag{4.40d}$$

For a boundary condition of the second kind at $x = x_1$ and a symmetry condition at $x = 0$, we consider the solution (4.20) which remains unchanged. That is,

$$T_p(x) = 0 \tag{4.40e}$$

Example 4.1   Obtain an expression for the distribution of the potential $T(x, t)$ satisfying the following boundary value problem

$$\frac{\partial T(x, t)}{\partial t} = \frac{\partial^2 T(x, t)}{\partial x^2} \quad \text{in } 0 < x < 1, \qquad t > 0$$

$$(4.41a)$$

$$\alpha_0 T(0, t) - \beta_0 \frac{\partial T(0, t)}{\partial x} = \phi_0 \quad \text{at } x = 0, \qquad t > 0 \qquad (4.41b)$$

$$\alpha_1 T(1, t) + \beta_1 \frac{\partial T(1, t)}{\partial x} = 0 \quad \text{at } x = 1, \qquad t > 0 \qquad (4.41c)$$

$$T(x, 0) = 0 \quad \text{for } t = 0, \qquad 0 \le x \le 1 \qquad (4.41d)$$

Solution   A comparison of problem (4.41) with that given by Equations (4.1) reveals that

$$x_0 = 0, \qquad\quad x_1 = 1, \qquad\quad w(x) = 1, \qquad k(x) = 1, \qquad d(x) = 0$$

$$P(x, t) = 0, \qquad \phi(x_0, t) = \phi_0, \qquad \phi(x_1, t) = 0, \qquad f(x) = 0 \qquad (4.42)$$

The eigenvalue problem associated with the problem (4.41) is taken as

$$\psi''(\mu, x) + \mu^2 \psi(\mu, x) = 0 \quad \text{in } 0 \le x \le 1 \qquad (4.43a)$$

$$\alpha_k \psi(\mu, x_k) - (-1)^k \beta_k \psi'(\mu, x_k) = 0 \qquad k = 0, 1 \qquad (4.43b)$$

This eigenvalue problem is exactly the same as that considered in Example 3.1 for $x_0 = 0$, $x_1 = 1$. Therefore, the eigencondition, eigenfunctions, and normalization integral are immediately obtained from Equations (3.20a, b, c), respectively, as

$$(\alpha_0 \beta_1 + \alpha_1 \beta_0)\cos \mu + (\alpha_0 \alpha_1 - \beta_0 \beta_1 \mu^2) \frac{\sin \mu}{\mu} = 0 \qquad (4.44a)$$

$$\psi_i(x) = \beta_0 \cos(\mu_i x) + \frac{\alpha_0}{\mu_i} \sin(\mu_i x) \qquad (4.44b)$$

$$N_i = \frac{1}{2\mu_i^2} \left\{ (\alpha_0^2 + \beta_0^2 \mu_i^2) \left[1 + \frac{\alpha_1 \beta_1}{\alpha_1^2 + \beta_1^2 \mu_i^2} \right] + \alpha_0 \beta_0 \right\} \qquad (4.44c)$$

The solution of the problem (4.41) is now split up into the solution of simpler problems according to Equation (4.8a) as

$$T(x, t) = T_0(x) + T_t(x, t) \tag{4.45a}$$

since the functions $T_\phi(x)$ and $T_p(x)$ are not needed and $j = 0$.

The solution for the steady-state function $T_0(x)$ is immediately obtained from Equation (4.27a) as

$$T_0(x) = \frac{\alpha_1(1 - x) + \beta_1}{\alpha_0\beta_1 + \alpha_1\beta_0 + \alpha_0\alpha_1}\phi_0 \tag{4.45b}$$

and the transient solution $T_t(x, t)$ is determined from Equation (4.35) as

$$T_t(x, t) = -\phi_0 \sum_{i=1}^{\infty} \frac{1}{N_i} e^{-\mu_i^2 t}\psi_i(x)\frac{1}{\mu_i^2} \tag{4.45c}$$

since the functions $\tilde{f}_i$, $\phi_e(x_k)$, $P_e(x)$, and $P_j(x)$ are not needed and $j = 0$ for the present problem. Introducing Equations (4.45b) and (4.45c) into Equation (4.45a) together with the foregoing expressions for $\psi_i(x)$ and N_i, the solution for $T(x, t)$ becomes

$$\frac{T(x, t)}{\phi_0} = \frac{\alpha_1(1 - x) + \beta_1}{\alpha_0\beta_1 + \alpha_1\beta_0 + \alpha_0\alpha_1}$$

$$- \sum_{i=1}^{\infty} 2\mu_i^2 \left\{ (\alpha_0^2 + \beta_0^2\mu_i^2)\left[1 + \frac{\alpha_1\beta_1}{\alpha_1^2 + \beta_1^2\mu_i^2}\right] + \alpha_0\beta_0 \right\}^{-1}$$

$$\times \left\{ \beta_0\cos(\mu_i x) + \frac{\alpha_0}{\mu_i}\sin(\mu_i x) \right\} e^{-\mu_i^2 t} \tag{4.45d}$$

where the μ_i are the positive roots of the transcendental Equation (4.44a).

Semi-Infinite Region, $x_0 \leq x < \infty$

In the foregoing analysis we considered the solutions of one-dimensional problems of Class I for a finite region $x_0 \leq x \leq x_1$. If the region is not finite but extends to infinity in one direction, say, $x_1 \rightarrow \infty$, the boundary condition at infinity is not suitable to establish a relation for the determination of a set of discrete eigenvalues μ_i $(i = 1, 2, \ldots, \infty)$. In such situations the eigenvalues assume all values from zero to infinity; that is, the discrete eigenvalues are replaced by continuum eigenvalues as now described.

We consider the eigenvalue problem (4.5a, b, c) which is the same as that given by Equations (3.2a, b, c). The asymptotic formula for the determination of the eigenvalues μ_i is given by Equation (3.71b) as

$$\mu_i = \frac{\pi i + \epsilon}{\int_{x_0}^{x_1} \sqrt{w(x)/k(x)}\,dx} \qquad (4.46a)$$

where the constant ϵ depends on the boundary condition parameters α_0, α_1, β_0, and β_1 of the eigenvalue problem (4.5a, b, c).

The difference between two consecutive eigenvalues μ_{i+1} and μ_i is obtained from Equation (4.45a) as

$$\mu_{i+1} - \mu_i = \frac{\pi}{\int_{x_0}^{x_1} \sqrt{w(x)/k(x)}\,dx} \qquad (4.46b)$$

It is apparent from this expression that $(\mu_{i+1} - \mu_i) \to 0$ as $x_1 \to \infty$; that is, μ_i assumes all values from 0 to infinity as $x_1 \to \infty$. In such situations, the summation over all discrete eigenvalues in Equation (4.2) should be replaced by an integration according to the formula

$$\sum_{i=1}^{\infty} F(\mu_i, x, t) \to \int_0^{\infty} F(\mu, x, t)\,di \quad \text{as } x_1 \to \infty \qquad (4.46c)$$

where di is the mean number of eigenvalues in the differential length $d\mu$. From the expression (4.46a) we obtain

$$di = \frac{1}{\pi}\left[\int_{x_0}^{\infty} \sqrt{\frac{w(x)}{k(x)}}\,dx\right] d\mu \qquad (4.46d)$$

When this result is introduced into Equation (4.46c) we obtain, for $x_1 \to \infty$,

$$\sum_{i=1}^{\infty} F(\mu_i, x, t) \to \frac{1}{\pi}\int_{\mu=0}^{\infty}\left[\int_{x_0}^{\infty} \sqrt{\frac{w(x)}{k(x)}}\,dx\right] F(\mu, x, t)\,d\mu \qquad (4.47)$$

which is *the desired relation to transform the solution for a finite region* $x_0 \le x \le x_1$ *to the solution for a semi-infinite region* $x_0 \le x < \infty$.

The formula (4.47) is now applied to transform the solution (4.2) for a finite region $x_0 \le x \le x_1$ to that for a semi-infinite region $x_0 \le x < \infty$ by letting $x_1 \to \infty$; we obtain

$$T(x, t) = \frac{1}{\pi}\int_{\mu=0}^{\infty} G(\mu)e^{-\mu^2 t}\psi(\mu, x)\left[\tilde{f}(\mu) + \int_0^t g(\mu, t')e^{\mu^2 t'}\,dt'\right]d\mu$$

$$(4.48)$$

where

$$G(\mu) = \lim_{x_1 \to \infty} \left\{ \frac{\int_{x_0}^{x_1} \sqrt{w(x)/k(x)} \, dx}{N_i} \right\} = \lim_{x_1 \to \infty} \left\{ \frac{\int_{x_0}^{x_1} \sqrt{w(x)/k(x)} \, dx}{\int_{x_0}^{x_1} w(x)\psi^2(\mu, x) \, dx} \right\}$$

(4.49a)

$$\tilde{f}(\mu) = \int_{x'=x_0}^{\infty} w(x')\psi(\mu, x')f(x') \, dx'$$

(4.49b)

$$g(\mu, t') = \left[\phi(x_0, t')\Omega(\mu, x_0) + \int_{x'=x_0}^{\infty} \psi(\mu, x')P(x', t') \, dx' \right]$$

(4.49c)

and

$$\Omega(\mu, x_0) = \frac{\psi(\mu, x_0) + k(x_0)\psi'(\mu, x_0)}{\alpha_0 + \beta_0} = \frac{\psi(\mu, x_0)}{\beta_0} = k(x_0)\frac{\psi'(\mu, x_0)}{\alpha_0}$$

(4.49d)

Clearly, Equation (4.48) is the solution of the following one-dimensional boundary value problem of Class I for $x_1 \to \infty$:

$$w(x)\frac{\partial T(x, t)}{\partial t} = \frac{\partial}{\partial x}\left[k(x)\frac{\partial T(x, t)}{\partial x} \right] - d(x) T(x, t) + P(x, t)$$

$$\text{in } x_0 < x < \infty, \qquad t > 0 \quad (4.50a)$$

$$\alpha_0 T(x_0, t) - \beta_0 k(x_0)\frac{\partial T(x_0, t)}{\partial x} = \phi(t)$$

$$\text{at } x = x_0, \qquad t > 0 \quad (4.50b)$$

$$T(x, 0) = f(x) \quad \text{for } t = 0 \tag{4.50c}$$

The solution of this problem given by Equation (4.48) is now rearranged in a more convenient form as

$$T(x, t) = \frac{1}{\pi}\int_{x'=x_0}^{\infty} w(x')f(x')\int_{\mu=0}^{\infty} e^{-\mu^2 t}G(\mu)\psi(\mu, x)\psi(\mu, x') \, d\mu \, dx'$$

$$+ \frac{1}{\pi}\int_{t'=0}^{t} \phi(t')\int_{\mu=0}^{\infty} e^{-\mu^2(t-t')}G(\mu)\Omega(\mu, x_0)\psi(\mu, x) \, d\mu \, dt'$$

$$+ \frac{1}{\pi}\int_{t'=0}^{t}\int_{x'=x_0}^{\infty} P(x', t')$$

$$\times \int_{\mu=0}^{\infty} e^{-\mu^2(t-t')}G(\mu)\psi(\mu, x)\psi(\mu, x') \, d\mu \, dx' \, dt' \tag{4.51}$$

where $G(\mu)$ and $\Omega(\mu, x_0)$ are defined previously and $\psi(\mu, x)$ is the solution of the eigenvalue problem (4.5a, b, c) for $x_1 \to \infty$.

Example 4.2 Obtain an expression for the distribution of the potential $T(x, t)$ satisfying the following boundary value problem

$$\frac{\partial^2 T(x, t)}{\partial x^2} + g(x, t) = \frac{\partial T(x, t)}{\partial t} \quad \text{in } x_0 < x < \infty, \qquad t > 0 \quad (4.52a)$$

$$T(x, 0) = f(x) \quad \text{for } t = 0, \quad \text{in } x_0 \le x < \infty \qquad (4.52b)$$

$$\alpha_0 T(x_0, t) - \beta_0 \frac{\partial T(x_0, t)}{\partial x} = \phi(t) \quad \text{at } x = x_0, \qquad t > 0 \quad (4.52c)$$

$$\frac{\partial T(\infty, t)}{\partial x} = 0 \quad \text{as } x \to \infty, \qquad t > 0 \qquad (4.52d)$$

Solution A comparison of problem (4.52) with that given by Equations (4.50) reveals that

$$k(x) = 1, \qquad w(x) = 1, \qquad d(x) = 0, \qquad P(x, t) = g(x, t) \quad (4.53)$$

and the eigenvalue problem associated with the problem (4.52) is obtained from that given by Equations (4.5a, b, c) by letting $x_1 \to \infty$ and taking into account (4.53). We obtain

$$\psi''(\mu, x) + \mu^2 \psi(\mu, x) = 0 \quad \text{in } x_0 < x < \infty \qquad (4.54a)$$

$$\alpha_0 \psi(\mu, x_0) - \beta_0 \psi'(\mu, x_0) = 0 \quad \text{at } x = x_0 \qquad (4.54b)$$

In this problem μ takes all values from zero to infinity continuously. We note that the eigenvalue problem (4.54) is obtainable from that given by Equations (3.17) by letting in the latter, $x_1 \to \infty$. Therefore, the eigenfunctions $\psi(\mu, x)$ and functions $G(\mu), \Omega(\mu, x_0)$ are obtainable from the results (3.20b) and utilizing the definitions of $G(\mu)$ and $\Omega(\mu, x_0)$ given by Equations (4.49a) and (4.49d), respectively. We find

$$\psi(\mu, x) = \beta_0 \cos[\mu(x - x_0)] + \frac{\alpha_0}{\mu} \sin[\mu(x - x_0)] \qquad (4.55a)$$

$$G(\mu) = \frac{2\mu^2}{\alpha_0^2 + \beta_0^2 \mu^2} \qquad (4.55b)$$

$$\Omega(\mu, x_0) = 1 \qquad (4.55c)$$

Equations (4.53) and (4.55) are introduced into Equation (4.51) and the following relations are utilized

$$2\cos[\mu(x - x_0)]\cos[\mu(x' - x_0)]$$
$$= \cos[\mu(x - x')] + \cos[\mu(x + x' - 2x_0)] \qquad (4.56a)$$

$$2\sin[\mu(x - x_0)]\sin[\mu(x' - x_0)]$$
$$= \cos[\mu(x - x')] - \cos[\mu(x + x' - 2x_0)] \qquad (4.56b)$$

$$\cos[\mu(x - x_0)]\sin[\mu(x' - x_0)] + \sin[\mu(x - x_0)]\cos[\mu(x' - x_0)]$$
$$= \sin[\mu(x + x' - 2x_0)] \qquad (4.56c)$$

Then the solution of the problem (4.52) becomes

$$T(x, t) = \frac{1}{\pi} \int_{x'=x_0}^{\infty} f(x') \int_{\mu=0}^{\infty} e^{-\mu^2 t}$$

$$\times \left\{ \cos[\mu(x - x_0)] + \frac{\beta_0^2 \mu^2 - \alpha_0^2}{\beta_0^2 \mu^2 + \alpha_0^2} \cos[\mu(x + x' - 2x_0)] \right.$$

$$\left. + \frac{2\alpha_0 \beta_0}{\beta_0^2 \mu^2 + \alpha_0^2} \mu \sin[\mu(x + x' - 2x_0)] \right\} d\mu \, dx'$$

$$+ \frac{1}{\pi} \int_{t'=0}^{t} \phi(t') \int_{\mu=0}^{\infty} e^{-\mu^2(t-t')} \frac{2\mu}{\beta_0^2 \mu^2 + \alpha_0^2}$$

$$\times \{ \beta_0 \mu \cos[\mu(x - x_0)] + \alpha_0 \sin[\mu(x - x_0)] \} \, d\mu \, dt'$$

$$+ \frac{1}{\pi} \int_{t'=0}^{t} \int_{x'=x_0}^{\infty} g(x', t') \int_{\mu=0}^{\infty} e^{-\mu^2(t-t')}$$

$$\times \left\{ \cos[\mu(x - x')] + \frac{\beta_0^2 \mu^2 - \alpha_0^2}{\beta_0^2 \mu^2 + \alpha_0^2} \cos[\mu(x + x' - 2x_0)] \right.$$

$$\left. + \frac{2\alpha_0 \beta_0}{\beta_0^2 \mu^2 + \alpha_0^2} \mu \sin[\mu(x + x' - 2x_0)] \right\} d\mu \, dx' \, dt' \qquad (4.57)$$

We now consider some special cases of the problem (4.52).

1. The boundary condition (4.52c) is given as $-[\partial T(x_0, t)/\partial x] = \phi(t)$. This special case is obtainable from the general boundary condition considered previously by setting $\alpha_0 = 0$ and $\beta_0 = 1$. Then the solution (4.57) reduces to

$$T(x, t) = \frac{1}{\pi} \int_{x'=x_0}^{\infty} f(x') \int_{\mu=0}^{\infty} e^{-\mu^2 t} \{ \cos[\mu(x - x')]$$

$$+ \cos[\mu(x + x' - 2x_0)] \} \, d\mu \, dx'$$

$$+ \frac{2}{\pi} \int_{t'=0}^{t} \phi(t') \int_{\mu=0}^{\infty} e^{-\mu^2(t-t')} \cos[\mu(x - x_0)] \, d\mu \, dt'$$

$$+ \frac{1}{\pi} \int_{t'=0}^{t} \int_{x'=x_0}^{\infty} g(x', t') \int_{\mu=0}^{\infty} e^{-\mu^2(t-t')}$$

$$\times \{ \cos[\mu(x - x')] + \cos[\mu(x + x' - 2x_0)] \} \, d\mu \, dx' \, dt'$$

$$(4.58)$$

The integrations with respect to μ are now performed by making use of the following relation [1, #861.20]

$$\int_0^{\infty} e^{-ax^2} \cos bx \, dx = \frac{1}{2} \sqrt{\frac{\pi}{a}} \, e^{-(b^2/4a)}, \qquad a > 0, \qquad b > 0 \qquad (4.59)$$

Then Equation (4.58) becomes

$$T(x, t) = \frac{1}{2\sqrt{\pi t}} \int_{x'=x_0}^{\infty} f(x') \left\{ e^{-[(x-x')^2/4t]} + e^{-[(x+x'-2x_0)^2/4t]} \right\} dx'$$

$$+ \frac{1}{\sqrt{\pi}} \int_{t'=0}^{t} \frac{\phi(t')}{\sqrt{t - t'}} e^{-[(x-x_0)^2/4(t-t')]} \, dt'$$

$$+ \frac{1}{2\sqrt{\pi}} \int_{t'=0}^{t} \frac{1}{\sqrt{t - t'}} \int_{x'=x_0}^{\infty} g(x', t')$$

$$\times \left\{ e^{-[(x-x')^2/4(t-t')]} + e^{-[(x+x'-2x_0)^2/4(t-t')]} \right\} dx' \, dt' \qquad (4.60)$$

2. The boundary condition (4.52c) is given as $T(0, t) = \phi(t)$. This special case is obtainable from the preceding general boundary condition by setting

$\alpha_0 = 1$, $\beta_0 = 0$, and $x_0 = 0$. Then solution (4.57) reduced to

$$T(x, t) = \frac{1}{\pi} \int_{x'=0}^{\infty} f(x') \int_{\mu=0}^{\infty} e^{-\mu^2 t} \{ \cos[\mu(x - x')] - \cos[\mu(x + x')] \} \, d\mu \, dx'$$

$$+ \frac{2}{\pi} \int_{t'=0}^{t} \phi(t') \int_{\mu=0}^{\infty} e^{-\mu^2 (t-t')} \mu \sin(\mu x) \, d\mu \, dt'$$

$$+ \frac{1}{\pi} \int_{t'=0}^{t} \int_{x'=0}^{\infty} g(x', t') \int_{\mu=0}^{\infty} e^{-\mu^2 (t-t')}$$

$$\times \{ \cos[\mu(x - x')] - \cos[\mu(x + x')] \} \, d\mu \, dx' \, dt' \qquad (4.61)$$

The integrations with respect to μ are performed using Equation (4.59) and the following relation [1, #861.21]

$$\int_0^{\infty} x e^{-ax^2} \sin bx \, dx = \frac{b\sqrt{\pi}}{4a^{3/2}} e^{-(b^2/4a)}, \qquad a > 0, \qquad b > 0 \qquad (4.62)$$

Then Equation (4.61) becomes

$$T(x, t) = \frac{1}{2\sqrt{\pi t}} \int_{x'=0}^{\infty} f(x') \left\{ e^{-[(x-x')^2/4t]} - e^{-[(x+x')^2/4t]} \right\} dx'$$

$$+ \frac{x}{2\sqrt{\pi}} \int_{t'=0}^{t} \frac{\phi(t')}{(t - t')^{3/2}} e^{-[x^2/4(t-t')]} \, dt'$$

$$+ \frac{1}{2\sqrt{\pi}} \int_{t'=0}^{t} \frac{1}{\sqrt{t - t'}} \int_{x'=0}^{\infty} g(x', t')$$

$$\times \left\{ e^{-[(x-x')^2/4(t-t')]} - e^{-[(x+x')^2/4(t-t')]} \right\} dx' \, dt' \qquad (4.63)$$

3.   We now restrict special case 2 to $f(x) = f_0$, $\phi(t) = \phi_0$, and $g(x, t) = 0$, where f_0 and ϕ_0 are constants. Then the resulting integrations are performed and the solution (4.63) becomes

$$T(x, t) = f_0 \mathrm{erf}\left(\frac{x}{2\sqrt{t}} \right) + \phi_0 \mathrm{erfc}\left(\frac{x}{2\sqrt{t}} \right) \qquad (4.64)$$

where $\mathrm{erf}(z)$ is called the *error function* of argument z and defined as

$$\mathrm{erf}(z) = \frac{2}{\sqrt{\pi}} \int_0^z e^{-\eta^2} \, d\eta \qquad (4.65a)$$

and erfc(z) is called the *complementary error function* of argument z and defined as

$$\text{erfc}(z) = 1 - \text{erf}(z) = \frac{2}{\sqrt{\pi}} \int_z^\infty e^{-\eta^2} d\eta \qquad (4.65b)$$

The reader should consult References 2 and 3 for extensive tabulations of this function.

Example 4.3 Obtain an expression for the distribution of the potential $T(x, t)$ satisfying the following boundary value problem

$$\frac{\partial T(x, t)}{\partial t} = \frac{\partial^2 T(x, t)}{\partial x^2} + \frac{1}{x} \frac{\partial T(x, t)}{\partial x} + g(x, t)$$

$$\text{in } 0 < x < \infty, \qquad t > 0 \quad (4.66a)$$

$$T(x, 0) = f(x) \qquad (4.66b)$$

$$\frac{\partial T(0, t)}{\partial x} = 0, \qquad \frac{\partial T(\infty, t)}{\partial x} = 0 \qquad (4.66c, d)$$

Solution Equation (4.66a) is written in an alternative form in order to compare it readily with Equation (4.50a)

$$x \frac{\partial T(x, t)}{\partial t} = \frac{\partial}{\partial x}\left[x \frac{\partial T(x, t)}{\partial x} \right] + xg(x, t) \quad \text{in } 0 < x < \infty, \qquad t > 0$$

$$(4.66e)$$

A comparison of system (4.66) with that given by Equations (4.50) reveals that

$$x_0 = 0, \qquad k(x) = x, \qquad w(x) = x, \qquad d(x) = 0$$

$$P(x, t) = xg(x, t), \qquad \alpha_0 = 0, \qquad \beta_0 = 1, \qquad \phi(t) = 0 \quad (4.67)$$

The eigenvalue problem associated with the solution of the problem (4.66) is given as

$$\frac{d}{dx}\left[x \frac{d\psi(\mu, x)}{dx} \right] + x\mu^2 \psi(\mu, x) = 0 \quad \text{in } 0 < x < \infty \qquad (4.68a)$$

$$\psi'(\mu, x) = 0 \quad \text{at } x = 0 \qquad (4.68b)$$

In this problem μ takes all values from 0 to infinity continuously. By comparing Equation (4.68a) with that given by Equation (3.10), we write

$$a = 0, \quad m = 0, \quad c = 1, \quad \text{and} \quad \nu = 0 \tag{4.69}$$

Then from Equation (3.11) we have

$$\psi(\mu, x) = C_1 J_0(\mu x) + D Y_0(\mu x) \tag{4.70a}$$

The function $Y_0(\mu x)$ becomes infinite as $x \to 0$. Since the region includes the origin $x = 0$, the function $Y_0(\mu x)$ is excluded from solution (4.70a) and the eigenfunction $\psi(\mu, x)$ is taken as

$$\psi(\mu, x) = J_0(\mu x) \tag{4.70b}$$

The normalization integral N is determined from Equation (3.16a) as

$$N = \int_0^{x_1} x J_0^2(\mu x)\, dx = \frac{x_1^2}{2} \left[J_0^2(\mu x_1) - J_{-1}(\mu x_1) J_1(\mu x_1) \right] \tag{4.70c}$$

Here the Bessel functions $J_{-1}(x)$ and $J_1(x)$ are related to each other by [2, p. 358]

$$J_{-1}(x) = -J_1(x) \tag{4.70d}$$

The functions $J_0(x)$ and $J_1(x)$ have oscillatory behavior as do the trigonometric functions; for large values of x and as $x \to \infty$ they are represented as [2, p. 364]

$$J_n \simeq \sqrt{\frac{2}{\pi x}} \cos\left(x - \frac{\pi}{4} - \frac{n\pi}{2} \right), \quad n = 0, 1, \ldots \tag{4.70e}$$

Introducing Equations (4.70d, e) into Equation (4.70c), we obtain the normalization integral N as

$$N = \frac{x_1}{\pi\mu} \quad \text{for large } x_1 \tag{4.71a}$$

and $G(\mu)$, according to its definition (4.49a) becomes

$$G(\mu) = \lim_{x_1 \to \infty} \left[\frac{\int_0^{x_1} dx}{N} \right] = \lim_{x_1 \to \infty} \left[\frac{x_1}{N} \right] = \pi\mu \tag{4.71b}$$

Introducing the results given by Equations (4.70b) and (4.71) into (4.51), and noting that $\phi(t){=}0$, $P(x, t) = xg(x, t)$, and $w(x) = x$, the solution of the problem (4.66) becomes

$$T(x, t) = \int_{x'=0}^{\infty} x'f(x') \int_{\mu=0}^{\infty} e^{-\mu^2 t} \mu J_0(\mu x) J_0(\mu x') \, d\mu \, dx'$$

$$+ \int_{t'=0}^{t} \int_{x'=0}^{\infty} x'g(x', t') \int_{\mu=0}^{\infty} e^{-\mu^2(t-t')} \mu J_0(\mu x) J_0(\mu x') \, d\mu \, dx' \, dt'$$

$$(4.72)$$

The integral with respect to μ is evaluated by making use of the following integral [4, #6.633.2]

$$\int_{\mu=0}^{\infty} e^{-\mu^2 t} \mu J_0(\mu x) J_0(\mu x') \, d\mu = \frac{1}{2t} e^{-(x^2+x'^2)/4t} I_0\left(\frac{xx'}{2t}\right) \qquad (4.73)$$

Then the solution (4.72) becomes

$$T(x, t) = \frac{1}{2t} \int_{x'=0}^{\infty} x' e^{-(x^2+x'^2)/4t} I_0\left(\frac{xx'}{2t}\right) f(x') \, dx'$$

$$+ \frac{1}{2} \int_{t'=0}^{t} \frac{1}{t-t'} \int_{x'=0}^{\infty} x'g(x', t') e^{-(x^2+x'^2)/4(t-t')} I_0\left(\frac{xx'}{2(t-t')}\right) dx' \, dt'$$

$$(4.74)$$

Infinite Region, $-\infty < x < \infty$

We now consider the one-dimensional form of the Class I problem for an infinite region $-\infty < x < \infty$ given as

$$w(x)\frac{\partial T(x, t)}{\partial t} = \frac{\partial}{\partial x}\left[k(x)\frac{\partial T(x, t)}{\partial x}\right] - d(x) T(x, t) + P(x, t)$$

$$\text{in } -\infty < x < \infty, \qquad t > 0 \quad (4.75a)$$

$$\frac{\partial T(-\infty, t)}{\partial x} = 0, \qquad \frac{\partial T(+\infty, t)}{\partial x} = 0 \qquad (4.75b)$$

$$T(x, 0) = f(x) \qquad (4.75c)$$

The solution of this problem is immediately obtainable from solution (4.51) of problem (4.50) for a semi-infinite medium, by letting in Equation (4.51),

$x_0 \to -\infty$ and $\phi(t) = 0$. We find

$$T(x, t) = \frac{1}{\pi} \int_{x' = -\infty}^{\infty} w(x')f(x') \int_{\mu = 0}^{\infty} e^{-\mu^2 t} G(\mu)\psi(\mu, x)\psi(\mu, x')\, d\mu\, dx'$$

$$+ \frac{1}{\pi} \int_{t' = 0}^{t} \int_{x' = -\infty}^{\infty} P(x', t')$$

$$\times \int_{\mu = 0}^{\infty} e^{-\mu^2 (t - t')} G(\mu)\psi(\mu, x)\psi(\mu, x')\, d\mu\, dx'\, dt' \qquad (4.76a)$$

For the special case of $k(x) = 1$, $w(x) = 1$, $d(x) = 0$, and $P(x, t) = g(x, t)$ the solution of problem (4.75) is obtainable from that for a semi-infinite medium $x_0 \le x < \infty$ given by Equation (4.60) by setting in that equation $x_0 \to -\infty$ and $\phi(t) = 0$. We find

$$T(x, t) = \frac{1}{2\sqrt{\pi t}} \int_{x' = -\infty}^{\infty} f(x') e^{-[(x - x')^2 / 4t]}\, dx'$$

$$+ \frac{1}{2\sqrt{\pi}} \int_{t' = 0}^{t} \frac{1}{\sqrt{t - t'}} \int_{x' = -\infty}^{\infty} g(x', t') e^{-(x - x')^2 / 4(t - t')}\, dx'\, dt'$$

$$(4.76b)$$

Example 4.4 Obtain an expression for the distribution of the potential $T(x, t)$ satisfying the following bondary value problem

$$\frac{\partial T(x, t)}{\partial t} = \frac{\partial^2 T(x, t)}{\partial x^2} \quad \text{in } -\infty < x < \infty, \qquad t > 0 \qquad (4.77a)$$

$$T(x, 0) = \begin{cases} T_0 & \text{in } -\infty < x < x_0 \\ T_1 & \text{in } x_0 < x < x_1 \\ T_2 & \text{in } x_1 < x < \infty \end{cases} \qquad (4.77b)$$

Solution By comparing the present problem (4.77) with that given by Equations (4.75) we write

$$w(x) = 1, \qquad k(x) = 1, \qquad d(x) = 0, \qquad g(x, t) = 0$$

and

$$f(x) = \begin{cases} T_0 & \text{in } -\infty < x < x_0 \\ T_1 & \text{in } x_0 < x < x_1 \\ T_2 & \text{in } x_1 < x < \infty \end{cases} \qquad (4.78)$$

Therefore, the solution of problem (4.77) is obtainable from solution (4.76b). Introducing the results (4.78) into Equation (4.76b) and after performing the integrations, we obtain the solution of the problem (4.77) as

$$T(x, t) = T_0 + \frac{1}{2} T_1 \left[\text{erf}\left(\frac{x_1 - x}{2\sqrt{t}} \right) - \text{erf}\left(\frac{x_0 - x}{2\sqrt{t}} \right) \right]$$

$$+ \frac{1}{2} T_2 \text{erfc}\left(\frac{x_1 - x}{2\sqrt{t}} \right) - \frac{1}{2} T_0 \text{erfc}\left(\frac{x_0 - x}{2\sqrt{t}} \right) \tag{4.79}$$

4.2 ONE-DIMENSIONAL PROBLEMS OF CLASS II

The mathematical formulation of the boundary value problems of Class II for the general three-dimensional case was given in Chapter 2 by Equations (2.70). In this section we examine the solution of these problems for the one-dimensional finite region.

We consider n parallel layers that are in contact and confined to the regions $x_{k-1} \le x \le x_k$, $k = 1, 2, \ldots, n$. For the one-dimensional case considered here, the differential operator L_k appearing in Equation (2.70a) takes the form

$$L_k \equiv -\frac{\partial}{\partial x} \left[k(x) \frac{\partial}{\partial x} \right] + d(x) \tag{4.80a}$$

where $x_{k-1} \le x \le x_k$, $k = 1, 2, \ldots, n$, and $x = x_0$, $x = x_n$ denote the outer boundaries of the region.

Letting $\alpha(x_0) = \alpha_0$, $\beta(x_0) = \beta_0$, $\alpha(x_n) = \alpha_n$, and $\beta(x_n) = \beta_n$, the boundary condition operator B_k defined by Equation (2.70f) becomes

$$B_0 \equiv \alpha_0 - \beta_0 k(x_0) \frac{\partial}{\partial n} \quad \text{at } x = x_0 \tag{4.80b}$$

$$B_n \equiv \alpha_n + \beta_n k(x_n) \frac{\partial}{\partial n} \quad \text{at } x = x_n \tag{4.80c}$$

Then the one-dimensional form of the boundary value problems of Class II becomes

$$w_k(x) \frac{\partial T_k(x, t)}{\partial t} = \frac{\partial}{\partial x} \left[k_k(x) \frac{\partial T_k(x, t)}{\partial x} \right] - d_k(x) T_k(x, t) + P_k(x, t)$$

$$\text{in } x_{k-1} < x < x_k, \quad k = 1, 2, \ldots, n \quad \text{for } t > 0 \tag{4.81a}$$

$$\alpha_0 T_1(x_0, t) - \beta_0 k_1(x_0) \frac{\partial T_1(x_0, t)}{\partial x} = \phi(x_0, t) \tag{4.81b}$$

$$k_k(x_k) \frac{\partial T_k(x_k, t)}{\partial x} = k_{k+1}(x_k) \frac{\partial T_{k+1}(x_k, t)}{\partial x}$$

$$= h_{k, k+1} [T_{k+1}(x_k, t) - T_k(x_k, t)] \tag{4.81c}$$

$$\alpha_n T_n(x_n, t) + \beta_n k_n(x_n) \frac{\partial T_n(x_n, t)}{\partial x} = \phi(x_n, t) \tag{4.81d}$$

$$T_k(x, 0) = f_k(x) \tag{4.81e}$$

The solution of problem (4.81) is immediately obtained from the general solution (2.81) by restricting it to the one-dimensional case. We find

$$T_k(x, t) = \sum_{i=1}^{\infty} \frac{1}{N_i} e^{-\mu_i^2 t} \psi_{ki}(x) \left[\tilde{f}_i + \int_0^t g_i(t') e^{\mu_i^2 t'} dt' \right]$$

$$\text{in } x_{k-1} \le x \le x_k, \qquad k = 1, 2, \ldots, n, \quad (4.82a)$$

where N_i, $\tilde{f}_i$, and $g_i(t')$ are defined by Equations (2.74), (2.80), and (2.79b), respectively; for the one-dimensional case considered here they take the form

$$N_i = \sum_{k=1}^{n} \int_{x_{k-1}}^{x_k} w_k(x) \psi_k^2(\mu_i, x) \, dx \qquad (4.82b)$$

$$\tilde{f}_i = \sum_{k=1}^{n} \int_{x_{k-1}}^{x_k} w_k(x) \psi_k(\mu_i, x) f_k(x) \, dx \qquad (4.82c)$$

$$g_i(t) = \phi(x_0, t) \Omega_i(x_0) + \phi(x_n, t) \Omega_i(x_n)$$

$$+ \sum_{k=1}^{n} \int_{x_{k-1}}^{x_k} \psi_k(\mu_i, x) P_k(x, t) \, dx \qquad (4.82d)$$

where

$$\Omega_i(x_0) = \frac{\psi_1(\mu_i, x_0) + k_1(x_0) \psi_1'(\mu_i, x_0)}{\alpha_0 + \beta_0} \qquad (4.82e)$$

$$\Omega_i(x_n) = \frac{\psi_n(\mu_i, x_n) - k_n(x_n) \psi_n'(\mu_i, x_n)}{\alpha_n + \beta_n} \qquad (4.82f)$$

and the eigenfunctions $\psi_k(\mu, x)$ are defined by the one-dimensional eigenvalue problem (3.96).

Splitting Up the Solution

When the nonhomogeneous terms $\phi(x_0, t)$, $\phi(x_n, t)$, and $P(x, t)$ of problem (4.81) are exponentials and q-order polynomials of time t in the form

$$\phi(x_0, t) = \phi_e(x_0) e^{-d_\phi t} + \sum_{j=0}^{q} \phi_j(x_0) t^j \qquad (4.83a)$$

$$\phi(x_n, t) = \phi_e(x_n) e^{-d_\phi t} + \sum_{j=0}^{q} \phi_j(x_n) t^j \qquad (4.83b)$$

$$P_k(x, t) = P_{ke}(x) e^{-d_p t} + \sum_{j=0}^{q} P_{kj}(x) t^j \qquad (4.83c)$$

Then the solution of problem (4.81) can be split up into the solution of simpler problems as given by Equation (2.83) in the form

$$T_k(x, t) = T_{k\phi}(x)e^{-d_\phi t} + T_{kp}(x)e^{-d_p t} + \sum_{j=0}^{q} T_{kj}(x)t^j + T_{kt}(x, t)$$

$$\text{in } x_{k-1} < x < x_k, \qquad k = 1, 2, \ldots, n \quad (4.84)$$

Here the functions $T_{k\phi}(x)$, $T_{kp}(x)$, and $T_{kj}(x)$ are the solutions of the one-dimensional form of the steady-state problems given by Equations (2.84)–(2.87). that is, the functions $T_{k\phi}(x)$ satisfy the one-dimensional form of problem (2.84) given as

$$\frac{d}{dx}\left[k_k(x)\frac{dT_{k\phi}(x)}{dx}\right] + \left[d_\phi w_k(x) - d(x)\right]T_{k\phi}(x) = 0$$

$$\text{in } x_{k-1} < x < x_k, \qquad k = 1, 2, \ldots, n \quad (4.85a)$$

$$\alpha_0 T_{1\phi}(x_0) - \beta_0 k_1(x_0)T'_{1\phi}(x_0) = \phi_e(x_0) \qquad (4.85b)$$

$$k_k(x_k)T'_{k\phi}(x_k) = k_{k+1}(x_k)T'_{k+1,\phi}(x_k) = h_{k,k+1}\left[T_{k+1,\phi}(x_k) - T_{k,\phi}(x_k)\right]$$

$$(4.85c)$$

$$\alpha_n T_{n\phi}(x_n) + \beta_n k_n(x_n)T'_{n\phi}(x_n) = \phi_e(x_n) \qquad (4.85d)$$

The functions $T_{kp}(x)$ satisfy the one-dimensional form of Equations (2.85) given as

$$\frac{d}{dx}\left[k_k(x)\frac{dT_{kp}(x)}{dx}\right] + \left[d_p w_k(x) - d(x)\right]T_{kp}(x) + P_{ke}(x) = 0$$

$$\text{in } x_{k-1} < x < x_k, \qquad k = 1, 2, \ldots, n \quad (4.86a)$$

$$\alpha_0 T_{1p}(x_0) - \beta_0 k_1(x_0)T'_{1p}(x_0) = 0 \qquad (4.86b)$$

$$k_k(x_k)T'_{kp}(x_k) = k_{k+1}(x_k)T'_{k+1,p}(x_k) = h_{k,k+1}\left[T_{k+1,p}(x_k) - T_{k,p}(x_k)\right]$$

$$(4.86c)$$

$$\alpha_n T_{np}(x_n) + \beta_n k_n(x_n)T'_{np}(x_n) = 0 \qquad (4.86d)$$

and the functions $T_{kj}(x)$ satisfy the one-dimensional form of Equations (2.87) given as

$$\frac{d}{dx}\left[k_k(x)\frac{dT_{kj}(x)}{dx}\right] - d(x)T_{kj}(x) + P_{kj}(x) = (j+1)w_k(x)T_{k,j+1}(x)$$

$$\text{in } x_{k-1} < x < x_k, \qquad k = 1, 2, \ldots, n \quad (4.87\text{a})$$

$$\alpha_0 T_{1j}(x_0) - \beta_0 k_1(x_0)T'_{1j}(x_0) = \phi_j(x_0) \qquad (4.87\text{b})$$

$$k_k(x_k)T'_{kj}(x_k) = k_{k+1}(x_k)T'_{k+1,j}(x_k) = h_{k,k+1}\left[T_{k+1,j}(x_k) - T_{k,j}(x_k)\right]$$

$$(4.87\text{c})$$

$$\alpha_n T_{nj}(x_n) + \beta_n k_n(x_n)T'_{nj}(x_n) = \phi_j(x_n) \qquad (4.87\text{d})$$

where $j = q - 1, q - 2, \ldots, 1, 0$.

Finally, the function $T_{kt}(x, t)$, $k = 1, 2, \ldots, n$ satisfies the one-dimensional version of the transient homogeneous problem (2.88). The general solution of problem (2.88) is given by Equation (2.89). Therefore, the solution for the functions $T_{kt}(x, t)$ is readily obtainable from Equation (2.89) by restricting it to the one-dimensional case. We find

$$T_{kt}(x, t) = \sum_{i=1}^{\infty} \frac{\psi_{ki}(x)}{N_i} e^{-\mu_i^2 t}$$

$$\times \left\{ \tilde{f}_i - \frac{1}{\mu_i^2 - d_\phi}\left[\phi_e(x_0)\Omega_i(x_0) + \phi_e(x_n)\Omega_i(x_n)\right] \right.$$

$$- \frac{1}{\mu_i^2 - d_p}\sum_{k=1}^{n}\int_{x_{k-1}}^{x_k}\psi_{ki}(x)P_{ke}(x)\,dx - \sum_{j=0}^{q}(-1)^j\frac{j!}{\mu_i^{2(j+1)}}$$

$$\times \left[\phi_j(x_0)\Omega_i(x_0) + \phi_j(x_n)\Omega_i(x_n)\right.$$

$$\left.\left. + \sum_{k=1}^{n}\int_{x_{k-1}}^{x_k}\psi_{ki}(x)P_{kj}(x)\,dx\right]\right\} \qquad (4.88)$$

Here $\Omega_i(x_k)$ is as defined by Equation (4.82e, f).

Boundary Condition of the Second Kind at Both Outer Boundaries

If $\alpha_0 = \alpha_n = 0$, $\beta_0 = \beta_n = 1$, that is, the boundary conditions at $x = x_0$ and $x = x_n$ are both of the second kind and, in addition, $d(x) = 0$, the solution of the problem (4.81) is obtainable from the general solution (2.91) by restricting it to the one-dimensional case. We find

$$T_k(x, t) = T_{av}(t) + \sum_{i=1}^{\infty} \frac{1}{N_i} e^{-\mu_i^2 t} \psi_{ki}(x) \left[\tilde{f}_i + \int_0^t g_i(t') e^{\mu_i^2 t'} dt' \right] \quad (4.89a)$$

where $T_{av}(t)$ is the mean potential over the entire region given by

$$T_{av}(t) = \left\{ \sum_{k=1}^{n} \int_{x_{k-1}}^{x_k} w_k(x) f_k(x) \, dx \right.$$

$$\left. + \int_0^t \left[\phi(x_0, t') + \phi(x_n, t') + \sum_{k=1}^{n} \int_{x_{k-1}}^{x_k} P_k(x, t') \, dx \right] dt' \right\}$$

$$\times \left[\sum_{k=1}^{n} \int_{x_{k-1}}^{x_k} w_k(x) \, dx \right]^{-1} \quad (4.89b)$$

We now consider the problem (4.81) for the case of $\alpha_0 = \alpha_n = 0$, $\beta_0 = \beta_n = 1$, and $d(x) = 0$, but assume that the nonhomogeneous terms $\phi(x_0, t)$, $\phi(x_n, t)$, and $P_k(x, t)$ are exponentials and q-order polynomials of time as represented by Equations (4.83). Then the solution $T_k(x, t)$ of the problem (4.81) can be split up into the solutions of simpler problems as given by Equation (2.94) in the form

$$T_k(x, t) = T_{av}(t) + T_{k\phi}(x) e^{-d_\phi t} + T_{kp}(x) e^{-d_p t} + \sum_{j=0}^{q} T_{kj}(x) t^j + T_{kt}(x, t)$$

$$(4.90)$$

where the functions $T_{k\phi}(x)$, $T_{kp}(x)$, and $T_{kj}(x)$ are the solutions of the one-dimensional form of the steady-state problems (2.95). That is, the functions $T_{k\phi}(x)$ satisfy the one-dimensional form of the steady-state problem (2.95a) given as

$$\frac{d}{dx} \left[k_k(x) \frac{dT_{k\phi}(x)}{dx} \right] + d_\phi w_k(x) T_{k\phi}(x) = w_k(x) \frac{\phi_e(x_0) + \phi_e(x_n)}{\sum_{k=1}^{n} \int_{x_{k-1}}^{x_k} w_k(x) \, dx}$$

$$\text{in } x_{k-1} < x < x_k, \quad k = 1, 2, \ldots, n \quad (4.91)$$

subject to the boundary conditions (4.85b, c, d) with $\alpha_0 = \alpha_n = 0$, $\beta_0 = \beta_n = 1$, and the additional condition

$$\sum_{k=1}^{n} \int_{x_{k-1}}^{x_k} w_k(x) T_{k\phi}(x) \, dx = 0 \quad (4.91a)$$

The functions $T_{kp}(x)$ satisfy the one-dimensional form of the steady-state problem (2.95b) given as

$$\frac{d}{dx}\left[k_k(x)\frac{dT_{kp}(x)}{dx}\right] + d_p w_k(x)T_{kp}(x) + P_{ke}(x)$$

$$= w_k(x)\frac{\displaystyle\sum_{k=1}^{n}\int_{x_{k-1}}^{x_k} P_{ke}(x)\,dx}{\displaystyle\sum_{k=1}^{n}\int_{x_{k-1}}^{x_k} w_k(x)\,dx}$$

$$\text{in } x_{k-1} < x < x_k, \qquad k = 1,2,\dots,n \quad (4.92)$$

subject to the boundary conditions (4.86b, c, d) with $\alpha_0 = \alpha_n = 0$, $\beta_0 = \beta_n = 1$, and the additional condition of (4.91a).

The functions $T_{kj}(x)$ satisfy the one-dimensional form of the steady-state problem (2.95d) given as

$$\frac{d}{dx}\left[k_k(x)\frac{dT_{kj}(x)}{dx}\right] + P_{kj}(x)$$

$$= \frac{\phi_j(x_0) + \phi_j(x_n) + \displaystyle\sum_{k=1}^{n}\int_{x_{k-1}}^{x_k} P_{kj}(x)\,dx}{\displaystyle\sum_{k=1}^{n}\int_{x_{k-1}}^{x_k} w_k(x)\,dx} w_k(x)$$

$$+ (j+1)w_k(x)T_{k,j+1}(x)$$

$$\text{in } x_{k-1} \le x \le x_k, \qquad k = 1,2,\dots,n \quad (4.93)$$

and subject to the boundary conditions (4.87b, c, d) with $\alpha_0 = \alpha_n = 0$, $\beta_0 = \beta_n = 1$, and the additional condition

$$\sum_{k=1}^{n}\int_{x_{k-1}}^{x_k} w_k(x)T_{kj}(x)\,dx = 0$$

Also, in the problem (4.93) we have $j = q - 1, q - 2, \dots, 1, 0$ and $T_{k,q+1}(x) = 0$.

Finally, the transient part of the solution, the function $T_{kt}(x, t)$, is exactly the same function given by Equation (4.88).

4.3   ONE-DIMENSIONAL PROBLEMS OF CLASS III

The mathematical formulation of the boundary value problems of Class III for the general three-dimensional case was given in Chapter 2 by Equations (2.96). In this section we examine the solution of such problems for the one-dimensional finite region. Taking the one-dimensional form of the operators L, B_k, B_{k1}, and B_{k2} as given by Equations (3.114), the one-dimensional form of the problem (2.96) becomes

$$w_k(x)\frac{\partial T_k(x, t)}{\partial t} = \frac{\partial}{\partial x}\left[k_k(x)\frac{\partial T_k(x, t)}{\partial x}\right] - d_k(x) T_k(x, t) + P_k(x, t)$$

$$\text{in } x_0 < x < x_1, \qquad t > 0 \quad (4.94\text{a})$$

$$\alpha_k T_k(x_0, t) - \beta_k k_k(x_0)\frac{\partial T_k(x_0, t)}{\partial x} = \phi_k(x_0, t) \qquad (4.94\text{b})$$

$$\alpha_{k1} T_1(x_1, t) + \alpha_{k2} T_2(x_2, t) + \beta_{k1} k_1(x_1)\frac{\partial T_1(x_1, t)}{\partial x}$$

$$+ \beta_{k2} k_2(x_1)\frac{\partial T_2(x_1, t)}{\partial x} = \phi_k(x_1, t) \qquad (4.94\text{c})$$

$$T_k(x, 0) = f_k(x) \qquad (4.94\text{d})$$

where $k = 1, 2$.

The solution of problem (4.94) is readily obtained from the general solution (2.109) by restricting it to the one-dimensional case. We find

$$T_k(x, t) = \sum_{i=1}^{\infty} \frac{1}{N_i} e^{-\mu_i^2 t} \psi_{ki}(x)\left[\tilde{f}_i + \int_0^t g_i(t') e^{\mu_i^2 t'}\, dt'\right], \qquad k = 1, 2 \quad (4.95\text{a})$$

where N_i is obtained from the one-dimensional form of Equation (2.101b) as

$$N_i = \frac{1}{2\mu_i} \sum_{k=1}^{2} \sigma_k \left\{ k_k(x_1) \left| \begin{array}{ccc} \left[\dfrac{\partial \psi_k(\mu, x_1)}{\partial \mu}\right]_{\mu=\mu_i} & \left[\dfrac{\partial^2 \psi_k(\mu, x_1)}{\partial x\, \partial \mu}\right]_{\mu=\mu_i} \\[2mm] \psi_k(\mu_i, x_1) & \dfrac{\partial \psi_k(\mu_i, x_1)}{\partial x} \end{array} \right| \right.$$

$$\left. - k_k(x_0) \left| \begin{array}{cc} \left[\dfrac{\partial \psi_k(\mu, x_0)}{\partial \mu}\right]_{\mu=\mu_i} & \left[\dfrac{\partial^2 \psi_k(\mu, x_0)}{\partial x\, \partial \mu}\right]_{\mu=\mu_i} \\[2mm] \psi_k(\mu_i, x_0) & \dfrac{\partial \psi_k(\mu_i, x_0)}{\partial x} \end{array} \right| \right\}$$

$$(4.95\text{b})$$

and σ_k, defined by Equation (2.100b), is written more compactly as

$$\sigma_k = \alpha_{3-k, k}\beta_{k, k} - \alpha_{k, k}\beta_{3-k, k} \qquad (4.95\text{c})$$

$\tilde{f}_i$ is obtained from Equation (2.108) as

$$\tilde{f}_i = \sum_{k=1}^{2} \sigma_k \int_{x_0}^{x_1} w_k(x)\psi_{ki}(x)f(x)\,dx \qquad (4.95d)$$

and in obtaining $g_i(t)$ from Equation (2.107b) we note that the boundary $x = x_1$ is associated with the boundary surface s_1 and the boundary $x = x_0$ with the boundary surface s_2 of the general three-dimensional problem. We find

$$g_i(t) = \begin{vmatrix} \phi_1(x_1, t) & \alpha_{11}\psi_{1i}(x_1) + \beta_{11}k_1(x_1)\psi_{1i}'(x_1) \\ \phi_2(x_1, t) & \alpha_{21}\psi_{1i}(x_1) + \beta_{21}k_1(x_1)\psi_{1i}'(x_1) \end{vmatrix}$$

$$+ \sum_{k=1}^{2} \sigma_k \left[\phi_k(x_0, t)\Omega_{ki}(x_0) + \int_{x_0}^{x_1} \psi_{ki}(x)P_k(x, t)\,dx \right] \qquad (4.95e)$$

Here $\Omega_{ki}(x_0)$ is defined as

$$\Omega_{ki}(x_0) = \frac{\psi_{ki}(x_0) + k_k(x_0)\psi_{ki}'(x_0)}{\alpha_k + \beta_k} \qquad (4.95f)$$

Splitting Up the Solution

When the nonhomogeneous terms $\phi_k(x_0, t)$, $\phi_k(x_1, t)$, and $P_k(x, t)$ of the problem (4.94) are exponentials and q-order polynomials of time in the form

$$\phi_k(x_0, t) = \phi_{ke}(x_0)e^{-d_\phi t} + \sum_{j=0}^{q} \phi_{kj}(x_0)t^j \qquad (4.96a)$$

$$\phi_k(x_1, t) = \phi_{ke}(x_1)e^{-d_\phi t} + \sum_{j=0}^{q} \phi_{kj}(x_1)t^j \qquad (4.96b)$$

$$P_k(x, t) = P_{ke}e^{-d_p t} + \sum_{j=0}^{q} P_{kj}(x)t^j \qquad (4.96c)$$

the solution of the problem (4.94) can be split up into the solution of simpler problems as given by Equation (2.111) in the form

$$T_k(x, t) = T_{k\phi}(x)e^{-d_\phi t} + T_{kp}(x)e^{-d_p t} + \sum_{j=0}^{q} T_{kj}(x)t^j + T_{kt}(x, t) \qquad (4.97)$$

where the functions $T_{k\phi}(x)$, $T_{kp}(x)$, and $T_{kj}(x)$ are the solutions of the one-dimensional form of the steady-state problems given by Equations (2.112)–(2.115). That is, the functions $T_{k\phi}(x)$ satisfy the one-dimensional form of the problem (2.112) given as

$$\frac{d}{dx}\left[k_k(x)\frac{dT_{k\phi}(x)}{dx}\right] + \left[d_\phi w_k(x) - d_k(x)\right]T_{k\phi}(x) = 0$$

$$\text{in } x_0 < x < x_1, \qquad k = 1,2 \quad (4.98a)$$

$$\alpha_k T_{k\phi}(x_0) - \beta_k k_k(x_0)T'_{k\phi}(x_0) = \phi_{ke}(x_0) \qquad (4.98b)$$

$$\alpha_{k1}T_{1\phi}(x_1) + \alpha_{k2}T_{2\phi}(x_1) + \beta_{k1}k_1(x_1)T'_{1\phi}(x_1)$$

$$+ \beta_{k2}k_2(x_1)T'_{2\phi}(x_1) = \phi_{ke}(x_1) \qquad (4.98c)$$

The functions $T_{kp}(x)$ satisfy the one-dimensional form of Equations (2.113) given as

$$\frac{d}{dx}\left[k_k(x)\frac{dT_{kp}(x)}{dx}\right] + \left[d_p w_k(x) - d_k(x)\right]T_{kp}(x) + P_{ke}(x) = 0$$

$$\text{in } x_0 \leq x \leq x_1, \qquad k = 1,2, \quad (4.99a)$$

$$\alpha_k T_{kp}(x_0) - \beta_k k_k(x_0)T'_{kp}(x_0) = 0 \qquad (4.99b)$$

$$\alpha_{k1}T_{kp}(x_1) + \alpha_{k2}T_{2p}(x_1) + \beta_{k1}k_1(x_1)T'_{1p}(x_1) + \beta_{k2}k_2(x_1)T'_{2p}(x_1) = 0$$

$$(4.99c)$$

and the functions $T_{kj}(x)$ satisfy the one-dimensional form of Equations (2.115) and (2.114) given as

$$\frac{d}{dx}\left[k_k(x)\frac{dT_{kj}(x)}{dx}\right] - d_k(x)\,T_{kj}(x) + P_{kj}(x) = (j+1)w_k(x)T_{k,\,j+1}(x)$$

$$\text{in } x_0 < x < x_1, \qquad k = 1,2 \quad (4.100a)$$

$$\alpha_k T_{kj}(x_0) - \beta_k k_k(x_0)T'_{kj}(x_0) = \phi_{kj}(x_0) \qquad (4.100b)$$

$$\alpha_{k1}T_{1j}(x_1) + \alpha_{k2}T_{2j}(x_1) + \beta_{k1}k_1(x_1)T'_{1j}(x_1)$$

$$+ \beta_{k2}k_2(x_1)T'_{2j}(x_1) = \phi_{kj}(x_1) \qquad (4.100c)$$

where $j = q - 1, q - 2, \ldots, 1, 0$, and $T_{k,\,q+1}(x) = 0$.

Finally, the transient part of the solution (4.97), the function $T_{kt}(x, t)$, is obtained from the one-dimensional form of solution (2.117); we find

$$T_{kt}(x, t) = \sum_{i=1}^{\infty} \frac{1}{N_i} e^{-\mu_i^2 t} \psi_{ki}(x)$$

$$\times \left\{ \tilde{f}_i - \frac{1}{\mu_i^2 - d_\phi} \begin{vmatrix} \phi_{1e}(x_1) & \alpha_{11}\psi_{1i}(x_1) + \beta_{11}k_1(x_1)\psi'_{1i}(x_1) \\ \phi_{2e}(x_1) & \alpha_{21}\psi_{1i}(x_1) + \beta_{21}k_1(x_1)\psi'_{1i}(x_1) \end{vmatrix} \right.$$

$$- \frac{1}{\mu_i^2 - d_\phi} \sum_{k=1}^{2} \sigma_k \phi_{ke} \Omega_{ki}(x_0)$$

$$- \frac{1}{\mu_i^2 - d_p} \sum_{k=1}^{2} \sigma_k \int_{x_0}^{x_1} \psi_{ki}(x) P_{ke}(x) \, dx$$

$$- \sum_{j=0}^{q} (-1)^j \frac{j!}{\mu_i^{2(j+1)}} \left[\begin{vmatrix} \phi_{1j}(x_1) & \alpha_{11}\psi_{1i}(x_1) + \beta_{11}k_1(x_1)\psi'_{1i}(x_1) \\ \phi_{2j}(x_1) & \alpha_{21}\psi_{1i}(x_1) + \beta_{21}k_1(x_1)\psi'_{1i}(x_1) \end{vmatrix} \right.$$

$$+ \left. \left. \sum_{k=1}^{2} \sigma_k \left| \phi_{kj}(x_0)\Omega_{ki}(x_0) + \int_{x_0}^{x_1} \psi_{ki}(x) P_{kj}(x) \, dx \right| \right] \right\} \qquad (4.101)$$

where N_i, $\tilde{f}_i$, and $\Omega_{ki}(x_0)$ are defined, respectively, by Equations (4.95b), (4.95d), and (4.95f).

Boundary Condition of the Second Kind at All Boundaries

If $\alpha_k = 0$, $\beta_k = 1$ for $k = 1, 2$ and $\alpha_{11}\alpha_{22} - \alpha_{12}\alpha_{21} = 0$, that is, the boundary conditions at $x = x_0$ and $x = x_1$ are both of the second kind, and, in addition, $d(x) = 0$, the solution of the problem (4.94) is obtainable from the general solution (2.119) by restricting it to the one-dimensional case. We find

$$T_k(x, t) = \frac{\psi_{k0}}{\sum\limits_{k=1}^{2} \sigma_k \int_{x_0}^{x_1} \psi_{k0}^2 w_k(x) \, dx}$$

$$\times \left\{ \sum_{k=1}^{2} \sigma_k \psi_{k0} \int_{x_0}^{x_1} w_k(x) f_k(x) \, dx + \int_0^t \left[\psi_{10} \begin{vmatrix} \phi_1(x_1, t') & \alpha_{11} \\ \phi_2(x_1, t') & \alpha_{21} \end{vmatrix} \right. \right.$$

$$+ \left. \left. \sum_{k=1}^{2} \sigma_k \psi_{k0} \left(\phi_k(x_0) + \int_{x_0}^{x_1} P_k(x, t') \, dx \right) \right] dt' \right\}$$

$$+ \sum_{i=1}^{\infty} \frac{1}{N_i} e^{-\mu_i^2 t} \psi_{ki}(x) \left[\tilde{f}_i + \int_0^t g_i(t') e^{\mu_i^2 t'} \, dt' \right] \qquad (4.102a)$$

where the constant eigenfunctions ψ_{10} and ψ_{20} corresponding to the zero eigenvalue ($\mu = 0$) must satisfy the relation given by Equation (2.118), that is,

$$\alpha_{k1}\psi_{10} + \alpha_{k2}\psi_{20} = 0, \qquad k = 1, 2 \qquad (4.102b)$$

and N_i and $\tilde{f}_i$ are defined by Equations (4.95b) and (4.95d), respectively.

4.4   ONE-DIMENSIONAL PROBLEMS OF CLASS IV

The mathematical formulation of the boundary value problems of Class IV for the general three-dimensional case was given in Chapter 2 by Equations (2.120). Taking the one-dimensional form of the differential operators L_k and B_k, the one-dimensional form of problem (2.120) becomes

$$w_k(x)\frac{\partial T_k(x, t)}{\partial t} = \frac{\partial}{\partial x}\left[k_k(x)\frac{\partial T_k(x, t)}{\partial x}\right] - d_k(x)\,T_k(x, t) + P_k(x, t)$$

$$+ b(x)\sum_{p=1}^{n}\alpha_{kp}\left[T_p(x, t) - T_k(x, t)\right]$$

$$\text{in } x_0 < x < x_1, \qquad t > 0 \quad (4.103a)$$

$$\alpha_{0k}T_k(x_0, t) - \beta_{0k}k_k(x_0)\frac{\partial T_k(x_0, t)}{\partial x} = \phi_k(x_0, t) \qquad (4.103b)$$

$$\alpha_{1k}T_k(x_1, t) + \beta_{1k}k_k(x_1)\frac{\partial T_k(x_1, t)}{\partial x} = \phi_k(x_1, t) \qquad (4.103c)$$

$$T_k(x, 0) = f_k(x) \qquad (4.103d)$$

where $k = 1, 2, \ldots, n$ and $\alpha_{kp} = \alpha_{pk}$.

The solution of problem (4.103) is obtainable from the general solution (2.128) by restricting it to the one-dimensional case. We find

$$T_k(x, t) = \sum_{i=1}^{\infty}\frac{1}{N_i}e^{-\mu_i^2 t}\psi_{ki}(x)\left[\tilde{f}_i + \int_0^t g_i(t')e^{\mu_i^2 t'}\,dt'\right], \qquad k = 1, 2, \ldots, n$$

$$(4.104a)$$

where N_i, $\tilde{f}_i$, and $g_i(t)$ are obtained, respectively, from the one-dimensional form of Equations (2.123b), (2.127), and (2.126c) as

$$N_i = \sum_{k=1}^{n} \int_{x_0}^{x_1} w_k(x)\psi_{ki}^2(x)\, dx \qquad (4.104b)$$

$$\tilde{f}_i = \sum_{k=1}^{n} \int_{x_0}^{x_1} w_k(x)\psi_{ki}(x)f_k(x)\, dx \qquad (4.104c)$$

$$g_i(t) = \sum_{k=1}^{n} \left[\phi_k(x_0, t)\Omega_{ki}(x_0) + \phi_k(x_1, t)\Omega_{ki}(x_1) \right.$$

$$\left. + \int_{x_0}^{x_1} \psi_{ki}(x')P_k(x', t)\, dx' \right] \qquad (4.104d)$$

here $\Omega_{ki}(x_0)$, $\Omega_{ki}(x_1)$ are defined as

$$\Omega_{ki}(x_0) = \frac{\psi_{ki}(x_0) + k_k(x_0)\psi_{ki}'(x_0)}{\alpha_{0k} + \beta_{0k}} \qquad (4.104e)$$

$$\Omega_{ki}(x_1) = \frac{\psi_{ki}(x_1) - k_k(x_1)\psi_{ki}'(x_1)}{\alpha_{1k} + \beta_{1k}} \qquad (4.104f)$$

The eigenfuntions $\psi_{ki}(x)$ and the eigenvalues μ_i needed in solution (4.104) are the eigenfunctions and eigenvalues of the eigenvalue problem (3.129) or of the one-dimensional form of the eigenvalue problem (2.121). The determination of the eigenfunctions and the eigenvalues of this eigenvalue problem was discussed in Section 3.4.

Boundary Condition of the Second Kind at All Boundaries

If $\alpha_{0k} = \alpha_{1k} = 0$, $\beta_{0k} = \beta_{1k} = 1$, that is, the boundary conditions at $x = x_0$ and $x = x_1$ are both of the second kind and $d(x) = 0$, the solution of problem (4.103) is obtainable from the one-dimensional form of the general solution (2.130) as

$$T_k(x, t) = \frac{\psi_{k0}}{\displaystyle\sum_{k=1}^{n} \psi_{k0}^2 \int_{x_0}^{x_1} w_k(x)\, dx}$$

$$\times \sum_{k=1}^{n} \psi_{k0} \left\{ \int_{x_0}^{x_1} w_k(x)f_k(x)\, dx + \int_0^t \left[\phi(x_0, t') + \phi(x_1, t') \right. \right.$$

$$\left. \left. + \int_{x_0}^{x_1} P_k(x, t')\, dx \right] dt' \right\}$$

$$+ \sum_{i=1}^{\infty} \frac{1}{N_i} e^{-\mu_i^2 t} \psi_{ki}(x) \left[\tilde{f}_i + \int_0^t g_i(t') e^{\mu_i^2 t'}\, dt' \right],$$

$$k = 1, 2, \ldots, n \quad (4.105a)$$

where the constant eigenfunctions ψ_{k0} corresponding to the zero eigenvalue $(\mu = 0)$ must satisfy Equations (2.129), that is,

$$\sum_{p=1}^{n} \alpha_{kp}(\psi_{p0} - \psi_{k0}) = 0, \qquad k = 1, 2, \ldots, n \qquad (4.105b)$$

and N_i, $\tilde{f}_i$, and $g_i(t)$ are defined by Equations (4.104b, c, d), respectively.

4.5   ONE-DIMENSIONAL PROBLEMS OF CLASS V

The mathematical formulation of the boundary value problems of Class V for the general three-dimensional case was given in Chapter 2 by Equations (2.131). Taking the one-dimensional form of the operators L_k and B_k, the one-dimensional form of problem (2.131) becomes

$$w_k(x)\frac{\partial T_k(x,t)}{\partial t} = \frac{\partial}{\partial x}\left[k_k(x)\frac{\partial T_k(x,t)}{\partial x}\right] - d_k(x)\,T_k(x,t) + P_k(x,t)$$

$$+ (-1)^k b(x)\left[\sigma_{k-1}T_1(x,t) - \sigma_{k+1}T_2(x,t)\right]$$

$$\text{in } x_0 < x < x_1, \qquad t > 0 \quad (4.106a)$$

$$\alpha_{0k}T_k(x_0,t) - \beta_{0k}k_k(x_0)\frac{\partial T_k(x_0,t)}{\partial x} = \phi_k(x_0,t) \qquad (4.106b)$$

$$\alpha_{1k}T_k(x_1,t) + \beta_{1k}k_k(x_1)\frac{\partial T_k(x_1,t)}{\partial x} = \phi_k(x_1,t) \qquad (4.106c)$$

$$T_k(x,0) = f_k(x) \qquad (4.106d)$$

where $k = 1, 2$.

The solution of problem (4.106) is obtainable from the general three-dimensional solution (2.140) by restricting it to the one-dimensional case. We find

$$T_k(x,t) = \sum_{i=1}^{\infty} \frac{1}{N_i}e^{-\mu_i^2 t}\psi_{ki}(x)\left[\tilde{f}_i + \int_0^t g_i(t')e^{\mu_i^2 t'}\,dt'\right], \qquad k = 1, 2 \quad (4.107a)$$

where N_i, $\tilde{f}_i$, and $g_i(t')$ are obtained, respectively, from the one-dimensional forms of equations (2.135b), (2.139c), and (2.139b). We find

$$
N_i = \frac{1}{2\mu_i} \sum_{k=1}^{2} \sigma_k \left\{ k_k(x_1) \left| \begin{array}{cc} \left[\dfrac{\partial \psi_k(\mu, x_1)}{\partial \mu} \right]_{\mu=\mu_i} & \left[\dfrac{\partial^2 \psi_k(\mu, x_1)}{\partial x\, \partial \mu} \right]_{\mu=\mu_i} \\[2ex] \psi_k(\mu_i, x_1) & \dfrac{\partial \psi_k(\mu_i, x_1)}{\partial x} \end{array} \right| \right.
$$

$$
\left. - k_k(x_0) \left| \begin{array}{cc} \left[\dfrac{\partial \psi_k(\mu, x_0)}{\partial \mu} \right]_{\mu=\mu_i} & \left[\dfrac{\partial^2 \psi_k(\mu, x_0)}{\partial x\, \partial \mu} \right]_{\mu=\mu_i} \\[2ex] \psi_k(\mu_i, x_0) & \dfrac{\partial \psi_k(\mu_i, x_0)}{\partial x} \end{array} \right| \right\}
$$

$$ \text{(4.107b)} $$

$$
\tilde{f}_i = \sum_{k=1}^{2} \sigma_k \int_{x_0}^{x_1} w_k(x) \psi_{ki}(x) f_k(x)\, dx \qquad \text{(4.107c)}
$$

$$
g_i(t) = \sum_{k=1}^{2} \sigma_k \left[\phi_k(x_0, t) \Omega_{ki}(x_0) + \phi_k(x_1, t) \Omega_{ki}(x_1) \right.
$$

$$
\left. + \int_{x_0}^{x_1} \psi_{ki}(x) P_k(x, t)\, dx \right] \qquad \text{(4.107d)}
$$

here $\Omega_{ki}(x_0)$ and $\Omega_{ki}(x_1)$ are defined as

$$
\Omega_{ki}(x_0) = \frac{\psi_{ki}(x_0) + k_k(x_0) \psi'_{ki}(x_0)}{\alpha_{0k} + \beta_{0k}} \qquad \text{(4.107e)}
$$

$$
\Omega_{ki}(x_1) = \frac{\psi_{ki}(x_1) - k_k(x_1) \psi'_{ki}(x_1)}{\alpha_{1k} + \beta_{1k}} \qquad \text{(4.107f)}
$$

The eigenfunctions $\psi_{ki}(x)$ and the eigenvalues μ_i needed for solution (4.107) are the eigenfunctions and eigenvalues of the eigenvalue problem (3.134) or the one-dimensional form of the eigenvalue problem (2.132). The determination of the eigenfunctions and eigenvalues of this eigenvalue problem was discussed in Section 3.5.

Boundary Condition of the Second Kind at All Boundaries

If $\alpha_{0k} = \alpha_{1k} = 0$, $\beta_{0k} = \beta_{1k} = 1$ for $k = 1, 2$, that is, the boundary conditions at $x = x_0$ and $x = x_1$ are both of the second kind and $d(x) = 0$, the solution of problem (4.106) is obtainable from the one-dimensional form of the general solution (2.142) as

$$T_k(x, t) = \frac{\psi_{k0}}{\sum\limits_{k=1}^{2} \sigma_k \psi_{k0}^2 \int_{x_0}^{x_1} w_k(x)\, dx}$$

$$\times \sum_{k=1}^{2} \sigma_k \psi_{k0} \left\{ \int_{x_0}^{x_1} w_k(x) f_k(x)\, dx + \int_{0}^{t} \left[\phi_k(x_0, t') \right.\right.$$

$$\left.\left. + \phi_k(x_1, t') + \int_{x_0}^{x_1} P_k(x', t')\, dx' \right] dt' \right\}$$

$$+ \sum_{i=1}^{\infty} \frac{1}{N_i} e^{-\mu_i^2 t}\, \psi_{ki}(x) \left[\tilde{f}_i + \int_{0}^{t} g_i(t') e^{\mu_i^2 t'}\, dt' \right], \qquad k = 1, 2 \quad (4.108a)$$

where the constant eigenfunctions ψ_{k0} corresponding to the zero eigenvalue ($\mu = 0$) must satisfy Equation (2.141), that is,

$$\sigma_{k-1}\psi_{10} - \sigma_{k+1}\psi_{20} = 0, \qquad k = 1, 2 \qquad (4.108b)$$

and N_i, $\tilde{f}_i$, and $g_i(t')$ are defined by Equations (4.107b, c, d), respectively.

4.6   ONE-DIMENSIONAL PROBLEMS OF CLASS VI

The mathematical formulation of the boundary value problems of Class VI for the one-dimensional case is immediately obtainable from the general three-dimensional Equations (2.143) as

$$w_k(x)\frac{\partial T_k(x, t)}{\partial t} = \frac{\partial}{\partial x}\left[k_k(x)\frac{\partial T_k(x, t)}{\partial x} \right] + P_k(x, t)$$

$$\text{in } x_0 < x < x_1, \qquad t > 0, \qquad k = 1, 2, \ldots, n \quad (4.109a)$$

$$\frac{\partial T_k(x_0, t)}{\partial x} = 0 \qquad (4.109b)$$

$$\alpha_k T_k(x_1, t) + \beta_k k_k(x_1)\frac{\partial T_k(x_1, t)}{\partial x} = \phi(t) \qquad (4.109c)$$

$$\frac{d\phi(t)}{dt} + \sum_{k=1}^{n} \gamma_k k_k(x_1)\frac{\partial T_k(x_1, t)}{\partial x} = Q(t) \qquad (4.109d)$$

$$T_k(x, 0) = f_k(x) \qquad (4.109e)$$

$$\phi(0) = \phi_0 \qquad (4.109f)$$

where γ_k (one dimensional) $\equiv \gamma_k$ (three dimensional) $\times S_k$.

The solution of problem (4.109) is obtainable from Equation (2.160) by restricting it to the one-dimensional case. We find

$$
T_k(x, t) = \left[\alpha_k + \alpha_k \sum_{k=1}^{n} \frac{\gamma_k}{\alpha_k} \int_{x_0}^{x_1} w_k(x)\, dx \right]^{-1}
$$

$$
\times \left\{ \phi_0 + \int_0^t Q(t')\, dt' + \sum_{k=1}^{n} \gamma_k \left[\int_{x_0}^{x_1} w_k(x) f_k(x)\, dx \right. \right.
$$

$$
\left. \left. + \int_0^t \int_{x_0}^{x_1} P_k(x, t')\, dx\, dt' \right] \right\}
$$

$$
+ \sum_{i=1}^{\infty} \frac{1}{N_i} e^{-\mu_i^2 t} \psi_{ki}(x) \left\{ \sum_{k=1}^{n} \alpha_k \gamma_k \left[\int_{x_0}^{x_1} w_k(x) \psi_{ki}(x) f_k(x)\, dx \right. \right.
$$

$$
\left. + \int_0^t e^{\mu_i^2 t'} \int_{x_0}^{x_1} \psi_{ki}(x') P_k(x, t')\, dx\, dt' \right]
$$

$$
\left. + \left[\phi_0 + \int_0^t e^{\mu_i^2 t'} Q(t')\, dt' \right] \left[\sum_{k=1}^{n} \gamma_k \frac{1}{\mu_i^2} k_k(x_1) \psi_k'(\mu_i, x_1) \right] \right\},
$$

$$
k = 1, 2, \ldots, n \quad (4.110)
$$

Here we utilized Equation (3.141d) to obtain the last term, $(1/\mu_i^2)$ $k_k(x_1)\psi_k'(\mu_i, x_1)$. The eigenfunctions $\psi_{ki}(x)$ and the eigenvalues μ_i needed for solution (4.110) are the eigenfunctions and eigenvalues of the eigenvalue problem (3.141). A discussion of the determination of these eigenvalues and eigenfunctions is given in Section 3.6.

4.7   ONE-DIMENSIONAL PROBLEMS OF CLASS VII

The mathematical formulation of the boundary value problems of Class VII for the general three-dimensional case was given by Equations (2.161). For the one-dimensional case the system becomes

$$
w(x) \left[\frac{\partial T_1(x, t)}{\partial t} + \frac{\partial T_2(x, t)}{\partial t} \right] = \frac{\partial}{\partial x} \left[k(x) \frac{\partial T_1(x, t)}{\partial x} \right] + P(x, t)
$$

$$
\text{in } x_0 < x < x_1, \qquad t > 0 \quad (4.111a)
$$

$$
\frac{\partial T_2(x, t)}{\partial t} = \sigma_1 T_1(x, t) - \sigma_2 T_2(x, t) \quad \text{in } x_0 < x < x_1, \quad t > 0
$$

$$
(4.111b)
$$

$$\frac{\partial T_1(x_0, t)}{\partial x} = 0 \qquad (4.111c)$$

$$\alpha T_1(x_1, t) + \beta k(x_1) \frac{\partial T_1(x_1, t)}{\partial x} = \phi(t) \qquad (4.111d)$$

$$\frac{d\phi(t)}{dt} + \gamma k(x_1) \frac{\partial T_1(x_1, t)}{\partial x} = 0 \qquad (4.111e)$$

$$T_k(x, 0) = f_k(x), \qquad k = 1, 2 \quad \text{in } x_0 \le x \le x_1 \qquad (4.111f)$$

$$\phi(0) = \phi_0 \qquad (4.111g)$$

The solution of problem (4.111) is obtainable from the one-dimensional form of the general solution (2.179) as

$$T_k(x, t) = \left\{ \phi_0 + \gamma \int_{x_0}^{x_1} w(x)[f_1(x) + f_2(x)] \, dx + \gamma \int_0^t \int_{x_0}^{x_1} P(x, t') \, dx \, dt' \right\}$$

$$\times \left(\frac{\sigma_1 \sigma_2}{\sigma_k} \right) \times \left\{ \alpha \sigma_2 + \gamma(\sigma_1 + \sigma_2) \int_{x_0}^{x_1} w(x) \, dx \right\}^{-1}$$

$$+ \sum_{i=1}^{\infty} \frac{1}{N_i} e^{-\mu_i^2 t} \psi_{ki}(x)$$

$$\times \left\{ \alpha \sum_{k=1}^{2} \sigma_k \int_{x_0}^{x_1} w(x) \psi_{ki}(x) f_k(x) \, dx \right.$$

$$- \phi_0 \sigma_1 \int_{x_0}^{x_1} w(x)[\psi_{1i}(x) + \psi_{2i}(x)] \, dx$$

$$\left. + \alpha \sigma_1 \int_0^t \int_{x_0}^{x_1} \psi_{1i}(x) P(x, t') e^{\mu_i^2 t'} \, dx \, dt' \right\}, \qquad k = 1, 2 \quad (4.112)$$

where N_i is defined by Equation (2.172b), that is,

$$N_i = \alpha \sum_{k=1}^{2} \sigma_k \int_{x_0}^{x_1} w(x) \psi_{ki}^2(x) \, dx + \gamma \sigma_1 \left\{ \int_{x_0}^{x_1} w(x)[\psi_{1i}(x) + \psi_{2i}(x)] \, dx \right\}^2$$

$$(4.113)$$

The eigenfunctions $\psi_{ki}(x)$ and the eigenvalues μ_i needed for solution (4.112) are the eigenvalues and eigenfunctions of the eigenvalue problem (3.149).

REFERENCES

1. H. B. Dwight, *Tables of Integrals and Other Mathematical Data*, MacMillan, New York, 1961.

2. M. Abramowitz and I. A. Stegun, *Handbook of Mathematical Functions*, National Bureau of Standards, Applied Mathematic Series 55, U.S. Government Printing Office, Washington, D.C., 1964.

3. E. Jahnke and F. Emde, *Tables of Functions*, 2nd. ed., Dover, New York, 1945.

4. I. S. Gradshteyn and I. M. Ryzhik, *Table of Integrals, Series, and Products*, Academic, New York, 1965.

CHAPTER FIVE

Class I Solutions Applied to Steady-State Diffusion

Example is always more efficacious than precept.

Samuel Johnson
1709–1784

In this and the following three chapters we practice what we preached on the general solution of Class I problems. The one and multidimensional steady-state diffusion in the rectangular, cylindrical, and spherical coordinates subject to different types of boundary conditions is the principal topic of this chapter. All the examples considered here are special cases of Class I problems; therefore, the solutions are obtained directly from the general solutions already developed by appropriate simplifications. The objective of this chapter is to illustrate how the general solutions given previously can be utilized to provide answers to a vast number of simpler problems.

5.1 STATEMENT OF THE STEADY-STATE CLASS I PROBLEMS

The steady-state Class I problem is defined by Equation (2.50) as follows

$$\nabla \cdot [k(\mathbf{x}) \nabla T(\mathbf{x})] - d(\mathbf{x}) T(\mathbf{x}) + P(\mathbf{x}) = 0, \qquad \mathbf{x} \in V \qquad (5.1a)$$

$$\alpha(\mathbf{x}) T(\mathbf{x}) + \beta(\mathbf{x}) k(\mathbf{x}) \frac{\partial T(\mathbf{x})}{\partial \mathbf{n}} = \phi(\mathbf{x}), \qquad \mathbf{x} \in S \qquad (5.1b)$$

When $k(\mathbf{x}) = k = $ constant, $d(\mathbf{x}) = 0$, and $P(\mathbf{x}) = 0$, problem (5.1) becomes

$$\nabla^2 T(\mathbf{x}) = 0, \qquad \mathbf{x} \in V \qquad (5.2a)$$

$$\left\{ \alpha(\mathbf{x}) + \beta^*(\mathbf{x}) \frac{\partial}{\partial \mathbf{n}} \right\} T(\mathbf{x}) = \phi(\mathbf{x}), \qquad \mathbf{x} \in S \qquad (5.2b)$$

where $\beta^*(x) = \beta(x)k$ and ∇^2 is the Laplacian operator in general in the three-dimensional domain.

We now present Equations (5.2) in the explicit form in the three-dimensional domain for the rectangular, cylindrical, and spherical coordinate systems.

The Rectangular Coordinates

For a finite region confined to the domain bounded by the planes $x = x_k$, $y = y_k$, and $z = z_k$, $k = 0,1$ and subject to the boundary conditions of the third kind at all boundaries, the mathematical formulation is given by

$$\left\{\frac{\partial^2}{\partial x^2} + \frac{\partial^2}{\partial y^2} + \frac{\partial^2}{\partial z^2}\right\}T(x, y, z) = 0$$

$$\text{in } x_0 < x < x_1, \quad y_0 < y < y_1, \quad z_0 < z < z_1 \quad (5.3a)$$

$$\left\{\alpha_{kx} - (-1)^k \beta_{kx}\frac{\partial}{\partial x}\right\}T(x_k, y, z) = \phi(x_k, y, z), \quad k = 0,1$$

$$(5.3b, c)$$

$$\left\{\alpha_{ky} - (-1)^k \beta_{ky}\frac{\partial}{\partial y}\right\}T(x, y_k, z) = \phi(x, y_k, z), \quad k = 0,1$$

$$(5.3d, e)$$

$$\left\{\alpha_{kz} - (-1)^k \beta_{kz}\frac{\partial}{\partial z}\right\}T(x, y, z_k) = \phi(x, y, z_k), \quad k = 0,1$$

$$(5.3f, g)$$

where $\alpha_{k\ell}, \beta_{k\ell}$ ($\ell = x, y, z$ and $k = 0,1$) are the prescribed boundary surface coefficients.

The Cylindrical Coordinates

For a finite region in the cylindrical coordinate system (r, ϕ, z) confined to the domain $r_0 \le r \le r_1$, $0 \le \phi \le 2\pi$, and $z_0 \le z \le z_1$, and the boundary conditions of the third kind for all boundaries, the mathematical formulation is given by

$$\left\{ \frac{1}{r}\frac{\partial}{\partial r}\left(r\frac{\partial}{\partial r} \right) + \frac{1}{r^2}\frac{\partial^2}{\partial \phi^2} + \frac{\partial^2}{\partial z^2} \right\} T(r, \phi, z) = 0$$

$$\text{in } r_0 < r < r_1, \qquad 0 \le \phi \le 2\pi, \qquad z_0 < z < z_1 \quad (5.4a)$$

$$\left\{ \alpha_{kr} - (-1)^k \beta_{kr}\frac{\partial}{\partial r} \right\} T(r_k, \phi, z) = \phi(r_k, \phi, z), \qquad k = 0,1 \quad (5.4b, c)$$

$$\left\{ \alpha_{kz} - (-1)^k \beta_{kz}\frac{\partial}{\partial z} \right\} T(r, \phi, z_k) = \phi(r, \phi, z_k), \qquad k = 0,1 \quad (5.4d, e)$$

where $\alpha_{k\ell}, \beta_{k\ell}(\ell \equiv r, z$ and $k = 0,1)$ are prescribed constant coefficients.

The Spherical Coordinates

For a finite region in the spherical coordinate system (r, θ, ϕ) confined to the domain $0 \le r \le r_1$, $0 \le \theta \le \pi$, and $0 \le \phi \le 2\pi$ and subject to the boundary condition of the third kind at the surface $r = r_1$, the mathematical formulation is given by

$$\left\{ \frac{\partial}{\partial r}\left(r^2\frac{\partial}{\partial r} \right) + \frac{1}{\sin \theta}\frac{\partial}{\partial \theta}\left(\sin \theta \frac{\partial}{\partial \theta} \right) + \frac{1}{\sin^2\theta}\frac{\partial^2}{\partial \phi^2} \right\} T(r, \theta, \phi) = 0$$

$$\text{in } 0 < r < r_1, \qquad 0 \le \theta \le \pi, \qquad 0 \le \phi \le 2\pi \quad (5.5a)$$

This equation is put into a more convenient form by defining a new variable z as

$$z = \cos \theta \qquad (5.5b)$$

Then Equation (5.5a) becomes

$$\left\{ \frac{\partial}{\partial r}\left(r^2\frac{\partial}{\partial r} \right) + \frac{\partial}{\partial z}\left[(1 - z^2)\frac{\partial}{\partial z} \right] + \frac{1}{1 - z^2}\frac{\partial^2}{\partial \phi^2} \right\} T(r, z, \phi) = 0$$

$$\text{in } 0 < r < r_1, \qquad -1 < z < 1, \qquad 0 \le \phi \le 2\pi \quad (5.5c)$$

and the boundary condition for Equation (5.5c) at $r = r_1$ is taken as

$$\left\{ \alpha_{r_1} + \beta_{r_1}\frac{\partial}{\partial r} \right\} T(r_1, z, \phi) = \phi(r_1, z, \phi) \qquad (5.5d)$$

Special Cases

For the one-dimensional case, Equations (5.3a), (5.4a), and (5.5a), with $k =$ constant, respectively, reduce to

$$\frac{d^2T(x)}{dx^2} = 0 \quad \text{in } x_0 < x < x_1 \tag{5.6a}$$

$$\frac{d}{dr}\left[r\frac{dT(r)}{dr}\right] = 0 \quad \text{in } r_0 < r < r_1 \tag{5.6b}$$

$$\frac{d}{dr}\left[r^2\frac{dT(r)}{dr}\right] = 0 \quad \text{in } r_0 < r < r_1 \tag{5.6c}$$

The one-dimensional form of Equations (5.1), for the case of $k(\mathbf{x}) \equiv k(x)$ and $P(\mathbf{x}) \equiv g(x)$ takes the form

$$\frac{d}{dx}\left\{x^nk(x)\frac{dT(x)}{dx}\right\} + x^ng(x) = 0 \quad \text{in } a < x < b \tag{5.7a}$$

$$\delta_a T(a) - \gamma_a k(a)\frac{dT(a)}{dx} = \phi_a \tag{5.7b}$$

$$\delta_b T(b) + \gamma_b k(b)\frac{dT(b)}{dx} = \phi_b \tag{5.7c}$$

where

$$n = \begin{cases} 0 & \text{slab} \\ 1 & \text{cylinder} \\ 2 & \text{sphere} \end{cases}$$

In the preceding formulation we used the notation $T(x)$ and $k(x)$ to characterize steady-state temperature distribution and thermal conductivity, respectively. The boundary conditions of the first, second, and third kinds are readily obtainable by proper choice of the values of the coefficients δ_i and γ_i ($i = a$ or b) as specified by Equations (5.7b, c).

5.2   ONE-DIMENSIONAL STEADY-STATE DIFFUSION

We first present a general solution of one-dimensional steady-state problems for a slab, cylinder, and sphere subject to the boundary conditions of the third kind at both boundaries, and then illustrate its application to the solution of specific problems.

The problem given by Equations (5.7) is a special case of the steady-state problem (4.23). The correspondence of various terms between these two problems is given by:

$$T_j(x) = T(x), \quad k(x) = x^n k(x), \quad d(x) = 0$$

$$P_j(x) = x^n g(x), \quad w(x) = 0, \quad x_0 = a, \quad x_1 = b$$

$$\alpha_0 = \delta_a, \quad \alpha_1 = \delta_b, \quad \beta_0 = a^{-n}\gamma_a, \quad \beta_1 = b^{-n}\gamma_b$$

$$\phi_j(x_0) = \phi_a, \quad \text{and} \quad \phi_j(x_1) = \phi_b \qquad (5.8)$$

Then the solution of problem (5.7) is immediately obtained from solution (4.27) of problem (4.23) as

$$T(x) = \left\{ \left[b^{-n}\gamma_b + \delta_b \int_x^b \frac{x'^{-n}}{k(x')} dx' \right] \phi_a + \left[a^{-n}\gamma_a + \delta_a \int_a^x \frac{x'^{-n}}{k(x')} dx' \right] \right.$$

$$\times \left[\phi_b + b^{-n}\gamma_b \int_a^b x^n g(x)\, dx + \delta_b \int_a^b \frac{x^{-n}}{k(x)} \int_a^x x'^n g(x')\, dx'\, dx \right] \right\}$$

$$\times \left\{ b^{-n}\delta_a\gamma_b + a^{-n}\delta_b\gamma_a + \delta_a\delta_b \int_a^b \frac{x^{-n}}{k(x)} dx \right\}^{-1}$$

$$- \int_a^x \frac{x'^{-n}}{k(x')} \int_a^{x'} x''^n g(x'')\, dx''\, dx' \qquad (5.9)$$

The heat flux $q(x)$ is determined from its definition

$$q(x) = -k(x)\frac{dT(x)}{dx} \qquad (5.10)$$

and we find

$$x^n q(x) = \left\{ \delta_b\phi_a - \delta_a \left[\phi_b + b^{-n}\gamma_b \int_a^b x^n g(x)\, dx \right. \right.$$

$$\left. \left. + \delta_b \int_a^b \frac{x^{-n}}{k(x)} \int_a^x x'^n g(x')\, dx'\, dx \right] \right\}$$

$$\times \left\{ b^{-n}\delta_a\gamma_b + a^{-n}\delta_b\gamma_a + \delta_a\delta_b \int_a^b \frac{x^{-n}}{k(x)} dx \right\}^{-1} + \int_a^x x^n g(x)\, dx$$

$$(5.11)$$

The total heat flow rate Q is determined by multiplying the heat flux $q(x)$ by the total area. We now consider the total heat flow rate Q through an area A of a slab, through an area at any radius $r = x$ of a cylinder of length H, and through an area at any radial position $r = x$ of a sphere, given by

$$\text{slab:} \qquad Q = q(x)A \qquad (5.12a)$$

$$\text{cylinder:} \qquad Q = q(x)2\pi x H \qquad (5.12b)$$

$$\text{sphere:} \qquad Q = q(x)4\pi x^2 \qquad (5.12c)$$

These three expressions given by Equations (5.12) can be combined into a single expression as

$$Q_n = q(x)x^n F_n \qquad (5.13)$$

where

$$F_0 = A, \qquad F_1 = 2\pi H, \qquad F_2 = 4\pi \qquad (5.14)$$

$n = 0, 1, 2$ refer to slab, cylinder, and sphere, respectively, and the quantity $x^n q(x)$ is given by Equation (5.11). Then the total heat transfer rate Q is readily determined according to Equation (5.13).

The general solutions given by Equations (5.9) and (5.11) for the temperature $T(x)$ and the heat flux $q(x)$, respectively, are applicable to any of the nine combinations of the boundary conditions at $x = a$ and $x = b$, except the case when *both boundary conditions are of the second kind*, that is,

$$\delta_a = \delta_b = 0, \qquad \gamma_a = \gamma_b = 1$$

Then ϕ_a and ϕ_b represent the externally applied heat fluxes at the boundary surfaces $x = a$ and $x = b$, respectively. For this particular case the solutions (5.9) and (5.11) are not defined. The implications of this particular situation is better envisioned if we consider, say, a sphere in which heat is generated continuously while the boundary surface is kept insulated (i.e., $\phi_a = \phi_b = 0$). The steady-state condition can never be established in such a problem because heat generated in the medium cannot escape through the boundaries.

Examples of One-Dimensional Problems

Example 5.1 Determine the steady-state temperature distribution in a slab, cylinder, and sphere confined to the region $a \le x \le b$ for the following situation: There is no energy generation in the medium and the boundary surfaces at $x = a$ and $x = b$ are subjected to convection with heat transfer coefficients h_a and h_b into environments at temperatures $T_{a\infty}$ and $T_{b\infty}$, respectively. The thermal conductivity of the material is assumed to depend on the position, that is, $k(x)$.

Solution   The mathematical formulation of this problem is given by

$$\frac{d}{dx}\left\{x^n k(x)\frac{dT(x)}{dx}\right\} = 0 \qquad (5.15a)$$

$$T(a) - \frac{1}{h_a}k(a)\frac{dT(a)}{dx} = T_{a\infty} \qquad (5.15b)$$

$$T(b) + \frac{1}{h_b}k(b)\frac{dT(b)}{dx} = T_{b\infty} \qquad (5.15c)$$

Then the solution is immediately obtained from Equation (5.9) by setting the various parameters as

$$\delta_a = \delta_b = 1, \qquad \gamma_a = \frac{1}{h_a}, \qquad \gamma_b = \frac{1}{h_b}$$

$$\phi_a = T_{a\infty}, \qquad \phi_b = T_{b\infty}, \qquad g(x) = 0 \qquad (5.16)$$

The solution of the problem (5.15) is determined as

$$T(x) = \left\{\left[\frac{1}{b^n h_b} + \int_x^b \frac{dx'}{x'^n k(x')}\right]T_{a\infty} + \left[\frac{1}{a^n h_a} + \int_a^x \frac{dx'}{x'^n k(x')}\right]T_{b\infty}\right\}$$

$$\times \left\{\frac{1}{a^n h_a} + \frac{1}{b^n h_b} + \int_a^b \frac{dx}{x^n k(x)}\right\}^{-1} \qquad (5.17)$$

The total heat flow rate Q_n through the medium is determined by utilizing the relations given by Equations (5.11) and (5.13) and setting various parameters as specified by Equations (5.16). We find

$$Q_n = \frac{T_{a\infty} - T_{b\infty}}{(1/a^n F_n h_a) + (1/F_n)\int_a^b [1/x^n k(x)]\,dx + (1/b^n F_n h_b)} \qquad (5.18)$$

By utilizing the *thermal resistance* concept with analogy between the flow of heat and the flow of electric current, this expression can be written more compactly in the form

$$Q_n = \frac{T_{a\infty} - T_{b\infty}}{R_a + R_m + R_b} \qquad (5.19a)$$

where various thermal resistances are defined as

$$R_a = \frac{1}{a^n F_n h_a}, \qquad R_b = \frac{1}{b^n F_n h_b} \qquad (5.19b, c)$$

$$R_m = \frac{1}{F_n}\int_a^b \frac{dx}{x^n k(x)} \qquad (5.19d)$$

and $n = 0, 1, 2$ refers, respectively, to slab, cylinder, and sphere; the coefficients F_n are given by Equation (5.14); that is,

$$F_0 = A, \qquad F_1 = 2\pi H, \qquad F_2 = 4\pi \qquad (5.19e)$$

Clearly, the terms R_a, R_m, and R_b refer, respectively, to the thermal resistances to heat flow at the boundary surface $x = a$, through the medium itself, and at the boundary surface $x = b$. We now summarize the thermal resistances for a slab, cylinder, and sphere.

Slab

The thermal resistances for heat flow through an area A of a slab of thickness L, subjected to convection at the boundaries $x = 0$ and $x = a$ are obtained from Equations (5.19b, c, d) as

$$R_a = \frac{1}{h_a A}, \qquad R_0 \equiv R_{\text{slab}} = \frac{1}{A} \int_0^L \frac{dx}{k(x)}, \qquad R_b = \frac{1}{h_b A}$$

$$(5.20a, b, c)$$

For uniform thermal conductivity $k(x) = k = $ constant, the thermal resistance of the slab reduces to

$$R_{\text{slab}} = \frac{L}{kA} \qquad (5.20d)$$

For a composite region composed of n *different parallel layers of slabs* in perfect thermal contact, each having a thickness L_i and a uniform thermal conductivity k_i $(i = 1, 2, \ldots, n)$, Equation (5.20b) becomes

$$R_{\text{slab}} = \frac{1}{A} \sum_{i=1}^{n} \frac{L_i}{k_i} \qquad (5.20e)$$

Hollow Cylinder

The thermal resistances for heat flow through a hollow cylinder of inner radius $r = a$, outer radius $r = b$, length H, subjected to convection at the boundaries $r = a$ and $r = b$ are obtained from Equations (5.19b, c, d) as

$$R_a = \frac{1}{2\pi a H h_a}, \qquad R_1 \equiv R_{\text{cyl}} = \frac{1}{2\pi H} \int_a^b \frac{dr}{r k(r)}, \qquad R_b = \frac{1}{2\pi b H h_b}$$

$$(5.21a, b, c)$$

For uniform thermal conductivity $k(r) = k = $ constant, the thermal resistance of the cylinder reduces to

$$R_{cyl} = \frac{\ln(b/a)}{2\pi Hk}$$ (5.21d)

For a composite region composed of n *coaxial layers of cylinders* in perfect thermal contact each having a thickness $L_i = r_i - r_{i-1}$ and thermal conductivity k_i ($i = 1, 2, \ldots, n$), Equation (5.21b) becomes

$$R_{cyl} = \frac{1}{2\pi H} \sum_{i=1}^{n} \frac{1}{k_i} \ln\left(\frac{r_i}{r_{i-1}}\right)$$ (5.21e)

Hollow Sphere

In the case of a hollow sphere of inner radius $r = a$, outer radius $r = b$, and subjected to convection at both boundaries, various thermal resistances are obtained from Equations (5.19) as

$$R_a = \frac{1}{4\pi a^2 h_a}, \qquad R_2 \equiv R_{sph} = \frac{1}{4\pi} \int_a^b \frac{dr}{r^2 k(r)}, \qquad R_b = \frac{1}{4\pi b^2 h_b}$$

(5.22a, b, c)

For uniform thermal conductivity $k(r) = k = $ constant, the thermal resistance of the sphere reduces to

$$R_{sph} = \frac{1}{4\pi k}\left(\frac{1}{a} - \frac{1}{b}\right)$$ (5.22d)

For a composite region composed of n *concentric layers of spheres* in perfect thermal contact, each having a thickness $L_i = r_i - r_{i-1}$ and thermal conductivity k_i ($i = 1, 2, \ldots, n$), Equation (5.22b) becomes

$$R_{sph} = \frac{1}{4\pi} \sum_{i=1}^{n} \frac{1}{k_i}\left(\frac{1}{r_{i-1}} - \frac{1}{r_i}\right)$$ (5.22e)

To illustrate the further application of the general solution (5.9) to obtain solutions for simpler problems, we consider the following example.

Example 5.2 Determine the steady-state temperature distribution in a slab of thickness b for uniform thermal conductivity k_0, and uniform energy generation at a rate of g_0 (i.e., W/m^3). The boundary surface at $x = 0$ is kept at a uniform temperature T_a and the boundary surface at $x = b$ dissipates heat by convection into an environment at constant temperature $T_{b\infty}$ with a heat transfer coefficient h_b.

Solution The mathematical formulation of this problem is given in the dimensionless form as

$$\frac{d^2\theta(X)}{dX^2} + G_0 = 0 \quad \text{in } 0 < X < 1 \tag{5.23a}$$

$$\theta(0) = 0 \tag{5.23b}$$

$$\theta(1) + \frac{1}{\text{Bi}}\frac{d\theta(1)}{dX} = 1 \tag{5.23c}$$

where various dimensionless quantities are defined as

$$X = \frac{x}{b}, \quad \theta(x) = \frac{T(x) - T_a}{T_{b\infty} - T_a}, \quad G = \frac{g_0 b^2}{k_0(T_{b\infty} - T_a)}, \quad \text{Bi} = \frac{h_b b}{k_0}$$

$$\tag{5.24a}$$

A comparison of problem (5.23) with (5.7) yields

$$x = X, \quad k(x) = 1, \qquad n = 0, \quad T = \theta, \qquad g(x) = G_0, \quad a = 0$$

$$b = 1, \qquad \delta_a = \delta_b = 1, \quad \gamma_a = 0, \quad \gamma_b = \frac{1}{\text{Bi}}, \qquad \phi_a = 0, \quad \phi_b = 1 \tag{5.24b}$$

Then the solution of problem (5.23) is immediately obtained from Equation (5.9) as

$$\theta(X) = \frac{X}{1 + (1/\text{Bi})} + \frac{1}{2}G_0\left[\frac{1 + (2/\text{Bi})}{1 + (1/\text{Bi})}X - X^2\right] \tag{5.25a}$$

For $h_b \to \infty$ (i.e., the surface at $x = b$ is kept at temperature $T_{b\infty}$), we have $\text{Bi} \to \infty$, and the solution (5.25a) reduces to

$$\theta(X) = X + \tfrac{1}{2}G_0(X - X^2) \tag{5.25b}$$

For $h_b \to 0$ (i.e., the boundary surface at $x = b$ is insulated), we have $\text{Bi} \to 0$ and the solution (5.25a) simplifies to

$$\theta(X) = \tfrac{1}{2}G_0(2X - X^2) \tag{5.25c}$$

Example 5.3 Determine the steady-state temperature distribution $T(r)$ in a solid cylinder in the region $0 \le r \le b$ in which energy is generated at a constant rate of g_0 while the boundary surface at $r = b$ is kept at a uniform temperature T_1.

Solution The mathematical formulation of this problem is given in dimensionless form as

$$\frac{d}{dR}\left[R\frac{d\theta(R)}{dR}\right] + R = 0 \quad \text{in } 0 \le R \le 1 \tag{5.26a}$$

$$\frac{d\theta(0)}{dR} = 0, \qquad \theta(1) = 0 \tag{5.26b, c}$$

where various dimensionless quantities are defined as

$$R = \frac{r}{b}, \qquad \theta(R) = \frac{T(r) - T_1}{g_0 b^2/k} \tag{5.27a, b}$$

A comparison of problem (5.26) with (5.7) yields

$$x = R, \qquad k(x) = 1, \qquad n = 1, \qquad T = \theta, \qquad g(x) = 1, \qquad a \to 0,$$

$$b = 1, \qquad \delta_a = \gamma_b = 0, \qquad \gamma_a = \delta_b = 1, \qquad \phi_a = \phi_b = 0 \tag{5.28}$$

Then the solution of problem (5.27) is immediately obtained from Equation (5.9) as

$$\theta(R) = \int_a^1 \frac{1}{R}\left(\int_a^R R'\, dR'\right) dR - \int_a^R \frac{1}{R'}\left(\int_a^{R'} R''\, dR''\right) dR' \tag{5.29}$$

For $a = 0$ integration of Equation (5.29) gives

$$\theta(R) = \frac{1 - R^2}{4} \tag{5.30}$$

which shows that the temperature distribution in the solid is parabolic.

Example 5.4   Determine the steady-state temperature distribution $T(r)$ and the radial heat-flow rate Q for a length H in a hollow cylinder in a region $a \le r \le b$ in which energy is generated at a constant rate of g_0 while the boundary surfaces at $r = a$ and $r = b$ are kept at uniform temperature T_a and T_b, respectively.

Solution   The mathematical formulation of this problem is given as

$$\frac{d}{dR}\left[R\frac{d\theta(R)}{dR}\right] + RG = 0 \quad \text{in } 1 < R < b/a \tag{5.31a}$$

$$T(1) = 0, \qquad T\left(\frac{b}{a}\right) = 1 \tag{5.31b, c}$$

where

$$R = \frac{r}{a}, \qquad \theta = \frac{T(r) - T_0}{T_1 - T_0}, \qquad G = \frac{g_0 a^2}{k(T_1 - T_0)} \qquad (5.32)$$

A comparison of problem (5.31) with (5.7) yields

$$x = R, \qquad k(x) = 1, \qquad n = 1, \qquad g(x) = G, \qquad a = 1, \qquad b = \frac{b}{a}$$

$$\delta_a = \delta_b = 1, \qquad \gamma_a = \gamma_b = 0, \qquad \phi_a = 0, \qquad \phi_b = 1 \qquad (5.33)$$

Then the solution is obtained as a special case from Equation (5.9) as

$$\theta(R) = \frac{\ln R}{\ln b/a} \left\{ 1 + \frac{G}{4} \left[\left(\frac{b}{a} \right)^2 - 1 \right] \right\} + \frac{G}{4} (1 - R^2) \qquad (5.34)$$

The radial heat-flow rate at any position r through the cylinder for a length H of the cylinder is determined from

$$Q(r) = -k \frac{dT(r)}{dr} 2\pi r H = -k \frac{d\theta(R)}{dR} 2\pi R H (T_1 - T_0) \qquad (5.35a)$$

When $dT(r)/dr$ is substituted from Equation (5.34) we find

$$Q(R) = \pi H R^2 Gk(T_1 - T_0) - \frac{2\pi H k (T_1 - T_0) G \left\{ 1 + G/4 \left[(b/a)^2 - 1 \right] \right\}}{\ln(b/a)}$$

$$(5.35b)$$

For the case of no energy generation, by setting $G = 0$ the temperature distribution $T(r)$ and the heat flow rate $Q(R)$ can be obtained, respectively, from Equations (5.34) and (5.35b).

We note that the total heat-flow rate Q for the case of no energy generation is independent of the radial position in the cylinder.

5.3   MULTIDIMENSIONAL STEADY-STATE PROBLEMS IN THE RECTANGULAR COORDINATE SYSTEM

In presenting the examples in this section we consider first the solution of the general problem (5.3) for a three-dimensional finite rectangular parallelepiped and then illustrate how this general result can be utilized to obtain solutions of a variety of specific situations.

Three-Dimensional Finite Region

By the principle of superposition we split up the problem (5.3) into three simpler problems by setting

$$T(x, y, z) = T^{(x)}(x, y, z) + T^{(y)}(x, y, z) + T^{(z)}(x, y, z) \quad (5.36)$$

where the functions $T^{(\ell)}$ ($\ell \equiv x, y, z$) are the solutions of the following three simpler problems each with two nonhomogeneous boundary conditions

$$\left\{ \frac{\partial^2}{\partial x^2} + \frac{\partial^2}{\partial y^2} + \frac{\partial^2}{\partial z^2} \right\} T^{(\ell)}(x, y, z) = 0$$

$$\text{in } x_0 < x < x_1, \quad y_0 < y < y_1, \quad z_0 < z < z_1 \quad (5.37a)$$

$$\left\{ \alpha_{kx} - (-1)^k \beta_{kx} \frac{\partial}{\partial x} \right\} T^{(\ell)}(x_k, y, z) = \delta_{x\ell} \phi(x_k, y, z),$$

$$k = 0, 1 \quad (5.37b, c)$$

$$\left[\alpha_{ky} - (-1)^k \beta_{ky} \frac{\partial}{\partial y} \right] T^{(\ell)}(x, y_k, z) = \delta_{y\ell} \phi(x, y_k, z), \quad k = 0, 1$$

$$(5.37d, e)$$

$$\left[\alpha_{kz} - (-1)^k \beta_{kz} \frac{\partial}{\partial z} \right] T^{(\ell)}(x, y, z_k) = \delta_{z\ell} \phi(x, y, z_k), \quad k = 0, 1$$

$$(5.37f, g)$$

where $\ell \equiv x, y, z$ and $\delta_{x\ell}, \delta_{y\ell}, \delta_{z\ell}$ are the Kronecker delta.

We now consider the solution of the problem (5.37) for the case $\ell = x$. By comparing this problem with that given by Equations (5.3) we write

$$t = x, \quad w(x) = 1, \quad \delta_k = \alpha_{kx}, \quad \gamma_k = \beta_{kx}$$

$$L_t = -\frac{\partial^2}{\partial x^2}, \quad L = -\left\{ \frac{\partial^2}{\partial y^2} + \frac{\partial^2}{\partial z^2} \right\}$$

$$B = \alpha_{ky} - (-1)^k \beta_{ky} \frac{\partial}{\partial y} \quad \text{at } y = y_k$$

$$B = \alpha_{kz} - (-1)^k \beta_{kz} \frac{\partial}{\partial z} \quad \text{at } z = z_k \quad (5.38)$$

Then the eigenvalue problem (2.2), when restricted to the y and z variables, is a special case of the eigenvalue problem (3.49). The eigenvalues are obtained as a double index set in the form $\mu_{mn}^2 \equiv \nu_m^2 + \eta_n^2$ $(m, n = 1, 2, \ldots, \infty)$ and the two-dimensional eigenfunctions are given as a double product in the form $\psi_{mn} \equiv \psi_m(y)\psi_n(z)$. The normalization integral becomes $N_{mn} \equiv N_m N_n$ [see Equations (3.54c, e, f)]. The eigenvalues ν_m, η_n, the eigenfunctions $\psi_m(y)$, $\psi_n(z)$, and the normalization integrals N_m, N_n are immediately determined according to expressions (3.20a, b, c), respectively. Substituting these results into solution (2.60a), we obtain the solution of problem (5.37) for the case $\ell = x$ as

$$T^{(x)}(x, y, z) = \sum_{m=1}^{\infty} \sum_{n=1}^{\infty} \frac{\psi_m(y)}{N_m} \frac{\psi_n(z)}{N_n}$$

$$\times \left\{ \sum_{k=0}^{1} (-1)^{1-k} \left[\int_{y_0}^{y_1} \int_{z_0}^{z_1} \psi_m(y')\psi_n(z')\phi(x_{1-k}, y', z') \, dy' \, dz' \right] \right.$$

$$\left. \times \frac{\alpha_{kx}\sinh[\mu_{mn}(x_k - x)] - (-1)^k \beta_{kx}\mu_{mn}\cosh[\mu_{mn}(x_k - x)]}{D} \right\}$$

$$\text{(5.39a)}$$

where

$$D = (\alpha_{0x}\alpha_{1x} + \beta_{0x}\beta_{1x}\mu_{mn}^2)\sinh[\mu_{mn}(x_1 - x_0)]$$

$$+ (\alpha_{0x}\beta_{1x} + \alpha_{1x}\beta_{0x})\mu_{mn}\cosh[\mu_{mn}(x_1 - x_0)]$$

$$\psi_m(y) = \beta_{0y}\cos[\nu_m(y - y_0)] + \frac{\alpha_{0y}}{\nu_m}\sin[\nu_m(y - y_0)] \qquad \text{(5.39b)}$$

$$\psi_n(z) = \beta_{0z}\cos[\eta_n(z - z_0)] + \frac{\alpha_{0z}}{\eta_n}\sin[\eta_n(z - z_0)] \qquad \text{(5.39c)}$$

$$N_m = \frac{1}{2\nu_m^2} \left\{ \left[\alpha_{0y}^2 + (\beta_{0y}\nu_m)^2 \right] \left[\frac{\alpha_{1y}\beta_{1y}}{\alpha_{1y}^2 + (\beta_{1y}\nu_m)^2} + y_1 - y_0 \right] + \alpha_{0y}\beta_{0y} \right\}$$

$$\text{(5.39d)}$$

$$N_n = \frac{1}{2\eta_n^2} \left\{ \left[\alpha_{0z}^2 + (\beta_{0z}\eta_n)^2 \right] \left[\frac{\alpha_{1z}\beta_{1z}}{\alpha_{1z}^2 + (\beta_{1z}\eta_n)^2} + z_1 - z_0 \right] + \alpha_{0z}\beta_{0z} \right\}$$

$$\text{(5.39e)}$$

$$\mu_{mn}^2 = \nu_m^2 + \eta_n^2 \qquad \text{(5.39f)}$$

and ν_m and η_n are the roots of the following transcendental equations

$$\left(\alpha_{0y}\beta_{1y} + \alpha_{1y}\beta_{0y}\right)\cos\left[\nu(y_1 - y_0)\right]$$

$$+ \left(\alpha_{0y}\alpha_{1y} - \beta_{0y}\beta_{1y}\nu^2\right)\frac{\sin\left[\nu(y_1 - y_0)\right]}{\nu} = 0 \qquad (5.39\text{g})$$

$$\left(\alpha_{0z}\beta_{1z} + \alpha_{1z}\beta_{0z}\right)\cos\left[\eta(z_1 - z_0)\right]$$

$$+ \left(\alpha_{0z}\alpha_{1z} - \beta_{0z}\beta_{1z}\eta^2\right)\frac{\sin\left[\eta(z_1 - z_0)\right]}{\eta} = 0 \qquad (5.39\text{h})$$

For the boundary condition of the second kind at both boundaries, we have

$$\alpha_{0y} = \alpha_{1y} = \alpha_{0z} = \alpha_{1z} = 0, \qquad \beta_{0y} = \beta_{1y} = \beta_{0z} = \beta_{1z} = 1 \quad (5.40\text{a})$$

Then Equations (5.39b)–(5.39h), respectively, take the form

$$\psi_m(y) = \cos\left[\nu_m(y - y_0)\right] \qquad (5.40\text{b})$$

$$\psi_n(z) = \cos\left[\eta_n(z - z_0)\right] \qquad (5.40\text{c})$$

$$N_m = \begin{cases} \dfrac{y_1 - y_0}{2} & \text{for } m \neq 0 \\ y_1 - y_0 & \text{for } m = 0 \end{cases} \qquad (5.40\text{d})$$

$$N_n = \begin{cases} \dfrac{z_1 - z_0}{2} & \text{for } n \neq 0 \\ z_1 - z_0 & \text{for } n = 0 \end{cases} \qquad (5.40\text{e})$$

$$\mu_{mn}^2 = \pi^2\left[\left(\frac{m}{y_1 - y_0}\right)^2 + \left(\frac{n}{z_1 - z_0}\right)^2\right] \qquad (5.40\text{f})$$

$$\nu_m = \frac{m\pi}{y_1 - y_0}, \qquad m = 0,1,2,\ldots \qquad (5.40\text{g})$$

$$\eta_n = \frac{n\pi}{z_1 - z_0}, \qquad n = 0,1,2,\ldots \qquad (5.40\text{h})$$

The solutions for the functions $T^{(y)}(x, y, z)$ and $T^{(z)}(x, y, z)$ are similar to the solutions (5.39) and (5.40) when appropriate changes are made in the notations.

Three-Dimensional Semi-Infinite Region

For a semi-infinite solid $x_0 \leq x < \infty$ we have

$$x_1 \to \infty, \qquad \alpha_{1x} = 0, \qquad \beta_{1x} = 1, \qquad \phi(x_1, y, z) = 0 \qquad (5.41\text{a})$$

Then solution (5.39a) becomes

$$T^{(x)}(x, y, z) = \sum_{m=0}^{\infty} \sum_{n=0}^{\infty} \frac{\psi_m(y)\psi_n(z)}{N_m N_n} \int_{y_0}^{y_1} \int_{z_0}^{z_1} \psi_m(y)\psi_n(z)\phi(x_0, y, z)\, dy\, dz$$

$$\times \frac{e^{-\mu_{nm}(x-x_0)}}{\beta_{0x}\mu_{mn} + \alpha_{0x}} \tag{5.41b}$$

where the eigenvalues, eigenfunctions, and normalization integrals are defined by Equations (5.39b–h) for general boundary conditions, and by Equations (5.40b–h) for the boundary conditions of the second kind.

Examples of Three-Dimensional Problems

Example 5.5 Develop an expression for the steady-state diffusion in a three-dimensional finite region $0 \le x \le a$, $0 \le y \le b$, $0 \le z \le c$, subject to the boundary conditions

$$T_{x=0} = T_{y=0} = T_{y=b} = T_{z=0} = 0, \qquad \left(\frac{\partial T}{\partial x}\right)_{x=a} = 0, \qquad T_{z=c} = T_0$$

$$\tag{5.42a}$$

Solution By comparing this problem with the one given by Equations (5.3) we write

$$
\begin{array}{lll}
\alpha_{0x} = 1, & \beta_{0x} = 0, & \phi(0, y, z) = 0 \\
\alpha_{1x} = 0, & \beta_{1x} = 1, & \phi(a, y, z) = 0 \\
\alpha_{0y} = 1, & \beta_{0y} = 0, & \phi(x, 0, z) = 0 \\
\alpha_{1y} = 1, & \beta_{1y} = 0, & \phi(x, b, z) = 0 \\
\alpha_{0z} = 1, & \beta_{0z} = 0, & \phi(x, y, 0) = 0 \\
\alpha_{1z} = 1, & \beta_{1z} = 0, & \phi(x, y, c) = T_0 \quad (5.42b)
\end{array}
$$

The solution is obtained as a special case of Equation (5.39) after making appropriate changes in the notations ($y \to x, z \to y, x \to z$). The final result is

$$T = T_0 \frac{16}{\pi^2} \sum_{m=1}^{\infty} \sum_{l=1}^{\infty} \frac{\sin\left(\dfrac{2m-1}{2a}\pi x\right) \sin\left(\dfrac{2l-1}{2b}\pi y\right)}{2m-1 \qquad\qquad 2l-1}$$

$$\times \frac{\sinh\left\{\pi z \sqrt{[(2m-1)/2a]^2 + [(2l-1)/b]^2}\right\}}{\sinh\left\{\pi c \sqrt{[(2m-1)/2a]^2 + [(2l-1)/b]^2}\right\}} \tag{5.43}$$

In the derivation of solution (5.43) the relation

$$\int_0^a \int_0^b \sin\left(\frac{2m-1}{2a}\pi x\right)\sin\left(\frac{n}{b}\pi y\right) dx\, dy$$

$$= \begin{cases} 0 & \text{for } n = 2l \\ \dfrac{4ab}{(2m-1)(2l-1)\pi^2} & \text{for } n = 2l-1 \end{cases} \qquad (5.44)$$

was used.

Example 5.6   Consider the steady-state diffusion in a three-dimensional region $0 \le x \le a,\ 0 \le y \le b,\ 0 \le z \le \infty$ subject to the boundary conditions

$$\left(\frac{\partial T}{\partial x}\right)_{x=0} = \left(\frac{\partial T}{\partial y}\right)_{y=0} = 0, \qquad T_{y=b} = 0, \qquad \left(T + k\frac{\partial T}{\partial x}\right)_{x=a} = 0$$

$$T_{z=0} = \begin{cases} T_0 & \text{for } 0 \le x \le a, & 0 \le y \le \dfrac{b}{2} \\ 0 & \text{for } 0 \le x \le a, & \dfrac{b}{2} < y \le b \end{cases} \qquad (5.45a)$$

Solution   By comparison of Equations (5.45a) with boundary conditions (5.3) we find

$$\begin{aligned}
&\alpha_{0x} = 0, & &\beta_{0x} = 1, & &\phi(0, y, z) = 0 \\
&\alpha_{1x} = 1, & &\beta_{1x} = k, & &\phi(a, y, z) = 0 \\
&\alpha_{0y} = 0, & &\beta_{0y} = 1, & &\phi(x, 0, z) = 0 \\
&\alpha_{1y} = 1, & &\beta_{1y} = 0, & &\phi(x, b, z) = 0 \\
&\alpha_{0z} = 1, & &\beta_{0z} = 0, & &\phi(x, y, 0) = T_{z=0} \qquad (5.45b)
\end{aligned}$$

Then the solution is obtained as a special case of Equation (5.41b) after making appropriate changes in the notation ($x \to z,\ z \to y,\ y \to x$). The final result is

$$T = T_0 \frac{8}{\pi} \sum_{m=1}^{\infty} \sum_{n=1}^{\infty} \frac{\sin \gamma_n}{\gamma_n \left\{ 1 + \dfrac{k/a}{1 + (\gamma_n k/a)^2} \right\}}$$

$$\times \frac{1}{2n-1} \sin\left[(2n-1)\frac{\pi}{4}\right] \cos\left(\frac{\gamma_n x}{a}\right) \cos\left(\frac{2n-1}{2b}\pi y\right)$$

$$\times \exp\left[-z\sqrt{\left(\frac{\gamma_n}{a}\right)^2 + \left(\frac{2n-1}{2b}\pi\right)^2}\right] \qquad (5.46)$$

where the γ_n are the roots of the transcendental equation

$$\cos \gamma = \frac{k}{a} \gamma \sin \gamma \tag{5.47}$$

Example 5.7 Consider the steady-state diffusion in a three-dimensional region $0 \le x \le a$, $0 \le y \le b$, $0 \le z < \infty$, subject to the boundary conditions

$$\left(\frac{\partial T}{\partial x}\right)_{x=0} = \left(\frac{\partial T}{\partial y}\right)_{y=0} = \left(\frac{\partial T}{\partial x}\right)_{x=a} = \left(\frac{\partial T}{\partial y}\right)_{y=b} = 0$$

$$T_{z=0} = \begin{cases} T_0 & \text{for } 0 \le x \le \dfrac{a}{2}, & 0 \le y \le \dfrac{b}{2} \\[2mm] 0 & \text{for } \dfrac{a}{2} < x \le a, & 0 \le y \le \dfrac{b}{2} \\[2mm] 0 & \text{for } 0 \le x \le \dfrac{a}{2}, & \dfrac{b}{2} < y \le b \\[2mm] -T_0 & \text{for } \dfrac{a}{2} < x \le a, & \dfrac{b}{2} < y \le b \end{cases} \tag{5.48a}$$

Solution By comparing this problem with the ones given by Equations (5.3) we write

$$\alpha_{0x} = \alpha_{1x} = \alpha_{0y} = \alpha_{1y} = 0, \qquad \beta_{0x} = \beta_{1x} = \beta_{0y} = \beta_{1y} = 1$$

$$\alpha_{0z} = 1, \qquad \beta_{0z} = 0 \tag{5.48b}$$

Then the solution is obtained as a special case of Equation (5.41b) after making appropriate changes in the notations $(x \rightarrow z, z \rightarrow y, y \rightarrow x)$

$$T = \sum_{m=0}^{\infty} \sum_{n=0}^{\infty} \frac{\cos(\pi mx/a)}{N_m} \frac{\cos(\pi ny/b)}{N_n} f_{mn} e^{-\mu_{mn}z} \tag{5.49a}$$

where

$$N_m = \begin{cases} a & \text{for } m \ne 0 \\[2mm] \dfrac{a}{2} & \text{for } m = 0 \end{cases} \tag{5.49b}$$

$$N_n = \begin{cases} b & \text{for } n \ne 0 \\[2mm] \dfrac{b}{2} & \text{for } n = 0 \end{cases} \tag{5.49c}$$

$$\mu_{mn} = \pi \sqrt{\left(\frac{m}{a}\right)^2 + \left(\frac{n}{b}\right)^2} \tag{5.49d}$$

$$f_{mn} = \int_0^a \int_0^b T_{z=0} \cos\left(\frac{\pi mx}{z}\right) \cos\left(\frac{\pi ny}{b}\right) dx\, dy \tag{5.49e}$$

In the last equation the integrals are evaluated as follows:

$$f_{00} = 0 \tag{5.50a}$$

$$f_{m0} = \begin{cases} 0 & \text{for } m = 2l \\ T_0 \dfrac{ab}{(2l-1)\pi}(-1)^{l+1} & \text{for } m = 2l-1 \end{cases} \tag{5.50b}$$

$$f_{0n} = \begin{cases} 0 & \text{for } n = 2s \\ T_0 \dfrac{ab}{(2s-1)\pi}(-1)^{s+1} & \text{for } n = 2s-1 \end{cases} \tag{5.50c}$$

$$f_{mn} = 2T_0 \dfrac{ab}{(2l-1)(2s-1)\pi^2}(-1)^{l+s} \tag{5.50d}$$

Using Equations (5.50) in solution (5.49a) we finally obtain

$$\begin{aligned}
T = T_0 \frac{2}{\pi} \Bigg\{ &\sum_{l=1}^{\infty} \frac{(-1)^{l+1}}{2l-1} \cos\left(\frac{2l-1}{a}\pi x\right)\exp\left(-\frac{2l-1}{a}\pi z\right) \\
&+ \sum_{s=1}^{\infty} \frac{(-1)^{s+1}}{2s-1} \cos\left(\frac{2s-1}{b}\pi y\right)\exp\left(-\frac{2s-1}{b}\pi z\right) \\
&+ \frac{4}{\pi} \sum_{l=1}^{\infty} \sum_{s=1}^{\infty} \frac{(-1)^{l+s}}{(2l-1)(2s-1)} \cos\left(\frac{2l-1}{a}\pi x\right)\cos\left(\frac{2s-1}{b}\pi y\right) \\
&\times \exp\left[-\pi z \sqrt{\left(\frac{2l-1}{a}\right)^2 + \left(\frac{2s-1}{b}\right)^2}\right] \Bigg\}
\end{aligned} \tag{5.51}$$

Two-Dimensional Finite Region

We consider steady-state diffusion in a two-dimensional finite rectangular region $x_0 \le x \le x_1$, $y_0 \le y \le y_1$, subject to the boundary condition of the third kind at each of the four boundary surfaces. The mathematical formulation of this problem is obtainable from that given by Equations (5.25) for the three-dimensional case by setting

$$\phi(x_k, y, z) \equiv \phi(x_k, y), \qquad \phi(x, y_k, z) \equiv \phi(x, y_k)$$

$$\alpha_{kz} = 0, \qquad \beta_{kz} = 1, \qquad \phi(x, y, z_k) = 0 \tag{5.52a}$$

When these quantities are introduced into Equations (5.3) it follows that the variation of T occurs only in the variables x and y; that is,

$$T(x, y, z) = T(x, y) \tag{5.52b}$$

Then $\partial/\partial z = 0$ and the two-dimensional formulation of the problem immediately follows from Equations (5.3).

By the principle of superposition we split up the general problem into two simpler problems by setting

$$T(x, y) = T^{(x)}(x, y) + T^{(y)}(x, y) \tag{5.52c}$$

where the functions $T^{(\ell)}$ ($\ell \equiv x, y$) satisfy the following two simpler problems which are the two-dimensional version of problems (5.37)

$$\left\{ \frac{\partial^2}{\partial x^2} + \frac{\partial^2}{\partial y^2} \right\} T^{(\ell)}(x, y) = 0 \quad \text{in } x_0 < x < x_1, \quad y_0 < y < y_1$$

$$\tag{5.53a}$$

$$\left\{ \alpha_{kx} - (-1)^k \beta_{kx} \frac{\partial}{\partial x} \right\} T^{(\ell)}(x, y) = \delta_{x\ell} \phi(x_k, y),$$

$$k = 0,1 \quad (5.53b)$$

$$\left[\alpha_{ky} - (-1)^k \beta_{ky} \frac{\partial}{\partial y} \right] T^{(\ell)}(x, y) = \delta_{y\ell} \phi(x, y_k),$$

$$k = 0,1 \quad (5.53c)$$

where $\ell \equiv x, y$ and $\delta_{x\ell}, \delta_{y\ell}$ are the Kronecker delta.

Clearly, problem (5.53) is a special case of problem (5.3); therefore, its solution is readily obtainable from Equations (5.39) as a special case by making use of the following results

$$\psi_n(z) = \cos[\eta_n(z - z_0)], \quad n = 1, 2, \ldots \tag{5.54a}$$

where

$$\eta_n = \frac{\pi(n - 1)}{z_1 - z_0} \tag{5.54b}$$

and

$$\frac{\psi_n(z)}{N_n} \int_{z_0}^{z_1} \psi_n(z) \, dz = \begin{cases} 0 & \text{for } n = 2, 3, \ldots, \infty \\ 1 & \text{for } n = 1 \end{cases} \tag{5.54c}$$

Then the solution of the problem (5.53) for $\ell = x$ is immediately obtained from

Equations (5.39) as

$$T^{(x)}(x, y) = \sum_{m=1}^{\infty} \frac{\psi_m(y)}{N_m} \left\{ \sum_{k=0}^{1} (-1)^{1-k} \int_{y_0}^{y_1} \psi_m(y)\phi(x_{1-k}, y)\, dy \right.$$

$$\left. \times \frac{\alpha_{kx}\sinh[\nu_m(x_k - x)] - (-1)^k \beta_{kx}\nu_m\cosh[\nu_m(x_k - x)]}{D} \right\}$$

$$(5.55)$$

where

$$D = \left(\alpha_{0x}\alpha_{1x} + \beta_{0x}\beta_{1x}\nu_m^2\right)\sinh[\nu_m(x_1 - x_0)]$$

$$+ \left(\alpha_{0x}\beta_{1x} + \alpha_{1x}\beta_{0x}\right)\nu_m\cosh[\nu_m(x_1 - x_0)]$$

where the eigenfunctions $\psi_m(y)$, the normalization integral N_m, and the eigenvalues ν_m are determined according to the expressions (5.39b, d, g), respectively, and for second kind boundary conditions according to the expressions (5.40b, d, g), respectively.

Semi-Infinite Rectangular Strip $x_0 \le x < \infty, y_0 \le y \le y_1$

We now consider diffusion in a semi-infinite strip $x_0 \le x < \infty$, $y_0 \le y \le y_1$ with a boundary condition of the third kind at all boundaries.

We split up the problem ino two simpler problems as given by Equations (5.52c). The separated problems are similar to those given by Equation (5.53) except for this particular case we have $x_1 \to \infty$, $\alpha_{1x} = 0$, and $\phi(x_1, y) = 0$. Then the solution for the function $T^{(x)}(x, y)$ is immediately obtainable from Equation (5.55) by setting in that equation $\alpha_{1x} = 0$, $\phi(x_1, y) = 0$, and $x_1 \to \infty$. We find

$$T^{(x)}(x, y) = \sum_{m=1}^{\infty} \frac{\psi_m(y)}{N_m} \int_{y_0}^{y_1} \psi_m(y')\phi(x_0, y')\, dy' \frac{e^{\nu_m(x_0 - x)}}{\alpha_{0x} + \beta_{0x}\nu_m} \quad (5.56)$$

Semi-Infinite Rectangular Strip $x_0 \le x \le x_1, 0 \le y < \infty$

We now consider a semi-infinite region $x_0 \le x \le x_1, 0 \le y < \infty$ with boundary conditions of the third kind at the boundaries $x = x_{1-k}$, $y = 0$ with $k = 0, 1$.

We again assume that the solution of the problem is split up into two simpler problems involving the functions $T^{(x)}(x, y)$ and $T^{(y)}(x, y)$ as discussed previously. The problem for $T^{(y)}(x, y)$ involves one nonhomogeneous boundary condition at $y = 0$, the nonhomogeneous term being $\phi(x, 0)$.

The solution for $T^{(y)}(x, y)$ is immediately obtainable from Equation (5.56) by interchanging in that equation x and y, and then setting $y_0 = 0$.

The problem for the function $T^{(x)}(x, y)$ involves two nonhomogeneous boundary conditions at $x = x_0$ and $x = x_1$, the nonhomogeneous terms being $\phi(x_0, y)$ and $\phi(x_1, y)$. The solution for $T(x, y)$ can be obtained from Equation (5.55) by taking into account the fact that the region in the y variable is semi-infinite; that is, $0 \le y < \infty$. Therefore, the summation over the discrete eigenvalues in Equation (5.55) should be transformed to an integral over the eigenvalues ν which now takes all values from zero to infinity as in Equation (4.55). By utilizing the results in Equations (5.55) and (4.57), we write the solution for $T^{(x)}(x, y)$ as

$$T^{(x)}(x, y) = \frac{1}{\pi} \sum_{k=0}^{1} (-1)^{1-k} \int_{y'=0}^{\infty} \phi(x_{1-k}, y')$$

$$\times \int_{\nu=0}^{\infty} \left\{ \cos[\nu(y - y')] + \frac{(\beta_{0y}\nu)^2 - \alpha_{0y}^2}{(\beta_{0y}\nu)^2 + \alpha_{0y}^2} \cos[\nu(y - y')] \right.$$

$$+ \frac{2\alpha_{0y}\beta_{0y}}{(\beta_{0y}\nu)^2 + \alpha_{0y}^2} \nu \sin[\nu(y + y')] \Bigg\}$$

$$\times \frac{\alpha_{kx}\sinh[\nu(x_k - x)] - (-1)^k \beta_{kx}\nu \cosh[\nu(x_k - x)]}{D} d\nu \, dy'$$

$$(5.57)$$

where

$$D = (\alpha_{0x}\alpha_{1x} + \beta_{0x}\beta_{1x}\nu^2)\sinh[\nu(x_1 - x_0)]$$

$$+ (\alpha_{0x}\beta_{1x} + \alpha_{1x}\beta_{0x})\nu \cosh[\nu(x_1 - x_0)]$$

Examples of Two-Dimensional Problems

Example 5.8 Consider the steady-state diffusion in a two-dimensional region $0 \le x \le a, 0 \le y \le b$, subject to the boundary conditions

$$T_{x=0} = T_{x=a} = T_{y=0} = 0, \qquad T_{y=b} = T_0 f(x) \qquad (5.58a)$$

Solution By comparing this problem with the ones given by Equations (5.53) we write

$$\ell = y, \qquad T(x, y) = T^{(y)}(x, y), \qquad \alpha_{kx} = \alpha_{ky} = 1$$

$$\beta_{kx} = \beta_{ky} = 0, \qquad \phi(x, 0) = 0, \qquad \phi(x, b) = T_0 f(x) \qquad (5.58b)$$

Then the solution is obtained as a special case of Equations (5.55) after making appropriate changes in the notation $(x \rightarrow y, y \rightarrow x)$

$$T = T_0 \frac{2}{a} \sum_{m=1}^{\infty} \sin\left(\frac{\pi mx}{a}\right) \int_0^a f(x)\sin\left(\frac{\pi mx}{a}\right) dx \, \frac{\sinh(\pi my/a)}{\sinh(\pi mb/a)} \quad (5.59)$$

Example 5.9 Consider the steady-state diffusion in a two-dimensional region $0 \leq x \leq a, 0 \leq y \leq b$ subject to the boundary conditions

$$T_{x=0} = 0, \qquad \left(\frac{\partial T}{\partial x}\right)_{x=a} = \left(\frac{\partial T}{\partial y}\right)_{y=0} = 0, \qquad T_{y=b} = T_0 f(x) \quad (5.60a)$$

Compute the temperature distribution for the two cases of boundary-surface temperature-distribution function $f(x)$ given by

(a) $f(x) = 1$

(b) $f(x) = \begin{cases} 1 & \text{for } 0 \leq x \leq \dfrac{a}{2} \\ 0 & \text{for } \dfrac{a}{2} < x \leq a \end{cases}$ \qquad (5.60b)

Solution By comparing this problem with ones given by Equations (5.53) we write

$$\beta_{0x} = \alpha_{1x} = \alpha_{0y} = \beta_{1y} = 0$$

$$\alpha_{0x} = \beta_{1x} = \beta_{0y} = \alpha_{1y} = 1 \quad (5.60c)$$

Then the solution is obtained as a special case of Equations (5.55) after making appropriate change in the notation $(x \rightarrow y, y \rightarrow x)$

$$T = T_0 \frac{2}{a} \sum_{m=1}^{\infty} \sin\left(\frac{2m-1}{2a}\pi x\right) \int_0^a f(x)\sin\left(\frac{2m-1}{2a}\pi x\right) dx$$

$$\times \frac{\cosh\{[(2m-1)/2a]\pi y\}}{\cosh\{[(2m-1)/2a]\pi b\}} \quad (5.61a)$$

The temperature distribution for $f(x) = 1$ becomes

$$T = T_0 \frac{4}{\pi} \sum_{m=1}^{\infty} \frac{1}{2m-1}\sin\left(\frac{2m-1}{2a}\pi x\right)\frac{\cosh\{[(2m-1)/2a]\pi y\}}{\cosh\{[(2m-1)/2a]\pi b\}}$$

$$(5.61b)$$

The temperature distribution for $f(x) = 1$ for $0 \leq x \leq a/2$ and $f(x) = 0$ for $a/2 < x \leq a$ becomes

$$T = T_0 \frac{4}{\pi} \sum_{m=1} \frac{1 - \cos\{[(2m - 1)/4]\pi\}}{2m - 1} \sin\left(\frac{2m - 1}{2a}\pi x\right)$$

$$\times \frac{\cosh\{[(2m - 1)/2a]\pi y\}}{\cosh\{[(2m - 1)/2a]\pi b\}} \qquad (5.61c)$$

Example 5.10 Consider the steady-state diffusion in a two-dimensional region $0 \leq x \leq a$, $0 \leq y \leq b$, subject to the boundary conditions

$$\left(\frac{\partial T}{\partial x}\right)_{x=0} = \left(\frac{\partial T}{\partial x}\right)_{x=a} = 0, \qquad T_{y=0} = T_0 f_0(x), \qquad \left(\frac{\partial T}{\partial y}\right)_{y=b} = Q_0 f_1(x)$$

$$(5.62a)$$

Compute the temperature distribution for the case of zero temperature at $y = 0$, that is, $f_0(x) = 0$, and the following four boundary-surface functions $f_1(x)$ given by:

(a) $\quad f_1(x) = 1$

(b) $\quad f_1(x) = \cos\left(\dfrac{\pi x}{a}\right)$

(c) $\quad f_1(x) = \begin{cases} 1 & \text{for } 0 \leq x \leq \dfrac{a}{2} \\ 0 & \text{for } \dfrac{a}{2} < x \leq a \end{cases}$

(d) $\quad f_1(x) = \begin{cases} 0 & \text{for } 0 \leq x \leq \dfrac{a}{2} \\ 1 & \text{for } \dfrac{a}{2} < x \leq a \end{cases}$ $\qquad (5.62b)$

Solution By comparing this problem with the ones given by Equations (5.53) we write

$$\alpha_{0x} = \alpha_{1x} = \beta_{0y} = \beta_{1y} = 0$$

$$\beta_{0x} = \beta_{1x} = \alpha_{0y} = \alpha_{1y} = 1 \qquad (5.62c)$$

Then the solution is obtained as a special case of Equations (5.55) after making changes $x \rightarrow y$ and $y \rightarrow x$ in the notation

$$T = \left[T_0 \int_0^a f_0(x) \, dx + y Q_0 \int_0^a f_1(x) \, dx \right] \left(\frac{1}{a} \right) + 2 \sum_{m=1}^{\infty} \cos\left(\frac{\pi m x}{a} \right)$$

$$\times \left\{ \frac{T_0}{a} \int_0^a f_0(x) \cos\left(\frac{\pi m x}{a} \right) dx \frac{\cosh[(\pi m/a)(b-y)]}{\cosh[(\pi m/a)b]} \right.$$

$$\left. + Q_0 \int_0^a f_1(x) \cos\left(\frac{\pi m x}{a} \right) dx \frac{\sinh(\pi m y/a)}{m \cosh(\pi m b/a)} \right\} \qquad (5.63)$$

For the case a, the temperature distribution becomes

$$T = y Q_0 \qquad (5.64)$$

For the case b, the integral term in Equation (5.63) is given by

$$\int_0^a \cos\left(\frac{\pi x}{a} \right) \cos\left(\frac{\pi m x}{a} \right) dx = \begin{cases} 0 & \text{for } m \neq 0 \\ \dfrac{a}{2} & \text{for } m = 0 \end{cases} \qquad (5.65a)$$

When this result is substituted into solution (5.63) we obtain

$$T = \frac{Q_0 a}{\pi} \frac{\sinh(\pi y/a)}{\sinh(\pi b/a)} \cos\left(\frac{\pi x}{a} \right) \qquad (5.65b)$$

For the case c, the integral term in the Equation (5.63) becomes

$$\int_0^{a/2} \cos\left(\frac{\pi m x}{a} \right) dx = \begin{cases} 0 & \text{for } m = 2l \\ \dfrac{a}{\pi m}(-1)^{l+1} & \text{for } m = 2l - 1 \end{cases} \qquad (5.66a)$$

where $l = 1, 2, 3, \ldots$.

Then all the even terms of the summation in the solution vanish and we obtain

$$T = Q_0 \left(\frac{y}{2} + \frac{2a}{\pi^2} \sum_{l=1}^{\infty} \frac{(-1)^{l+1}}{(2l-1)^2} \frac{\sinh\{[(2l-1)/a]\pi y\}}{\cosh\{[(2l-1)/a]\pi b\}} \cos\left(\frac{2l-1}{a} \pi x \right) \right)$$

$$(5.66b)$$

Finally, *for the case* d the solution (5.63) becomes

$$T = Q_0 \left(\frac{y}{2} + \left(\frac{2}{\pi} \right) \sum_{l=1}^{\infty} \frac{(-1)^l}{2l-1} \frac{\sinh\{[(2l-1)/a]\pi y\}}{\sinh\{[(2l-1)/a]\pi b\}} \cos\left(\frac{2l-1}{a} \pi x \right) \right)$$

$$(5.67)$$

Example 5.11 Consider a semi-infinite rectangular strip $0 \le x \le a$, $0 \le y < \infty$ with boundary surface at $y = 0$ kept at temperature $T_0 f(x)$ and the remaining boundaries at $x = 0$ and $x = a$ are insulated. Compute the distribution of temperature for the case $f(x) = 1$.

Solution By comparing this problem with the one given by Equations (5.53) we write

$$\beta_{0x} = \beta_{1x} = \beta_{0y} = \alpha_{1y} = 0$$

$$\alpha_{0x} = \alpha_{1x} = \alpha_{0y} = \beta_{1y} = 1$$

$$\phi(x,0) = T_0 f(x), \qquad \phi(x, \infty) = 0 \qquad (5.68)$$

Then the solution is obtained as a special case of Equation (5.56) after making changes $x \to y$ and $y \to x$ in the notation

$$T = T_0 \frac{2}{a} \sum_{m=1}^{\infty} \int_0^a f(x) \sin\left(\frac{\pi m x}{a}\right) dx \, e^{-(\pi m y/a)} \sin\left(\frac{\pi m x}{a}\right) \qquad (5.69)$$

The integral term in this equation for $f(x) = 1$ becomes

$$\int_0^1 \sin\left(\frac{\pi m x}{a}\right) dx = \begin{cases} 0 & \text{for } m \text{ even} \\ 2 & \text{for } m \text{ odd} \end{cases} \qquad (5.70)$$

When the result in Equation (5.70) is substituted into the solution given by Equation (5.69), all the even terms of the summation vanish and we obtain

$$T = T_0 \frac{4}{\pi} \sum_{l=1}^{\infty} \frac{1}{2l-1} \exp\left(-\frac{2l-1}{a}\pi y\right) \sin\left(\frac{2l-1}{a}\pi x\right) \qquad (5.71a)$$

The sum is evaluated by making use of the following relation

$$\sum_{k=0}^{\infty} \frac{e^{-(2k+1)a}}{2k+1} \sin[(2k+1)b] = \frac{1}{2} \text{arctg}\left(\frac{\sin b}{\sinh a}\right) \qquad (5.72)$$

Then the solution (5.71a) becomes

$$T = T_0\left(\frac{2}{\pi}\right) \text{arctg} \frac{\sin(\pi x/a)}{\sinh(\pi y/a)} \qquad (5.71b)$$

Example 5.12 Consider a semi-infinite rectangular strip $0 \le x \le a$, $0 \le y < \infty$ with the boundary conditions

$$\left(\frac{\partial T}{\partial x}\right)_{x=0} = \left(\frac{\partial T}{\partial x}\right)_{x=a} = 0, \qquad \left(T - \kappa\frac{\partial T}{\partial y}\right)_{y=0} = T_0 f(x) \quad (5.73a)$$

Compute the distribution of temperature for the cases

$$f(x) = \begin{cases} 1 & \text{for } 0 \le x \le \dfrac{a}{2} \\ -1 & \text{for } \dfrac{a}{2} < x \le a \end{cases} \tag{5.73b}$$

Solution   By comparing this problem with the one given by Equations (5.53) we write

$$\alpha_{0x} = \alpha_{1x} = \alpha_{1y} = 0, \qquad \alpha_{0y} = 1$$

$$\beta_{0x} = \beta_{1x} = \beta_{1y} = 1, \qquad \beta_{0y} = \kappa$$

$$\phi(x,0) = T_0 f(x), \qquad \phi(x,\infty) = 0 \tag{5.73c}$$

Then the solution is obtained as a special case of Equation (5.56)

$$T = \frac{T_0}{a}\left[\int_0^a f(x)\,dx + 2\sum_{m=1}^{\infty} \cos(\pi mx/a)\frac{e^{-mx/a}}{1+\pi\kappa m/a}\int_0^a f(x)\cos\left(\pi\frac{mx}{a}\right)dx\right] \tag{5.74a}$$

Substituting Equation (5.73b) into the integral of Equation (5.74a) we obtain

$$T = T_0\left(\frac{4}{\pi}\right)\sum_{l=1}^{\infty}\frac{(-1)^{l+1}}{2l-1}\frac{\cos\{[(2l-1)/a]\pi x\}}{1+\kappa\pi(2l-1)/a}\exp\left(-\frac{2l-1}{a}\pi y\right) \tag{5.74b}$$

5.4   MULTIDIMENSIONAL STEADY-STATE PROBLEMS IN THE CYLINDRICAL COORDINATE SYSTEM

In this section we first obtain the solution for a three-dimensional, finite, hollow cylinder subject to boundary conditions of the third kind at all boundaries. We then determine the solutions of a number of specific problems as special cases from this general solution.

Three-Dimensional Finite, Hollow Cylinder

We consider a three-dimensional problem in the cylindrical coordinate system (r, ϕ, z) for a region specified by the coordinates $r_0 \le r \le r_1$, $0 \le \phi \le \pi$, and

$z_0 \leq z \leq z_1$. We assume boundary conditions of the third kind for all boundaries. The mathematical formulation of the problem is given by Equations (5.4).

To solve this problem, we split up the function $T(r, \phi, z)$ by the principle of superposition into the solution of two simpler problems in the form

$$T(r, \phi, z) = T^{(r)}(r, \phi, z) + T^{(z)}(r, \phi, z) \tag{5.75}$$

where the $T^{(\ell)}$ ($\ell \equiv r, z$) are the solutions of the following simpler problems each with two nonhomogeneous boundary conditions

$$\left\{ \frac{1}{r}\frac{\partial}{\partial r}\left(r\frac{\partial}{\partial r}\right) + \frac{1}{r^2}\frac{\partial^2}{\partial\phi^2} + \frac{\partial^2}{\partial z^2} \right\} T^{(\ell)}(r, \phi, z) = 0$$

$$\text{in } r_0 < r < r_1, \quad 0 \leq \phi \leq 2\pi, \quad z_0 < z < z_1 \tag{5.76a}$$

$$\left\{ \alpha_{kr} - (-1)^k \beta_{kr}\frac{\partial}{\partial r} \right\} T^{(\ell)}(r_k, \phi, z) = \delta_{r\ell}\phi(r_k, \phi, z), \quad k = 0, 1$$

$$\tag{5.76b, c}$$

$$\left\{ \alpha_{kz} - (-1)^k \beta_{kz}\frac{\partial}{\partial z} \right\} T^{(\ell)}(r, \phi, z_k) = \delta_{z\ell}\phi(r, \phi, z_k), \quad k = 0, 1$$

$$\tag{5.76d, e}$$

where $\ell \equiv r, z$ and $\delta_{r\ell}$, $\delta_{z\ell}$ are the Kronecker delta.

We now present the solutions for the functions $T^{(z)}(r, \phi, z)$ and $T^{(r)}(r, \phi, z)$.

Solution for $T^{(z)}(r, \phi, z)$

We set in Equations (5.76), $\ell = z$, and by comparing the resulting problem with that given by Equations (2.52) we write

$$t = z, \quad w(x) = 1, \quad \delta_k = \alpha_{kz}, \quad \gamma_k = \beta_{kz}$$

$$L_t = -\frac{\partial^2}{\partial z^2}, \quad L = -\left[\frac{1}{r}\frac{\partial}{\partial r}\left(r\frac{\partial}{\partial r}\right) + \frac{1}{r^2}\frac{\partial^2}{\partial\phi^2} \right]$$

$$B = \alpha_{kr} - (-1)^k \beta_{kr}\frac{\partial}{\partial r} \quad \text{at } r = r_k \tag{5.77}$$

The eigenvalue problem appropriate for system (5.76) is now a special case of the eigenvalue problem (3.55). Then the solution for $T^{(z)}(r, \phi, z)$ is immediately obtained according to the solution (2.60) as

$$
T^{(z)}(r, \phi, z) = \sum_{\ell=1}^{\infty} \sum_{m=0}^{\infty} \frac{\psi(\eta_{\ell m} r)}{N_{\ell m} N_m} \begin{Bmatrix} \cos(m\phi) \\ \sin(m\phi) \end{Bmatrix} \sum_{k=0}^{1} (-1)^{1-k}
$$

$$
\times \int_{r_0}^{r_1} \int_0^{2\pi} \psi(\eta_{\ell m} r') \begin{Bmatrix} \cos(m\phi') \\ \sin(m\phi') \end{Bmatrix} \phi(r', \phi', z_{1-k}) \, d\phi' \, dr'
$$

$$
\times \frac{\alpha_{kz} \sinh[\eta_{\ell m}(z_k - z)] - (-1)^k \beta_{kz} \eta_{\ell m} \cosh[\eta_{\ell m}(z_k - z)]}{D}
$$

$$
k = 0,1 \quad (5.78a)
$$

where

$$
D = (\alpha_{0z}\alpha_{1z} + \beta_{0z}\beta_{1z}\eta_{\ell m}^2)\sinh[\eta_{\ell m}(z_1 - z_0)]
$$

$$
+ (\alpha_{0z}\beta_{1z} + \alpha_{1z}\beta_{0z})\eta_{\ell m}\cosh[\eta_{\ell m}(z_1 - z_0)]
$$

and the eigenfunctions $\psi(\beta_{\ell m} r)$, the normalization integrals $N_{\ell m}$ and N_m, and the eigencondition for the determination of $\eta_{\ell m}$ are given, respectively, by

$$
\psi(\eta_{\ell m} r) = J_m(\eta_{\ell m} r)
$$

$$
- \frac{(\alpha_{0r} r_0 + \beta_{0r} m) J_m(\eta_{\ell m} r_0) - \beta_{0r} r_0 \eta_{\ell m} J_{m-1}(\eta_{\ell m} r_0)}{(\alpha_{0r} r_0 + \beta_{0r} m) Y_m(\eta_{\ell m} r_0) - \beta_{0r} r_0 \eta_{\ell m} Y_{m-1}(\eta_{\ell m} r_0)} Y_m(\beta_{0m} r)
$$

$$
(5.78b)
$$

$$
N_{\ell m} = \frac{2}{(\pi \eta_{\ell m})^2} \sum_{k=0}^{1} (-1)^{k+1}
$$

$$
\times \frac{(\alpha_{kr} r_k)^2 + (\eta_{\ell m} \beta_{kr} r_k)^2 - (\beta_{kr} m)^2}{\left[(\alpha_{kr} r_k + (-1)^k \beta_{kr} m) Y_m(\eta_{\ell m} r_k) - (-1)^k \beta_{kr} \eta_{\ell m} r_k Y_{m-1}(\eta_{\ell m} r_k) \right]^2}
$$

$$
k = 0,1 \quad (5.78c)
$$

$$
N_m = \begin{cases} \pi & \text{for } m = 1,2,3,\dots \\ 2\pi & \text{for } m = 0 \end{cases} \quad (5.78d)
$$

and the $\eta_{\ell m}$ are the roots of the following eigencondition

$$
\sum_{k=0}^{1} (-1)^k \frac{\left[\alpha_{kr} r_k + (-1)^k \beta_{kr} m \right] J_m(\eta r_k) - (-1)^k \beta_{kr} r_k \beta J_{m-1}(\eta r_k)}{\left[\alpha_{kr} r_k + (-1)^k \beta_{kr} m \right] Y_m(\eta r_k) - (-1)^k \beta_{kr} r_k \eta Y_{m-1}(\eta r_k)} = 0
$$

$$
(5.78e)
$$

Before discussing the solution for the function $T^{(r)}$, we examine some special cases of the solution (5.78).

Special Case 1 If the source function $\phi(r, \phi, z_k)$ is independent of the ϕ variable, then the solution (5.78) becomes independent of ϕ and simplifies to

$$T^{(z)}(r, z) = \sum_{i=1}^{\infty} \frac{\psi(\mu_i r)}{N_i} \sum_{k=0}^{1} (-1)^{1-k} \int_{r_0}^{r_1} \psi(\mu_i r') \phi(r', z_{1-k}) \, dr'$$

$$\times \frac{\alpha_{kz} \sinh[\mu_i(z_k - z)] - (-1)^k \beta_{kz} \mu_i \cosh[\mu_i(z_k - z)]}{D}$$

$$(5.79a)$$

where

$$D = \left(\alpha_{0z}\alpha_{1z} + \beta_{0z}\beta_{1z}\mu_i^2\right) \sinh[\mu_i(z_1 - z_0)]$$

$$+ \left(\alpha_{0z}\beta_{1z} + \alpha_{1z}\beta_{0z}\right) \mu_i \cosh[\mu_i(z_1 - z_0)]$$

and we set $\beta_{\ell 0} \equiv \mu_i$ and $\psi(\beta_{\ell 0} r) \equiv \psi(\mu_i r)$.

The eigenfunctions $\psi(\mu_i r)$, the normalization integral N_i, and the eigencondition for the determination of μ_i are given, respectively, by

$$\psi(\mu_i r) = J_0(\mu_i r) - \frac{\alpha_{0r} J_0(\mu_i r_0) + \beta_{0r} \mu_i J_1(\mu_i r_0)}{\alpha_{0r} Y_0(\mu_i r_0) + \beta_{0r} \mu_i Y_1(\mu_i r_0)} Y_0(\mu_i r) \quad (5.79b)$$

$$N_i = \frac{2}{(\pi\mu_i)^2} \sum_{k=0}^{1} (-1)^{k+1} \frac{\alpha_{kr}^2 + (\beta_{kr}\mu_i)^2}{\left[\alpha_{kr} Y_0(\mu_i r_k) + (-1)^k \beta_{kr}\mu_i Y_1(\mu_i r_k)\right]^2}$$

$$(5.79c)$$

and the μ_i are the roots of the following eigencondition

$$\sum_{k=0}^{1} (-1)^k \frac{\alpha_{kr} J_0(\mu r_k) + (-1)^k \beta_{kr}\mu J_1(\mu r_k)}{\alpha_{kr} Y_0(\mu r_k) + (-1)^k \beta_{kr}\mu Y_1(\mu r_k)} = 0 \quad (5.79d)$$

Examples of Hollow Cylinder

Example 5.13 Consider the steady-state diffusion in a hollow cylinder $a \leq r \leq b$, $0 \leq z \leq c$, subject to the boundary conditions

$$T_{r=a} = 0, \quad \left(\frac{\partial T}{\partial r}\right)_{r=b} = 0, \quad \left(\frac{\partial T}{\partial z}\right)_{z=0} = q_0, \quad T_{z=c} = 0 \quad (5.80a)$$

Solution By comparing this problem with the ones given by Equations (5.76) we write

$$\beta_{0r} = \alpha_{1r} = \alpha_{0z} = \beta_{1z} = 0$$

$$\alpha_{0r} = \beta_{1r} = \beta_{0z} = \alpha_{1z} = 1 \text{ and } \phi(r,0) = -q_0 \qquad (5.80b)$$

Then the solution is obtained as a special case of Equations (5.79), namely,

$$T = q_0 a \pi \sum_{i=1}^{\infty} \frac{\sinh\{\vartheta_i(c-z)/a\}[Y_0(\vartheta_i)J_0(\vartheta_i r/a) - J_0(\vartheta_i)Y_0(\vartheta_i r/a)]}{\vartheta_i \cosh(\vartheta_i c/a)[J_0^2(\vartheta_i)/J_1^2(\vartheta_i b/a) - 1]}$$

$$(5.81a)$$

where the ϑ_i are the roots of the transcendental equation

$$J_0(\nu)Y_1\left(\frac{\nu b}{a}\right) - Y_0(\nu)J_1\left(\frac{\nu b}{a}\right) = 0 \qquad (5.81b)$$

Example 5.14   Consider the steady-state diffusion in a hollow semi-infinite cylinder $a \le r \le b$, $0 \le z < \infty$, subject to the boundary conditions

$$T_{r=a} = T_{r=b} = 0, \qquad T_{z=0} = T_0 \qquad (5.82a)$$

Solution By comparing this problem with the ones given by Equations (5.76) we write

$$\beta_{0r} = \beta_{1r} = \beta_{0z} = \alpha_{1z} = 0, \qquad \phi(r,0) = T_0$$

$$\alpha_{0r} = \alpha_{1r} = \alpha_{0z} = \beta_{1z} = 1 \qquad \phi(r,\infty) = 0 \qquad (5.82b)$$

Then the solution is obtained as a special case of Equations (5.79) after using $c \to \infty$

$$T = T_0 \pi \sum_{i=1}^{\infty} \frac{Y_0(\vartheta_i)J_0(\vartheta_i r/a) - J_0(\vartheta_i)Y_0(\vartheta_i r/a)}{1 + J_0(\vartheta_i)/J_0(\vartheta_i b/a)} e^{-\vartheta_i z/a} \qquad (5.83a)$$

where the ϑ_i are the roots of the transcendental equation

$$J_0(\nu)Y_0\left(\frac{\nu b}{a}\right) - Y_0(\nu)J_0\left(\frac{\nu b}{a}\right) = 0 \qquad (5.83b)$$

Special Case 2   We now restrict the solution (5.79) for a solid cylinder confined to a finite region $0 \le r \le r_1$, $z_0 \le z \le z_1$. This condition implies that we set in Equations (5.79a) $r_0 = 0$, $\alpha_{0r} = 0$, and $Y_1(0) = \infty$. Then the eigenfunctions, normalization integral, and eigencondition given by Equations (5.79b, c, d), respectively, simplify to

$$\psi(\mu_i r) = J_0(\mu_i r) \tag{5.84a}$$

$$N_i = \frac{r_1^2}{2}\left[J_0^2(\mu_i r_1) + J_1^2(\mu_i r_1) \right] \tag{5.84b}$$

$$\frac{J_0(\mu r_1)}{J_1(\mu r_1)} = \frac{\beta_{1r}}{\alpha_{1r}}\mu \tag{5.84c}$$

We now proceed with the analysis to solve problem (5.76) for the case $\ell = r$. For simplicity, we consider the problem for a solid cylinder, $0 \le r \le r_1$, instead of a hollow cylinder as given previously. Furthermore, in order to illustrate the application to specific cases, we consider solutions for the two-dimensional situations involving the functions $T^{(r)}(r, z)$ and $T^{(r)}(r, \phi)$ instead of the three-dimensional case $T^{(r)}(r, \phi, z)$.

Solution for $T^{(r)}(r, z)$

First we set in Equations (5.76) $\ell = r$; for a solid cylinder with azimuthal symmetry we have $r_0 = 0$ and $\phi(r_k, \phi, z) \equiv \phi(r_k, z)$. Then Equations (5.76) are simplified to

$$\left[\frac{1}{r}\frac{\partial}{\partial r}\left(r\frac{\partial}{\partial r} \right) + \frac{\partial^2}{\partial z^2} \right] T^{(r)}(r, z) = 0 \quad \text{in } 0 < r < r_1, \qquad z_0 < z < z_1$$

$$\tag{5.85a}$$

$$\left[\alpha_{kz} - (-1)^k \beta_{kz}\frac{\partial}{\partial z} \right] T^{(r)}(r, z) = 0, \qquad k = 0, 1 \tag{5.85b, c}$$

$$\frac{\partial}{\partial r} T^{(r)}(0, z) = 0 \tag{5.85d}$$

$$\left(\alpha_{1r} + \beta_{1r}\frac{\partial}{\partial r} \right) T^{(r)}(r_1, z) = \phi(r_1, z) \tag{5.85e}$$

A comparison of this problem with that given by Equations (2.52) reveals that

$$t = r, \qquad w(\mathbf{x}) = 1, \qquad L_t = -\frac{1}{r}\frac{\partial}{\partial r}\left(r\frac{\partial}{\partial r} \right), \qquad L = -\frac{\partial^2}{\partial z^2} \tag{5.86}$$

and the eigenvalue problem for this particular case coincides with the eigenvalue problem (3.17) considered in Example (3.1). Hence, the eigenfunctions, eigenvalues, and eigencondition are obtainable from Equations (3.20). The solution of the problem (5.85) is immediately obtained, according to solution (2.63), as

$$T^{(r)}(r, z) = \sum_{i=1}^{\infty} \frac{\psi_i(z)}{N_i} \int_{z_0}^{z_1} \psi_i(z')\phi(r, z')\, dz' \frac{I_0(\mu_i r)}{\alpha_{1r} I_0(\mu_i r_1) + \beta_{1r}\mu_i K_0(\mu_i r_1)}$$

(5.87a)

where $\psi_i(z)$, N_i, and the eigencondition are determined from Equations (3.20) as

$$\psi_i(z) = \beta_{0z}\cos[\mu_i(z - z_0)] + \frac{\alpha_{0z}}{\mu_i}\sin[\mu_i(z - z_0)]$$

(5.87b)

$$N_i = \frac{1}{2\mu_i^2}\left\{\left[\alpha_{0z}^2 + (\beta_{0z}\mu_i)^2\right]\left[\frac{\alpha_{1z}\beta_{1z}}{\alpha_{1z}^2 + (\beta_{1z}\mu_i)^2} + z_1 - z_0\right] + \alpha_{0z}\beta_{0z}\right\}$$

(5.87c)

and the eigenvalues μ_i are the roots of the following eigencondition

$$(\alpha_{0z}\beta_{1z} + \alpha_{1z}\beta_{0z})\cos[\mu(z_1 - z_0)] + (\alpha_{0z}\alpha_{1z} - \beta_{0z}\beta_{1z}\mu^2)\frac{\sin \mu(z_1 - z_0)}{\mu} = 0$$

(5.87d)

Solution for $T^{(r)}(r, \phi)$

We set $\ell = r$ in Equations (5.76) and consider the resulting problem for a solid cylinder with no dependence on the z variable; that is, $r_0 = 0$ and $\phi(r_k, \phi, z) \equiv \phi(r_k, \phi)$. Then Equations (5.76) are simplified to

$$\left[\frac{1}{r}\frac{\partial}{\partial r}\left(r\frac{\partial}{\partial r}\right) + \frac{1}{r^2}\frac{\partial^2}{\partial \phi^2}\right]T^{(r)}(r, \phi) = 0 \quad \text{in } 0 < r < r_1, \qquad 0 \le \phi \le 2\pi$$

(5.88a)

$$\left(\alpha_{1r} + \beta_{1r}\frac{\partial}{\partial r}\right)T^{(r)}(r, \phi) = \phi(r_1, \phi)$$

(5.88b)

In order to compare this problem with that given by Equations (2.52) we write Equation (5.88a) in the form

$$\left[r\frac{\partial}{\partial r}\left(r\frac{\partial}{\partial r}\right) + \frac{\partial^2}{\partial \phi^2}\right]T^{(r)}(r, \phi) = 0$$

(5.88c)

A comparison of the problem (5.88) with that given by Equations (2.52) reveals that

$$t = r, \qquad w(\mathbf{x}) = 1, \qquad L_t = -r\frac{\partial}{\partial r}\left(r\frac{\partial}{\partial r}\right), \qquad L = -\frac{\partial^2}{\partial\phi^2} \qquad (5.89)$$

Then the eigenvalue problem coincides with the system (3.57a) for $0 \le \phi \le 2\pi$ (i.e., $\phi_0 = 0$ and $\phi_1 = 2\pi$). The eigenvalues for this eigenvalue problem are $\mu_i = i$ ($i = 0, 1, 2, \ldots$) [see Equation (3.61d)] and every eigenvalue i corresponds to two independent eigenfunctions $\cos(i\phi)$ and $\sin(i\phi)$. The normalization integrals are defined by Equations (3.61f, g). Finally, the solution of problem (5.88) is obtainable from Equation (2.66b).

Substituting these results in Equation (2.66b), the solution for $T^{(r)}(r, \phi)$ is determined as

$$T^{(r)}(r, \phi) = \frac{1}{\pi}\sum_{i=0}^{\infty}\left\{\begin{matrix}\cos(\phi i)\\ \sin(\phi i)\end{matrix}\right\}\int_0^{2\pi}\left\{\begin{matrix}\cos(\phi' i)\\ \sin(\phi' i)\end{matrix}\right\}\phi(r_1, \phi')\,d\phi'\frac{(r/r_1)^i}{\alpha_{1r} + (\beta_{1r}/r_1)i}$$

$$(5.90a)$$

where π should be replaced by 2π for $i = 0$.

After rearranging, Equation (5.90a) can be expressed in the form

$$T^{(r)}(r, \phi) = \frac{1}{\pi}\sum_{i=0}^{\infty}\frac{(r/r_1)^i}{\alpha_{1r} + (\beta_{1r}/r_1)i}\int_0^{2\pi}\cos[(\phi - \phi')i]\phi(r_1, \phi')\,d\phi'$$

$$(5.90b)$$

where π should be replaced by 2π when $i = 0$ (see [1a], p. 183).

Examples of a Solid Cylinder

Example 5.15   Consider the steady-state diffusion in a solid cylinder $0 \le x \le a$, $0 \le z \le c$, subject to the boundary conditions

$$T_{r=a} = T_{z=0} = 0, \qquad T_{z=c} = T_0 f(r) \qquad (5.91a)$$

Compute the temperature distribution for the cases of $f(x)$ given by

(a) $f(r) = 1$

(b) $f(r) = \begin{cases} 0 & \text{for } 0 \le r \le a_1 \\ 1 & \text{for } a_1 < r \le a \end{cases}$ $\qquad (5.91b)$

Solution   By comparing the boundary conditions (5.91a) with boundary conditions (5.76) we find

$$\beta_{1r} = \beta_{0z} = \beta_{1z} = 0$$

$$\alpha_{1r} = \alpha_{0z} = \alpha_{1z} = 1$$

$$\phi(r,0) = 0, \qquad \phi(r,c) = T_0 f(r) \qquad (5.91c)$$

Then from Equations (5.79a), after using Equations (5.84), we obtain

$$T = T_0 \left(\frac{2}{a^2} \right) \sum_{i=1}^{\infty} \frac{J_0(\nu_i r/a)}{J_1^2(\nu_i)} \frac{\sinh(\nu_i z/a)}{\sinh(\nu_i c/a)} \int_0^1 r J_0\left(\frac{\nu_i r}{a} \right) f(r) \, dr \quad (5.92a)$$

where the ν_i are the roots of the equation

$$J_0(\nu) = 0 \qquad (5.92b)$$

The temperature distribution *for case* a becomes

$$T = T_0 2 \sum_{i=1}^{\infty} \frac{J_0(\nu_i r/a)}{\nu_i J_1(\nu_i)} \frac{\sinh(\nu_i z/a)}{\sinh(\nu_i c/a)} \qquad (5.92c)$$

The temperature distribution *for case* b becomes

$$T = T_0 2 \sum_{i=1}^{\infty} \frac{J_0(\nu_i r/a)}{\nu_i J_1^2(\nu_i)} \left[J_1(\nu_i) - \frac{a_1}{a} J_1\left(\nu_i \frac{a_1}{a} \right) \right] \frac{\sinh(\nu_i z/a)}{\sinh(\nu_i c/a)} \qquad (5.92d)$$

Example 5.16   Develop an expression for the steady-state temperature distribution in a cylinder $0 \leq r \leq a$, $0 \leq z \leq c$, subject to the boundary conditions

$$\left(T + k \frac{\partial T}{\partial r} \right)_{r=r_1} = 0, \qquad \left(\frac{\partial T}{\partial z} \right)_{z=0} = -q_0, \qquad T_{z=c} = 0 \quad (5.93a)$$

Solution   For the case considered here we have

$$\beta_{1r} = k, \qquad \alpha_{0z} = \beta_{1z} = 0$$

$$\alpha_{1r} = 1, \qquad \beta_{0z} = \alpha_{1z} = 1 \qquad (5.93b)$$

$$\phi(r,0) = q_0$$

Then from Equations (5.79a), after using Equations (5.84), we obtain

$$T = q_0 a 2 \sum_{i=1}^{\infty} \frac{J_0(\nu_i r/a) J_1(\nu_i)}{\nu_i^2 \{ J_0^2(\nu_i) + J_1^2(\nu_i) \}} \frac{\sinh\{ \nu_i [(c-z)/a] \}}{\cosh[\nu_i(c/a)]} \qquad (5.94a)$$

where the ν_i are the roots of the equation

$$J_0(\nu) - \left(\frac{k}{a}\right)\nu J_1(\nu) = 0 \tag{5.94b}$$

Example 5.17   Consider the steady-state temperature distribution in a cylinder $0 \le r \le a$, $0 \le z \le c$, subject to the boundary conditions

$$T_{z=0} = T_{z=c} = 0, \qquad T_{r=r_1} = T_0 f(z) \tag{5.95a}$$

Compute the temperature distribution for the cases

(a) $f(z) = 1$

(b) $f(z) = \begin{cases} 1 & \text{for } 0 \le z \le \dfrac{c}{2} \\ 0 & \text{for } \dfrac{c}{2} < z \le c \end{cases}$

(c) $f(z) = \sin\left(\dfrac{\pi z}{c}\right)$ \qquad\qquad (5.95b)

Solution   By the comparison of Equations (5.76) with boundary conditions (5.95a) we find

$$\beta_{0z} = \beta_{1z} = \beta_{1r} = 0$$

$$\alpha_{0z} = \alpha_{1z} = \alpha_{1r} = 1$$

$$\phi(r_1, z) = T_0 f(z) \tag{5.95c}$$

Then from Equations (5.87) we find

$$T = T_0 \frac{2}{c} \sum_{i=1}^{\infty} \frac{I_0(\pi i r/c)}{I_0(\pi i a/c)} \sin\left(\frac{\pi i z}{c}\right) \int_0^c f(z) \sin\left(\frac{\pi i z}{c}\right) dz \tag{5.96a}$$

The temperature distribution *for case* a becomes

$$T = T_0 \frac{4}{\pi} \sum_{i=1}^{\infty} \frac{1}{2i-1} \frac{I_0\{[(2i-1)/c]\pi r\}}{I_0\{[(2i-1)/c]\pi a\}} \sin\left(\frac{2i-1}{c}\pi z\right) \tag{5.96b}$$

The temperature distribution *for case* b becomes

$$T = T_0 \frac{2}{\pi} \sum_{i=1}^{\infty} \frac{1 - \cos(\pi i/2)}{i} \frac{I_0(\pi i r/c)}{I_0(\pi i a/c)} \sin\left(\frac{\pi i z}{c}\right) \tag{5.96c}$$

The temperature distribution *for case* c becomes

$$T = T_0 \frac{I_0(\pi r/c)}{I_0(\pi a/c)} \sin\left(\frac{\pi z}{c}\right) \tag{5.96d}$$

5.5   MULTIDIMENSIONAL STEADY-STATE PROBLEMS IN THE SPHERICAL COORDINATE SYSTEM

In this section we first consider the solution of a three-dimensional problem for a finite spherical region in the spherical coordinate system (r, θ, ϕ) and then obtain solutions for some specific situations as special cases from the general solution.

Three-Dimensional Finite Solid Sphere

We consider steady-state diffusion in a three-dimensional spherical finite region specified by the coordinates $0 \le r \le r_1$, $0 \le \theta \le \pi$, $0 \le \phi \le 2\pi$ and subject to a nonhomogeneous boundary condition of the third kind at the boundary surface $r = r_1$. The mathematical formulation of the problem is given by Equations (5.5). We consider the form given by Equation (5.5c) and write

$$\left\{ \frac{\partial}{\partial r}\left(r^2 \frac{\partial}{\partial r} \right) + \frac{\partial}{\partial z}\left[(1 - z^2)\frac{\partial}{\partial z} \right] + \frac{1}{1 - z^2}\frac{\partial^2}{\partial \phi^2} \right\} T(r, z, \phi) = 0$$

$$\text{in } 0 \le r \le r_1, \quad -1 \le z \le 1, \quad 0 \le \phi \le 2\pi \tag{5.97a}$$

where

$$z = \cos \theta \tag{5.97b}$$

and the boundary condition is taken the same as given by Equation (5.5d), that is,

$$\left\{ \alpha_{r1} + \beta_{r1}\frac{\partial}{\partial r} \right\} T(r_1, z, \phi) = \phi(r_1, z, \phi) \tag{5.97c}$$

By comparing Equation (5.97a) with Equation (2.52) we write

$$t = r, \quad w(\mathbf{x}) \equiv 1, \quad L_t = -\frac{\partial}{\partial r}\left(r^2 \frac{\partial}{\partial r} \right)$$

$$L \equiv -\left\{ \frac{\partial}{\partial z}\left[(1 - z^2)\frac{\partial}{\partial z} \right] + \frac{1}{1 - z^2}\frac{\partial^2}{\partial \phi^2} \right\}$$

$$B = \alpha_{r1} + \beta_{r1}\frac{\partial}{\partial r} \quad \text{at } r = r_1 \tag{5.98}$$

The eigenvalue problem for this particular problem is a special case of problem (3.62) and the solution of problem (5.97) is obtainable according to Equation (2.69). Then the solution for $T(r, z, \phi)$ is determined as

$$T(r, z, \phi) = \frac{1}{\pi} \sum_{n=0}^{\infty} \sum_{m=0}^{n} \frac{2n+1}{2} \frac{(n-m)!}{(n+m)!} P_n^m(z) \int_{z=-1}^{1} P_n^m(z')$$

$$\times \int_{\phi'=0}^{2\pi} \cos[m(\phi - \phi')] \phi(r_1, z, \phi') \, d\phi' \, dz' \frac{(r/r_1)^n}{\alpha_{r1} + \beta_{r1}(n/r_1)}$$

$$(5.99)$$

where π should be replaced by 2π when $m = 0$.

Finite, Solid Hemisphere

We now consider steady-state diffusion in a hemispherical finite region specified by the coordinates $0 \le r \le r_1$, $0 \le z \le 1$, and $0 \le \phi \le 2\pi$. The boundary condition at the spherical surface $r = r_1$ is a nonhomogeneous boundary condition of the third kind as given by Equation (5.5d).

The solution of this problem can be expressed in the same form as that given by Equation (5.99) for the full sphere, provided that the degree n on the Legendre function is so chosen as to satisfy the requirement of boundary condition at $z = 0$. With this consideration we examine the following cases.

If the source function $\phi(r_1, z, \phi)$ is independent of the ϕ variable, that is, $\phi(r_1, z, \phi) = \phi(r_1, z)$, we set $m = 0$ and the integration with respect to ϕ is performed. Then Equation (5-99) simplify to

(1) $$T(r, z) = \sum_{n=0}^{\infty} (4n + 3) P_{2n+1}(z)$$

$$\times \int_{z=0}^{1} P_{2n+1}(z') \phi(r_1, z') \frac{(r/r_1)^{2n+1}}{\alpha_{r1} + \beta_{r1}[(2n+1)/r_1]} dz' \quad (5\text{-}100)$$

for $T(r, 0) = 0$, where we replaced n by $(2n + 1)$ in order to satisfy the boundary condition at $z = 0$.

(2) $$T(r, z) = \sum_{n=0}^{\infty} (4n + 1) P_{2n}(z) \int_{z=0}^{1} P_{2n}(z') \phi(r_1, z') \frac{(r/r_1)^{2n}}{\alpha_{r1} + \beta_{r1}(2n/r_1)} dz'$$

$$(5\text{-}101)$$

for

$$\frac{\partial T(r,0)}{\partial r} = 0$$

where we replaced n by $2n$ in order to satisfy the boundary condition at $z = 0$.

Example 5.18 Develop an expression for the steady-state temperature distribution in a solid hemisphere, confined to the region $0 \le r \le a$, $0 \le \theta \le \pi/2$, and the base of which (i.e., $\theta = \pi/2$) is kept at zero temperature while the spherical surface $r = a$ is subjected to the boundary condition of the third kind, given by

$$\left(T + \kappa \frac{\partial T}{\partial r}\right)_{r=a} = T_0 \tag{5-102}$$

Solution   For the case considered

$$r_1 = a, \qquad \alpha_{1r} = 1, \qquad \beta_{1r} = \kappa, \qquad \phi(r_1, z) = T_0 \tag{5.103}$$

and the solution (5-100) becomes

$$T(r, z) = \frac{T_0}{\sqrt{\pi}} \sum_{n=0}^{\infty} \frac{(4n + 3)(-1)^n \Gamma[(2n + 3)/2]}{(n + 1)!(2n + 1)(1 + \kappa[(2n + 1)/a])} \left(\frac{r}{a}\right)^{2n+1} P_{2n+1}(z)$$

$$\tag{5.104}$$

Example 5.19   Develop an expression for the steady-state temperature distribution in a solid sphere of radius $r = a$, subject to the boundary condition shown in Figure 5.1a.

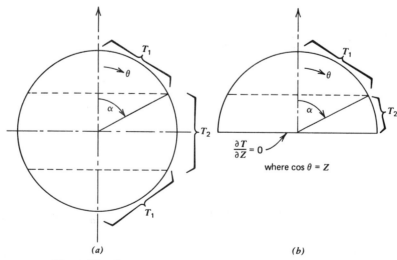

(a)   *(b)*

Figure 5.1   Geometry and boundary conditions for Example 5.19.

Solution For the boundary condition assumed, the temperature distribution possesses even symmetry with respect to the plane $\theta = \pi/2$. Therefore, one can formulate an equivalent model shown on Figure 5.1*b*. For this case we have

$$r_1 = a, \qquad \alpha_{r1} = 1, \qquad \beta_{r1} = 0$$

$$\phi(r_1, z) = \phi(a, \theta) = \begin{cases} T_1 & \text{for } 0 \le \theta \le \alpha \\ T_2 & \text{for } \alpha < \theta \le \dfrac{\pi}{2} \end{cases} \qquad (5.105)$$

and solution (5-101) becomes

$$T = T_1 + (T_2 - T_1)$$

$$\times \left\{ \cos\alpha + \sum_{n=1}^{\infty} \left(\frac{r}{a}\right)^{2n} \left[P_{2n+1}(\cos\alpha) - P_{2n-1}(\cos\alpha) \right] P_{2n}(\cos\theta) \right\}$$

$$(5.106)$$

where $\cos\theta \equiv z$.

REFERENCES

1. (a) M. N. Özişik, *Boundary Value Problems of Heat Conduction*, International Textbook Co., Scranton, PA, 1968. (b) M. N. Özişik, *Heat Conduction*, Wiley, New York, 1980.

2. H. S. Carslaw and J. C. Jaeger, *Conduction of Heat in Solids*, Clarendon Press, London, 1959.

3. J. Crank, *The Mathematics of Diffusion*, second ed., Clarendon Press, London, 1975.

4. M. D. Mikhailov and M. N. Özişik, An Alternative General Solution of the Steady-State Heat Diffusion Equation, *Int. J. Heat and Mass Transfer*, **23**, 609–612 (1980).

5. Yu. Ia. Iossel, *Calculation of Potential Field in Power Engineering*, Energiya, Moscow, 1978.

CHAPTER SIX

Class I Solutions Applied
to Heat Flow Through Fins

Science is built up with facts, as a house is with stones.
But a collection of facts is no more a science than a
heap of stones is a house.

Jules Henry Poincaré
1854–1912

In Chapter 4, we developed a number of solutions to the Class I problem. In this Chapter we illustrate the application of such results to obtain solutions to the problems of temperature distribution and heat flow through extended surfaces. Fins or the extended surfaces are extensively used in engineering applications to increase the heat transfer efficiency of surfaces. A wide variety of fin geometries is available and the determination of the fin efficiency for each specific fin geometry is of interest. Once the temperature distribution through the fin is known, the heat transfer rate and the fin efficiency can be readily determined. In this section we present the generalized fin equation and develop its solution as a special case of the Class I problem. The general solution to the fin problem obtained in this manner is then utilized to determine the temperature distribution, heat flow rate, and fin efficiency for a number of specific fin geometries.

6.1 GENERAL ANALYSIS OF HEAT FLOW THROUGH FINS

A large variety of fin geometries are used in heat transfer applications; typical fin profiles are illustrated in Figure 6.1. A comprehensive treatment of heat flow through fins is given in Reference 1.

To determine heat transfer through a fin, the temperature distribution through the fin is needed. We now develop the generalized fin equation and its

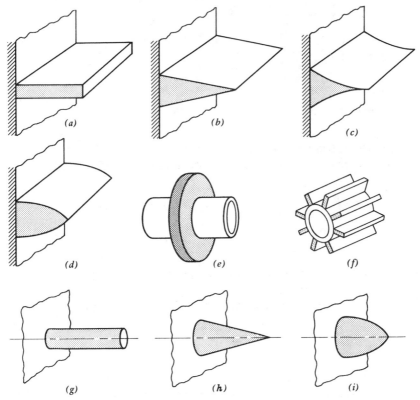

Figure 6.1   Typical fin profiles: (a) Longitudinal fin of rectangular profile; (b) longitudinal fin of triangular profile; (c) longitudinal fin of concave parabolic profile; (d) longitudinal fin of convex parabolic profile; (e) radial fin of rectangular profile; (f) cylindrical tube with fins of radial profile; (g) cylindrical spine; (h) conical spine; (i) spine with convex parabolic profile.

general formal solution. The temperature distribution and heat flow through a large variety of fin geometries are then determined by utilizing this solution.

Generalized Fin Equation

Basic assumptions pertinent to heat transfer through a fin include the following: the temperature varies in one direction only (i.e., the temperature at any cross-section is uniform); the temperature remains constant with time (i.e., steady-state); the material is homogeneous; the thermal conductivity is constant; there are no energy sources or sinks within the fin; the temperature of the surrounding fluid is uniform.

To derive the fin equation, let x be the coordinate measured along the fin as illustrated in Figure 6.2. We consider a volume element of the fin of length Δx, having a cross-sectional area $A(x)$ normal to the x direction. The steady-state temperature distribution within the fin is governed by the Laplace equation,

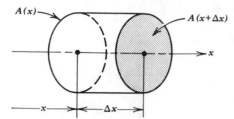

Figure 6.2   Control volume for the derivation of fin equation.

$\nabla^2 T = 0$. The integration of this Laplace equation over this volume element and then the conversion of the volume integral to the surface integral yields

$$\int_{A(x+\Delta x)} \frac{\partial T}{\partial x} dA - \int_{A(x)} \frac{\partial T}{\partial x} dA + \int_{\Delta s} \frac{\partial T}{\partial n} ds = 0 \qquad (6.1)$$

where

$A(x)$, $A(x + \Delta x)$ = cross-sectional area at the positions x and $x + \Delta x$, respectively

Δs = lateral surface area of the volume element

$\dfrac{\partial}{\partial n}$ = normal derivative in the outward direction at the lateral surface Δs

Various terms in Equation (6.1) are determined as now described.

The lateral surface Δs is in contact with a fluid at temperature T_∞. Then an energy balance for this surface yields

$$-k \int_{\Delta s} \frac{\partial T}{\partial n} ds = h(x)[T_s(x) - T_\infty] \Delta s \qquad (6.2a)$$

where

$h(x)$ = heat transfer coefficient between the lateral surface Δs and the fluid

k = thermal conductivity of the fin material

$T_s(x)$ = temperature of the fin surface

T_∞ = temperature of the surrounding fluid

The average temperature $T_{\text{fin}}(x)$ over the cross-section area $A(x)$ is defined as

$$T_{\text{fin}}(x) = \frac{1}{A(x)} \int_{A(x)} T \, dA \qquad (6.2b)$$

and from the assumption that the temperature at any cross-section is uniform, it follows

$$T_s(x) = T_{\text{fin}}(x) \qquad (6.2c)$$

Introducing Equations (6.2) into (6.1) we obtain

$$A(x + \Delta x)\frac{dT_{\text{fin}}(x + \Delta x)}{dx} - A(x)\frac{dT_{\text{fin}}(x)}{dx} - \frac{h(x)}{k}\left[T_{\text{fin}}(x) - T_{\infty}\right]\Delta s = 0$$

$$(6.3)$$

Dividing Equation (6.3) by Δx and considering the limit $\Delta x \to 0$, we obtain the *generalized fin equation* as

$$\frac{d}{dx}\left[A(x)\frac{dT_{\text{fin}}(x)}{dx}\right] - \frac{h(x)}{k}\frac{dS(x)}{dx}\left[T_{\text{fin}}(x) - T_{\infty}\right] = 0 \qquad (6.4)$$

This equation holds from the fin base at $x = x_b$ to the fin tip at $x = x_t$.

Boundary Conditions

To solve the fin Equation (6.4) two boundary conditions are needed: one for the fin base $x = x_b$ and the other for the fin tip $x = x_t$.

For most applications the temperature at the fin base is uniform and known, hence this boundary condition for the fin base is taken as

$$T_{\text{fin}}(x) = T_b \quad \text{at } x = x_b \qquad (6.5a)$$

where T_b is a specified fin base temperature.

If heat transferred through the fin tip is negligible compared with that through the lateral surface of the fin, as is the case for many applications, the boundary condition for the fin tip is taken as

$$\frac{dT_{\text{fin}}(x)}{dx} = 0 \quad \text{at } x = x_t \qquad (6.5b)$$

A more general boundary condition, allowing for convection from the fin tip, can also be considered. The general fin Equation (6.4) is not dependent upon the boundary conditions; therefore, it can be used for any combination of suitable boundary conditions at the fin base and fin tip.

In the following analysis we consider the fin Equation (6.4) subject to the boundary conditions (6.5), and obtain its solution by appropriate simplification of the general solutions developed in Chapter 4. The general solution given in Chapter 4 is also capable of yielding, as a special case, the solution of the generalized fin Equation (6.4) for other combinations of boundary conditions.

The Fin Problem

For convenience in the analysis, we introduce the following dimensionless variables

$$X = \frac{x}{\ell_0}, \qquad \theta(X) = \frac{T_{\text{fin}}(x) - T_\infty}{T_b - T_\infty}, \qquad K(X) = \frac{A(x)}{A_0}$$

$$W(X) = \frac{h(x)}{p_0 h_{\text{av}}} \frac{dS(x)}{dx}, \qquad M = m\ell_0 \qquad\qquad (6.6a, b, c, d, e)$$

where

$A_0 =$ a reference cross-section area
$h_{\text{av}} =$ an average heat transfer coefficient
$\ell_0 =$ a reference length
$m^2 = \dfrac{h_{\text{av}} p_0}{kA_0}$
$p_0 =$ perimeter of the reference cross-section area A_0

and the lateral area $S(x)$ is related to the perimeter $p(x)$ of the cross-section area $A(x)$ by

$$\frac{dS(x)}{dx} = p(x) \qquad\qquad (6.6f)$$

As pointed out by Gardner [2], the relation (6.6f) is valid for thin fins and spins; that is, when the square of the slope of the fin sides is negligible compared to unity.

With various dimensionless quantities as previously defined, the mathematical formulation of the fin problem comprising the differential Equation (6.4) and the boundary conditions (6.5a, b) takes the form

$$\frac{d}{dX}\left[K(X)\frac{d\theta(X)}{dX}\right] - M^2 W(X)\theta(X) = 0 \quad \text{in } X_b < X < X_t \quad (6.7a)$$

$$\theta(X) = 1 \quad \text{at } X = X_b \qquad\qquad (6.7b)$$

$$\frac{d\theta(X)}{dX} = 0 \quad \text{at } X = X_t \qquad\qquad (6.7c)$$

Here the quantity M^2 by definition, is

$$M^2 = \frac{h_{\text{av}} p_0}{kA_0}\ell_0^2 \qquad\qquad (6.8a)$$

$K(X)$ is given by

$$K(X) = \frac{A(x)}{A_0} \tag{6.8b}$$

and for $dS(x)/dx$ as given by Equation (6.6f), the quantity $W(X)$, defined by Equation (6.6d), takes the form

$$W(X) = \frac{h(x)}{h_{av}} \frac{p(x)}{p_0} \tag{6.8c}$$

Formal Solution of the Fin Problem

The solution to the fin problem given by Equations (6.7) can readily be obtained from the general solutions developed in Chapter 4. That is, the fin problem (6.7) is a special case of the more general problem (4.9); a comparison of these two problems reveals that

$$x = X, \qquad T_\phi = \theta, \quad k(x) = K(X), \quad w(x) = W(X)$$

$$d_\phi = -M^2, \qquad x_0 = X_b, \qquad x_1 = X_t, \qquad \alpha_0 = 1$$

$$\beta_0 = 0, \qquad \phi_e(x_0) = 1, \qquad \alpha_1 = 0, \qquad \beta_1 = 1, \qquad \phi_e(x_1) = 0 \tag{6.9}$$

Then the solution of the fin problem (6.7) is immediately obtained from the solution (4.12), where the functions U and V are specified by Equations (3.4c–f). We find

$$\theta(X) = \frac{v'(M, X_t)u(M, X) - u'(M, X_t)v(M, X)}{v'(M, X_t)u(M, X_b) - u'(M, X_t)v(M, X_b)} \tag{6.10}$$

where $u(M, X)$ and $v(M, X)$ are two independent solutions of Equation (6.7a).

Up to this point our solution is formal because we have yet to determine the two elementary solutions $u(M, X)$ and $v(M, X)$ which depend on the choice of the functional forms of the parameters $K(X)$ and $W(X)$. This matter is examined in the following.

A Generalized Solution of the Fin Problem

We now examine a situation in which the functions $K(X)$ and $W(X)$ are chosen as

$$K(X) = X^{1-2m} \tag{6.11a}$$

$$W(X) = c^2 n^2 X^{2c-2} K(X) \tag{6.11b}$$

Then the fin Equation (6.7a) becomes

$$\frac{d^2\theta(X)}{dX^2} + \frac{1-2m}{X}\frac{d\theta(X)}{dX} - M^2n^2c^2X^{2c-2}\theta(X) = 0 \qquad (6.12)$$

which is a special case of the generalized Bessel Equation (3.10) with

$$a = 0, \qquad \nu = \frac{m}{c}, \qquad \mu = inM, \qquad x = X \qquad (6.13)$$

where $i = \sqrt{-1}$. Then the solutions $u(M, X)$ and $v(M, X)$ of Equation (6.12) are obtained from Equation (3.11) as

$$u(M, X) = X^m I_{m/c}(nMX^c) \qquad (6.14a)$$

and

$$v(M, X) = X^m K_{m/c}(nMX^c) \qquad (6.14b)$$

or

$$v(M, X) = X^m I_{-(m/c)}(nMX^c) \quad \text{for } \frac{m}{c} \text{ nonintegral} \qquad (6.14c)$$

Here $I_\nu(x)$ and $K_\nu(x)$ are modified Bessel functions of order ν of the first and second kind, respectively. For detailed treatment of Bessel functions the reader should consult the standard texts on the subject [3, 4]. We note that $K_\nu(x)$ becomes infinite as x goes to zero, whereas $I_\nu(x)$ becomes infinite as x goes to infinity for $\nu \geq 0$.

The solutions (6.14) are differentiated with respect to X,

$$u'(M, X) = cnMX^{m+c-1}I_{(m/c)-1}(nMX^c) \qquad (6.15a)$$

and

$$v'(M, X) = -cnMX^{m+c-1}K_{(m/c)-1}(nMX^c) \qquad (6.15b)$$

or

$$v'(M, X) = cnMX^{m+c-1}I_{-(m/c)+1}(nMX^c) \quad \text{for } \frac{m}{c} \text{ nonintegral} \qquad (6.15c)$$

When Equations (6.14) and (6.15) are introduced into Equation (6.10), the solution $\theta(X)$ for the fin Equation (6.12), subject to the boundary conditions (6.7b, c), becomes

$$\theta(X)$$

$$= \left(\frac{X}{X_b}\right)^m \frac{K_{(m/c)-1}(nMX_t^c)I_{m/c}(nMX^c) + I_{(m/c)-1}(nMX_t^c)K_{m/c}(nMX^c)}{K_{(m/c)-1}(nMX_t^c)I_{m/c}(nMX_b^c) + I_{(m/c)-1}(nMX_t^c)K_{m/c}(nMX_b^c)}$$

$$(6.16a)$$

for the *integral values* of m/c; it is written as

$$\theta(X)$$

$$= \left(\frac{X}{X_b}\right)^m \frac{I_{-(m/c)+1}(nMX_t^c)I_{m/c}(nMX^c) - I_{(m/c)-1}(nMX_t^c)I_{-m/c}(nMX^c)}{I_{-(m/c)+1}(nMX_t^c)I_{m/c}(nMX_b^c) - I_{(m/c)-1}(nMX_t^c)I_{-(m/c)}(nMX_b^c)}$$

(6.16b)

for the *nonintegral values* of m/c.

Special case for cn = 1 and c = 0

For the special case of $cn = 1$ and $c = 0$, the fin Equation (6.12) reduces to

$$\frac{d^2\theta(X)}{dX^2} + \frac{1 - 2m}{X}\frac{d\theta(X)}{dX} - \frac{M^2}{X^2}\theta(X) = 0 \qquad (6.17)$$

which is now a homogeneous differential equation (i.e., the Euler equation). To solve this equation in any interval *not containing the origin*, we substitute into Equation (6.17)

$$\theta(X) = X^r \qquad (6.18a)$$

and compute the roots r_1 and r_2 of equation

$$r(r - 2m) - M^2 = 0 \qquad (6.18b)$$

The roots of this equation are

$$r_k = m + (-1)^{1+k}\sqrt{m^2 + M^2}, \qquad k = 1, 2 \qquad (6.18c)$$

Then the two independent solutions $u(M, X)$ and $v(M, X)$ of Equation (6.17) become

$$u(M, X) = X^{r_1}, \qquad v(M, X) = X^{r_2} \qquad (6.19a, b)$$

and the differentiation of these functions yields

$$u'(M, X) = r_1 X^{r_1-1}, \qquad v'(M, X) = r_2 X^{r_2-1} \qquad (6.19c, d)$$

Introducing the results given by Equations (6.19) into Equation (6.10), we obtain the solution of the fin Equation (6.17) as

$$\theta(X) = \frac{r_2 X^{r_1} - r_1 X_t^{r_1-r_2} X^{r_2}}{r_2 X_b^{r_1} - r_1 X_t^{r_1-r_2} X_b^{r_2}} \qquad (6.20)$$

Solution for the Region $0 \leq X \leq 1$

The solutions given by Equations (6.16) and (6.20) are for situations in which the region does not contain the origin. In the analysis of fin problems it is convenient to choose, without the loss of generality, $X_t = 0$ and $X_b = 1$, so that the fin is considered to lie in the region $0 \leq X \leq 1$. Then for this special case solutions (6.16a, b), respectively, simplify to

$$\theta(X) = X^m \frac{I_{m/c}(nMX^c)}{I_{m/c}(nM)} \quad \text{for } \frac{m}{c} \text{ integral} \tag{6.21a}$$

$$\theta(X) = X^m \frac{I_{-(m/c)}(nMX^c)}{I_{-m/c}(nM)} \quad \text{for } \frac{m}{c} \text{ nonintegral}, \quad \frac{m}{c} < 1 \tag{6.21b}$$

and

$$\theta(X) = X^m \frac{I_{m/c}(nMX^c)}{I_{m/c}(nM)} \quad \text{for } \frac{m}{c} \text{ nonintegral}, \quad \frac{m}{c} > 1 \tag{6.21c}$$

The solution (6.20) for the case of $cn = 1$ and $c = 0$ reduces to

$$\theta(X) = X^{m + \sqrt{m^2 + M^2}} \tag{6.22}$$

Table 6.1.　Values of c, m, and n for some special cases of $K(X)$ and $W(X)$ with $cn = 1$

	$K(X)$	$W(X)$	m	c	n	$\frac{m}{c}$
Longitudinal fin of rectangular profile	1	1	$\frac{1}{2}$	1	1	$\frac{1}{2}$
Longitudinal fin of triangular profile	X	1	0	$\frac{1}{2}$	2	0
Longitudinal fin of concave parabolic profile	X^2	1	$-\frac{1}{2}$	0	∞	$-\infty$
Longitudinal fin of convex parabolic profile	$X^{1/2}$	1	$\frac{1}{4}$	$\frac{3}{4}$	$\frac{4}{3}$	$\frac{1}{3}$
Radial fin of rectangular profile	X	X	0	1	1	0
Radial fin of hyperbolic profile	1	X	$\frac{1}{2}$	$\frac{3}{2}$	$\frac{2}{3}$	$\frac{1}{3}$
Conical spine	X^2	X	$-\frac{1}{2}$	$\frac{1}{2}$	2	-1
Spine fin of concave parabolic profile	X^4	X^2	$-\frac{3}{2}$	0	∞	$-\infty$
Spine of convex parabolic profile	X	$X^{1/2}$	0	$\frac{3}{4}$	$\frac{4}{3}$	0

In the derivation of the solutions given by Equations (6.21) and (6.22), it is assumed that the fin lies in the region $0 \leq X \leq 1$, and the coefficients $K(X)$ and $W(X)$ appearing in the fin Equation (6.7a) are defined by Equations (6.11a, b), that is,

$$K(X) = X^{1-2m}, \qquad W(x) = c^2 n^2 X^{2c-2} K(X)$$

Clearly, the numerical values of the coefficients c, n, and m establish the functional forms of $K(X)$ and $W(X)$. For convenience in the subsequent analysis, we present in Table 6.1 the numerical values of c, m, and n corresponding to some special cases of $K(X)$ and $W(X)$ with $cn = 1$.

6.2   HEAT FLOW THROUGH FINS—APPLICATIONS

Having established in the previous section the solution of the fin problem, we now examine the temperature distribution and heat flow through fins and the fin efficiency for some specific fin geometries of practical interest.

Longitudinal Fins

Consider a longitudinal fin with arbitrary cross-section as illustrated in Figure 6.3. Let x be the axial coordinate with its origin $x = 0$ at the fin tip. The fin profile consists of two symmetrical curves about the axis $0x$. Then the fin thickness $\delta(x)$ varies with the distance x. Let ℓ be the depth of the fin and $\delta_0 \equiv \delta(b)$ the fin thickness at the fin base $x = b$ as illustrated in Figure 6.3. The cross-section area $A(x)$ and the perimeter $p(x)$ at any location are given

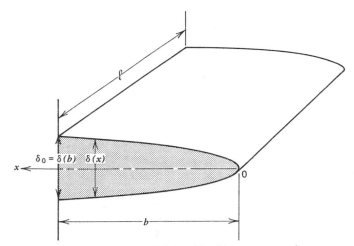

Figure 6.3   Longitudinal fin with arbitrary cross-section.

by

$$A(x) = \delta(x)\ell, \qquad A_0 = \delta_0\ell \qquad (6.23a, b)$$

$$p(x) = 2\ell\left[1 + \frac{\delta(x)}{\ell}\right] \approx 2\ell, \qquad p_0 = 2\ell \qquad (6.23c, d)$$

where A_0 and p_0 are the cross-section area and perimeter at the fin base, respectively.

The results given by Equations (6.23) are introduced into Equations (6.8) and the reference length ℓ_0 is chosen as $\ell_0 = b$. Then the parameters $K(X)$, $W(X)$, and M are determined as

$$K(X) = \frac{\delta(x)}{\delta_0}, \qquad W(X) = \frac{h(x)}{h_0} \qquad (6.24a, b)$$

$$M = \left(\frac{2h_{av}}{k\delta_0}\right)^{1/2} b \qquad (6.24c)$$

and the dimensionless coordinates $X = x/b$ of the fin tip X_t, and the fin base X_b, respectively, become

$$X_t = 0 \quad \text{and} \quad X_b = 1 \qquad (6.24d)$$

If we further assume that heat transfer coefficient $h(x) = h_0$, constant, $W(X)$ simplifies to

$$W(X) = 1 \qquad (6.24e)$$

Then the distribution of temperature $\theta(X)$ through the fin is readily determined from Equations (6.21), once the functional form of the fin thickness $\delta(x)$ is specified.

Knowing the temperature distribution $\theta(X)$, the heat flow through the fin, Q_{fin}, is determined according to the relation

$$Q_{fin} = kA(b)\frac{dT(b)}{dx} = k\delta_0\ell(T_b - T_\infty)\frac{1}{b}\frac{d\theta(1)}{dX} \qquad (6.25)$$

The *fin efficiency* η is defined as

$$\eta = \frac{Q_{fin}}{Q_{ideal}} = \frac{\text{Actual heat transfer through fin}}{\text{Ideal heat transfer through fin if the entire fin surface were at fin-base temperature } T_b} \qquad (6.26)$$

where Q_{ideal} is given by

$$Q_{ideal} = h_{av}2\ell b(T_b - T_\infty) \qquad (6.27)$$

Substituting Equations (6.25) and (6.27) into (6.26), the fin efficiency is determined as

$$\eta = \frac{1}{M^2} \frac{d\theta(1)}{dX} \qquad (6.28)$$

where M is defined by Equation (6.24c). The preceding results are now applied to determine the temperature distribution and fin efficiency for longitudinal fins having rectangular, triangular, and parabolic profiles.

Longitudinal Fin of Rectangular Profile

For a fin having a constant thickness δ_0 as illustrated in Fig. 6.4 we have

$$\delta(X) = \delta_0 \qquad (6.29a)$$

and by Equations (6.24a, e) we write

$$K(X) = 1, \qquad W(X) = 1 \qquad (6.29b, c)$$

For $K(X)$ and $W(X)$ as specified by Equation (6.29b, c), from Table 6.1 we obtain

$$m = \frac{1}{2}, \qquad c = 1, \qquad n = 1, \qquad \frac{m}{c} = \frac{1}{2} \qquad (6.29d)$$

Then the temperature distribution $\theta(X)$ through the fin is determined from Equation (6.21b) as

$$\theta(X) = X^{1/2} \frac{I_{-(1/2)}(MX)}{I_{-(1/2)}(M)} \qquad (6.30a)$$

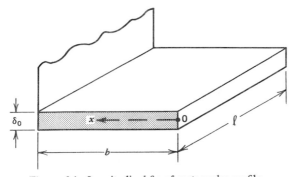

Figure 6.4   Longitudinal fin of rectangular profile.

By making use of the relation

$$I_{-(1/2)}(X) = \sqrt{\frac{2}{\pi X}} \cosh X \tag{6.30b}$$

The solution (6.30a) is written as

$$\theta(X) = \frac{\cosh(MX)}{\cosh M} \tag{6.30c}$$

and the fin efficiency is determined according to Equation (6.28) as

$$\eta = \frac{\tanh M}{M} \tag{6.31}$$

where M is defined by Equation (6.24c).

Figure 6.5, curve A, shows the fin efficiency η determined from Equation (6.31), plotted as a function of M.

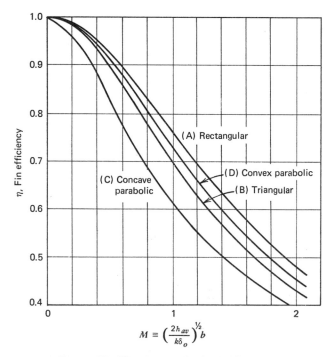

Figure 6.5   Fin efficiency for longitudinal fins.

Longitudinal Fin of Triangular Profile

For a longitudinal fin of triangular profile as illustrated in Figure 6.6, the fin profile $\delta(x)$ is given by

$$\delta(x) = \delta_0 \frac{x}{b} = \delta_0 X \tag{6.32a}$$

and by Equations (6.24a, e) we have

$$K(X) = \frac{\delta(x)}{\delta_0} = X, \qquad W(X) = 1 \tag{6.32b, c}$$

For $K(X)$ and $W(X)$ as given by Equations (6.32b, c), from Table 6.1 we obtain

$$m = 0, \qquad c = \frac{1}{2}, \qquad n = 2, \qquad \frac{m}{c} = 0 \tag{6.32d}$$

Then the temperature distribution $\theta(X)$ is determined from Equation (6.21a) as

$$\theta(X) = \frac{I_0(2M\sqrt{X})}{I_0(2M)} \tag{6.33}$$

and the fin efficiency (6.28) becomes

$$\eta = \frac{I_1(2M)}{MI_0(2M)} \tag{6.34}$$

where M is defined by Equation (6.24c).

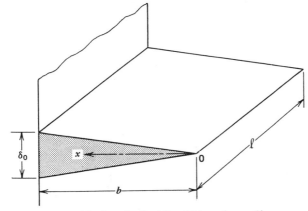

Figure 6.6   Longitudinal fin of triangular profile.

Figure 6.5, curve B, shows the fin efficiency η determined from Equation (6.34), plotted as a function of M.

Longitudinal Fin of Concave Parabolic Profile

For a longitudinal fin of concave parabolic profile as illustrated in Figure 6.7, the profile is given by

$$\delta(x) = \delta_0\left(\frac{x}{b}\right)^2 \quad \text{or} \quad K(X) = \frac{\delta(x)}{\delta_0} = X^2 \qquad (6.35a)$$

For $K(X) = X^2$ and $W(X) = 1$, from Table 6.1 we have

$$m = -\frac{1}{2}, \quad c = 0, \quad n = \infty, \quad \frac{m}{c} = -\infty \qquad (6.35b)$$

For this particular case the fin Equation (6.12) reduces to an homogeneous Equation (6.17), the solution of which is given by Equation (6.22). By setting $m = -\frac{1}{2}$ in Equation (6.22) we obtain

$$\theta(X) = X^{-(1/2)+\sqrt{(1/4)+M^2}} \qquad (6.36)$$

The fin efficiency η is determined according to its definition (6.28) as

$$\eta = \frac{2}{1 + \sqrt{1 + (2M)^2}} \qquad (6.37)$$

where M is defined by Equation (6.24c).

Figure 6.5, curve C, shows the fin efficiency η for a concave parabolic fin plotted as a function of the parameter M.

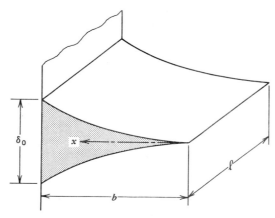

Figure 6.7 Longitudinal fin of concave parabolic profile.

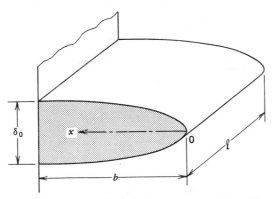

Figure 6.8 Longitudinal fin of convex parabolic profile.

Longitudinal Fin of Convex Parabolic Profile

For a longitudinal fin of convex parabolic profile shown in Figure 6.8, the profile is given by

$$\delta(x) = \delta_0\left(\frac{x}{b}\right)^{1/2} \quad \text{or} \quad K(X) = \frac{\delta(x)}{\delta_0} = X^{1/2} \tag{6.38a}$$

For $K(X) = X^{1/2}$ and $W(X) = 1$, from Table 6.1 we have

$$m = \frac{1}{4}, \qquad c = \frac{3}{4}, \qquad n = \frac{4}{3}, \qquad \frac{m}{c} = \frac{1}{3} \tag{6.38b}$$

Then the solution for $\theta(X)$ is obtained from Equation (6.21b) as

$$\theta(X) = X^{1/4}\frac{I_{-(1/3)}\left(\frac{4}{3}MX^{3/4}\right)}{I_{-(1/3)}\left(\frac{4}{3}M\right)} \tag{6.39}$$

The fin efficiency η is determined by Equation (6.28) as

$$\eta = \frac{1}{M}\frac{I_{2/3}\left(\frac{4}{3}M\right)}{I_{-(1/3)}\left(\frac{4}{3}M\right)} \tag{6.40}$$

where M is defined by Equation (6.24c).

Figure 6.5, curve D, shows the efficiency η for a convex parabolic fin as a function of the parameter M.

Radial Fins

We now consider radial fins in the form of a circular disc as illustrated in Figure 6.9, where the fin thickness $\delta(r)$ may vary radially. Let δ_0 be the fin

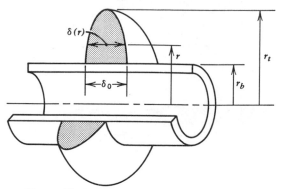

Figure 6.9 Radial fin of arbitrary cross-section.

thickness at the fin base $r = r_b$. The cross-section area $A(r)$ and the perimeter $p(r)$ of the fin at any radial location are given by

$$A(r) = 2\pi r \delta(r), \qquad A_0 = 2\pi r_b \delta_0 \qquad (6.41a, b)$$

$$p(r) = 4\pi r, \qquad p_0 = 4\pi r_b \qquad (6.41c, d)$$

where A_0 and p_0 are the cross-section area and perimeter of the fin at the fin base, respectively.

If the reference length ℓ_0 is chosen as $\ell_0 = r_t$, where r_t is the outer radius of the fin shown in Figure 6.9, the dimensionless space variables become

$$X \equiv R = \frac{r}{r_t}, \qquad X_b \equiv R_b = \frac{r_b}{r_t}, \qquad X_t \equiv R_t = 1 \qquad (6.42a, b, c)$$

With various quantities as defined by Equations (6.41) and (6.42), the fin problem (6.7) is modified accordingly and the parameters M^2, $K(X)$, and $W(X)$ defined by Equations (6.8) take the form

$$M = \left(\frac{2h_{av}}{k\delta_0}\right)^{1/2} r_t \qquad (6.43a)$$

$$K(R) = R\frac{\delta(r)}{\delta_0}, \qquad W(R) = R \qquad (6.43b, c)$$

where we assumed $h(x) = h_{av}$ and cancelled out the constant multiplier $1/R_b$ that would have appeared in the parameters $K(R)$ and $W(R)$ because the fin Equation (6.7a) is an homogeneous equation.

The heat flow rate through a radial fin is determined as

$$Q_{fin} = -k2\pi r_b \delta_0 \frac{dT(r_b)}{dr} = -k2\pi\delta_0(T_b - T_\infty)R_b\frac{d\theta(R_b)}{dR} \qquad (6.44a)$$

and the ideal heat transfer rate through a radial fin is given by

$$Q_{ideal} = 2\pi h_{av} r_t^2 (1 - R_b^2)(T_b - T_\infty) \tag{6.44b}$$

Introducing Equations (6.44) into the definition of fin efficiency given by Equation (6.26), the fin efficiency η for radial fins is expressed in the form

$$\eta = -\frac{2}{M^2} \frac{R_b}{1 - R_b^2} \frac{d\theta(R_b)}{dR} \tag{6.45}$$

where M is defined by Equation (6.43a).

The foregoing results are now applied to determine the temperature distribution and fin efficiency for radial fins having rectangular and hyperbolic profiles.

Radial Fin of Rectangular Profile

For a radial fin having constant cross-section as illustrated in Figure 6.10, we have

$$\delta(r) = \delta_0 \tag{6.46a}$$

and by Equations (6.43b, c) we write

$$K(R) = R \quad \text{and} \quad W(R) = R \tag{6.46b, c}$$

For $K(R)$ and $W(R)$ as given by Equation (6.46b, c), from Table 6.1 we obtain

$$m = 0, \quad c = 1, \quad n = 1, \quad \frac{m}{c} = 0 \tag{6.46d}$$

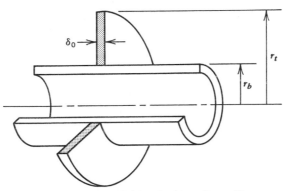

Figure 6.10   Radial fin of rectangular profile.

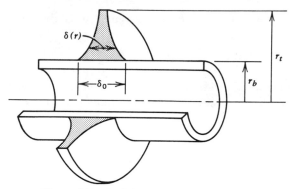

Figure 6.11   Radial fin of hyperbolic profile.

By making use of the relations

$$I_{-1}(x) = I_1(x), \qquad K_{-1}(x) = K_1(x) \tag{6.46e}$$

the temperature distribution $\theta(R)$ is determined from Equation (6.16a) as

$$\theta(R) = \frac{K_1(M)I_0(MR) + I_1(M)K_0(MR)}{K_1(M)I_0(MR_b) + I_1(M)K_0(MR_b)} \tag{6.47}$$

Introducing $\theta(R)$ into Equation (6.45), the fin efficiency becomes

$$\eta = \frac{2}{M} \frac{R_b}{1 - R_b^2} \frac{I_1(M)K_1(MR_b) - K_1(M)I_1(MR_b)}{I_1(M)K_0(MR_b) + K_1(M)I_0(MR_b)} \tag{6.48}$$

where R, R_b are defined by Equations (6.42) and the parameter M is defined by Equation (6.43a).

Radial Fin of Hyperbolic Profile

For a radial fin of hyperbolic profile as illustrated in Figure 6.11, the fin thickness $\delta(r)$ varies with the radial coordinate as

$$\delta(r) = \delta_0 \frac{r_b}{r} \tag{6.49}$$

and by Equations (6.43) we write

$$M^2 = \left(\frac{2h_{av}}{k\delta_0}\right)r_t^2 \tag{6.50a}$$

$$K(R) = R\frac{r_b}{r} = R_b, \qquad W(R) = R \tag{6.50b, c}$$

The definition of M^2 and $K(R)$ given by Equations (6.50) can be modified by dividing them by R_b because the fin Equation (6.7a) is homogeneous and remains invariant if the terms $K(R)$ and M^2 are divided by the same constant. With this consideration we prefer to write the results in Equation (6.50) in the alternative form as

$$M^* = \left(\frac{2h_{av}}{k\delta_0}\right)^{1/2} \frac{r_t}{\sqrt{R_b}} \tag{6.51a}$$

$$K(R) = 1, \qquad W(R) = R \tag{6.51b, c}$$

For $K(R)$ and $W(R)$ as given by Equations (6.51b, c), from Table 6.1 we obtain

$$m = \frac{1}{2}, \qquad c = \frac{3}{2}, \qquad n = \frac{2}{3}, \qquad \frac{m}{c} = \frac{1}{3} \tag{6.51d}$$

Then the solution for the temperature $\theta(R)$ is obtained from Equation (6.16b) as

$$\theta(R) = \left(\frac{R}{R_b}\right)^{1/2} \frac{I_{2/3}\left(\frac{2}{3}M^*\right)I_{1/3}\left(\frac{2}{3}M^*R^{3/2}\right) - I_{-2/3}\left(\frac{2}{3}M^*\right)I_{-1/3}\left(\frac{2}{3}M^*R^{3/2}\right)}{I_{2/3}\left(\frac{2}{3}M^*\right)I_{1/3}\left(\frac{2}{3}M^*R_b^{3/2}\right) - I_{-2/3}\left(\frac{2}{3}M^*\right)I_{-1/3}\left(\frac{2}{3}M^*R_b^{3/2}\right)} \tag{6.52}$$

To obtain the fin efficiency, the derivative of the following functions are needed

$$Y_{\pm} = R^{1/2}I_{\pm 1/3}\left(\frac{2}{3}MR^{3/2}\right) \tag{6.53}$$

We define a new independent variable x as

$$x = \frac{2}{3}MR^{3/2} \quad \text{or} \quad R^{1/2} = \left(\frac{3}{2}M\right)^{-1/3}x^{1/3} \tag{6.54a, b}$$

and we have $dx/dR = MR^{1/2}$ $\tag{6.54c}$

Then Equation (6.53) becomes

$$Y_{\pm} = \left(\frac{3}{2}M\right)^{-1/3}x^{1/3}I_{\pm 1/3}(x) \tag{6.55}$$

The differentiation of Equation (6.55) gives

$$\frac{dY_{\pm}}{dx} = \left(\frac{3}{2}M\right)^{-1/3}x^{1/3}I_{\mp 2/3}(x) = R^{1/2}I_{\mp 2/3}\left(\frac{2}{3}MR^{3/2}\right) \tag{6.56}$$

or

$$\frac{dY_\pm}{dR} = \frac{dY_\pm}{dx}\frac{dx}{dR} = MRI_{\mp 2/3}\left(\frac{2}{3}MR^{3/2}\right) \tag{6.57}$$

Finally, Equation (6.52) is introduced into Equation (6.45) and the Bessel functions are differentiated according to the rule of differentiation given by Equations (6.53) and (6.57). We obtain

$$\eta = \frac{2}{M^*}\frac{R_b}{1 - R_b^2}\frac{I_{2/3}\left(\frac{2}{3}M^*\right)I_{-2/3}\left(\frac{2}{3}M^*R_b^{3/2}\right) - I_{-2/3}\left(\frac{2}{3}M^*\right)I_{2/3}\left(\frac{2}{3}M^*R_b^{3/2}\right)}{I_{-2/3}\left(\frac{2}{3}M^*\right)I_{-1/3}\left(\frac{2}{3}M^*R_b^{3/2}\right) - I_{2/3}\left(\frac{2}{3}M^*\right)I_{1/3}\left(\frac{2}{3}M^*R_b^{3/2}\right)} \tag{6.58}$$

where M^* is defined by Equation (6.51a).

Spines

We now consider a spine of arbitrary profile as illustrated in Figure 6.12. Let x be the axial coordinate with its origin ($x = 0$) at the tip of the spine and the radius $r(x)$ of the spine varies with the distance along x. The cross-section area $A(x)$ and the perimeter $p(x)$ of the spine at any axial location are given by

$$A(x) = \pi[r(x)]^2, \qquad A_0 = \pi r_b^2 \tag{6.59a, b}$$

$$p(x) = 2\pi r(x), \qquad p_0 = 2\pi r_b \tag{6.59c, d}$$

where A_0 is the cross-section area, p_0 is the perimeter, and r_b is the radius of the spine at the base of the spine.

We now assume $h(x) = h_{av}$, choose the reference length ℓ_0 as $\ell_0 = b$ where b is the height of the spine, and define the dimensionless space variables as

$$R(X) = \frac{r(x)}{r_b}, \qquad X = \frac{x}{b} \tag{6.60a, b}$$

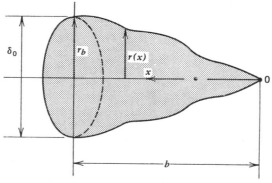

Figure 6.12 Spine of arbitrary cross-section.

Then the dimensionless coordinates of the spine tip and spine base become

$$X_t = 0, \qquad X_b = 1 \qquad\qquad (6.60c, d)$$

The results in Equations (6.59) are introduced into Equations (6.8); we obtain

$$M = \left(\frac{2h_{av}}{kr_b}\right)^{1/2} b \qquad\qquad (6.61a)$$

$$K(X) = [R(x)]^2, \qquad W(X) = R(X) \qquad\qquad (6.61b, c)$$

The heat flow rate Q_{fin} through the spine is determined from the relation

$$Q_{fin} = k\pi r_b^2 \frac{T_b - T_\infty}{b} \frac{d\theta(1)}{dX} \qquad\qquad (6.62a)$$

and the ideal heat transfer rate through the spine is given by

$$Q_{ideal} = 2\pi r_b b \int_0^1 R(X) h_{av}(T_b - T_\infty)\, dX \qquad\qquad (6.62b)$$

Introducing Equations (6.62) into the definition of fin efficiency given by Equation (6.26), the fin efficiency η for spine is expressed in the form

$$\eta = \frac{1}{M^2 \int_0^1 R(X)\, dX} \frac{d\theta(1)}{dX} \qquad\qquad (6.63)$$

where M is defined by Equation (6.61a).

We now examine the temperature distribution and efficiency for some specific spine profiles.

Cylindrical Spine

For a cylindrical spine having a constant radius as illustrated in Figure 6.13 we have

$$r(x) = r_b \quad \text{or} \quad R(X) = 1 \qquad\qquad (6.64a)$$

and by Equations (6.61b, c) we write

$$K(X) = 1, \qquad W(X) = 1 \qquad\qquad (6.64b, c)$$

We note that the values of $K(X)$ and $W(X)$ for a cylindrical spine are the same as those given by Equations (6.29b, c) for a longitudinal fin of rectangular profile. In addition, the spine lies in the region $0 \le X \le 1$. Therefore, the

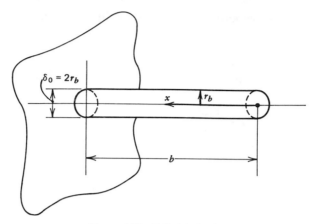

Figure 6.13   Cylindrical spine.

temperature distribution $\theta(X)$ and the fin efficiency η for a cylindrical spine are the same as those given by Equations (6.30c) and (6.31) for a longitudinal fin of rectangular profile. Thus we write

$$\theta(X) = \frac{\cosh(MX)}{\cosh M} \tag{6.65}$$

and

$$\eta = \frac{\tanh M}{M} \tag{6.66a}$$

where M is defined by Equation (6.61a), that is,

$$M = \left(\frac{2h_{av}}{kr_b}\right)^{1/2} b \tag{6.66b}$$

We also note that the definition of M for a cylindrical spine given by Equation (6.66b) is based on the spine radius r_b, whereas for a rectangular longitudinal fin given by Equation (6.24c), the definition is based on the fin thickness δ_0. Therefore, the efficiency of a cylindrical spine can be obtained from Figure 6.5, curve A for a rectangular longitudinal fin by appropriate interpretation of the definition of M.

Conical Spine

For a conical spine as illustrated in Figure 6.14, the radius $r(x)$ is given by

$$r(x) = \frac{\delta_0}{2}\frac{x}{b} = r_b\frac{x}{b} \quad \text{or} \quad R(X) = \frac{r(x)}{r_b} = X \tag{6.67a}$$

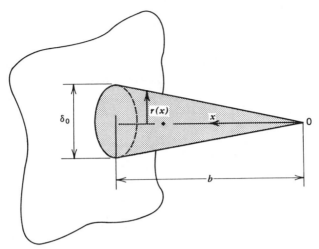

Figure 6.14   Conical spine.

and by Equations (6.61b, c) we write

$$K(X) = X^2, \qquad W(X) = X \tag{6.67b, c}$$

For $K(X)$ and $W(X)$ as given by Equations (6.67b, c), from Table 6.1 we obtain

$$m = -\frac{1}{2}, \qquad c = \frac{1}{2}, \qquad n = 2, \qquad \frac{m}{c} = -1 \tag{6.67d}$$

Then the solution for the temperature distribution $\theta(X)$ is determined from Equation (6.21a) as

$$\theta(X) = X^{-1/2} \frac{I_1(2M\sqrt{X})}{I_1(2M)} \tag{6.68}$$

where we made use of the relation $I_{-1}(x) = I_1(x)$. Introducing Equation (6.68) into Equation (6.63) we obtain the efficiency for a conical spine as

$$\eta = \frac{2}{M} \frac{I_2(2M)}{I_1(2M)} \tag{6.69}$$

where M is defined by Equation (6.61a).

Figure 6.15, curve A shows the efficiency η of a conical spine plotted as a function of the parameter M.

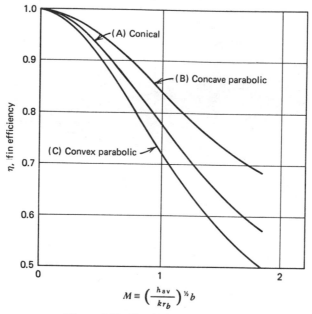

Figure 6.15   Fin efficiency for spines.

Spine Fin of Concave Parabolic Profile

For a spine fin of concave parabolic profile as shown in Figure 6.16, the radius $r(x)$ is given by

$$r(x) = r_b\left(\frac{x}{b}\right)^2 \quad \text{or} \quad R(X) = X^2 \tag{6.70a}$$

and by Equations (6.61b, c) we write

$$K(X) = X^4, \qquad W(X) = X^2 \tag{6.70b, c}$$

For $K(X)$ and $W(X)$ as given by Equations (6.70b, c), from Table 6.1 we obtain

$$m = -\frac{3}{2}, \qquad c = 0, \qquad n = \infty, \qquad \frac{m}{c} = -\infty \tag{6.70d}$$

Then the solution for the temperature $\theta(X)$ is obtained from Equation (6.22) as

$$\theta(X) = X^{-3/2 + \sqrt{(9/4) + M^2}} \tag{6.71}$$

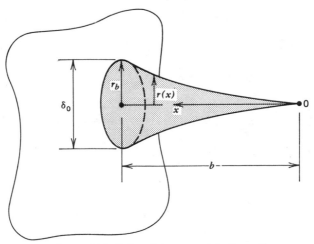

Figure 6.16   Spine of concave parabolic profile.

and the efficiency of a concave parabolic fin is determined according to Equation (6.63) as

$$\eta = \frac{2}{1 + \sqrt{1 + [(2/3)M]^2}} \tag{6.72}$$

where M is defined by Equation (6.61a).

Figure 6.15, curve **B** shows the efficiency η for a spine fin of concave parabolic profile.

Spine of Convex Parabolic Profile

For a spine of convex parabolic profile as illustrated in Figure 6.17, the radius $r(x)$ is given by

$$r(x) = r_b \left(\frac{x}{b} \right)^{1/2} \quad \text{or} \quad R(X) = X^{1/2} \tag{6.73a}$$

and by Equations (6.61b, c) we write

$$K(X) = X, \quad W(X) = X^{1/2} \tag{6.73b, c}$$

For $K(X)$ and $W(X)$ as defined by Equations (6.73b, c), from Table 6.1 we obtain

$$m = 0, \quad c = \frac{3}{4}, \quad n = \frac{4}{3}, \quad \frac{m}{c} = 0 \tag{6.73d}$$

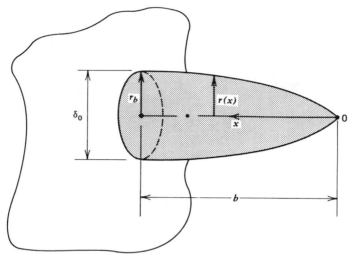

Figure 6.17 Spine of convex parabolic profile.

Then the solution for the temperature $\theta(X)$ is obtained from Equation (6.21a) as

$$\theta(X) = \frac{I_0\left(\frac{4}{3}MX^{3/4}\right)}{I_0\left(\frac{4}{3}M\right)} \qquad (6.74)$$

and the efficiency of a convex parabolic fin is determined from Equation (6.63) as

$$\eta = \frac{3}{2}\frac{1}{M}\frac{I_1\left(\frac{4}{3}M\right)}{I_0\left(\frac{4}{3}M\right)} \qquad (6.75)$$

where M is defined by Equation (6.21a).

Figure 6.15, curve C, shows the efficiency η of a convex parabolic fin.

6.3   HEAT TRANSFER THROUGH AN ARRAY OF FINS

In the previous sections we discussed heat transfer through single fins of various geometries. In many engineering applications such as those encountered in certain types of compact heat exchangers, the repeating section of a finned matrix is not a single fin, but is an array of fins. Consider for illustration purposes a complex finned matrix in the coolant passage between two hot surfaces as illustrated in Figure 6.18a. Let the temperature of the two hot surfaces be uniform and equal. Then from symmetry, the heat transfer characteristics of this fin matrix can be represented with that of the *repeating*

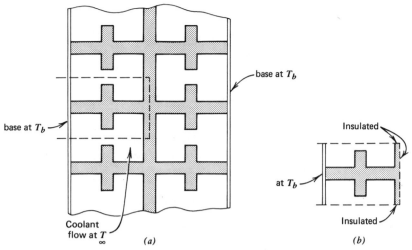

Figure 6.18   Array of fins. (*a*) Coolant passage with complex fin array. (*b*) The repeating section of a fin array.

section enclosed by the dashed lines. This repeating section, as shown in Figure 6.18b, is not a single fin, but is an array of fins.

The heat transfer analysis of an array of fins is a complicated matter; only a limited number of papers is available on this subject. The method of analysis proposed in Reference 1 involves a simultaneous solution of a set of ordinary differential equations associated with each of the fins in the array. If the differential equations can be integrated, the problem becomes one of solving a system of linear algebraic equations associated with the determination of the integration constants. The application of the method for the solution of practical problems is quite involved. To alleviate this difficulty, a procedure is described in Reference 5 which treats each fin as a lumped parameter and cascades all the fins in the array by means of operations.

We now present a very efficient method, described in References 6 and 7, for solving such problems by using the finite element assembly procedure. In this method each fin in the array is considered as a finite element whose characteristics are determined by the solution of a one-dimensional fin problem. Thus the analysis of the problem is reduced to the solution of a system of algebraic equations. In such a system, each element of the *global fin matrix* (i.e., coefficient matrix) is related to the fundamental solutions of the ordinary differential equation governing heat transfer through a single fin. Explicit expressions are developed for the determination of the elements of this matrix.

We consider the generalized fin Equation (6.12), which is the generalized Bessel equation, and take its two fundamental solutions given by Equations (6.14), that is,

$$u(X) = X^m I_{m/c}(nMX^c) \qquad (6.76a)$$

and

$$v(X) = X^m K_{m/c}(nMX^c) \tag{6.76b}$$

or

$$v(X) = X^m I_{-m/c}(nMX^c) \quad \text{for } \frac{m}{c} \text{ nonintegral} \tag{6.76c}$$

For the special case of $cn = 1$ and $c = 0$, the fin Equation (6.12) reduces to the Euler equation. Then the two fundamental solutions in any interval not containing the origin are

$$u(X) = X^{r_1}, \qquad v(X) = X^{r_2} \tag{6.77a, b}$$

where

$$r_k = m + (-1)^{1+k}(m^2 + M^2)^{1/2}, \qquad k = 1, 2 \tag{6.77c}$$

The solution for $\theta(X)$ can be constructed by taking any linear combination of the two fundamental solutions $u(X)$ and $v(X)$; we prefer to construct the solution in the form

$$\theta(X) = \theta_b U(X) + \theta_t V(X) \tag{6.78a}$$

where

$$\theta_b = \theta(X_b) \quad \text{and} \quad \theta_t = \theta(X_t). \tag{6.78b, c}$$

Then the functions $U(X)$ and $V(X)$, satisfying the requirement of Equations (6.78b, c) are determined as

$$U(X) = \frac{u(X)v(X_t) - u(X_t)v(X)}{u(X_b)v(X_t) - u(X_t)v(X_b)} \tag{6.79a}$$

$$V(X) = \frac{u(X_b)v(X) - u(X)v(X_b)}{u(X_b)v(X_t) - u(X_t)v(X_b)} \tag{6.79b}$$

We note that with the choice of the functions $U(X)$ and $V(X)$ as defined by Equations (6.79a, b), the following conditions are automatically satisfied

$$U(X_b) = 1, \qquad U(X_t) = 0 \tag{6.80a}$$

$$V(X_b) = 0, \qquad V(X_t) = 1 \tag{6.80b}$$

In the case of a single fin problem, the integration constants θ_b and θ_t are determined by constraining the solution (6.78a) to meet the two boundary conditions specified at $X = X_b$ and $X = X_t$, and the analysis is simple and straightforward.

In the case of a fin-array in which several fins are connected, it becomes an elaborate and complicated matter to solve the problem by writing the fin equation for each of the fins in the array and then trying to determine the resulting integration constants by matching the solutions to meet the boundary conditions. To alleviate such difficulty we now describe an efficient method of solving the heat transfer problems for a fin-array.

Consider the one-dimensional heat flow through a typical element of the fin-array as shown in Figure 6.19. Let Q_b and Q_t be the heat flow rates entering the element at the node $X = X_b$ and $X = X_t$ through the surfaces A_b and A_t, respectively. Let the temperature distribution $\theta(X)$ in the element be governed by Equation (6.78a). The heat flow rates Q_b and Q_t entering the elements are determined according to Fourier's law and the result is written in matrix notation as

$$kA_0 \frac{\Delta T}{L_0} \begin{bmatrix} -K(X_b)U'(X_b) & -K(X_b)V'(X_b) \\ K(X_t)U'(X_t) & K(X_t)V'(X_t) \end{bmatrix} \begin{Bmatrix} \theta_b \\ \theta_t \end{Bmatrix} = \begin{Bmatrix} Q_b \\ Q_t \end{Bmatrix} \quad (6.81)$$

where primes denote differentiation with respect to X.

Equation (6.81) is written in the *standard form* for the finite element method [9]. When the temperature of the surrounding fluid T_∞ is not constant and therefore has different values for the fins of the array, Equation (6.81) must be written in the form

$$\begin{bmatrix} k_{bb} & k_{bt} \\ k_{tb} & k_{tt} \end{bmatrix} \begin{Bmatrix} T_b \\ T_t \end{Bmatrix} = \begin{Bmatrix} Q_b \\ Q_t \end{Bmatrix} + \begin{Bmatrix} P_b \\ P_t \end{Bmatrix} \quad (6.82a)$$

or, more compactly, in the form

$$[K] \quad \{T\} = \{Q\} + \{P\} \quad (6.82b)$$

where $\{T\}$ is the column vector of the two *node temperature*, $\{Q\}$ is the column vector of the *two node heat transfer rates*, and $\{K\}$ is the *fin matrix* of thermal influence coefficients:

$$k_{bb} = -\left(\frac{k}{L_0}\right) A_0 K(X_b) U'(X_b) \quad (6.83a)$$

$$k_{bt} = -\left(\frac{k}{L_0}\right) A_0 K(X_b) V'(X_b) \quad (6.83b)$$

$$k_{tb} = \left(\frac{k}{L_0}\right) A_0 K(X_t) U'(X_t) \quad (6.83c)$$

$$k_{tt} = \left(\frac{k}{L_0}\right) A_0 K(X_t) V'(X_t) \quad (6.83d)$$

and $\{P\}$ is the column vector of the two components

$$P_b = (k_{bb} + k_{bt})\frac{T_\infty}{\Delta T} \tag{6.84a}$$

$$P_t = (k_{tb} + k_{tt})\frac{T_\infty}{\Delta T} \tag{6.84b}$$

For the fin equation defined by Equation (6.12), the functional forms of $U'(X)$ and $V'(X)$ can be determined by introducing the fundamental solutions, Equations (6.76), into Equations (6.79a, b) and then substituting the resulting expressions into Equations (6.83). Then when (m/c) *is integral*, Equations (6.83) become:

$$k_{bb} = -\frac{\kappa n M X_b^{c-2m} N(n M X_b^c, n M X_t^c)}{D} \tag{6.85a}$$

$$k_{bt} = \frac{\kappa}{D X_b^m X_t^m} \tag{6.85b}$$

$$k_{tb} = \frac{\kappa}{D X_b^m X_t^m} \tag{6.85c}$$

$$k_{tt} = -\frac{\kappa n M X_t^{c-2m} N(n M X_t^c, n M X_b^c)}{D} \tag{6.85d}$$

where

$$\kappa = \left(\frac{k}{L_0}\right)A_0 c \tag{6.85e}$$

$$N(X, Y) = I_{(m/c)-1}(X)K_{m/c}(Y) + I_{m/c}(Y)K_{(m/c)-1}(X) \tag{6.85f}$$

$$D = I_{m/c}(n M X_b^c)K_{m/c}(n M X_t^c) - I_{m/c}(n M X_t^c)K_{m/c}(n M X_b^c) \tag{6.85g}$$

and in Equation (6.85f), X and Y are dummy variables. When m/c *is nonintegral* Equations (6.83) take the form:

$$k_{bb} = -\frac{\kappa n M X_b^{c-2m} N(n M X_b^c, n M X_t^c)}{D} \tag{6.86a}$$

$$k_{bt} = -\frac{\dfrac{2}{\pi}\kappa\sin\left[\left(\dfrac{m}{c} - 1\right)\pi\right]}{D X_b^m X_t^m} \tag{6.86b}$$

$$k_{tb} = -\frac{\dfrac{2}{\pi}\kappa\sin\left[\left(\dfrac{m}{c} - 1\right)\pi\right]}{D X_b^m X_t^m} \tag{6.86c}$$

$$k_{tt} = -\frac{\kappa n M X_t^{c-2m} N(n M X_t^c, n M X_b^c)}{D} \tag{6.86d}$$

where

$$\kappa = \left(\frac{k}{L_0}\right) A_0 c \tag{6.86e}$$

$$N(X, Y) = I_{(m/c)-1}(X) I_{-m/c}(Y) - I_{m/c}(Y) I_{-(m/c)+1}(X) \tag{6.86f}$$

$$D = I_{m/c}(nMX_b^c) I_{-m/c}(nMX_t^c) - I_{m/c}(nMX_t^c) I_{-m/c}(nMX_b^c)$$

$$\tag{6.86g}$$

and in Equation (6.86) X and Y are dummy variables.

For the special case of $cn = 1$ and $c = 0$, we recall that the fin Equation (6.12) reduces to the Euler equation and its two fundamental solutions are given by Equations (6.71). In that case Equations (6.77a, b) are introduced into Equations (6.79) and the resulting expressions are substituted into Equations (6.83), which become

$$k_{bb} = -\frac{\kappa N(X_b, X_t)}{D} \tag{6.87a}$$

$$k_{bt} = \frac{\kappa(r_1 - r_2)}{D} \tag{6.87b}$$

$$k_{tb} = \frac{\kappa(r_1 - r_2)}{D} \tag{6.87c}$$

$$k_{tt} = -\frac{\kappa N(X_t, X_b)}{D} \tag{6.87d}$$

where

$$\kappa = \left(\frac{k}{L_0}\right) A_0 \tag{6.87e}$$

$$N(X, Y) = \frac{r_1 X^{r_1} Y^{r_2} - r_2 Y^{r_1} X^{r_2}}{X^{2m}} \tag{6.87f}$$

$$D = X_b^{r_1} X_t^{r_2} - X_t^{r_1} X_b^{r_2} \tag{6.87g}$$

and in Equation (6.87f), X and Y are dummy variables. The foregoing procedure establishes the fundamental properties of each element (i.e., fin) in a system composed of many individual elements connected through the nodes. The global relationships between the individual elements for the entire system can now be established by utilizing the finite element assembly procedure as now described.

Finite Element Assembly Procedure

Let us consider the heat flow network system composed of many individual fins ($e = 1, 2, \ldots, E$) connected through the nodes ($n = 1, 2, \ldots, N$). We focus our attention on an individual fin treated as a finite element similar to the one illustrated in Figure 6.19, except the end points are denoted as the nodes m and n, and the fin is referred to as the element e of the assembly. Let T_m and T_n be the temperatures, and $Q_m^{(e)}$ and $Q_n^{(e)}$ be the heat flow rates entering the element e through the nodes m and n, respectively. The relationship given by Equation (6.82b) is now applied to this element and written in the abbreviated form as

$$K^{(e)}T = Q^{(e)} + P^{(e)} \tag{6.88a}$$

or in the expanded form as

$$
\begin{bmatrix}
0 & \cdot & 0 & \cdot & 0 & \cdot & 0 & \cdot & 0 \\
0 & \cdot & k_{nn}^{(e)} & \cdot & 0 & \cdot & k_{nm}^{(e)} & \cdot & 0 \\
0 & \cdot & 0 & \cdot & 0 & \cdot & 0 & \cdot & 0 \\
0 & \cdot & k_{mn}^{(3)} & \cdot & 0 & \cdot & k_{mm}^{(e)} & \cdot & 0 \\
0 & \cdot & 0 & \cdot & 0 & \cdot & 0 & \cdot & 0
\end{bmatrix}
\begin{Bmatrix}
T_1 \\ \cdot \\ T_n \\ \cdot \\ \cdot \\ T_m \\ \cdot \\ T_N
\end{Bmatrix}
=
\begin{Bmatrix}
0 \\ \cdot \\ Q_n^{(e)} \\ \cdot \\ 0 \\ Q_m^{(e)} \\ \cdot \\ 0
\end{Bmatrix}
+
\begin{Bmatrix}
0 \\ \cdot \\ P_n^{(e)} \\ \cdot \\ 0 \\ P_m^{(e)} \\ \cdot \\ 0
\end{Bmatrix}
$$

$$\tag{6.88b}$$

We note that (1) at the nodes where the fins are connected the temperatJre is the same for all fins forming the node, and (2) the algebraic sum of the heat flow rates at each node must be equal to the external heat flow at that node.

Now Equation (6.88b) is written for each element and the results are summed up for all $e = 1, 2, \ldots, E$ where E is the total number of elements in the assembly. We obtain

$$[K] \quad \{T\} = \{Q\} + \{P\} \tag{6.89a}$$

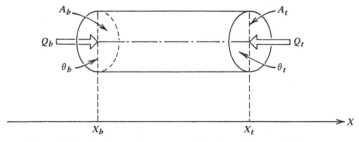

Figure 6.19 A schematic representation of an element of fin-array.

where $\{Q\}$ is a column vector for external nodal heat flow rates, and the global matrix $[K]$ and the column vector $\{P\}$ are given by

$$[K] = \sum_{e=1}^{E} K^{(e)} \tag{6.89b}$$

$$\{P\} = \sum_{e=1}^{E} P^{(e)} \tag{6.89c}$$

The computer assembly procedure forming the global matrix $[K]$ consists of the following steps: (1) Set up an $N \times N$ null master matrix (all zero entries). (2) Starting with one element, insert its coefficients $k_{nn}^{(e)}$, $k_{nm}^{(e)}$, $k_{mn}^{(e)}$, and $k_{mm}^{(e)}$ into the master matrix in the locations designated by their indices. Each time a term is placed in a location where another term has already been placed, it is added to whatever value is there. (3) Return to step 2 and repeat this procedure for one element after another until all elements have been treated. The result will be the global matrix $[K]$. This assembly procedure is an essential part of the finite element method.

The Boundary Conditions

The next step in the analysis is the inclusion of the boundary condition at each node into the problem.

At each node a boundary condition can be written as

$$A_n T_n + B_n Q_n = F_n, \qquad n = 1, 2, \ldots, N \tag{6.90a}$$

where Q_n is the external nodal heat flow rate; A_n, B_n, and F_n are known constants. The cases $A_n = 1$, $B_n = 0$, and $A_n = 0$, $B_n = 1$ correspond to a prescribed temperature and heat flow rate, respectively.

To preserve the sparse, banded, and symmetric properties of the global matrix $[K]$, the boundary conditions (6.90a) are rewritten as

$$\left(\frac{A_n}{B_n}\right) T_n + Q_n = \frac{F_n}{B_n}, \qquad n = 1, 2, \ldots, N \tag{6.90b}$$

To insert prescribed nodal temperature $A_n = 1$, B_n is replaced by a small number, say, 1×10^{-15}. Then Equation (6.83b) express the fact that $T_n = F_n$ since $1 \times 10^{15} \gg 1$.

The nodal boundary conditions (6.90b) can be written in the matrix form as

$$\left[\frac{A}{B}\right]\{T\} + \{Q\} = \left\{\frac{F}{B}\right\} \tag{6.90c}$$

where

$$\left[\frac{A}{B}\right] \equiv \begin{bmatrix} \dfrac{A_1}{B_1} & & & \\ & \dfrac{A_2}{B_2} & & \\ & & \ddots & \\ & & & \dfrac{A_n}{B_n} \end{bmatrix} \quad \text{and} \quad \left\{\frac{F}{B}\right\} \equiv \begin{Bmatrix} \dfrac{F_1}{B_1} \\ \dfrac{F_2}{B_2} \\ \vdots \\ \dfrac{F_n}{B_n} \end{Bmatrix} \quad (6.90\text{d, e})$$

and the elements of $[A/B]$ not shown are zero.

Finally, Equations (6.89a) and (6.90c) are combined to obtain the solution in the form

$$\{T\} = \left(\left[\frac{A}{B}\right] + [K]\right)^{-1}\left(\left\{\frac{F}{B}\right\} + \{P\}\right) \qquad (6.91)$$

Illustrative Example

Consider now as an illustrative example the fin matrix configuration of Figure (6.20a) to represent the repeating section of a plate fin compact heat exchanger. Let the elements 2 and 3 represent the splitter plates; then there is no heat transfer at the nodes and we have $Q_2 = Q_4 = 0$. The nodes 1 and 5 are subjected to prescribed heat flow rates, hence Q_1 and Q_5 are known. At the node 3 there is no externally applied heat flow.

The specific fin geometry considered for this example is a plate fin as illustrated in Figure 6.20b, for which the standard fin equation applies, that is,

$$\frac{d^2\theta(X)}{dX^2} - M^2\theta(X) = 0 \qquad (6.92)$$

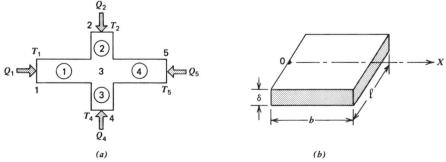

(a)   (b)

Figure 6.20 Nomenclature for: (a) Repeating element of the fin-array; (b) geometry of an element of the fin-array.

where $\theta(X)$ is the fin temperature in excess of the surrounding coolant temperature.

A comparison of Equation (6.92) with Equation (6.7a) reveals that

$$K(X) = 1 \quad \text{and} \quad W(X) = 1 \tag{6.93a}$$

and in view of the definitions, Equations (6.11a, b), we have

$$m = \frac{1}{2}, \quad c = 1, \quad \text{and} \quad n = 1 \tag{6.93b}$$

Then the thermal influence coefficients, Equations (6.86), reduce to

$$k_{bb} = k_{tt} = \frac{k}{L_0} A_0 M \frac{\cosh[M(X_t - X_b)]}{\sinh[M(X_t - X_b)]} \tag{6.94a}$$

$$k_{bt} = k_{tb} = \frac{k}{L_0} A_0 M \frac{1}{\sinh[M(X_t - X_b)]} \tag{6.94b}$$

where we utilized the relationships

$$I_{-1/2}(z) = \sqrt{\frac{2}{\pi z}} \cosh z, \quad I_{1/2}(z) = \sqrt{\frac{2}{\pi z}} \sinh z \tag{6.95}$$

As reference length we select $L_0 = b$ (see Figure 6.20b). The cross-section area is taken as $A_0 = \delta\ell$. Finally we select

$$X_b = 0, \quad X_t = 1$$

$$M = mb \quad \text{where } m = \left(\frac{2h}{k\delta}\right)^{1/2} \tag{6.96}$$

Then the resulting fin matrix becomes

$$\begin{vmatrix} k_{bb}^{(e)} & k_{tb}^{(e)} \\ k_{bt}^{(e)} & k_{tt}^{(e)} \end{vmatrix} = \frac{(k\delta\ell m)^{(e)}}{\sinh[m^{(e)}b^{(e)}]} \begin{vmatrix} \cosh[m^{(e)}b^{(e)}] & -1 \\ -1 & \cosh[m^{(e)}b^{(e)}] \end{vmatrix} \tag{6.97}$$

The pertinent dimensions of the elements and the magnitudes of various other quantities are taken as in [5]:

Fins 1 and 4: $b^{(1)} = b^{(4)} = 6.34$ mm, $\delta^{(1)} = \delta^{(4)} = 0.152$ mm
Fins 2 and 3: $b^{(2)} = b^{(3)} = 1.16$ mm, $\delta^{(2)} = \delta^{(3)} = 0.254$ mm
Depth of array: $\ell = 0.3048$ m
$k = 173$ W/m °C, $h = 56.77$ W/m² °C
Heat flow rates: $Q_1 = 2.93$ W, $Q_5 = 2.344$ W
There is no heat flow at the nodes 2 and 4; hence $Q_2 = Q_4 = 0$.

These numerical results are used to calculate the influence coefficients for each element in the fin matrix (6.97). Finally, the solution of the global matrix for the problem yielded the temperatures in excess of the surrounding temperature for each node as: $\theta_1 = 11.2010$, $\theta_2 = 9.7652$, $\theta_3 = 9.7825$, $\theta_4 = 9.7652$, and $\theta_5 = 10.7570$ where the subscripts correspond to the node number.

REFERENCES

1. D. A. Kern and A. D. Kraus, *Extended Surface Heat Transfer*, McGraw-Hill, New York, 1972.
2. K. A. Gardner, Efficiency of Extended Surfaces, *Trans. ASME*, **67**, 621–631 (1945).
3. G. N. Watson, *A Treatize of the Theory of Bessel Functions*, 2nd ed., Cambridge University Press, London, 1966.
4. N. W. McLachlan, *Bessel Functions for Engineers*, 2nd ed., Clarendon Press, London, 1961.
5. A. D. Kraus, A. D. Snider, and L. F. Doty, An Efficient Algorithm for Evaluating Arrays of Extended Surface, *J. Heat Transf.*, **100**, 288–293 (1978).
6. M. D. Mikhailov and M. N. Özişik, Finite Element Analysis of Heat Exchangers. In *Heat Exchangers*, S. Kakaç, A. E. Bergles, and F. Mayinger, Eds. Hemisphere, New York, 1981.
7. M. D. Mikhailov and M. N. Özişik, On the Solution of Heat Transfer Through an Array of Extended Surfaces, *Int. J. Heat Mass Transfer*, **27**, 893–899 (1984).
8. M. N. Özişik, *Basic Heat Transfer*, McGraw-Hill, New York, 1977.
9. K. H. Huebner, *The Finite Element Method for Engineers*, Wiley, 1975.

CHAPTER SEVEN

Class I Solutions Applied to Time Dependent Heat and Mass Diffusion

Numerical precision is the very soul of science.

Sir D'Arcy Wentworth Thompson
1860–1948

The material presented in this chapter is both analytical and numerical application of the general solutions of the problem of Class I developed in the previous chapters for the solution of unsteady heat or mass diffusion problems in the rectangular, cylindrical, and spherical coordinate systems, subject to different types of boundary conditions. To illustrate the similarity in the mathematical formulation of heat and mass diffusion problems, a brief discussion is given of the physical significance of the conservation equation and the pertinent boundary conditions in relation to heat and mass diffusion. Because of the analogy between the heat and mass diffusion processes, the mathematical formulation of a heat diffusion problem can be interpreted as that of an analogous mass diffusion process by proper interpretation of the notation. All the problems considered in this chapter belong to special cases of the problem of Class I; therefore, the solutions to the problems considered in this chapter are obtained from the general solutions developed previously by appropriate simplifications.

7.1 FORMULATION OF HEAT AND MASS DIFFUSION PROBLEMS OF CLASS I

In heat transfer, diffusion (i.e., *conduction*) is a process in which the energy is transferred from the region of high temperature to the region of low temperature as a result of random motion of molecules as in the case of fluid in rest, and due to the drift of electrons in the case of metals. In mass transfer, *diffusion* is a process in which mass is transferred from one part of the system to another as a result of random motion of molecules. For example, if iodine or dye is introduced to the lower portion of clear water in a container without setting up convective currents, the colored part gradually spreads towards the top as a result of diffusion of iodine or dye molecules through the water.

The mathematical theory of diffusion in an *isotropic* medium (i.e., a medium in which properties are independent of direction) is based on the hypothesis that the transfer of substance (i.e., heat or mass) per unit area, per unit time is proportional to the normal gradient of the concentration of the substance through the area. The equation governing heat diffusion, originally introduced by Joseph Fourier in his analytic theory of heat [1], is given by

$$\mathbf{q} = -k\nabla T(\mathbf{x}, t) \tag{7.1}$$

Here $\mathbf{q}$ is the *heat flux vector*, $T(\mathbf{x}, t)$ is the temperature, and $\nabla T(\mathbf{x}, t)$ is the gradient of temperature. The minus sign is included in order to make the heat flow a positive quantity when the heat flux vector $\mathbf{q}$ points in the direction of decreasing temperature. The proportionality constant k is called the *thermal conductivity* of the material and is a positive quantity. When the temperature gradient is measured in °C/m, the heat flux is in W/m^2, and the thermal conductivity has units W/m °C. There is a wide difference in the thermal conductivities of various engineering materials. The highest value is given by pure metals and the lowest value by gases and vapors; the amorphous insulating materials and inorganic liquids have thermal conductivities that lie in between. A comprehensive compilation of thermal conductivities of materials may be found in References 2–4.

In the case of mass diffusion in a binary system, that is, species A diffuses in the direction of decreasing concentration of A inside species B, the similarity between heat and mass diffusion was recognized by Fick [5] and the equation governing mass diffusion is given in an analogous fashion

$$\mathbf{J} = -D\nabla C(\mathbf{x}, t) \tag{7.2}$$

Here $\mathbf{J}$ is the *mass flux vector*, $C(\mathbf{x}, t)$ is the concentration of the diffusion species, $\nabla C(\mathbf{x}, t)$ is the gradient of the concentration, and D is called the *diffusion coefficient*. The minus sign is included for the same reason discussed previously. To give some idea of the units of various quantities in Equation (7.2), let the concentration $C(\mathbf{x}, t)$ be measured in kg/m^3, the distance in meters, the mass flux in kg/m^2 hr; then the diffusion coefficient D is in m^2/hr. Various models have been proposed for the determination of the *diffusion coefficient* for a binary mixture of gases. The reader should consult References 6–8 for discussion of various mathematical models. In the case of diffusion of a specie inside a liquid, a rigorous theory is yet to be developed for the prediction of the diffusion coefficient. A number of semiempirical relations have been proposed for the determination of diffusion coefficient in liquids. The reader should consult References 9 and 10 for a collection of such expressions. In the case of diffusion inside a solid containing pores and capillaries, the theoretical prediction of diffusion coefficient is an extremely complicated matter; no fundamental theory is yet available on the subject. Experimental approach is the only reliable means of determining the diffusion coefficient for a given solid structure and fluid combination.

Clearly, the expressions for the heat and mass flux, given by Equations (7.1) and (7.2), respectively, are identical to the *linear diffusion law* given by Equation (1.24). Having established the expressions governing the heat and mass fluxes, we can now proceed to the determination of the conservation equations for heat and mass diffusion.

Heat Diffusion Equation

The differential equation governing the diffusion of heat in an isotropic medium is immediately obtained from Equations (1.32) and (1.33) as

$$\rho C_p \frac{\partial T(\mathbf{x}, t)}{\partial t} = \nabla \cdot [k(\mathbf{x}) \nabla T(\mathbf{x}, t)] + g(\mathbf{x}, t) \quad \text{in } V \qquad t > 0 \quad (7.3a)$$

For the case of constant thermal conductivity $k(\mathbf{x}) = k$, Equation (7.3a) simplifies to

$$\frac{1}{\alpha} \frac{\partial T(\mathbf{x}, t)}{\partial t} = \nabla^2 T(\mathbf{x}, t) + \frac{1}{k} g(\mathbf{x}, t) \quad \text{in } V \qquad t > 0 \qquad (7.3b)$$

In the preceding equations $T(\mathbf{x}, t)$ is the temperature, C_p is the specific heat, ρ is the density, $k(\mathbf{x})$ is the thermal conductivity of the medium. The term $g(\mathbf{x}, t)$ represents the energy generation (or sink) when it is positive (or negative) (i.e., W/m³ hr). In addition we defined

$$\alpha = \frac{k}{\rho C_p} \qquad (7.3c)$$

The physical significance of thermal diffusivity of a material is associated with the speed of propagation of heat into the solid. The higher is the thermal diffusivity, the faster is the propagation of heat into the medium.

Mass Diffusion Equation

The differential equation governing the diffusion of mass in an isotropic medium is given by

$$\frac{\partial C(\mathbf{x}, t)}{\partial t} = \nabla \cdot [D(\mathbf{x}) \nabla C(\mathbf{x}, t)] + g(\mathbf{x}, t) \quad \text{in } V \qquad t > 0 \quad (7.4a)$$

For the case of the constant diffusion coefficient $D(\mathbf{x}) = D$, Equation (7.4a) simplifies to

$$\frac{1}{D}\frac{\partial C(\mathbf{x}, t)}{\partial t} = \nabla^2 C(\mathbf{x}, t) + \frac{1}{D}g(\mathbf{x}, t) \quad \text{in } V \quad t > 0 \qquad (7.4b)$$

In these equations $C(\mathbf{x}, t)$ is the mass concentration and $D(\mathbf{x})$ is the diffusion coefficient. The term $g(\mathbf{x}, t)$ represents the mass generation (or sink) when it is positive (or negative), that is, kg/m^3 hr. The physical significance of the diffusion coefficient is associated with the speed of diffusion of the mass into the medium. The higher is the value of diffusion coefficient, the faster is the diffusion of the mass into the medium.

The reader should consult References 8–10 for the measured values and the theoretical predictions of the values of the diffusion coefficient D.

A comparison of Equations (7.3) and (7.4) reveals their similarity: the diffusion coefficient for mass diffusion problems is analogous to the thermal diffusivity in heat diffusion problems.

Boundary Conditions

To solve the foregoing heat and mass diffusion equations, a set of boundary conditions are needed. Here we consider the following three different types of linear boundary conditions which are referred to as the boundary conditions of the *first*, the *second* and the *third* kind.

The boundary condition at a surface S of the region V is said to be of *the first kind* if the distribution of temperature or mass concentration is specified at that surface. This type of boundary condition for the heat and mass diffusion problems is formally written as

$$\text{Heat diffusion:} \quad T(\mathbf{x}, t) = \phi(\mathbf{x}, t), \quad \mathbf{x} \in S \qquad (7.5a)$$

$$\text{Mass diffusion:} \quad C(\mathbf{x}, t) = \phi(\mathbf{x}, t), \quad \mathbf{x} \in S \qquad (7.5b)$$

where $\phi(\mathbf{x}, t)$ denotes the distribution of temperature or mass concentration for the heat diffusion and mass diffusion problems, respectively, at the boundary surface S.

In the case of the mass diffusion, care must be exercized in the interpretation of the mass boundary condition Equation (7.5b). The physical situation defined by this equation implies that the concentration of the solute in the body just at the boundary surface is the same as that in the solution at the boundary surface. This may not be so if the partition factor discussed by Crank [11] is not equal to unity. That is, the solute concentration in the body just at the boundary surface is not necessarily equal to that within the solution at the boundary if the partition factor differs from unity. Therefore, the mass boundary condition Equation (7.5b) implies a partition factor of unity for the analysis.

The boundary condition at the surface S is said to be of *the second kind* if the distribution of the heat flux or the mass flux is prescribed at the surface S. This type of boundary condition for the heat and mass diffusion problems is formally written as

$$\text{Heat diffusion:} \quad k(\mathbf{x})\frac{\partial T(\mathbf{x}, t)}{\partial \mathbf{n}} = \phi(\mathbf{x}, t), \quad \mathbf{x} \in S \quad (7.6a)$$

$$\text{Mass diffusion:} \quad D(\mathbf{x})\frac{\partial C(\mathbf{x}, t)}{\partial \mathbf{n}} = \phi(\mathbf{x}, t), \quad \mathbf{x} \in S \quad (7.6b)$$

where $\phi(\mathbf{x}, t)$ denotes the prescribed heat or mass flux distribution, respectively, at the boundary surface S. The symbol $\partial/\partial \mathbf{n}$ denotes the derivative along the *outward drawn normal* $\hat{\mathbf{n}}$ to the boundary surface S.

The boundary condition at the surface S is said to be of *the third kind* if heat or mass transfer takes place between the boundary surface and the surrounding environment by convection characterized by a heat or mass transfer coefficient. The boundary condition for such a situation is determined by writing an energy or mass balance for the surface; the results for heat and mass transfer are expressed, respectively, in the forms

$$\text{Heat diffusion:} \quad T(\mathbf{x}, t) + \frac{k(\mathbf{x})}{h}\frac{\partial T(\mathbf{x}, t)}{\partial \mathbf{n}} = \phi(\mathbf{x}, t), \quad \mathbf{x} \in S$$
$$(7.7a)$$

$$\text{Mass diffusion:} \quad C(\mathbf{x}, t) + \frac{D(\mathbf{x})}{h}\frac{\partial C(\mathbf{x}, t)}{\partial \mathbf{n}} = \phi(\mathbf{x}, t), \quad \mathbf{x} \in S$$
$$(7.7b)$$

where h represents the *heat transfer coefficient* or the *mass transfer coefficient* for the heat diffusion or mass diffusion problem, respectively. The function $\phi(\mathbf{x}, t)$ is the prescribed temperature or mass concentration of the environment over the boundary surface S for the heat or mass diffusion problems, respectively. The symbol $\partial/\partial \mathbf{n}$ denotes the derivative along the *outward drawn normal* $\hat{\mathbf{n}}$ to the boundary surface S.

It is to be noted that in the writing of the mass diffusion boundary condition Equation (7.7b) the partition factor is assumed to be unity.

Analytical and empirical expressions are available for the heat and mass diffusion coefficient h for many situations encountered in practical applications [12–14].

If the right-hand side of the boundary conditions given by Equations (7.5), (7.6), and (7.7) is zero, they are called, respectively, *homogeneous boundary conditions of the first, second, and third kind.*

The three different types of boundary conditions given by Equations (7.5), (7.6), and (7.7) are readily obtainable from the following general expressions

$$\delta T(\mathbf{x}, t) + \gamma k(\mathbf{x}) \frac{\partial T(\mathbf{x}, t)}{\partial \mathbf{n}} = \phi(\mathbf{x}, t), \qquad \mathbf{x} \in S \qquad (7.8a)$$

$$\delta C(\mathbf{x}, t) + \gamma D(\mathbf{x}) \frac{\partial C(\mathbf{x}, t)}{\partial \mathbf{n}} = \phi(\mathbf{x}, t), \qquad \mathbf{x} \in S \qquad (7.8b)$$

if δ, γ, and $\phi(\mathbf{x}, t)$ are chosen as specified in the following.
Boundary condition of the first kind:

$$\left. \begin{array}{l} \delta = 1, \qquad \gamma = 0 \\[2mm] \phi(\mathbf{x}, t) = \text{temperature or mass concentration} \\ \qquad \text{at the boundary surface.} \end{array} \right\} \qquad (7.9a)$$

Boundary condition of the second kind:

$$\left. \begin{array}{l} \delta = 0, \qquad \gamma = 1 \\[2mm] \phi(\mathbf{x}, t) = \text{heat flux or mass flux at the boundary surface} \end{array} \right\} \qquad (7.9b)$$

Boundary condition of the third kind:

$$\left. \begin{array}{l} \delta = 1, \qquad \gamma = \dfrac{1}{h} \\[2mm] h = \text{heat or mass transfer coefficient} \\[2mm] \phi(\mathbf{x}, t) = \text{temperature or mass concentration} \\ \qquad \text{of the surrounding fluid} \end{array} \right\} \qquad (7.9c)$$

One-Dimensional Formulation of Heat and Mass Diffusion Problems

When temperature depends on time and one of the space variables only, the mathematical formulation of the heat diffusion problem is obtained from Equations (7.3b) and (7.8a) by making appropriate simplifications. The resulting system, valid for the rectangular, cylindrical, and spherical coordinate systems, is given by

$$\frac{1}{\alpha_0} \frac{\partial T(x, t)}{\partial t} = \frac{1}{x^n} \frac{\partial}{\partial x} \left[x^n k^*(x) \frac{\partial T(x, t)}{\partial x} \right] + \frac{1}{k_0} g(x, t)$$

$$\text{in } x_0 < x < x_1, \qquad t > 0 \quad (7.10a)$$

$$\delta_k T(x_k, t) - (-1)^k \gamma_k k(x_k) \frac{\partial T(x_k, t)}{\partial x} = \phi_k(t) \quad \text{at } x = x_k, \qquad k = 0, 1,$$

$$(7.10b)$$

$$T(x, 0) = f(x) \quad \text{in } x_0 \leq x \leq x_1 \qquad (7.10c)$$

where

$$k*(x) = \frac{k(x)}{k_0}, \qquad \alpha_0 = \frac{k_0}{\rho C_p} \qquad \text{(7.10d, e)}$$

and

$$n = \begin{cases} 0 & \text{slab} \\ 1 & \text{cylinder} \\ 2 & \text{sphere} \end{cases} \qquad \text{(7.10f)}$$

In the preceding equations k_0 is a reference thermal conductivity and α_0 is a reference thermal diffusivity based on the reference thermal conductivity k_0. Clearly, by appropriate choice of δ_k, γ_k, and $\phi_k(t)$ ($k = 0, 1$) as given by Equations (7.9), the boundary condition of the first, second, or third kind is readily obtainable for any of the boundary surfaces at $x = x_0$ and $x = x_1$. Thus nine different combinations of boundary conditions are possible for the preceding heat diffusion problem.

In the case of the mass diffusion problem, equations analogous to (7.10) are obtained from Equation (7.4b) and (7.8b) and given by

$$\frac{1}{D_0}\frac{\partial C(x,t)}{\partial t} = \frac{1}{x^n}\frac{\partial}{\partial x}\left[x^n D*(x)\frac{\partial C(x,t)}{\partial x}\right] + \frac{1}{D_0}g(x,t)$$

$$\text{in } x_0 < x < x_1, \qquad t > 0 \quad \text{(7.11a)}$$

$$\delta_k C(x_k, t) - (-1)^k \gamma_k D(x_k)\frac{\partial C(x_k, t)}{\partial x} = \phi_k(t) \quad \text{at } x = x_k, \qquad k = 0, 1$$

$$\text{(7.11b)}$$

$$C(x,0) = f(x) \quad \text{in } x_0 \le x \le x_1 \qquad \text{(7.11c)}$$

where $D*(x) = D(x)/D_0$ and $n = 0$, 1, or 2 for slab, cylinder, or sphere, respectively.

7.2   ONE-DIMENSIONAL, TIME DEPENDENT DIFFUSION IN SLAB, CYLINDER, AND SPHERE

In this section we illustrate how the general solution of one-dimensional problems of Class I can be applied to obtain solutions to specific heat diffusion problems in simple geometries such as slab, long solid cylinder, and solid sphere. The heat diffusion problems considered here can also be interpreted as mass diffusion problems with appropriate interpretation of the physical significance of various parameters.

We focus our attention on problems in the region $0 \le x \le \ell_0$, with symmetry boundary condition at $x = 0$; that is, a slab of thickness ℓ_0 with the boundary at $x = 0$ insulated, or a long solid cylinder and a solid sphere of radius ℓ_0. We assume that the medium is initially at a temperature $f(x)$. For times $t > 0$, heat is generated in the region at a rate of $g(x, t)$, W/m³, and heat is dissipated from the boundary surface at $x = \ell_0$ by convection into an environment whose temperature varies with time. The mathematical formulation of this heat diffusion problem, for slab, solid cylinder, and solid sphere is given in the dimensionless form as

$$X^{1-2m} \frac{\partial \theta (X, \tau)}{\partial \tau} = \frac{\partial}{\partial X} \left[X^{1-2m} \frac{\partial \theta (X, \tau)}{\partial X} \right] + X^{1-2m} G(X, \tau)$$

$$\text{in } 0 < X < 1, \qquad \tau > 0 \quad (7.12a)$$

$$\frac{\partial \theta (X, \tau)}{\partial X} = 0 \quad \text{at } X = 0, \qquad \tau > 0 \qquad (7.12b)$$

$$\alpha \theta (X, \tau) + \beta \frac{\partial \theta (X, \tau)}{\partial X} = \phi(\tau) \quad \text{at } X = 1, \qquad \tau > 0 \qquad (7.12c)$$

$$\theta (X, \tau) = F(X) \quad \text{for } \tau = 0, \qquad 0 \le X \le 1 \quad (7.12d)$$

where

$$m = \begin{cases} \frac{1}{2} & \text{for slab} \\ 0 & \text{for cylinder} \\ -\frac{1}{2} & \text{for sphere} \end{cases}$$

and various dimensionless quantities are defined as

$F(X) = \dfrac{f(x) - T_r}{\Delta T} = $ dimensionless initial temperature

$G(X, \tau) = \dfrac{g(x, t)\ell_0^2}{k \Delta T} = $ dimensionless energy generation

$T_r = $ a reference temperature

$\Delta T = $ a reference temperature difference

$X = \dfrac{x}{\ell_0} = $ dimensionless space variable

$\ell_0 = $ slab thickness; cylinder or sphere radius

$\alpha = 1, \qquad \beta = \dfrac{k}{h\ell_0}$ for boundary condition of the third kind

$\alpha = 0, \qquad \beta = 1$ for boundary condition of the second kind

$\alpha = 1, \qquad \beta = 0$ for boundary condition of the first kind

$\theta (X, \tau) = \dfrac{T(x, t) - T_r}{\Delta T} = $ dimensionless temperature

$\tau = \dfrac{\alpha t}{\ell_0^2} = $ dimensionless time or Fourier number. $\qquad (7.13)$

The problem (7.12) considered here is a special case of the general problem Equations (4.1), as follows:

$$T = \theta, \qquad x = X, \qquad t = \tau, \qquad d(x) = 0$$

$$w(x) = k(x) = X^{1-2m}, \qquad P(x, t) = X^{1-2m}G(X, \tau)$$

$$x_0 = 0, \qquad x_1 = 1, \qquad \alpha_0 = 0, \qquad \beta_0 = 1$$

$$\phi(x_0, t) = 0, \qquad \alpha_1 = \alpha, \qquad \beta_1 = \beta, \qquad \phi(x_1, t) = \phi(\tau) \quad (7.14)$$

The eigenvalue problem appropriate for the solution of the problem (7.12) has already been considered as Example 3.6, Equations (3.42). Once the eigenfunctions, eigenvalues, and normalization integral are obtained from the results given in Example 3.6, the solution of the problem (7.12) can be determined from the general solution (4.2) for the case $\alpha \neq 0$ and from the solution (4.6) for the case $\alpha = 0$ as now described.

The Case $\alpha \neq 0$

We consider the boundary condition of the third kind at $X = 1$; that is, the case $\alpha \neq 0$ and $\beta =$ finite. The eigenfunctions $\psi_i(X)$ and the normalization integral N_i for this case are obtained from Equations (3.43a) and (3.43c), respectively, as

$$\psi_i(X) = X^m J_{-m}(\mu_i X) \tag{7.15a}$$

$$N_i = \frac{1}{2}[J_{1-m}(\mu_i)]^2\left[1 + 2m\frac{\beta}{\alpha} + \left(\frac{\beta}{\alpha}\mu_i\right)^2\right] \tag{7.15b}$$

In writing these results we omitted the constant multipliers C and μ_i^m, since they will cancel out when they are introduced into the general solution of the problem.

When Equations (7.15) are introduced into the general solution (4.2) together with the values of various quantities as defined by Equations (7.14), the solution of the problem (7.12) is obtained as

$$\theta(X, \tau) = 2\sum_{i=1}^{\infty} \frac{e^{-\mu_i^2\tau}}{1 + 2m\dfrac{\beta}{\alpha} + \left(\dfrac{\beta}{\alpha}\mu_i\right)^2} X^m \frac{J_{-m}(\mu_i X)}{J_{1-m}(\mu_i)}$$

$$\times \left\{\int_0^1 X^{1-m}\frac{J_{-m}(\mu_i X)}{J_{1-m}(\mu_i)}F(X)\,dX\right.$$

$$\left. + \int_0^\tau e^{\mu_i^2\tau'}\left[\phi(\tau')\frac{\mu_i}{\alpha} + \int_0^1 X^{1-m}\frac{J_{-m}(\mu_i X)}{J_{1-m}(\mu_i)}G(X, \tau')\,dX\right]d\tau'\right\}$$

$$\tag{7.16a}$$

where the μ_i are the roots of the transcendental equation [see Equation (3.43b)]

$$\alpha J_{-m}(\mu) - \beta\mu J_{1-m}(\mu) = 0 \qquad (7.16b)$$

The Bessel functions appearing in solution (7.16) are listed in Table 3.1 for the values of $m = \tfrac{1}{2}, 0,$ and $-\tfrac{1}{2}$.

The mean temperature $\theta_{av}(\tau)$ is defined according to Equation (2.20) as

$$\theta_{av}(\tau) = 2(1-m)\int_0^1 X^{1-2m}\theta(X,\tau)\,dX \qquad (7.17)$$

Introducing solution (7.16a) into Equation (7.17), the mean temperature becomes

$$\theta_{av}(\tau) = 4(1-m)\sum_{i=1}^{\infty}\frac{e^{-\mu_i^2\tau}}{1+2m\beta/\alpha+(\beta\mu_i/\alpha)^2}\frac{1}{\mu_i}$$

$$\times\left\{\left[\int_0^1 X^{1-m}\frac{J_{-m}(\mu_i X)}{J_{1-m}(\mu_i)}F(X)\,dX\right.\right.$$

$$\left.\left.+\int_0^\tau e^{\mu_i^2\tau'}\left[\phi(\tau')\frac{\mu_i}{\alpha}+\int_0^1 X^{1-m}\frac{J_{-m}(\mu_i X)}{J_{1-m}(\mu_i)}G(X,\tau')\,dX\right]d\tau'\right\}$$

$$\qquad (7.18)$$

The solutions (7.16) and (7.18) are applicable for both the boundary condition of the third kind (i.e., $\alpha \neq 0$, β = finite) and the first kind (i.e., $\alpha = 1$, $\beta = 0$) at the boundary surface $X = 1$. The physical significance of $\beta = 0$ for the boundary conditions of the first kind is that the heat transfer coefficient h is infinite, hence the temperature of the boundary surface is the same as that of the environment.

The Case $\alpha = 0$

When $\alpha = 0$ and $\beta = 1$, the boundary condition (7.12c) represents a prescribed heat flux at the boundary surface $X = 1$. Then the boundary conditions at $X = 0$ and $X = 1$ of the eigenvalue problem are both of the second kind. For this special case, the eigenfunctions $\psi_i(X)$ are taken as

$$\psi_i(X) = X^m J_{-m}(\mu_i X) \qquad (7.19a)$$

the normalization integral given by Equation (3.43c) simplifies to

$$N_i = \tfrac{1}{2}[J_{-m}(\mu_i)]^2 \qquad (7.19b)$$

and the eigenvalues μ_i are the roots of [see Equation (7.16b) for $\alpha = 0$]

$$J_{1-m}(\mu) = 0 \tag{7.19c}$$

When the eigenfunctions (7.19a) and the normalization integral (7.19b) are introduced into the general solution (4.6), the solution of the problem (7.12) for the case $\alpha = 0$ and $\beta = 1$ becomes

$$\theta(X, \tau) = \theta_{av}(\tau) + 2 \sum_{i=1}^{\infty} e^{-\mu_i^2 \tau} X^m \frac{J_{-m}(\mu_i X)}{J_{-m}(\mu_i)} \left\{ \int_0^1 X^{1-m} \frac{J_{-m}(\mu_i X)}{J_{-m}(\mu_i)} F(X) \, dX \right.$$

$$\left. + \int_0^{\tau} e^{\mu_i^2 \tau'} \left[\phi(\tau') + \int_0^1 X^{1-m} \frac{J_{-m}(\mu_i X)}{J_{-m}(\mu_i)} G(X, \tau') \, dX \right] d\tau' \right\} \tag{7.20a}$$

where

$$\theta_{av}(\tau) = 2(1 - m)$$

$$\times \left\{ \int_0^1 X^{1-2m} F(X) \, dX + \int_0^{\tau} \left[\phi(\tau') + \int_0^1 X^{1-2m} G(X, \tau') \, dX \right] d\tau' \right\}$$

$$\tag{7.20b}$$

and the μ_i are the roots of the transcendental equation

$$J_{1-m}(\mu) = 0 \tag{7.20c}$$

Splitting up the Solution

When the forcing functions such as the initial condition function $F(X)$, the heat generation term $G(X, \tau)$, or the boundary condition function $\phi(\tau)$ can be expressed as polynomials or exponentials, it is convenient to split up the preceding solutions into the solutions of simpler problems as now described.

Let the functions $F(X)$, $\phi(\tau)$, and $G(X, \tau)$ be represented as

$$F(X) = \sum_{j=0}^{n_f} f_{2j} X^{2j} \tag{7.21a}$$

$$\phi(\tau) = \phi_e e^{-d_\phi \tau} + \sum_{j=0}^{n_\phi} \phi_j \tau^j \tag{7.21b}$$

$$G(X, \tau) = G_e e^{-d_g \tau} + \sum_{j=0}^{n_g} G_j(X) \tau^j \tag{7.21c}$$

where

$$G_j(X) = \sum_{k=0}^{n_j} g_{j,2k} X^{2k} \tag{7.21d}$$

Then the solution of problem (7.12) can be split up into the solutions of simpler problems as discussed in Chapter 4. The resulting solutions are summarized for the cases of both $\alpha \neq 0$ and $\alpha = 0$ as follows.

$$\theta(X, \tau) = \theta_\phi(X)e^{-d_\phi\tau} + \theta_g(X)e^{-d_g\tau} + \sum_{j=0}^{n_\phi} \theta_{\phi j}(X)\tau^j$$

$$+ \sum_{j=0}^{n_g} \theta_{gj}(X)\tau^j + \theta_t(X, \tau) \tag{7.22a}$$

for $\alpha \neq 0$, β = finite, by Equation (4.8a);

$$\theta(X, \tau) = \theta_{av}(\tau) + \theta_\phi(X)e^{-d_\phi\tau} + \theta_g(X)e^{-d_g\tau} + \sum_{j=0}^{n_\phi} \theta_{\phi j}(X)\tau^j$$

$$+ \sum_{j=0}^{n_g} \theta_{gj}(X)\tau^j + \theta_t(X, \tau) \tag{7.22b}$$

for $\alpha = 0$, $\beta = 1$, by Equation (4.8b).

In the solutions (7.22), the function $\theta_{av}(\tau)$ is defined by Equation (7.20b); the functions $\theta_\phi(X)$, $\theta_g(X)$, $\theta_{\phi j}(X)$, $\theta_{gj}(X)$, and $\theta_t(X, \tau)$, obtained from the solutions in Chapter 4, are summarized in the following.

Solution for $\theta_\phi(X)$

By Equation (4.12) we have

$$\theta_\phi(X) = \phi_e \frac{X^m J_{-m}\left(X\sqrt{d_\phi}\right)}{\alpha J_{-m}\left(\sqrt{d_\phi}\right) - \beta\sqrt{d_\phi} J_{1-m}\left(\sqrt{d_\phi}\right)}, \qquad (\alpha \neq 0) \tag{7.23a}$$

$$\lim_{d_\phi \to 0} \theta_\phi(X) = \frac{\phi_e}{\alpha}, \qquad (\alpha \neq 0) \tag{7.23b}$$

and by Equation (4.14) we have

$$\theta_\phi(X) = \phi_e\left\{ \frac{2(1-m)}{d_\phi} - \frac{X^m J_{-m}\left(X\sqrt{d_\phi}\right)}{\sqrt{d_\phi}\, J_{1-m}\left(\sqrt{d_\phi}\right)} \right\}, \qquad (\alpha = 0, \beta = 1)$$

$$(7.23c)$$

$$\lim_{d_\phi \to 0} \theta_\phi(X) = \phi_e \frac{1}{2}\left(X^2 - \frac{1-m}{2-m} \right), \qquad (\alpha = 0, \beta = 1) \qquad (7.23d)$$

Solution for $\theta_g(X)$

By Equation (4.18) we have

$$\theta_g(X) = \frac{G_e}{d_g}\left\{ \alpha \frac{X^m J_{-m}\left(X\sqrt{d_g}\right)}{\alpha J_{-m}\left(\sqrt{d_g}\right) - \beta\sqrt{d_g}\, J_{1-m}\left(\sqrt{d_g}\right)} - 1 \right\}, \qquad (\alpha \neq 0)$$

$$(7.24a)$$

$$\lim_{d_g \to 0} \theta_g(X) = 0, \qquad (\alpha \neq 0) \qquad (7.24b)$$

and by Equation (4.20) we have

$$\theta_g(X) = 0, \qquad (\alpha = 0, \beta = 1) \qquad (7.24c)$$

Solution for $\theta_{\phi j}(X)$

By setting in Equation (4.27a), $\phi_j(x_0) = 0$, $\phi_j(x_1) = \phi_j$, and $P_j(x) = 0$, and making appropriate simplifications for the symmetry condition at the boundary $x_0 = 0$, we obtain

$$\theta_{\phi j}(X) = \frac{\phi_j}{\alpha} - (j+1)\left\{ \frac{\beta}{\alpha}\int_0^1 X^{1-2m}\theta_{\phi, j+1}(X)\, dX \right.$$

$$\left. + \int_X^1 \frac{1}{X'^{1-2m}} \int_0^{X'} X''^{1-2m}\theta_{\phi, j+1}(X'')\, dX''\, dX' \right\},$$

$$(\alpha \neq 0) \quad (7.25a)$$

where

$$\theta_{\phi, n_\phi+1}(X) = 0, \qquad j = n_\phi, \quad n_\phi - 1, \ldots, 2, 1, 0$$

For the case of $\alpha = 0$, $\beta = 1$, we make use of Equation (4.33a). By setting in this equation $\phi_j(x_0) = 0$, $\phi_j(x_1) = \phi_j$, and $P_j(x) = 0$, and making simplifications appropriate for the symmetry condition at $x_0 = 0$, we obtain

$$\theta_{\phi j}(X) = \frac{1}{2}\phi_j\left(X^2 - \frac{1-m}{2-m}\right) - (j+1)$$

$$\times \int_X^1 \frac{1}{X'^{1-2m}} \int_0^{X'} X''^{1-2m}\theta_{\phi, j+1}(X'') \, dX'' \, dX'$$

$$+ (j+1)\int_0^1 X' \int_0^{X'} X^{1-2m}\theta_{\phi, j+1}(X) \, dX \, dX', \qquad (\alpha = 0, \beta = 1)$$

$$(7.25\text{b})$$

where

$$\theta_{\phi, n+1}(X) = 0, \qquad j = n_\phi, \quad n_\phi - 1, \ldots, 2, 1, 0$$

Solution for $\theta_{gj}(X)$

By setting in Equation (4.27a), $\phi_j(x_0) = \phi_j(x_1) = 0$, $w(x) = k(x) = X^{1-2m}$, $P_j(x) = X^{1-2m}G_j(X)$, $T_j(x) = \theta_{gj}(X)$, $x_0 = 0$, and $x_1 = 1$, and noting the symmetry boundary condition at $x_0 = 0$, we obtain

$$\theta_{gj}(X) = \frac{\beta}{\alpha}\int_0^1 X^{1-2m}\left[G_j(X) - (j+1)\theta_{g, j+1}(X)\right] dX$$

$$+ \int_X^1 \frac{1}{X'^{1-2m}}\int_0^{X'} X''^{1-2m}\left[G_j(X'') - (j+1)\theta_{g, j+1}(X'')\right] dX'' \, dX',$$

$$(\alpha \neq 0) \quad (7.26\text{a})$$

where

$$\theta_{g, n_g+1}(X) = 0, \qquad j = n_g, n_g - 1, \ldots, 2, 1, 0$$

For the case of $\alpha = 0$, $\beta = 1$, we make use of Equation (4.33a). By setting in this equation $\phi_j(x_0) = \phi_j(x_1) = 0$, $P_j(x) = X^{1-2m}G_j(X)$, $w(x) = k(x) = X^{1-2m}$, $x_0 = 0$, and $x_1 = 1$, and noting the symmetry condition at $x_0 = 0$, we obtain

$$\theta_{gj}(X) = \frac{1}{2}\left[X^2 - \frac{1-m}{2-m}\right]\int_0^1 X^{1-2m}G_j(X)\,dX$$

$$+ \int_X^1 \frac{1}{X'^{1-2m}}\int_0^{X'} X''^{1-2m}\left[G_j(X'') - (j+1)\theta_{g,j+1}(X'')\right]dX''\,dX'$$

$$- \int_0^1 X'\int_0^{X'} X^{1-2m}\left[G_j(X) - (j+1)\theta_{g,j+1}(X)\right]dX\,dX',$$

$$(\alpha = 0, \beta = 1)\quad(7.26b)$$

where

$$\theta_{g,n_g+1}(X) = 0, \qquad j = n_g, n_g - 1,\ldots,2,1,0$$

Solution for $\theta_t(X,\tau)$

The solution for $\theta_t(X,\tau)$ is determined from Equation (4.35) and the cases $\alpha \neq 0$ and $\alpha = 0$ are considered separately.

1. For the case $\alpha \neq 0$, the eigenfunctions $\psi_i(x)$ and the normalization integral N_i are given by Equations (7.15a) and (7.15b), respectively, and the function $\Omega_i(x_k)$ is determined according to the third equality in Equations (4.4d). Introducing these results into Equation (4.35) we obtain

$$\theta_t(X,\tau) = 2\sum_{i=1}^\infty \frac{e^{-\mu_i^2\tau}}{1 + 2m\beta/\alpha + (\beta\mu_i/\alpha)^2} X^m \frac{J_{-m}(\mu_i X)}{J_{1-m}(\mu_i)}$$

$$\times \left\{ \sum_{j=0}^{n_f} f_{2j}\int_0^1 X^{1-m+2j}\frac{J_{-m}(\mu_i X)}{J_{1-m}(\mu_i)}\,dX - \frac{\phi_e}{\alpha}\frac{\mu_i}{\mu_i^2 - d_\phi} \right.$$

$$- \sum_{j=0}^{n_g} (-1)^j \frac{\phi_j}{\alpha}\frac{j!}{\mu_i^{2j+1}} - \frac{G_e}{\mu_i(\mu_i^2 - d_g)}$$

$$\left. - \sum_{j=0}^{n_g} (-1)^j \frac{j!}{\mu_i^{2(j+1)}} \sum_{k=0}^{n_j} g_{j,2k}\int_0^1 X^{1-m+2k}\frac{J_{-m}(\mu_i X)}{J_{1-m}(\mu_i)}\,dX \right\},$$

$$(\alpha \neq 0)\quad(7.27a)$$

and the eigenvalues μ_i are the roots of the transcendental Equation (7.16b).

2.   For the case of $\alpha = 0$, $\beta = 1$, the eigenfunctions $\psi_i(x)$ and the normalization integral N_i are given by Equations (7.19a) and (7.19b), respectively. In addition, the function $\Omega_i(x_k)$ is determined according to the second equality in Equation (4.4d), that is, $\Omega_i(x_k) = \psi_i(x_k)/\beta_k$. Introducing these results into Equation (4.35) we obtain

$$\theta_t(X, \tau) = 2 \sum_{i=1}^{\infty} e^{-\mu_i^2 \tau} X^m \frac{J_{-m}(\mu_i X)}{J_{-m}(\mu_i)}$$

$$\times \left\{ \sum_{j=0}^{n_f} f_{2j} \int_0^1 X^{1-m+2j} \frac{J_{-m}(\mu_i X)}{J_{-m}(\mu_i)} dX - \frac{\phi_e}{\mu_i^2 - d_\phi} \right.$$

$$- \sum_{j=0}^{n_g} (-1)^j \phi_j \frac{j!}{\mu_i^{2(j+1)}} - \frac{G_e}{\mu_i(\mu_i^2 - d_g)}$$

$$\left. - \sum_{j=0}^{n_g} (-1)^j \frac{j!}{\mu_i^{2(j+1)}} \sum_{k=0}^{n_j} g_{j,2k} \int_0^1 X^{1-m+2k} \frac{J_{-m}(\mu_i X)}{J_{-m}(\mu_i)} dX \right\},$$

$$(\alpha = 0, \beta = 1) \quad (7.27b)$$

and the eigenvalues μ_i are the roots of the transcendental equation (7.19c).

The integrals in Equation (7.27) can be evaluated by integration by parts. The result for boundary conditions of the first and third kind is

$$\int_0^1 X^{1-m+2j} \frac{J_{-m}(\mu_i X)}{J_{1-m}(\mu_i)} dX = \frac{1}{\mu_i} \left(1 + 2j\frac{\beta}{\alpha}\right) - \frac{4j(j-m)}{\mu_i^2}$$

$$\times \int_0^1 X^{2j-m-1} \frac{J_{-m}(\mu_i X)}{J_{1-m}(\mu_i)} dX \quad (7.28a)$$

and for the second kind becomes

$$\int_0^1 X^{1-m+2j} \frac{J_{-m}(\mu_i X)}{J_{-m}(\mu_i)} dX = \frac{2j}{\mu_i^2} - \frac{4j(j-m)}{\mu_i^2} \int_0^1 X^{2j-m-1} \frac{J_{-m}(\mu_i X)}{J_{-m}(\mu_i)} dX$$

$$(7.28b)$$

Applications

Solutions to the one-dimensional, time dependent heat and mass diffusion problems are given in References 14, 15, and 16. In the following examples we illustrate the application of the expressions given in this section to obtain solutions to one-dimensional, time dependent temperature distribution in a slab, long solid cylinder, and solid sphere with symmetry condition at $X = 0$ and the boundary condition of the first, second, and third kind at $X = \ell_0$.

Example 7.1   Determine the dimensionless temperature distribution $\theta(X, \tau)$ for a slab of thickness ℓ_0, a long solid cylinder, and a solid sphere of radius ℓ_0 with symmetry at $x = 0$ for the following conditions. The solid is initially at a uniform temperature T_i; for times $t > 0$ the boundary surface at $x = \ell_0$ is kept at constant temperature T_e and there is no energy generation in the medium.

Solution   We choose T_e as the reference temperature, $\Delta T = T_i - T_e$ as the reference temperature difference, ℓ_0 as the reference length, and define the dimensionless variables as

$$\theta(X, \tau) = \frac{T(x, t) - T_e}{T_i - T_e}, \qquad \tau = \frac{\alpha t}{\ell_0^2}, \qquad X = \frac{x}{\ell_0} \qquad (7.29a)$$

Then the boundary condition is reduced to a homogeneous boundary condition of the first kind, the dimensionless initial condition becomes unity, and we write

$$F(X) = 1, \qquad \phi(\tau) = 0, \qquad G(X, \tau) = 0, \qquad \alpha = 1, \qquad \beta = 0$$

$$(7.29b)$$

The solution of this problem is immediately obtained from Equation (7.16a) by making simplifications according to Equations (7.29b) and performing the resulting integrals. We find

$$\theta(X, \tau) = 2 \sum_{i=1}^{\infty} X^m \frac{J_{-m}(\mu_i X)}{\mu_i J_{1-m}(\mu_i)} e^{-\mu_i^2 \tau} \qquad (7.30a)$$

where the μ_i are the roots of the transcendental equation obtained from Equation (7.16b) for $\beta = 0$

$$J_{-m}(\mu) = 0 \qquad (7.30b)$$

The solution (7.30) is valid for a slab, long solid cylinder, and solid sphere depending on the choice of the value of m; by making use of the relations given in Table 3.1, we present the resulting solutions for these three cases.

Slab $(m = \frac{1}{2})$

$$\theta(X, \tau) = 2 \sum_{i=1}^{\infty} (-1)^{i+1} \frac{\cos \mu_i X}{\mu_i} e^{-\mu_i^2 \tau} \qquad (7.31a)$$

where

$$\mu_i = (2i - 1)\frac{\pi}{2} \tag{7.31b}$$

Long Solid Cylinder ($m = 0$)

$$\theta(X, \tau) = 2 \sum_{i=1}^{\infty} \frac{J_0(\mu_i X)}{\mu_i J_1(\mu_i)} e^{-\mu_i^2 \tau} \tag{7.32a}$$

where the μ_i are the roots of

$$J_0(\mu) = 0 \tag{7.32b}$$

Solid Sphere ($m = -\frac{1}{2}$)

$$\theta(X, \tau) = 2 \sum_{i=1}^{\infty} \frac{j_0(\mu_i X)}{\mu_i j_1(\mu_i)} e^{-\mu_i^2 \tau} \tag{7.33a}$$

where the μ_i are the roots of

$$j_0(\mu) = 0 \tag{7.33b}$$

If the spherical Bessel functions $j_0(z)$ and $j_1(z)$ are transformed to trigonometrical functions according to the relations (3.36d, e), the solution (7.33) is written in the alternative form as

$$\theta(X, \tau) = 2 \sum_{i=1}^{\infty} (-1)^{i+1} \frac{\sin(\mu_i X)}{\mu_i X} e^{-\mu_i^2 \tau} \tag{7.34a}$$

where

$$\mu_i = \pi i \tag{7.34b}$$

Figures 7.1a, b, and c show the dimensionless temperature distribution, $\theta(X, \tau)$, for a slab, long solid cylinder, and solid sphere, respectively, as obtained from the preceding solutions as a function of the position in the region $0 \leq X \leq 1$ for several different values of the dimensionless time τ.

Example 7.2 Determine the dimensionless temperature distribution $\theta(X, \tau)$ for the problem considered in Example 7-1 if the initial temperature distribution were

$$T(0, x) \equiv f(x) = T_i - (T_i - T_e)\left(\frac{x}{\ell_0}\right)^2$$

where T_e is the temperature at the boundary surface $x = \ell_0$.

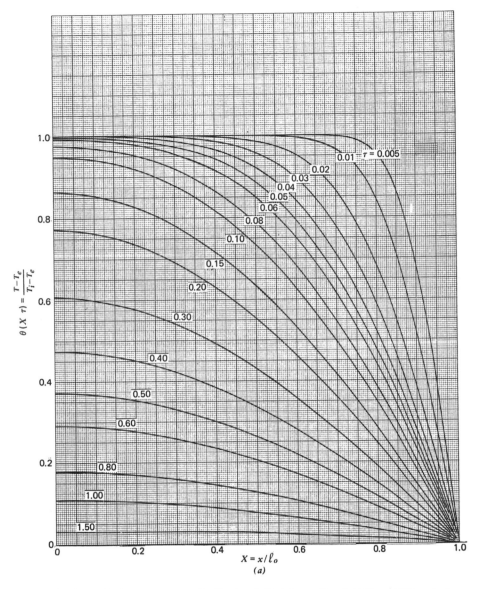

Figure 7.1 Dimensionless temperature distribution $\theta(X, \tau)$ for medium initially at $F(X) = 1$, the boundary at $X = 1$ kept at $\phi(\tau) = 0$. (a) Slab. (b) Long solid cylinder. (c) Solid sphere.

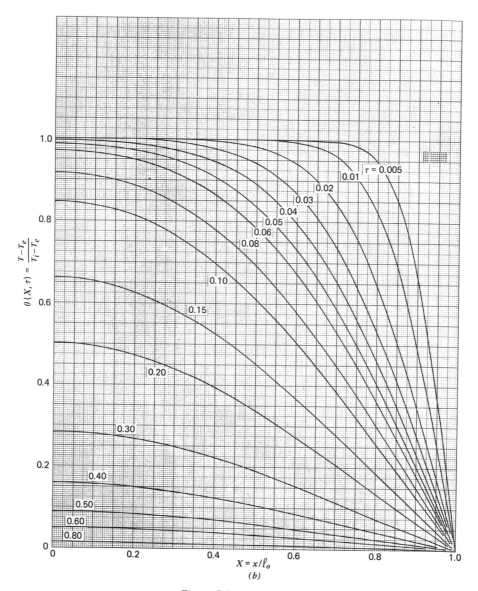

Figure 7.1 (*Continued*).

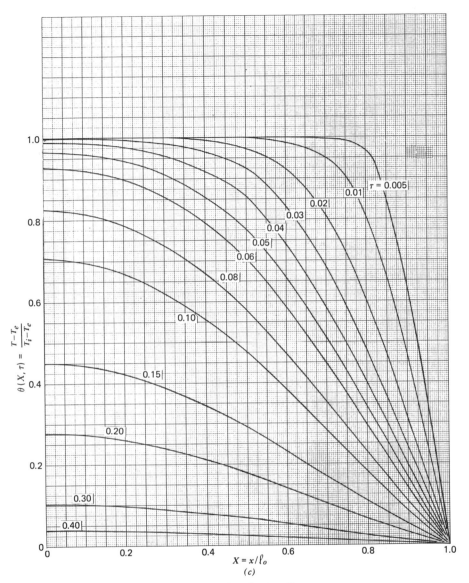

Figure 7.1 (*Continued*).

291

Solution We choose the dimensionless variables as

$$\theta(X,\tau) = \frac{T(x,t) - T_e}{T_i - T_e}, \qquad \tau = \frac{\alpha t}{\ell_0^2}, \qquad X = \frac{x}{\ell_0} \qquad (7.35\text{a})$$

Then the boundary condition at $x = \ell_0$ becomes an homogeneous boundary condition of the first kind and the dimensionless initial temperature distribution is given by

$$F(X) = 1 - X^2 \quad \text{or} \quad f_0 = 1, \qquad f_2 = -1 \quad \text{by Equation (7.21a)}$$

$$(7.35\text{b})$$

Other parameters are taken as

$$G(X,\tau) = 0, \qquad \phi(\tau) = 0, \qquad \alpha = 1, \qquad \beta = 0 \qquad (7.35\text{c})$$

The solution for this problem is immediately obtained from Equation (7.16a) by introducing the simplifications given by Equations (7.35b, c). The integrals in the resulting equation are performed by making use of the integral given by Equation (7.28a). Finally, we obtain the solution as

$$\theta(X,\tau) = 8(1-m) \sum_{i=1}^{\infty} X^m \frac{J_{-m}(\mu_i X)}{\mu_i^3 J_{1-m}(\mu_i)} e^{-\mu_i^2 \tau} \qquad (7.36\text{a})$$

where the μ_i are the roots of

$$J_{-m}(\mu) = 0 \qquad (7.36\text{b})$$

This result is valid for a slab, long solid cylinder, and solid sphere by the appropriate choice of the value of m. Also, the solutions obtained in this manner can be expressed in the alternative form by making use of the relations in Table 3.1 as discussed in the Example 7.1.

Figures 7.2a, b, and c show the dimensionless temperature distribution $\theta(X,\tau)$ given by Equation (7.36), for a slab, long solid cylinder, and solid sphere, respectively, as a function of position or several different values of the dimensionless time τ.

Example 7.3 Determine the dimensionless temperature distribution for a slab, long solid cylinder, and solid sphere with symmetry at $x = 0$, for the following conditions.

The medium is initially at a uniform temperature T_i. For times $t > 0$, the temperature of the boundary surface at $x = \ell_0$ varies with time as

$$T(\ell_0, t) = T_e - (T_e - T_i)e^{-bt}$$

where T_e and b are constants, and there is no energy generation in the medium.

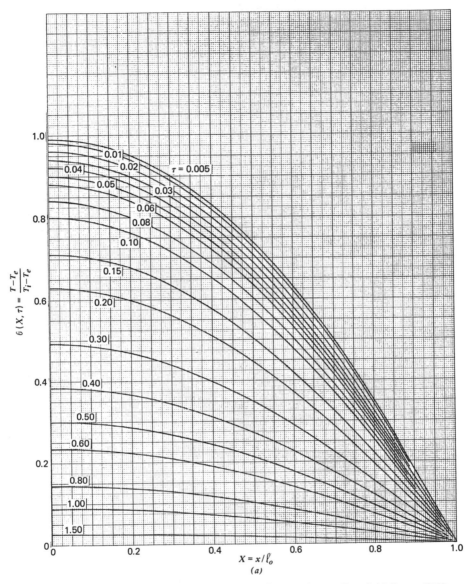

Figure 7.2 Dimensionless temperature distribution $\theta(X, \tau)$ for medium initially at $F(X) = 1 - X^2$, the boundary at $X = 1$ kept at $\theta(\tau) = 0$. (a) slab. (b) Long solid cylinder. (c) Solid sphere.

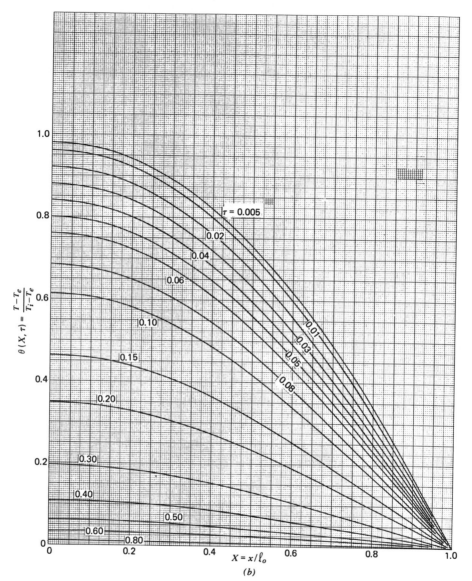

$$\theta(X, \tau) = \frac{T - T_e}{T_i - T_e}$$

$\tau = 0.005$

0.02

0.04

0.06

0.01

0.03

0.05

0.08

0.10

0.15

0.20

0.30

0.40

0.50

0.60

0.80

$X = x/\ell_o$

(b)

Figure 7.2 (*Continued*).

294

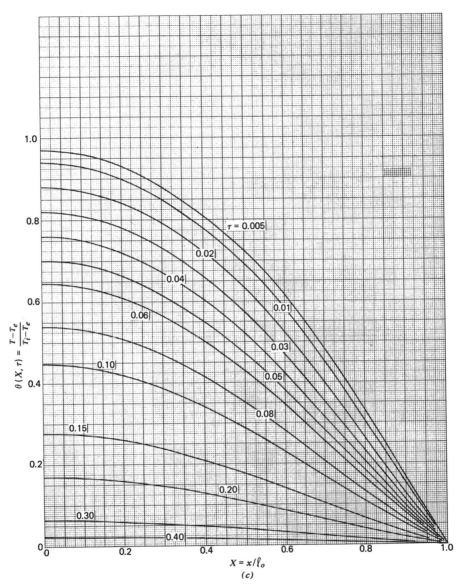

Figure 7.2 (*Continued*).

295

Solution   We choose the dimensionless variables as

$$\theta(X, \tau) = \frac{T(x, t) - T_i}{T_e - T_i}, \qquad \tau = \frac{\alpha t}{\ell_0^2}, \qquad X = \frac{x}{\ell_0} \qquad (7.37a)$$

We note that we have chosen the initial temperature T_i as the reference temperature in order to make the dimensionless initial temperature zero. Other dimensionless quantities are taken as

$$\phi(\tau) \equiv \frac{T(\ell_0, t) - T_i}{T_e - T_i} = 1 - e^{-p\tau} \quad \text{where } p = \frac{b\ell_0^2}{\alpha} \qquad (7.37b)$$

$$F(X) = 0, \qquad G(X, \tau) = 0, \qquad \alpha = 1, \qquad \beta = 0 \qquad (7.37c)$$

The solution of this problem is obtained from Equation (7.22a) by noting that the functions $\theta_g(X)$ and $\theta_{gi}(X)$ are zero, and the functions $\theta_\phi(X)$, $\theta_{\phi 0}(X)$, and $\theta_t(X, \tau)$ are determined by Equations (7.23a), (7.25a), and (7.27a), respectively, with appropriate simplification. Finally, the solution for $\theta(X, \tau)$ is determined as

$$\theta(X, \tau) = 1 - X^m \frac{J_{-m}(X\sqrt{p})}{J_{-m}(\sqrt{p})} e^{-P\tau}$$

$$-2 \sum_{i=1}^{\infty} \frac{1}{\mu_i[1 - (\mu_i^2/p)]} X^m \frac{J_{-m}(\mu_i X)}{J_{1-m}(\mu_i)} e^{-\mu_i^2 \tau} \qquad (7.38a)$$

where the μ_i are the roots of

$$J_{-m}(\mu) = 0 \qquad (7.38b)$$

The average temperature $\theta_{av}(\tau)$ is obtained by introducing the solution (7.38a) into Equation (7.17) and performing the integrations. We obtain

$$\theta_{av}(\tau) = 1 - 2(1 - m) \frac{J_{1-m}(\sqrt{p})}{\sqrt{p} J_{-m}(\sqrt{p})} e^{-p\tau}$$

$$- 4(1 - m) \sum_{i=1}^{\infty} \frac{1}{\mu_i^2[1 - (\mu_i^2/p)]} e^{-\mu_i^2 \tau} \qquad (7.39)$$

Figures 7.3*a*, *b*, and *c* show the average dimensionless temperature for a slab, cylinder, and sphere, respectively, as a function of dimensionless time for several different values of the parameter *p*.

Example 7.4   Determine the dimensionless temperature distribution for a slab of thickness ℓ_0, a long solid cylinder, and a solid sphere of radius ℓ_0 with symmetry at $x = 0$ for the following condition.

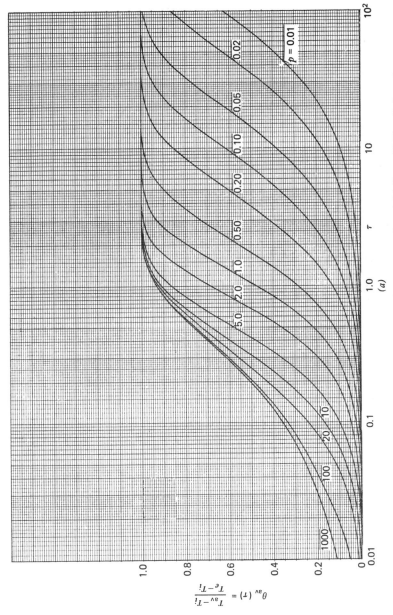

Figure 7.3 Dimensionless average temperature $\theta_{av}(\tau)$ for medium initially at $F(X) = 0$, boundary at $X = 1$ kept at $\phi(\tau) = 1 - e^{-p\tau}$. (a) Slab. (b) Long solid cylinder. (c) Solid sphere.

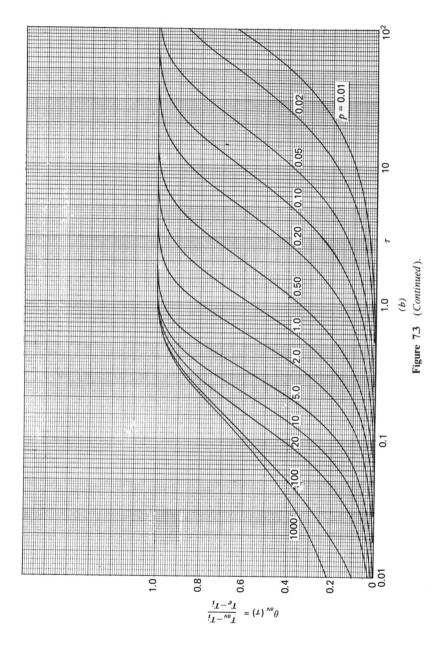

Figure 7.3 (*Continued*).

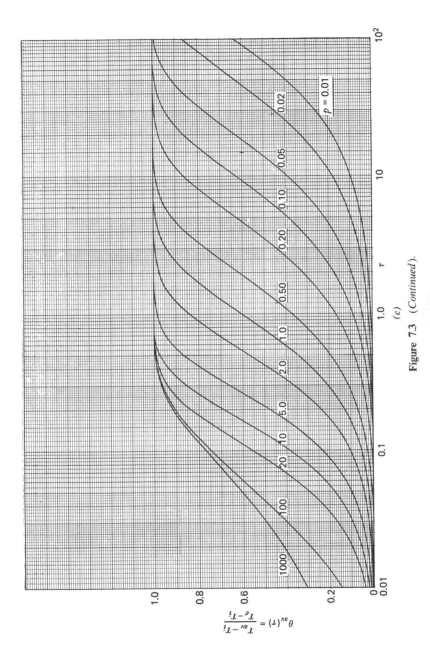

$$\theta_{av}(\tau) = \frac{T_{av} - T_i}{T_e - T_i}$$

$p = 0.01$

τ

(c)

Figure 7.3 (*Continued*).

Initially the medium is at uniform temperature T_i. For times $t > 0$, heat is supplied into the medium from the boundary surface $x = \ell_0$ at a constant rate of q per unit area, per unit time. There is no energy generation in the medium.

Solution   We choose the dimensionless variables as

$$\theta(X, \tau) = \frac{T(x, t) - T_i}{\Delta T}, \qquad \tau = \frac{\alpha t}{\ell_0^2}, \qquad X = \frac{x}{\ell_0} \qquad (7.40a)$$

where ΔT is a reference temperature. Then the boundary condition at $X = 1$ becomes

$$\frac{\partial \theta(1, \tau)}{\partial X} = \frac{q\ell_0}{k \Delta T} \equiv Q \text{ or } \phi(\tau) = Q \qquad (7.40b)$$

and in addition we have

$$F(X) = 0, \qquad G(X, \tau) = 0, \qquad \alpha = 0, \qquad \beta = 1 \qquad (7.40c)$$

The solution for this problem is written by Equation (7.22b) as

$$\theta(X, \tau) = \theta_{av}(\tau) + \theta_\phi(X) + \theta_t(X, \tau) \qquad (7.41a)$$

where $\theta_{av}(\tau)$ is determined from Equation (7.20b) for $\phi(\tau) = Q$ as

$$\theta_{av}(\tau) = Q2(1 - m)\tau \qquad (7.41b)$$

$\theta_\phi(X)$ is determined from Equation (7.23d) as

$$\theta_\phi(X) = \frac{1}{2}\left(X^2 - \frac{1 - m}{2 - m} \right)Q \qquad (7.41c)$$

and $\theta_t(X, \tau)$ is determined from Equation (7.27b) for $\phi_e = Q$ as

$$\theta_t(X, \tau) = -Q2\sum_{i=1}^{\infty} X^m \frac{J_{-m}(\mu_i X)}{\mu_i^2 J_{-m}(\mu_i)} e^{-\mu_i^2 \tau} \qquad (7.41d)$$

Introducing Equations (7.41b, c, d) into Equation (7.41a) we obtain the solution for the problem in the form

$$\frac{\theta(X, \tau)}{Q} = 2(1 - m)\tau + \frac{1}{2}\left(X^2 - \frac{1 - m}{2 - m} \right) - 2\sum_{i=1}^{\infty} X^m \frac{J_{-m}(\mu_i X)}{\mu_i^2 J_{-m}(\mu_i)} e^{-\mu_i^2 \tau}$$

$$(7.42a)$$

where the μ_i are the roots of [see Equation (7.16b) for $\alpha = 0$]

$$J_{1-m}(\mu) = 0 \tag{7.42b}$$

Figures 7.4a, b, and c show a plot of $[\theta(X, \tau)]/Q = [T(x, t) - T_i]/(q\ell_0/k)$ for a slab, cylinder, and sphere, respectively, as a function of the position X in the region $0 \le X \le 1$ for several different values of the dimensionless time τ.

Example 7.5 Solve the problem considered in Example 7.4 for a heat supply at the boundary surface $x = \ell_0$, at a rate of qe^{-bt} per unit area per unit time, where q and b are constants.

Solution The boundary condition for this case is written in the dimensionless form as

$$\frac{\partial \theta(1, \tau)}{\partial X} = Qe^{-p\tau} \tag{7.43a}$$

where

$$Q \equiv \frac{q\ell_0}{k\,\Delta T} = \text{constant}, \qquad p = \frac{b\ell_0^2}{\alpha} \tag{7.43b}$$

and the other dimensionless quantities are taken the same as given by Equations (7.40a, c).

The solution of this problem is written by Equation (7.22b) as

$$\theta(X, \tau) = \theta_{av}(\tau) + \theta_\phi(X)e^{-p\tau} + \theta_t(X, \tau) \tag{7.44a}$$

where $\theta_{av}(\tau)$ is determined from Equation (7.20b) for $\phi(\tau) = Qe^{-p\tau}$ as

$$\theta_{av}(\tau) = Q2(1 - m)\frac{1 - e^{-p\tau}}{p} \tag{7.44b}$$

$\theta_\phi(X)$ is determined from (7.23c) for $\phi_e = Q, d_\phi = p$ as

$$\theta_\phi(\tau) = Q\left[\frac{2(1 - m)}{p} - X^m\frac{J_{-m}(X\sqrt{p})}{\sqrt{p}\,J_{1-m}(\sqrt{p})}\right] \tag{7.44c}$$

and $\theta_t(X, t)$ is determined from Equation (7.27b) for $\phi_e \equiv Q, d_\phi = p$ as

$$\theta_t(X, \tau) = -Q2\sum_{i=1}^{\infty}\frac{1}{\mu_i^2 - p}X^m\frac{J_{-m}(\mu_i X)}{J_{-m}(\mu_i)}e^{-\mu_i^2\tau} \tag{7.44d}$$

Introducing Equations (7.44b, c, d) into (7.44a), we obtain the solution for the

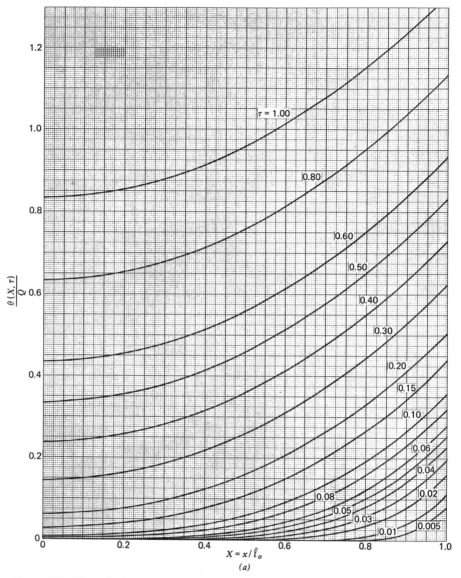

Figure 7.4 Dimensionless temperature distribution $\theta(X, \tau)/Q$ for medium initially at $F(X) = 0$, the boundary at $X = 1$ subjected to heat flux $[\partial\theta(1, \tau)]/\partial X = Q$. (*a*) Slab. (*b*) Long solid cylinder. (*c*) Solid sphere.

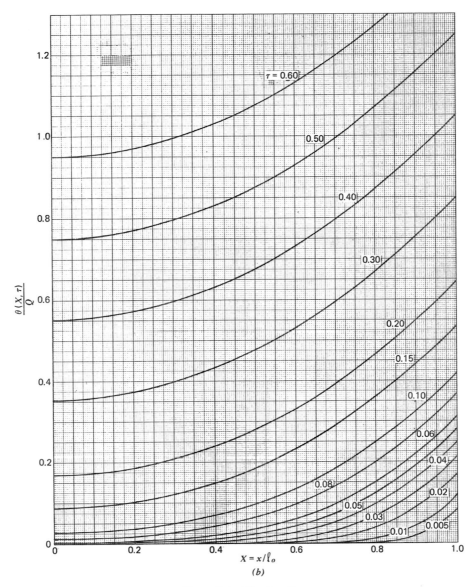

$\frac{\theta(X,\tau)}{Q}$

$X = x/\ell_o$

(b)

Figure 7.4 (*Continued*).

303

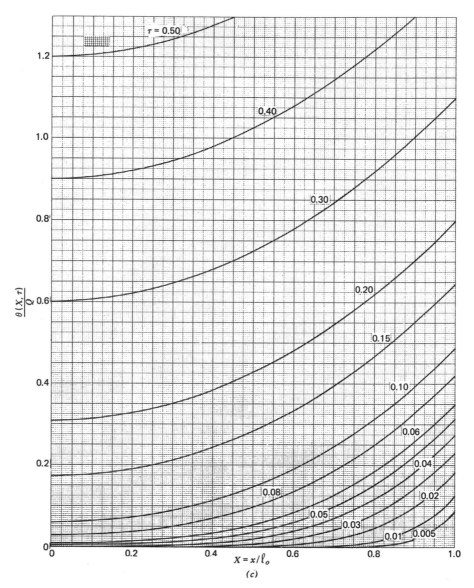

Figure 7.4 (*Continued*).

problem as

$$\frac{\theta(X,\tau)}{Q} = \frac{2(1-m)}{p} - X^m \frac{J_{-m}(X\sqrt{p})}{\sqrt{p}\,J_{1-m}(\sqrt{p})} e^{-p\tau}$$

$$-2\sum_{i=1}^{\infty} \frac{1}{\mu_i^2 - p} X^m \frac{J_{-m}(\mu_i X)}{J_{-m}(\mu_i)} e^{-\mu_i^2 \tau} \qquad (7.45)$$

where the μ_i are the roots of Equation (7.42b).

For the special case of $p \to 0$, it can be shown by making use of the results in Equations (7.23c, d) that the solution (7.45) reduces to (7.42a).

Figures 7.5a, b, and c show a plot of

$$\frac{p\theta(X,\tau)}{2(1-m)Q} = \frac{T(x,t) - T_i}{2(1-m)[q\alpha/b\ell_0 k]}$$

for a slab ($m = \frac{1}{2}$), cylinder ($m = 0$), and sphere ($m = -\frac{1}{2}$), respectively, as a function of the dimensionless time τ for several different values of the parameter $p = b\ell_0^2/\alpha$ at the center of the body (i.e., $X = 0$).

Figures 7.6a, b, and c are plots for temperature at the surface of the body (i.e., $X = 1$).

Example 7.6   Determine the dimensionless temperature distribution for a slab of thickness ℓ_0, a long solid cylinder, and a solid sphere of radius ℓ_0 with symmetry at $x = 0$ for the following conditions.

Initially the medium is at a uniform temperature T_i. For times $t > 0$, heat transfer at the boundary surface $x = \ell_0$ is by convection, with a heat transfer coefficient h, into a medium at temperature T_∞. There is no energy generation in the medium.

Solution   We choose the dimensionless variables as

$$\theta(X,\tau) = \frac{T(x,t) - T_i}{T_\infty - T_i}, \qquad \tau = \frac{\alpha t}{\ell_0^2}, \qquad X = \frac{x}{\ell_0} \qquad (7.46a)$$

Then the boundary condition at the surface $X = 1$ becomes

$$\theta(1,\tau) + \frac{1}{\text{Bi}} \frac{\partial \theta(1,\tau)}{\partial X} = 1 \qquad (7.46b)$$

where

$$\text{Bi} \equiv \frac{h\ell_0}{k} = \text{Biot number}$$

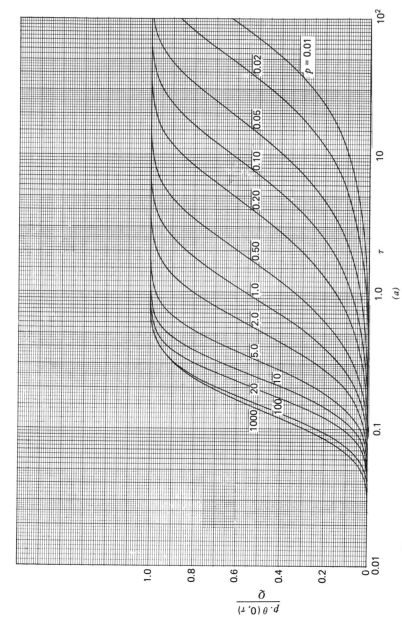

Figure 7.5 Dimensionless center temperature $p\theta(0,\tau)/Q$ for medium initially at $F(X) = 0$, the boundary at $X = 1$ subjected to heat flux $[\partial\theta(1,\tau)]/\partial X = Qe^{-p\tau}$. (a) Slab. (b) Long solid cylinder. (c) Solid sphere.

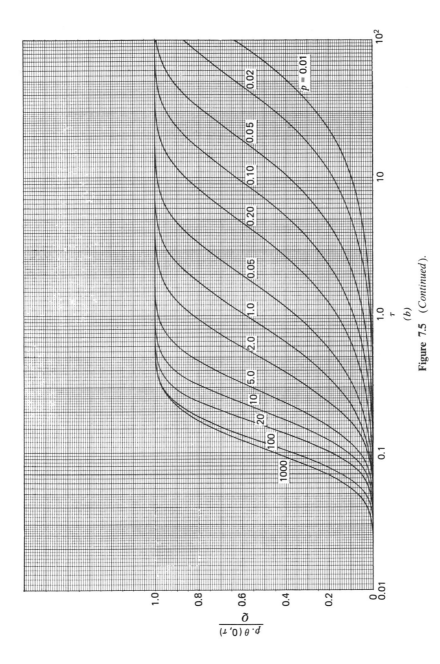

Figure 7.5 (*Continued*).

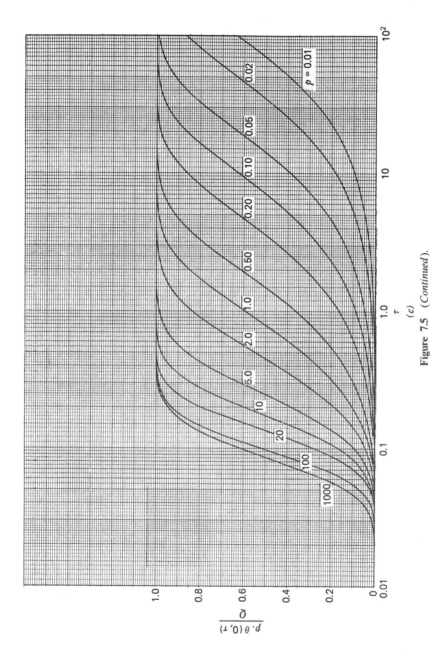

Figure 7.5 (*Continued*).

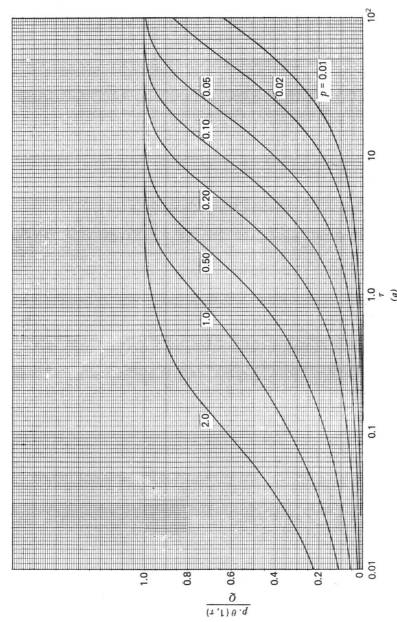

Figure 7.6 Dimensionless boundary surface temperature $p\theta(1, \tau)/Q$ for medium initially at $F(X) = 0$, the boundary $X = 1$ subjected to heat flux $[\partial\theta(1, \tau)]/\partial X = Qe^{-p\tau}$. (a) Slab. (b) Long solid cylinder. (c) Solid sphere.

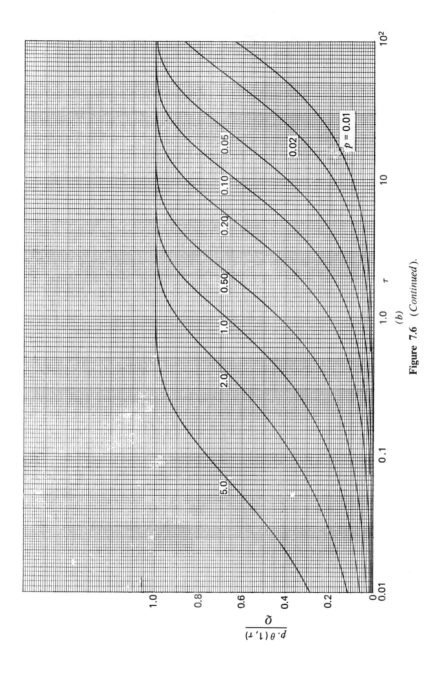

Figure 7.6 (*Continued*).

$\dfrac{p \cdot \theta (1, \tau)}{Q}$

τ

(b)

$p = 0.01$

0.02

0.05

0.10

0.20

0.50

1.0

2.0

5.0

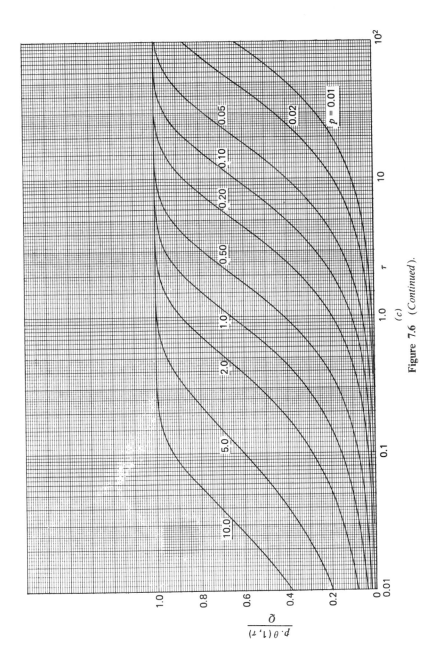

Figure 7.6 (Continued).

(c)

311

and in addition we have

$$F(X) = 0, \qquad G(X, \tau) = 0, \qquad \alpha = 1, \qquad \beta = \frac{1}{\text{Bi}} \qquad (7.46c)$$

The solution of this problem is written by Equation (7.22a) in the form

$$\theta(X, \tau) = \theta_\phi(X) + \theta_t(X, \tau) \qquad (7.47a)$$

where the function $\theta_\phi(X)$ is determined from Equation (7.23b) as

$$\theta_\phi(X) = 1 \qquad (7.47b)$$

and the function $\theta_t(X, \tau)$ is determined from Equation (7.27a) as

$$\theta_t(X, \tau) = -2 \sum_{i=1}^{\infty} \frac{1}{\mu_i} \frac{e^{-\mu_i^2 \tau}}{1 + (2m/\text{Bi}) + (\mu_i/\text{Bi})^2} X^m \frac{J_{-m}(\mu_i X)}{J_{1-m}(\mu_i)}$$

$$(7.47c)$$

Introducing Equations (7.47b, c) into (7.47a), the solution for the problem becomes

$$\theta(X, \tau) = 1 - 2 \sum_{i=1}^{\infty} \frac{1}{\mu_i} \left[1 + \frac{2m}{\text{Bi}} + \left(\frac{\mu_i}{\text{Bi}} \right)^2 \right]^{-1} X^m \frac{J_{-m}(\mu_i X)}{J_{1-m}(\mu_i)} e^{-\mu_i^2 \tau}$$

$$(7.48a)$$

where the μ_i are the roots of the transcendental equation [see Equation (7.16b) for $\alpha = 1$, $\beta = 1/\text{Bi}$].

$$\frac{J_{-m}(\mu)}{J_{1-m}(\mu)} = \frac{\mu}{\text{Bi}} \qquad (7.48b)$$

Figures 7.7a, b, and c show a plot of the dimensionless center temperature $\theta(0, \tau) = [T(0, \tau) - T_i]/(T_\infty - T_i)$ for a slab, cylinder, and sphere, respectively, as a function of dimensionless time τ for several different values of the parameter $1/\text{Bi}$.

Figures 7.8a, b, and c show similar plots for the dimensionless surface temperature $\theta(1, \tau) = [T(\ell_0, t) - T_i]/(T_\infty - T_i)$.

Example 7.7 Solve the problem considered in Example 7.6 for a convection boundary condition in which the ambient temperature varies with time exponentially, that is,

$$k \frac{\partial T}{\partial x} + hT = h\left[T_i + (T_\infty - T_i)e^{-bt} \right] \quad \text{at } x = \ell_0$$

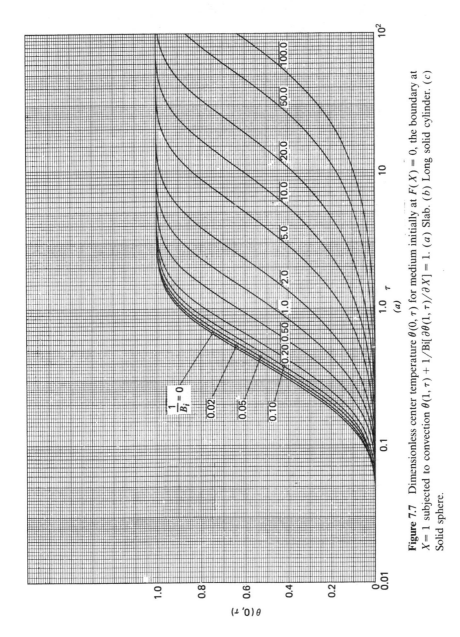

Figure 7.7 Dimensionless center temperature $\theta(0, \tau)$ for medium initially at $F(X) = 0$, the boundary at $X = 1$ subjected to convection $\theta(1, \tau) + 1/\text{Bi}[\partial\theta(1, \tau)/\partial X] = 1$. (*a*) Slab. (*b*) Long solid cylinder. (*c*) Solid sphere.

313

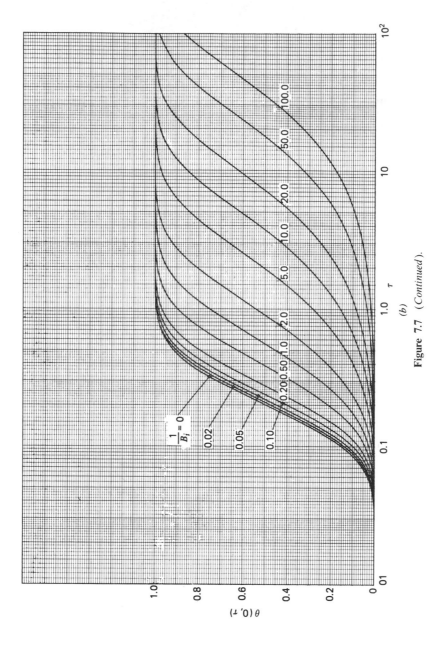

Figure 7.7 (*Continued*).

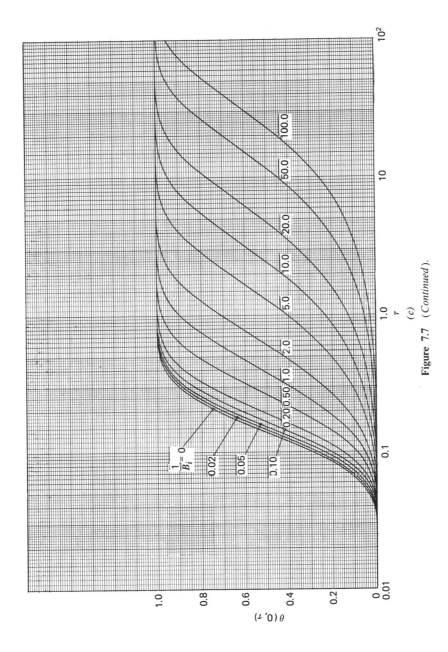

Figure 7.7 (*Continued*).

315

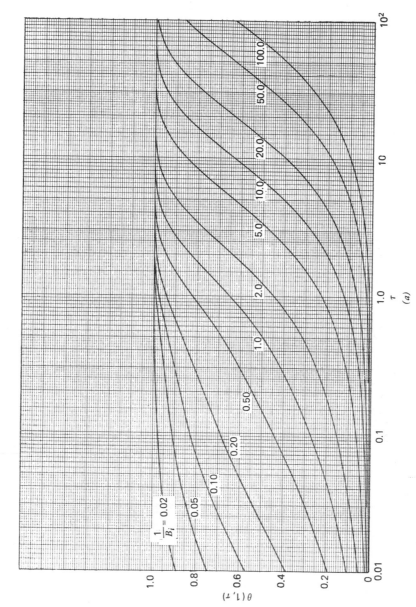

Figure 7.8 Dimensionless boundary surface temperature $\theta(1, \tau)$ for medium initially at $F(X) = 0$, the boundary $X = 1$ subjected to convection $\theta(1, \tau) + 1/\text{Bi}[\partial\theta(X, \tau)/\partial X] = 1$. (*a*) Slab. (*b*) Long solid cylinder. (*c*) Solid sphere.

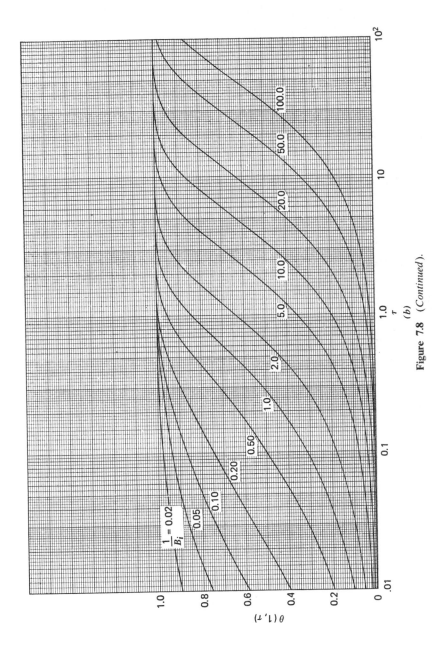

Figure 7.8 (Continued).

317

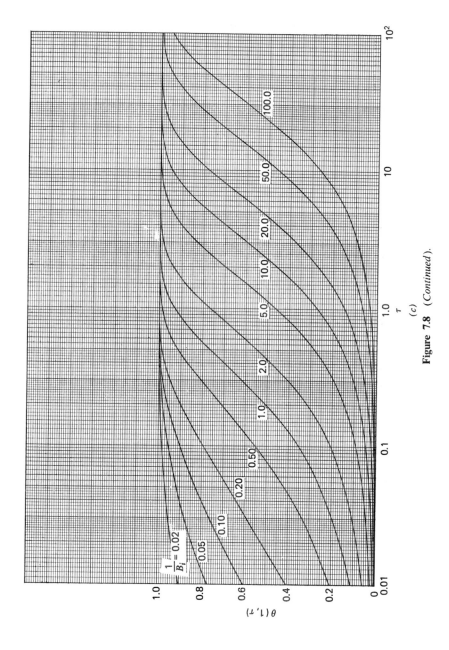

Figure 7.8 (*Continued*).

Solution The boundary condition for this case is given in the dimensionless form as

$$\theta(1, \tau) + \frac{1}{\text{Bi}} \frac{\partial \theta(1, \tau)}{\partial X} = e^{-p\tau} \qquad (7.49a)$$

where

$$p = \frac{b\ell_0^2}{\alpha} \qquad (7.49b)$$

and various other dimensionless parameters are as defined by Equations (7.46a, c).

The solution of the problem is written by Equation (7.22a) in the form

$$\theta(X, \tau) = \theta_\phi(X) e^{-p\tau} + \theta_t(X, \tau) \qquad (7.50a)$$

where $\theta_\phi(X)$ is determined from Equation (7.23a) by setting $\phi_e = 1$, $\alpha = 1$, $\beta = 1/\text{Bi}$, and $d_\phi = p$

$$\theta_\phi(X) = \frac{X^m J_{-m}(X\sqrt{p})}{J_{-m}(\sqrt{p}) - (\sqrt{p}/\text{Bi}) J_{1-m}(\sqrt{p})} \qquad (7.50b)$$

and $\theta_t(X, \tau)$ is determined from Equation (7.27a) as

$$\theta_t(X, \tau) = -2 \sum_{i=1}^{\infty} \frac{\mu_i}{\mu_i^2 - p} \frac{e^{-\mu_i^2 \tau}}{1 + (2m/\text{Bi}) + (\mu_i/\text{Bi})^2} X^m \frac{J_{-m}(\mu_i X)}{J_{1-m}(\mu_i)} \qquad (7.50c)$$

Introducing Equations (7.50b, c) into (7.50a), the solution for the problem is obtained as

$$\theta(X, \tau) = \frac{X^m J_{-m}(X\sqrt{p})}{J_{-m}(\sqrt{p}) - (\sqrt{p}/\text{Bi}) J_{1-m}(\sqrt{p})} e^{-p\tau}$$

$$- 2 \sum_{i=1}^{\infty} \frac{\mu_i}{\mu_i^2 - p} \left[1 + \frac{2m}{\text{Bi}} + \left(\frac{\mu_i}{\text{Bi}}\right)^2 \right]^{-1} X^m \frac{J_{-m}(\mu_i X)}{J_{1-m}(\mu_i)} e^{-\mu_i^2 \tau} \qquad (7.51)$$

where the μ_i are the roots of the transcendental Equation (7.48b).

Example 7.8 Solve the problem considered in Example 7.6 for a convection boundary condition in which the environment temperature fluctuates cosinosoidally with time, that is,

$$k\frac{\partial T}{\partial x} + hT = h\big[T_i + (T_\infty - T_i)\cos(bt)\big] \quad \text{at } x = \ell_0$$

Solution The boundary condition is given in the dimensionless form as

$$\theta(1,\tau) + \frac{1}{\text{Bi}}\frac{\partial\theta(1,\tau)}{\partial X} = \cos(p\tau) \tag{7.52a}$$

where

$$p = \frac{b\ell_0^2}{\alpha} \tag{7.52b}$$

and various other dimensionless parameters are defined by Equations (7.46a, c).

In order to express the nonhomogeneous term in the boundary condition in terms of exponentials, we make use of complex variables and rewrite Equation (7.52a) in the form

$$\theta(1,\tau) + \frac{1}{\text{Bi}}\frac{\partial\theta(1,\tau)}{\partial X} = \frac{1}{2}\big(e^{ip\tau} + e^{-ip\tau}\big) \tag{7.52c}$$

Then the problem considered in this example becomes similar to that considered in Example 7.7 except that p should be replaced by $\pm ip$ and ϕ_e taken as $\frac{1}{2}$. The solution is immediately obtained from Equation (7.51) as

$$\theta(X,\tau) = \frac{1}{2}\big\{A^+ e^{ip\tau} + A^- e^{-ip\tau}\big\}$$

$$-2\sum_{i=1}^{\infty}\frac{\mu_i^3}{\mu_i^4 + p^2}\left\{1 + \frac{2m}{\text{Bi}} + \left(\frac{\mu_i}{\text{Bi}}\right)^2\right\}^{-1} X^m \frac{J_{-m}(\mu_i X)}{J_{1-m}(\mu_i)} e^{-\mu_i^2\tau} \tag{7.53a}$$

where

$$A^\pm = \frac{X^m I_{-m}\big(X\sqrt{\pm ip}\,\big)}{I_{-m}\big(\sqrt{\pm ip}\,\big) + \dfrac{\sqrt{\pm ip}}{\text{Bi}} I_{1-m}\big(\sqrt{\pm ip}\,\big)} \tag{7.53b}$$

where the μ_i are the roots of the transcendental Equation (7.48b). The functions $I_\nu(z)$ are the modified Bessel functions of imaginary argument.

We note that the solution (7.53) consists of the sum of periodic and transient terms. As time goes to infinity (i.e., $\tau \to \infty$), the transient term decays out and the remaining solution that characterizes the *sustained temperature oscillations* or the *quasi-steady-state solution* can be written in the form

$$\theta(X, \tau) = (A^+ A^-)^{1/2} \cos(p\tau - \delta) \tag{7.54a}$$

where

$$\delta = \tan^{-1}\left(i\frac{A^+ - A^-}{A^+ + A^-}\right) \tag{7.54b}$$

We now examine some special cases of the quasi-steady-state solution (7.54) for a slab, long solid cylinder, and solid sphere.

Slab ($m = \frac{1}{2}$)

By making use of the relations

$$I_{-1/2}(z) = \sqrt{\frac{2}{\pi}} \frac{\cosh z}{\sqrt{z}} \quad \text{and} \quad I_{1/2}(z) = \sqrt{\frac{2}{\pi}} \frac{\sinh z}{\sqrt{z}} \tag{7.55}$$

we write Equation (7.53b) for $m = \frac{1}{2}$ as

$$A_{\pm} = \frac{\cosh(X\sqrt{\pm ip})}{\cosh(\sqrt{\pm ip}) + (\sqrt{\pm ip}/\text{Bi})\sinh(\sqrt{\pm ip})} \tag{7.56}$$

The hyperbolic sine and cosine function of imaginary argument in Equation (7.56) are split up into real and imaginary parts by expressing the term $\sqrt{\pm ip}$ in the form

$$\sqrt{\pm ip} = (1 \pm i)\sqrt{\frac{p}{2}} = \sqrt{\frac{p}{2}} \pm i\sqrt{\frac{p}{2}} \tag{7.57a}$$

and by making use of the following expansions

$$\cosh(\alpha + i\beta) = \cosh\alpha\cos\beta + i\sinh\alpha\sin\beta \tag{7.57b}$$

$$\sinh(\alpha + i\beta) = \sinh\alpha\cos\beta + i\cosh\alpha\sin\beta \tag{7.57c}$$

The resulting expressions for A^+ and A^- determined in this manner are introduced into Equations (7.54) and the quasi-steady-state temperature distribution or a slab with *boundary condition* of the first kind at $X = 1$ (i.e., $1/\text{Bi} = 0$) is obtained as

$$\theta(X, \tau) = \left[\frac{\cos^2(X\sqrt{p/2}) + \sinh^2(X\sqrt{p/2})}{\cos^2(\sqrt{p/2}) + \sinh^2(\sqrt{p/2})}\right]^{1/2} \cos(p\tau - \delta)$$

$$\tag{7.58a}$$

where

$$\delta = \tan^{-1}\left(\frac{C_1 - C_2}{D_1 + D_2}\right) \tag{7.58b}$$

$$C_2 = \cosh\left(\sqrt{\frac{p}{2}}\right)\cos\left(\sqrt{\frac{p}{2}}\right)\sinh\left(X\sqrt{\frac{p}{2}}\right)\sin\left(X\sqrt{\frac{p}{2}}\right) \tag{7.58c}$$

$$C_1 = \sinh\left(\sqrt{\frac{p}{2}}\right)\sin\left(\sqrt{\frac{p}{2}}\right)\cosh\left(X\sqrt{\frac{p}{2}}\right)\cos\left(X\sqrt{\frac{p}{2}}\right) \tag{7.58d}$$

$$D_1 = \cosh\left(\sqrt{\frac{p}{2}}\right)\cos\left(\sqrt{\frac{p}{2}}\right)\cosh\left(X\sqrt{\frac{p}{2}}\right)\cos\left(X\sqrt{\frac{p}{2}}\right) \tag{7.58e}$$

$$D_2 = \sinh\left(\sqrt{\frac{p}{2}}\right)\sin\left(\sqrt{\frac{p}{2}}\right)\sinh\left(X\sqrt{\frac{p}{2}}\right)\sin\left(X\sqrt{\frac{p}{2}}\right) \tag{7.58f}$$

Long Solid Cylinder ($m = 0$)

For this particular case we set $m = 0$ in Equation (7.53b) and obtain

$$A^{\pm} = \frac{I_0\left(X\sqrt{\pm ip}\right)}{I_0\left(\sqrt{\pm ip}\right) + \left(\sqrt{\pm ip}/\text{Bi}\right)I_1\left(\sqrt{\pm ip}\right)} \tag{7.59}$$

The modified Bessel functions of imaginary argument appearing in this equation can be split up into real and imaginary parts by making use of the following relations

$$I_n(z) = i^{-n}J_n(zi) \tag{7.60a}$$

and

$$J_n(zi^{\pm 3/2}) = ber_n(z) \pm ibei_n(z) \tag{7.60b}$$

where n is integer and the functions $ber_n(z)$ and $bei_n(z)$ are called *Kelvin functions* of order n, argument z, which are real functions that have been well tabulated [17, p. 430; 22]. The reader should consult References 19 and 20 for further discussion of these functions.

In view of the relations (7.60), we can expand the functions $I_0(X\sqrt{ip})$ and $I_1(X\sqrt{ip})$ as

$$I_0\left(X\sqrt{ip}\right) = ber\left(X\sqrt{p}\right) + ibei\left(X\sqrt{p}\right) \tag{7.61a}$$

$$I_1\left(X\sqrt{ip}\right) = bei_1\left(X\sqrt{p}\right) - iber_1\left(X\sqrt{p}\right) \tag{7.61b}$$

where $ber \equiv ber_0$ and $bei \equiv bei_0$; the functions with argument $-X\sqrt{ip}$ can be expanded in a similar manner.

Now the modified Bessel functions of imaginary argument appearing in Equation (7.59) are expanded into real and imaginary parts as described previously and the resulting expressions are introduced into the quasi-steady-state solution for temperature distribution in a long solid cylinder for *boundary condition of the first kind* at $X = 1$ (i.e., $1/Bi = 0$) is determined as

$$\theta(X, \tau) = \left\{ \frac{ber^2(X\sqrt{p}) + bei^2(X\sqrt{p})}{ber^2(\sqrt{p}) + bei^2(\sqrt{p})} \right\}^{1/2} \cos(p\tau - \delta) \qquad (7.62a)$$

where

$$\delta = \tan^{-1}\left[\frac{bei(\sqrt{p})ber(X\sqrt{p}) - ber(\sqrt{p})bei(X\sqrt{p})}{ber(\sqrt{p})ber(X\sqrt{p}) + bei(\sqrt{p})bei(X\sqrt{p})} \right] \qquad (7.62b)$$

Solid Sphere ($m = -\frac{1}{2}$)

For $m = -\frac{1}{2}$, Equation (7.53b) becomes

$$A^{\pm} = \frac{X^{-1/2}I_{1/2}(X\sqrt{\pm ip})}{I_{1/2}(\sqrt{\pm ip}) + (\sqrt{\pm ip}/Bi) I_{3/2}(\sqrt{\pm ip})} \qquad (7.63a)$$

where

$$I_{1/2}(z) = \sqrt{\frac{2}{\pi}} \frac{\sinh z}{\sqrt{z}} \qquad (7.63b)$$

and $I_{3/2}(z)$ is obtained by making use of the recurrence relation for Bessel functions and the relations (7.55) as

$$I_{3/2}(z) = \sqrt{\frac{2}{\pi}}\left[\frac{\cosh z}{\sqrt{z}} - \frac{\sinh z}{z\sqrt{z}} \right] \qquad (7.63c)$$

Then the functions $A^{\pm}$ are written as

$$A^{\pm} = \frac{Bi}{X} \frac{\sinh(X\sqrt{\pm ip})}{(Bi - 1)\sinh(\sqrt{\pm ip}) + \sqrt{\pm ip}\cosh(\sqrt{\pm ip})} \qquad (7.64)$$

For a *boundary condition of the first kind* at $X = 1$, (i.e., $1/Bi = 0$), we obtain

$$A^{\pm} = \frac{1}{X} \frac{\sinh(X\sqrt{\pm ip})}{\sinh(\sqrt{\pm ip})} \qquad (7.65)$$

The hyperbolic sine functions of imaginary argument appearing in Equation

(7.65) are split up into real and imaginary parts according to the relations given by Equations (7.57), and the resulting expressions for $A^{\pm}$ are introduced into Equation (7.54). After some manipulation, the quasi-steady-state solution for a solid sphere for a boundary condition of the first kind becomes

$$\theta(X,\tau) = \left[\frac{1}{X^2} \frac{\sin^2(X\sqrt{p/2}) + \sinh^2(X\sqrt{p/2})}{\sin^2(\sqrt{p/2}) + \sinh^2(\sqrt{p/2})} \right]^{1/2} \cos(p\tau - \delta)$$

(7.66a)

where

$$\delta = \tan^{-1}\left[\frac{C_1 - C_2}{D_1 + D_2} \right]$$

(7.66b)

$$C_2 = \sinh\left(\sqrt{\frac{p}{2}}\right)\cos\left(\sqrt{\frac{p}{2}}\right)\cosh\left(X\sqrt{\frac{p}{2}}\right)\sin\left(X\sqrt{\frac{p}{2}}\right)$$

(7.66c)

$$C_1 = \cosh\left(\sqrt{\frac{p}{2}}\right)\sin\left(\sqrt{\frac{p}{2}}\right)\sinh\left(X\sqrt{\frac{p}{2}}\right)\cos\left(X\sqrt{\frac{p}{2}}\right)$$

(7.66d)

$$D_1 = \sinh\left(\sqrt{\frac{p}{2}}\right)\cos\left(\sqrt{\frac{p}{2}}\right)\sinh\left(X\sqrt{\frac{p}{2}}\right)\cos\left(X\sqrt{\frac{p}{2}}\right)$$

(7.66e)

$$D_2 = \cosh\left(\sqrt{\frac{p}{2}}\right)\sin\left(\sqrt{\frac{p}{2}}\right)\cosh\left(X\sqrt{\frac{p}{2}}\right)\sin\left(X\sqrt{\frac{p}{2}}\right)$$

(7.66f)

Example 7.9 Determine the dimensionless temperature distribution for a slab of thickness ℓ_0, a long solid cylinder, and a solid sphere of radius ℓ_0 with symmetry at $x = 0$ for the following conditions.

Initially the medium is at a uniform temperature T_i. For times $t > 0$, energy is generated in the medium at a constant rate g_0, W/m^3 while the boundary surface at $x = \ell_0$ is kept at the temperature T_i.

Solution We choose the dimensionless variables as

$$\theta(X,\tau) = \frac{T(x,t) - T_i}{\Delta T}, \qquad \tau = \frac{\alpha t}{\ell_0^2}, \qquad X = \frac{x}{\ell_0}$$

(7.67a)

and, in addition, we have

$$F(X) = 0, \qquad \frac{g_0 \ell_0^2}{k\,\Delta T} \equiv G_0, \qquad \phi(\tau) = 0, \qquad \alpha = 1, \qquad \beta = 0$$

(7.67b)

where ΔT is a reference temperature difference.

The solution for this problem is written by Equation (7.22a) as

$$\theta(X, \tau) = \theta_{g0}(X) + \theta_t(X, \tau) \tag{7.68a}$$

where $\theta_{g0}(X)$ is determined from Equation (7.26a) for $j = 0$ with $G_j(X)|_{j=0} = G_0 = $ constant, as

$$\theta_{g0}(X) = \frac{G_0}{4(1 - m)}(1 - X^2) \tag{7.68b}$$

and $\theta_t(X, \tau)$ is determined from Equation (7.27a) as

$$\theta_t(X, \tau) = -2G_0 \sum_{i=1}^{\infty} \frac{X^m}{\mu_i^3} \frac{J_{-m}(\mu_i X)}{J_{1-m}(\mu_i)} e^{-\mu_i^2 \tau} \tag{7.68c}$$

Introducing Equations (7.68b, c) into (7.68a), the solution for the problem is determined as

$$\frac{\theta(X, \tau)}{G_0} = \frac{1 - X^2}{4(1 - m)} - 2 \sum_{i=1}^{\infty} \frac{1}{\mu_i^3} X^m \frac{J_{-m}(\mu_i X)}{J_{1-m}(\mu_i)} e^{-\mu_i^2 \tau} \tag{7.69a}$$

where the μ_i are the roots of [see Equation (7.16b) for $\beta = 0$]

$$J_{-m}(\mu) = 0 \tag{7.69b}$$

The solution (7.69) is valid for slab ($m = \frac{1}{2}$), cylinder ($m = 0$), and sphere ($m = -\frac{1}{2}$); it can be expressed in the alternative, more convenient form by utilizing the results given in Table 3.1.

7.3   ONE DIMENSIONAL TIME DEPENDENT DIFFUSION IN SEMI-INFINITE REGIONS

Problems of heat diffusion in a semi-infinite medium have many practical engineering applications. This matter is better envisioned if we consider a slab of finite thickness ℓ_0 which is initially at uniform temperature T_i and for times $t > 0$, the boundary surface at $x = 0$ is kept at temperature T_s and the boundary at $x = \ell_0$ is kept insulated. The solution of this problem for very short times is equivalent to that for semi-infinite media, because the region affected by the temperature variation is quite small compared to the thickness ℓ_0 of the plate. For example, in the study of variation of the Earth's temperature due to periodic heating by the sun or in the calculation of temperature oscillations in the cylinder walls of internal combustion engines, both the Earth and the cylinder wall can be considered as a semi-infinite medium. The solution for one-dimensional, time dependent heat diffusion problems of semi-infinite media are obtainable by appropriate simplifications of the general solutions given in Chapter 4. We illustrate the application with the following examples.

Example 7.10 Determine the dimensionless temperature distribution in a semi-infinite medium, $0 \le x < \infty$, which is initially at uniform temperature T_i, and for times $t > 0$, the boundary surface at $x = 0$ is subjected to a prescribed temperature $f(t)$. There is no energy generation in the medium.

Solution   We choose the dimensionless variables as

$$\theta(X, \tau) = \frac{T(x, t) - T_i}{\Delta T}, \qquad \tau = \frac{\alpha t}{\ell_0^2}, \qquad X = \frac{x}{\ell_0} \qquad (7.70a)$$

where ℓ_0 is a reference length and ΔT a reference temperature difference. The initial and boundary conditions become

$$\theta(0, \tau) = \frac{f(t) - T_i}{\Delta T} \equiv \phi(\tau), \qquad \frac{\partial \theta(\infty, \tau)}{\partial X} = 0, \qquad \theta(X, 0) = 0$$

$$(7.70b)$$

The heat diffusion problem considered here is a special case of the general problem Equations (4.52) as:

$$T = \theta, \qquad x = X, \qquad t = \tau, \qquad g(x, t) = 0$$
$$f(x) = 0, \qquad \alpha_0 = 1, \qquad \beta_0 = 0, \qquad \phi(t) = \phi(\tau) \qquad (7.70c)$$

Then the solution is immediately obtained from Equation (4.63) as

$$\theta(X, \tau) = \frac{X}{2\sqrt{\pi}} \int_{\tau'=0}^{\tau} \frac{\phi(\tau')}{(\tau - \tau')^{3/2}} e^{-X^2/4(\tau - \tau')} \, d\tau' \qquad (7.71a)$$

This result can be put into different form by defining a new variable as

$$\eta = \frac{X}{2\sqrt{\tau - \tau'}}, \qquad d\tau = \frac{2}{\eta}(\tau - \tau') \, d\eta$$

Then Equation (7.71a) becomes

$$\theta(X, \tau) = \frac{2}{\sqrt{\pi}} \int_{X/2\sqrt{\tau}}^{\infty} e^{-\eta^2} \phi\left(\tau - \frac{X^2}{4\eta^2}\right) d\eta \qquad (7.71b)$$

For the special case of constant applied temperature at the boundary surface

$X = 0$, we set

$$f(t) = T_s = \text{constant}, \qquad \Delta T = T_s - T_i, \qquad \phi(\tau) = 1 \qquad (7.72)$$

Then Equation (7.71b) reduces to

$$\theta(X, \tau) = \frac{2}{\sqrt{\pi}} \int_{X/2\sqrt{\tau}}^{\infty} e^{-\eta^2} \, d\eta = \text{erfc}\left(\frac{X}{2\sqrt{\tau}}\right) \qquad (7.73a)$$

which can be written more compactly as

$$\theta(z) = \text{erfc}(z) \qquad (7.73b)$$

where

$$z = \frac{X}{2\sqrt{\tau}} \qquad (7.73c)$$

Figure 7.9 shows a plot of $\theta(z)$ versus z. The results in this figure can be interpreted in different ways.

1. For a given time, the figure represents the temperature distribution in a half-space as a function of position at that particular time,
2. For a given position, it represents the variation of temperature at that position as a function of time.

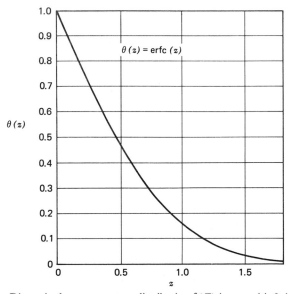

Figure 7.9 Dimensionless temperature distribution $\theta(Z)$ in a semi-infinite medium.

The heat flux at the boundary surface $X = 0$ is determined from its definition as

$$q(0, t) = -k\frac{\partial T(0, t)}{\partial X} \tag{7.74a}$$

By differentiating Equation (7.73a) with respect to X, we have

$$\frac{\partial T(0, t)}{\partial X} = \frac{T_s - T_i}{\sqrt{\pi\alpha t}} \tag{7.74b}$$

Introducing (7.74b) into (7.74a) we find

$$q(0, t) = \sqrt{kc_p\rho}\,\frac{T_s - T_i}{\sqrt{\pi t}} \tag{7.74c}$$

where T_s is the temperature of the boundary surface.

Example 7.11   Two semi-infinite regions are initially at different uniform temperatures T_{1i} and T_{2i}. At time $t = 0$ the solids are brought into contact and kept in contact for times $t > 0$. It is assumed that the surfaces are in perfect thermal contact (i.e., no contact resistance). Immediately after the contact, the two surfaces will instantaneously come to the same interface temperature T_s. Determine this interface temperature.

Solution   This problem can be solved by a straightforward application of the result obtained in Example 7.10. Immediately after the contact, the surface of each solid is at the same interface temperature T_s, and Equation (7.74c) is applicable for each solid. Hence we write the rate of heat flow *into* the first solid as

$$q_1(0, t) = \sqrt{(kc_p\rho)_1}\,\frac{T_s - T_{1i}}{\sqrt{\pi t}} \tag{7.75a}$$

and the rate of heat flow from the second solid as

$$q_2(0, t) = \sqrt{(kc_p\rho)_2}\,\frac{T_{2i} - T_s}{\sqrt{\pi t}} \tag{7.75b}$$

Since heat flow from the hot solid must be equal to that into the cold solid, we equate Equations (7.75a, b) to obtain

$$\frac{T_s - T_{1i}}{T_{2i} - T_s} = \sqrt{\frac{(kc_p\rho)_2}{(kc_p\rho)_1}} \tag{7.76a}$$

or solving for T_s

$$T_s = \frac{T_{1i}\sqrt{(kc_p\rho)_1} + T_{2i}\sqrt{(kc_p\rho)_2}}{\sqrt{(kc_p\rho)_1} + \sqrt{(kc_p\rho)_2}} \qquad (7.76b)$$

Thus the contact temperature T_s is equal to the weighted average of the initial temperatures T_{1i} and T_{2i} with the weight factors $\sqrt{(kc_p\rho)_1}$ and $\sqrt{(kc_p\rho)_2}$.

One practical application of this result is the estimation of contact temperature if one touches a finger to a hot object [21]. For very short times after the contact, both the object and the finger can be treated as semi-infinite solids since the depth of temperature penetration is quite small compared to the thickness of both the object and the finger. The contact temperature is a measure of how hot or how cold an object will feel upon touching it, hence the result given by Equation (7.76b) represents the contact temperature when there is no thermal resistance at the interface.

Example 7.12 Determine the dimensionless temperature distribution in a semi-infinite solid which is initially at uniform temperature T_i and for times $t > 0$, subjected to a prescribed heat flux $q(t)$ per unit time per unit area at the boundary surface $x = 0$.

Solution   We choose the dimensionless variables as

$$\theta(X, \tau) = \frac{T(x, t) - T_i}{\Delta T}, \qquad \tau = \frac{\alpha t}{\ell_0^2}, \qquad X = \frac{x}{\ell_0} \qquad (7.77a)$$

where ℓ_0 is a reference length and ΔT a reference temperature difference. The boundary and initial conditions become

$$-\frac{\partial\theta(0, \tau)}{\partial X} = \frac{\ell_0 q(t)}{k\,\Delta T} \equiv Q(\tau), \qquad \frac{\partial\theta(\infty, \tau)}{\partial X} = 0, \qquad \theta(X, 0) = 0$$

$$(7.77b)$$

The heat diffusion problem considered here is a special case of the general problem Equations (4.52) as follows:

$$T = \theta, \qquad x = X, \qquad t = \tau, \qquad g(x, t) = 0, \qquad x_0 = 0$$

$$f(X) = 0, \qquad \alpha_0 = 0, \qquad \beta_0 = 1, \qquad \phi(t) = Q(\tau) \qquad (7.78)$$

Then the solution is immediately obtained from the general solution Equation (4.60) as

$$\theta(X, \tau) = \frac{1}{\sqrt{\pi}}\int_{\tau'=0}^{\tau}\frac{Q(\tau')}{\sqrt{\tau - \tau'}}e^{-X^2/4(\tau - \tau')}\,d\tau' \qquad (7.79a)$$

or, by defining a new independent variable $\eta = \tau - \tau'$, this result can be written in the form

$$\theta(X, \tau) = \frac{1}{\sqrt{\pi}} \int_{\eta=0}^{\tau} \frac{Q(\tau - \eta)}{\sqrt{\eta}} e^{-X^2/4\eta} \, d\eta \qquad (7.79b)$$

For the special case of $Q = $ constant, we have

$$\theta(X, \tau) = \frac{Q}{\sqrt{\pi}} \int_{\eta=0}^{\tau} \frac{1}{\sqrt{\eta}} e^{-X^2/4\eta} \, d\eta \qquad (7.80a)$$

This result can be transformed into a more convenient form by defining a new variable $\xi = X/2\sqrt{\eta}$ and then integrating the result by parts. We find

$$\frac{\theta(X, \tau)}{Q} = 2\sqrt{\frac{\tau}{\pi}} e^{-X^2/4\tau} - X \, \mathrm{erfc}\left(\frac{X}{2\sqrt{\tau}}\right) \qquad (7.80b)$$

7.4   TIME DEPENDENT TEMPERATURE DISTRIBUTION IN A TRANSPIRATION-COOLED POROUS PLATE

The use of porous materials subjected to the flow of liquid or gas under pressure through the capillary passages inside the metal provides an efficient means for cooling surfaces that are in contact with a very high temperature gas. Such an approach, called the *transpiration cooling*, has applications in the cooling of heated structures encountered in the nozzles of intermittently operated rockets, components of vehicles in accelerated flight, rocket nozzles of booster engines, and tunnel walls of short run-time test facilities. The determination of temperature distribution in transpiration-cooled structures has been the subject of numerous investigations [22–24]. In this section we illustrate how the general solution of the one-dimensional Class I problem can be applied to determine the transient temperature distribution in transpiration-cooled structures.

We consider a porous plate of thickness L, confined to the region $0 \leq x \leq L$. A coolant at reservoir temperature T_f and confined to the region $x \leq 0$ enters the plate at $x = 0$ and flows through the porous matrix of the plate with a constant mass flow velocity of G_f kg/s · m². Figure 7.10 illustrates the geometry and coordinates. The density ρ_f, specific heat C_f, and the thermal conductivity k_f of the coolant fluid are constant. Similarly, the *effective* density ρ_w, specific heat C_w, and thermal conductivity k_w of the plate are also constant.

For times $t < 0$, a steady-state is established between the coolant fluid and the plate, hence the entire plate is at a uniform temperature T_f.

At time $t = 0$, the plate surface at $x = L$ is suddenly exposed to convective heating with a hot gas flowing at an elevated temperature T_g and a heat

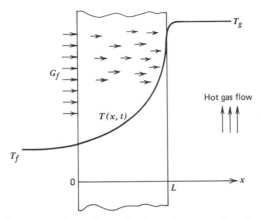

Figure 7.10 Geometry and coordinates for transpiration cooling of a porous plate.

transfer coefficient h. The thermal convective resistance at the coolant entry $x = 0$ is neglected.

During the transpiration cooling of the plate, the coolant and solid temperatures at any location within the plate $0 < x < L$ are assumed to be equal. This assumption implies a high ratio of wetted area to unit volume of wall material and a very high (i.e., infinite) surface conductance between the coolant and the porous wall material.

The mathematical formulation of this transpiration cooling problem is governed by the energy equation given in the dimensionless form as

$$\frac{\partial \theta(X, \tau)}{\partial \tau} = \frac{\partial^2 \theta(X, \tau)}{\partial X^2} - g \frac{\partial \theta(X, \tau)}{\partial X} \quad \text{in } 0 < X < 1, \quad \tau > 0$$

$$(7.81)$$

For convenience in the subsequent analysis, we rewrite this equation in the form

$$e^{-gX} \frac{\partial \theta(X, \tau)}{\partial \tau} = \frac{\partial}{\partial X} \left\{ e^{-gX} \frac{\partial \theta(X, \tau)}{\partial X} \right\} \quad \text{in } 0 < x < 1, \quad \tau > 0$$

$$(7.82a)$$

The boundary and the initial conditions for this problem are taken in the dimensionless form as

$$g\theta(0, \tau) - \frac{\partial \theta(0, \tau)}{\partial X} = 0 \qquad (7.82b)$$

$$\delta\theta(1, \tau) + \gamma \frac{\partial \theta(1, \tau)}{\partial X} = \phi \qquad (7.82c)$$

$$\theta(X, 0) = 0 \quad \text{in } 0 \le X \le 1 \qquad (7.82d)$$

Here various dimensionless quantities are defined as

$$g = \frac{C_f G_f L}{k_w}, \qquad X = \frac{x}{L},$$

$$\tau = \frac{\alpha_w t}{L^2}, \qquad \alpha_w = \frac{k_w \rho_w}{C_w}, \qquad \theta = \frac{T(x,t) - T_f}{\Delta T}$$

$$\Delta T = \text{reference temperature difference} \qquad (7.83)$$

The boundary condition (7.82c) is chosen sufficiently general to include several different physical situations as special cases. For example, the case

$$\phi = 1, \qquad \delta = 1$$

$$\gamma = \frac{k_w}{hL} = \frac{1}{\text{Biot number}} = \frac{1}{\text{Bi}} \qquad (7.84)$$

corresponds to that considered in Reference 23. The case

$$\delta = 0, \qquad \gamma = 1$$

$$\phi = \text{dimensionless heat flux at } x = L \qquad (7.85)$$

corresponds to that solved in Reference 24.

Clearly, the problem defined by Equations (7.82) is a special case of the general problem defined by Equations (4.1). The correspondence between the two problems is as follows:

From Equations (4.1a) and (7.82a):

$$x = X, \qquad t = \tau, \qquad T = \theta, \qquad w(x) = k(x) = e^{-gX},$$

$$d(x) = 0, \qquad P(x,t) = 0 \qquad (7.86a)$$

From Equations (4.1b) and (7.82b):

$$x_0 = 0, \qquad \alpha_0 = g, \qquad \beta_0 = 1, \qquad \phi(x_0, t) = 0 \qquad (7.86b)$$

From Equations (4.1c) and (7.82c):

$$x_1 = 1, \qquad \alpha_1 = \delta, \qquad \beta_1 = \gamma e^g, \qquad \phi(x_1, t) = \phi \qquad (7.86c)$$

From Equations (4.1d) and (7.82d):

$$f(x) = 0 \qquad (7.86d)$$

Having established the correspondence between the two problems, we can now utilize the solutions of the general problem given by Equations (4.8a), (4.27a), and (4.35). The procedure is as follows.

From Equation (4.8a) we write

$$\theta(X, \tau) = \theta_0(X) + \theta_t(X, \tau) \tag{7.87a}$$

The function $\theta_0(X)$ is obtained from Equation (4.27a) as

$$\theta_0(X) = \phi \frac{1 + g\int_0^X e^{gX'} \, dX'}{g\gamma e^g + \delta + g\delta\int_0^1 e^{gX'} \, dX'} = \phi \frac{e^{g(X-1)}}{\delta + \gamma g} \tag{7.87b}$$

The function $\theta_t(X, \tau)$ is obtained from Equation (4.35) as

$$\theta_t(X, \tau) = -\phi \sum_{i=1}^{\infty} \frac{1}{N_i} e^{-\mu_i^2 \tau} \psi_i(X) \frac{1}{\mu_i^2} \Omega_i(1) \tag{7.87c}$$

The quantity $\Omega_i(1)$ is given by Equation (4.4d)

$$\Omega_i(1) = \frac{\psi_i(1)}{\gamma e^g} \tag{7.87d}$$

The solution of the problem is now reduced to that of determining the eigencondition or evaluating the eigenvalues μ_i, the eigenfunctions $\psi_i(x)$, and the normalization integral N_i.

The eigenvalue problem appropriate for the problem considered here can be obtained as a special case from the one considered in Example 3.5. The correspondence between the eigenvalue problem defined by Equations (3.37) and the one appropriate for the present problem is as follows:

$$x_0 = 0, \qquad x_1 = 1, \qquad \delta_0 = g, \qquad \gamma_0 = 1, \qquad \delta_1 = \delta, \qquad \gamma_1 = \gamma \tag{7.88}$$

Then the eigencondition for the determination of the eigenvalues μ_i is obtained from Equation (3.41a) as

$$(\delta + \gamma g)\cos \lambda + \left\{ \frac{1}{2} g\left(\delta + \frac{1}{2}\gamma g \right) - \gamma\lambda^2 \right\} \frac{\sin \lambda}{\lambda} = 0 \tag{7.89a}$$

where λ_i is related to the eigenvalues μ_i by Equation (3.39d) as

$$\mu_i^2 = \lambda_i^2 + \left(\frac{g}{2} \right)^2 \tag{7.89b}$$

The eigenfunctions $\psi_i(X)$ are obtained from Equation (3.41b) as

$$\psi_i(X) = \left\{ \cos(\lambda_i X) + \frac{g}{2\lambda_i} \sin(\lambda_i X) \right\} e^{(g/2)X} \tag{7.89c}$$

The normalization integral N_i is determined from Equation (3.41c)

$$N_i = \frac{1}{2\lambda_i^2}\left(\left(\lambda_i^2 + \frac{g^2}{4}\right)\left\{\frac{\gamma(\delta + \gamma g/2)}{(\delta + \gamma g/2)^2 + (\gamma\lambda_i)^2} + 1\right\} + \frac{g}{2}\right) \quad (7.89d)$$

Then introducing Equation (7.89c) into (7.87d) the quantity $\Omega_i(1)$ is determined as

$$\Omega_i(1) = \frac{e^{g/2}}{\gamma e^g}\frac{\sin\lambda_i}{\lambda_i}\left(\frac{1}{2}g + \lambda_i\frac{\cos\lambda_i}{\sin\lambda_i}\right) \quad (7.90a)$$

The quantity inside the parentheses can be expressed in a more convenient form by utilizing the eigencondition, Equation (7.89a), that is,

$$\frac{1}{2}g + \lambda_i\frac{\cos\lambda_i}{\sin\lambda_i} = \frac{g}{2} - \frac{(g/2)[\delta + (1/2)\gamma g] - \gamma\lambda_i^2}{\delta + \gamma g}$$

$$= \frac{\gamma}{\delta + \gamma g}\left(\lambda_i^2 + \frac{1}{4}g^2\right) \quad (7.90b)$$

Then from Equations (7.90a) and (7.90b), we find

$$\Omega_i(1) = \frac{e^{g/2}}{\gamma e^g}\left(\lambda_i^2 + \frac{1}{4}g^2\right)\frac{\sin\lambda_i}{\lambda_i} \quad (7.90c)$$

Finally, Equations (7.89) and (7.90) are introduced into Equations (7.87) to obtain the solution of the problem given by Equations (7.82) as

$$\frac{\delta + \gamma g}{\phi}\theta(X, \tau)$$

$$= \exp\{g(X - 1)\} - 2\exp\left\{\frac{1}{2}g(X - 1)\right\}\sum_{i=1}^{\infty}\exp\left\{-\left[\lambda_i^2 + \left(\frac{g}{2}\right)^2\right]\tau\right\}$$

$$\times \frac{\lambda_i\sin\lambda_i\{\cos(\lambda_i X) + (g/2\lambda_i)\sin(\lambda_i X)\}}{\{\lambda_i^2 + (g/2)^2\}\left\{\frac{\gamma[\delta + (1/2)\gamma g]}{[\delta + (1/2)\gamma g]^2 + (\gamma\lambda)^2} + 1\right\} + (1/2)g} \quad (7.91)$$

We now examine the special cases of this solution.

Case 1 $\delta = 1$, $\gamma = 0$, $\phi = 1$: The eigencondition (7.89a) reduces to

$$\cos\lambda + \frac{g}{2\lambda}\sin\lambda = 0 \quad (7.92a)$$

By utilizing this relationship, the following term is simplified as

$$\cos(\lambda_i X) + \frac{g}{2\lambda_i}\sin(\lambda_i X) \equiv \frac{\sin[\lambda_i(1-X)]}{\sin\lambda_i} \qquad (7.92b)$$

Then the solution (7.91) takes the form

$$\theta(X,\tau) = \exp\{g(X-1)\} - 2\exp\left\{\frac{1}{2}g(X-1)\right\}$$

$$\times \sum_{i=1}^{\infty} \exp\left\{-\left[\lambda_i^2 + \frac{g^2}{4}\right]\tau\right\}\frac{\lambda_i\sin[\lambda_i(1-X)]}{\lambda_i^2 + (g/2)^2 + (g/2)} \qquad (7.92c)$$

Case 2 $\delta = 0,\ \gamma = 1,\ \phi = K$: The eigencondition (7.89a) reduces to

$$\cos\lambda + \left\{\frac{g}{4\lambda} - \frac{\lambda}{g}\right\}\sin\lambda = 0 \qquad (7.93a)$$

The solution (7.91) simplifies to

$$\frac{g}{K}\theta(X,\tau) = \exp\{g(X-1)\} - 2\exp\left\{\frac{1}{2}g(X-1)\right\}$$

$$\times \sum_{i=1}^{\infty} \exp\left\{-\left[\lambda_i^2 + \left(\frac{g}{2}\right)^2\right]\tau\right\}$$

$$\times \left\{\cos(\lambda_i X) + \frac{g}{2\lambda_i}\sin(\lambda_i X)\right\}\frac{\lambda_i\sin\lambda_i}{\lambda_i^2 + (g/2)^2 + g} \qquad (7.93b)$$

Case 3 $\delta = 1,\ \gamma = 1/\text{Bi},\ \phi = 1$: The eigencondition (7.89a) reduces to

$$(\text{Bi} + g)\cos\lambda + \left\{\frac{g}{2}\left(\text{Bi} + \frac{g}{2}\right) - \lambda^2\right\}\frac{\sin\lambda}{\lambda} = 0 \qquad (7.94a)$$

and the solution (7.91) simplifies to

$$\left(1 + \frac{1}{\text{Bi}}g\right)\theta(X,\tau) = \exp\{g(X-1)\} - 2\exp\left\{\frac{1}{2}g(X-1)\right\}$$

$$\times \sum_{i=1}^{\infty} \exp\left\{-\left[\lambda_i^2 + \left(\frac{g}{2}\right)^2\right]\tau\right\}$$

$$\times \frac{\lambda_i\sin\lambda_i\{\cos(\lambda_i X) + (g/2\lambda_i)\sin(\lambda_i X)\}}{\left(\text{Bi} + \frac{g}{2}\right)\left\{\dfrac{\lambda_i^2 + (g/2)^2}{\lambda_i^2 + [\text{Bi} + (g/2)]^2}\right\} + \lambda_i^2 + (g/2)^2 + g/2}$$

$$(7.94b)$$

REFERENCES

1. J. B. Fourier, *Theory Analytique de la Chaleur*, Paris, 1822 (English transl. A. Freeman, Dover, New York, 1955).

2. R. W. Powell, C. Y. Ho, and P. E. Liley, *Thermal Conductivity of Selected Materials*, NSRDS —NBS 8, U.S. Department of Commerce, National Bureau of Standards, Washington, D.C., 1966.

3. *Thermophysical Properties of Matter*, Vols. 1–3, 1F1/Plenum Data Corp., New York, 1969.

4. C. Y. Ho, R. W. Powell, and P. E. Liley, Thermal Conductivity of Elements, Vol. 1, First Supplement to *J. Phys. Chem. Ref. Data* (1972).

5. A. Fick, *Ann. Phys. Lpz.* **170**, 59, (1855).

6. R. B. Bird, W. E. Stewart, and E. N. Lightfoot, *Transport Phenomena*, Wiley, New York, 1960.

7. R. D. Prescnt, *Kinetic Theory of Gases*, McGraw-Hill, New York, 1958.

8. J. O. Hirschfelder, C. F. Curtiss, and R. B. Bird, *Molecular Theory of Gases and Liquids*, Wiley, New York, 1954.

9. A. H. P. Skelland, *Diffusional Mass Transfer*, Wiley, New York, 1974.

10. R. C. Reid and T. K. Sherwood, *The Properties of Gases and Liquids*, Chap. 11, McGraw-Hill, New York, 1966.

11. J. Crank, *The Mathematics of Diffusion*, 2nd ed., Clarendon Press, London, 1975.

12. W. M. Kays and M. E. Crawford, *Convective Heat and Mass Transfer*, McGraw-Hill, New York, 1980.

13. E. R. G. Eckert and R. M. Drake, *Analysis of Heat and Mass Transfer*, McGraw-Hill, New York, 1972.

14. M. N. Özişik, *Heat Transfer*, McGraw-Hill, New York, 1984.

15. H. S. Carslaw and J. C. Jaeger, *Conduction of Heat in Solids*, 2nd ed., Clarendon Press, London, 1959.

16. M. N. Özişik, *Heat Conduction*, Wiley, New York, 1980.

17. M. Abramowitz and I. A. Stegun, *Handbook of Mathematical Functions*, National Bureau of Standards, Applied Mathematics Series, 55, U.S. Government Printing Office, Washington, D.C., 1964.

18. W. Flügge, *Four Place Tables of Transcendental Functions*, McGraw-Hill, New York, 1954.

19. N. W. McLachlan, *Bessel Functions for Engineers*, Clarendon Press, London, 1955.

20. H. W. Reddick and F. H. Miller, *Advanced Mathematics for Engineers*, Wiley, New York, 1955.

21. G. E. Myers, *Analytical Methods in Conduction Heat Transfer*, McGraw-Hill, New York, 1971.

22. S. Weinbaum and H. L. Wheeler, Jr., Heat Transfer in Sweat-Cooled Porous Metals, *J. Appl. Phys.*, **20**, 113–122 (1949).

23. P. J. Schneider and J. J. Brogan, Temperature Response of a Transpiration-Cooled Plate, *ARS J.*, **32**, 233–236 (1962).

24. A. R. Mendelsohn, Transient Temperature of a Porous-Cooled Wall, *AIAA J.*, **1**, 1449–1451 (1963).

CHAPTER EIGHT

Class I Solutions Applied to Forced Convection in Conduits

Every cause produces more than one effect.

Herbert Spencer
1820–1903

In this chapter we illustrate the application of the one-dimensional Class I solutions developed in Chapter 4 for the analysis of another type of heat transfer problem. This time it is applied to solve forced convection inside conduits having simple geometries such as a parallel plate or a circular cylinder. It is shown that the steady and unsteady velocity distribution for fully developed flow and the steady-state temperature distribution for thermally developing, hydrodynamically developed laminar or turbulent flow are obtainable as special cases of the one-dimensional Class I solutions.

8.1. STEADY FULLY DEVELOPED FLOW THROUGH CONDUITS

The analysis of heat transfer in forced convection is complicated by the fact that the motion of the fluid plays an important part in heat transfer. Therefore, in the determination of the temperature field in fluid flow, a knowledge of the velocity distribution is essential. In this section we are concerned with the determination of the velocity distribution for *hydrodynamically fully developed* flow inside conduits. The flow is said to be hydrodynamically fully developed when the velocity distribution at any cross-section is independent of the axial coordinate z; that is, there is no component of fluid velocity normal to the axis of the conduit.

We consider fully developed, steady flow of an incompressible, constant property fluid inside a conduit having uniform cross-section and axial symmetry. Let z be the axial and r the radial (or normal) coordinates. The momentum equation for the flow is taken as

$$\frac{d}{dr}\left[r^n(\mu + \rho\epsilon_m)\frac{dw(r)}{dr}\right] - r^n\frac{dP}{dz} = 0 \quad \text{in } r_0 < r < r_1 \qquad (8.1a)$$

where μ is viscosity, ρ is density, $\epsilon_m \equiv \epsilon_m(r)$ is eddy viscosity, $dP/dz = $ constant is the pressure gradient along the conduit, and $w(r)$ is the velocity profile. The exponent n is defined as

$$n = \begin{cases} 0 & \text{for parallel plate channel} \\ 1 & \text{for circular or annular tube} \end{cases} \qquad (8.1b)$$

Here viscosity and density are properties of the fluid, whereas eddy viscosity $\epsilon_m(r)$ is a purely local function in the fluid and determined from the turbulent model. Eddy viscosity is equal to zero at the channel surface, that is, $\epsilon_m(r_0) = \epsilon_m(r_1) = 0$.

To nondimensionalize the preceding momentum equation, we choose a characteristic length ℓ_0 and define the dimensionless variables as

$$R = \frac{r}{\ell_0}, \qquad W(R) = \frac{\mu w(r)}{(-dP/dz)\ell_0^2}, \qquad E_m(R) = 1 + \frac{\rho}{\mu}\epsilon_m(r)$$

$$(8.2a, b, c)$$

By introducing these dimensionless quantities into Equation (8.1a) we obtain

$$\frac{d}{dR}\left\{ R^n E_m(R)\frac{dW(R)}{dR}\right\} + R^n = 0 \quad \text{in } R_0 < R < R_1 \qquad (8.3a)$$

The boundary conditions are taken as velocity is equal to zero at the wall, that is,

$$W(R_k) = 0, \qquad k = 0 \quad \text{and} \quad 1 \qquad (8.3b, c)$$

The system (8.3) is the complete mathematical formulation of the problem for the determination of the velocity field $W(R)$; it is a special case of the one-dimensional problem (4.23) as follows:

$$x = R, \qquad T_j(x) = W(R), \qquad T_{j+1}(x) = 0, \qquad k(x) = R^n E_m(R),$$

$$d(x) = 0, \qquad P_j(x) = R^n, \qquad \alpha_k = 1, \qquad \beta_k = 0, \qquad \phi_j(x_k) = 0,$$

$$E_m(R_k) = 1, \qquad x_k = R_k$$

Then the solution of problem (8.3) is immediately obtained from solution

(4.27) by appropriate change of notation and simplification; we find

$$W(R) = \frac{1}{n+1}\left\{ \frac{\int_{R_0}^{R}\frac{dR'}{R'^n E_m(R')}}{\int_{R_0}^{R_1}\frac{dR}{R^n E_m(R)}}\cdot\int_{R_0}^{R_1}\frac{R\,dR}{E_m(R)} - \int_{R_0}^{R}\frac{R'\,dR'}{E_m(R')}\right\} \quad (8.4)$$

The *average velocity* for the flow is defined as

$$w_{av} = \frac{\int_{r_0}^{r_1} r^n w(r)\,dr}{\int_{r_0}^{r_1} r^n\,dr} \quad (8.5a)$$

or given in the dimensionless form by

$$W_{av} = \frac{n+1}{R_1^{n+1} - R_0^{n+1}}\int_{R_0}^{R_1} R^n W(R)\,dR \quad (8.5b)$$

Introducing the solution (8.4) into Equation (8.5b), the dimensionless average velocity becomes

$$W_{av} = \frac{1}{(n+1)(R_1^{n+1} - R_0^{n+1})}\left\{ \int_{R_0}^{R_1}\frac{R^{n+2}\,dR}{E_m(R)} - \frac{\left[\int_{R_0}^{R_1}\frac{R\,dR}{E_m(R)}\right]^2}{\int_{R_0}^{R_1}\frac{dR}{R^n E_m(R)}}\right\} \quad (8.6)$$

Normalized Velocity Distribution

An alternative way of presenting the dimensionless velocity distribution is normalizing it with respect to the average velocity as

$$U(R) = \frac{w(r)}{w_{av}} = \frac{W(R)}{W_{av}} \quad (8.7a)$$

Introducing Equations (8.4) and (8.6) into Equation (8.7a), the normalized velocity distribution becomes

$$U(R) = \left(R_1^{n+1} - R_0^{n+1}\right)$$

$$\times \frac{\int_{R_0}^{R}\frac{dR'}{R'^n E_m(R')}\int_{R_0}^{R_1}\frac{R\,dR}{E_m(R)} - \int_{R_0}^{R}\frac{R'\,dR'}{E_m(R')}\int_{R_0}^{R_1}\frac{dR}{R^n E_m(R)}}{\int_{R_0}^{R_1}\frac{dR}{R^n E_m(R)}\int_{R_0}^{R_1}\frac{R^{n+2}\,dR}{E_m(R)} - \left[\int_{R_0}^{R_1}\frac{R\,dR}{E_m(R)}\right]^2}$$

$$(8.7b)$$

Velocity profiles have a maximum for a certain value of the radius. The radial location R_m where this maximum occurs is determined by differentiating Equation (8.4) or (8.7b) with respect to R and setting the resulting expression equal to zero. We find

$$R_m = \left\{ \frac{\int_{R_0}^{R_1} \dfrac{R\,dR}{E_m(R)}}{\int_{R_0}^{R_1} \dfrac{dR}{R^n E_m(R)}} \right\}^{1/(n+1)} \tag{8.8}$$

Velocity Distribution for Circular Tube and Parallel Plate Channel

As a special case of the foregoing analysis we now consider a parallel plate channel of spacing $2r_1$ and a circular tube of radius r_1. The origin of the coordinate axis r is located at the center and the characteristic dimension is chosen as $\ell_0 = r_1$; then we have $R_0 = 0$ and $R_1 = 1$. The velocity distribution in the region $0 \le R \le 1$ is governed by the differential Equation (8.3a) subject to the symmetry condition at $R_0 = 0$ and zero velocity at $R_1 = 1$; the mathematical formulation becomes

$$\frac{d}{dR}\left\{ R^n E_m(R) \frac{dW(R)}{dR} \right\} + R^n = 0 \quad \text{in } 0 < R < 1 \tag{8.9a}$$

$$\frac{dW(0)}{dR} = 0, \qquad W(1) = 0 \tag{8.9b, c}$$

This problem is a special case of the one-dimensional problem (4.23) as discussed previously, but with $\alpha_0 = 0$, $\beta_0 = 1$, $\alpha_1 = 1$, and $\beta_1 = 0$. Then the solution of the problem (8.9) is obtained from solution (4.27) as

$$W(R) = \frac{1}{n+1} \int_R^1 \frac{R'\,dR'}{E_m(R')} \tag{8.10}$$

and the definition of the average dimensionless velocity, Equation (8.5b), for $R_0 = 0$ and $R_1 = 1$, becomes

$$W_{av} = (n+1) \int_0^1 R^n W(R)\,dR \tag{8.11a}$$

Introducing Equation (8.10) into (8.11a) we find

$$W_{av} = \frac{1}{n+1} \int_0^1 \frac{R^{n+2}\,dR}{E_m(R)} \tag{8.11b}$$

The normalized velocity distribution $U(R)$ is determined by substituting Equations (8.10) and (8.11b) into Equation (8.7a).

$$U(R) = \frac{\int_R^1 R' \, dR'/E_m(R')}{\int_0^1 R^{n+2} \, dR/E_m(R)} \qquad (8.12)$$

Equation (8.12) is applicable only for shear stress composed of a viscous component and a turbulent component in the form

$$\tau = -(\mu + \rho \epsilon_m) \frac{dw(r)}{dr} \qquad (8.13a)$$

If the turbulent stress term in Equation (8.13a) is zero (i.e., $\epsilon_m = 0$), the resulting equation is known as *Newton's law of viscosity*. According to Newton's law of viscosity, a plot of τ versus $-(dw/dr)$ for a given fluid should give a straight line through the origin. Indeed, for all gases and ordinary fluids such as water and oils, this linear relationship holds and such fluids are called *Newtonian fluids*.

There are also some fluids for which the relation between the shear stress and the velocity gradient is more complicated and can be defined by

$$f(\tau) = -\frac{dw(r)}{dr} \qquad (8.13b)$$

where f is considered a known function. Such fluids are called *non-Newtonian fluids*. Colloidal suspensions and emulsions are examples of non-Newtonian fluids. Depending on the nature of the function $f(\tau)$, non-Newtonian fluids may be subdivided into the following distinct types [26]:

1. Bingham plastics.
2. Pseudoplastic fluids.
3. Dilatant fluids.

Common examples of Bingham plastics include slurries, drilling muds, oil paints, toothpaste, and sewage sludges. Examples of pseudoplastic fluids include suspensions of asymmetric particles or solutions of high polymers such as cellulose derivatives.

Equations (8.13) can be used to develop relations for shear stress in the fluid and a generalized formula for the velocity distribution for flow in a parallel-plate channel or a circular tube as now described.

Introducing Equation (8.13a) into Equation (8.1a) we obtain an alternative form of the momentum equation

$$\frac{d}{dr}(r^n \tau) + r^n \frac{dP}{dz} = 0 \quad \text{in } 0 \le r \le r_1 \qquad (8.14a)$$

The boundary condition for this equation is taken as shear stress equal to zero at tube axis

$$\tau = 0 \quad \text{at } r = 0 \tag{8.14b}$$

The solution of Equation (8.14a) subject to boundary condition (8.14b) gives

$$\tau = \frac{r}{n+1}\left(-\frac{dP}{dz}\right) \tag{8.15a}$$

Then the wall shear stress at $r = r_1$ is

$$\tau_w = \frac{r_1}{n+1}\left(-\frac{dP}{dz}\right) \tag{8.15b}$$

From Equations (8.15) we obtain

$$\tau = \tau_w R \tag{8.16}$$

Therefore, the shear stress distribution in a parallel plate channel or circular tube for steady fully developed flow is always linear.

To develop a general relationship for the velocity distribution, we integrate Equation (8.13b) to obtain

$$w(r) = r_1 \int_R^1 f(\tau_w R)\, dR \tag{8.17a}$$

Introducing Equation (8.17a) into Equation (8.5a), after integration by parts, we obtain

$$w_{av} = r_1 \int_0^1 R^{n+1} f(\tau_w R)\, dR \tag{8.17b}$$

Using Equations (8.17) in the definition (8.7a), we find a generalized formula for velocity distribution

$$U(R) = \frac{\displaystyle\int_R^1 f(\tau_w R)\, dR}{\displaystyle\int_0^1 R^{n+1} f(\tau_w R)\, dR} \tag{8.18}$$

Clearly, Equation (8.18) includes as a special case given by Equation (8.12) for

$$f(\tau) = \frac{\tau_w}{\mu} \frac{R}{E_m(R)}$$

We now illustrate with examples the application of the foregoing results for the determination of velocity distribution in laminar flow inside conduits.

Example 8.1 Determine the velocity distribution for Poiseuille flow inside a circular tube of radius r_1 and a parallel plate duct of spacing $2r_1$.

Solution Laminar flow through a circular tube or a parallel plate duct is referred to as Poiseuille flow or Hagen–Poiseuille flow after Hagen [1] and Poiseuille [2]. For laminar flow, we set $E_m(R) = 1$ and from Equations (8.10), (8.11b), and (8.12), respectively, obtain

$$W(R) = \frac{1 - R^2}{2(n + 1)} \tag{8.19a}$$

$$W_{av} = \frac{1}{(n + 1)(n + 3)} \tag{8.19b}$$

$$U(R) = \frac{n + 3}{2}(1 - R^2) \tag{8.19c}$$

where

$$n = \begin{cases} 0 & \text{for parallel plate} \\ 1 & \text{for circular tube} \end{cases}$$

Clearly, the velocity profile for the Poiseuille flow is parabolic.
    The wall shear stress τ_w is determined from its definition as

$$\tau_w = -\mu \frac{dw(r_1)}{dr} = -\mu \frac{W_{av}}{r_1} \frac{dU(1)}{dR} = \mu \frac{W_{av}}{r_1}(n + 3) \tag{8.19d}$$

Example 8.2 Determine the normalized velocity distribution $U(R)$ for the Poiseuille flow in a parallel plate duct $0 \le r \le h$ with the coordinate axis located at $r = 0$.

Solution We choose the characteristic length as $\ell_0 = h$. Then the solution of this problem is a special case of Equation (8.7b) for $E_m(R) = 1$, $n = 0$, $R_0 = 0$, and $R_1 = 1$. We find

$$U(R) = 6(R - R^2) \tag{8.20}$$

Example 8.3 Determine the normalized velocity distribution $U(R)$ for laminar flow inside a circular annulus $r_0 \le r \le r_1$ and the location r_m where the velocity profile has a maximum.

Solution $U(R)$ is obtained from Equation (8.7b), for $E_m(R) = 1$ and $n = 1$, as

$$U(R) = 2\frac{(R_1^2 - R_0^2)\ln(R/R_0) - (R^2 - R_0^2)\ln(R_1/R_0)}{(R_1^2 + R_0^2)\ln(R_1/R_0) - (R_1^2 - R_0^2)} \quad (8.21a)$$

The location R_m where $U(R)$ has a maximum is determined from Equation (8.8) for $E_m(R) = 1$. We find

$$R_m = \left(\frac{R_1^2 - R_0^2}{2\ln(R_1/R_0)}\right)^{1/2} \quad (8.21b)$$

Example 8.4   Determine the velocity distribution for non-Newtonian power law fluid in laminar flow inside conduits.

Solution   For power law fluid

$$\tau = a\left[-\frac{dw(r)}{dr}\right]^b \quad (8.22a)$$

where a and b are model parameters (i.e., material constants of the fluid and are both positive). Equation (8.22a) is rearranged as

$$\left(\frac{\tau}{a}\right)^{1/b} = -\frac{dw(r)}{dr} \quad (8.22b)$$

By comparison of Equations (8.22b) and (8.13b) we find

$$f(\tau) = \left(\frac{\tau_w}{a}\right)^{1/b} R^{1/b} \quad (8.22c)$$

Using this result in Equation (8.18) we obtain

$$U(R) = \frac{1 + (2 + n)b}{1 + b}[1 - R^{(1+b)/b}] \quad (8.23)$$

where

$$n = \begin{cases} 0 & \text{for parallel plate} \\ 1 & \text{for circular tube} \end{cases}$$

The velocity profile for a circular tube ($n = 1$) is shown in Figure 8.1 for *Newtonian fluid* ($b = 1$), *dilatant fluid* ($b = 3$), and *pseudoplastic fluid* ($b = \frac{1}{3}$).

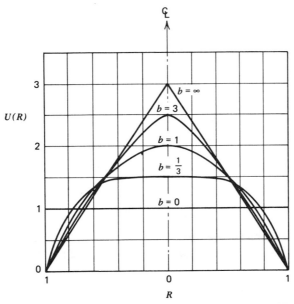

Figure 8.1 Velocity profile for flow inside a circular tube for typical non-Newtonian fluids.

In the same figure are shown the velocity profiles for the ideal dilatant ($b = \infty$) and ideal pseudoplastic ($b = 0$) materials [26].

8.2 UNSTEADY FULLY DEVELOPED FLOW THROUGH CONDUITS

In many instances it is of interest to know the time variation of velocity for forced flow through conduits resulting from time variation of the applied pressure gradient. For example, in the cooling channel of a nuclear reactor, a change in pumping pressure causes a transient in the flow velocity which, in turn, effects the heat transfer. In a rocket engine, the start of the firing is accompanied by a transient in the stream velocity. The circulation of blood in the blood vessels, the pumping of a fluid through a channel by a reciprocating device, and the motion of fluid resulting from flow instabilities are the examples of pulsating or oscillatory flow.

We now present an analysis of unsteady, fully developed velocity distribution in laminar flow inside a circular tube and a parallel plate channel and illustrate the application of the results with examples.

Consider fully developed laminar flow inside a circular tube or a parallel plate channel subjected to a pressure gradient dP/dz, which is independent of the space coordinates r and z, but is a prescribed function of time. Then the time dependent z-momentum equation is taken as

$$\rho\frac{\partial w\,(r,t)}{\partial t} = \mu\frac{1}{r^n}\frac{\partial}{\partial r}\left\{r^n\frac{\partial w\,(r,t)}{\partial r}\right\} - \frac{dP}{dz} \quad \text{in } 0 < r < r_1 \quad (8.24a)$$

We consider the solution of this equation subject to the boundary conditions

$$\frac{\partial w\,(0,t)}{\partial r} = 0, \qquad w(r_1,t) = 0 \qquad\qquad (8.24b,c)$$

and the initial condition

$$w(r,0) = w_0(r) \qquad\qquad (8.24d)$$

To nondimensionalize the problem (8.24), we choose a characteristic length $\ell_0 = r_1$, a reference average velocity w_{av}^*, and introduce the following dimensionless variables

$$R = \frac{r}{r_1}, \qquad \tau = \frac{\nu t}{r_1^2}, \qquad U(R,\tau) = \frac{w(r,t)}{w_{av}^*}, \qquad P(\tau) = \frac{-(dP/dz)r_1^2}{\mu w_{av}^*}$$

$$(8.25a,b,c,d)$$

Then the problem (8.24) is written in the dimensionless form as

$$R^n\frac{\partial U\,(R,\tau)}{\partial \tau} = \frac{\partial}{\partial R}\left\{R^n\frac{\partial U(R,\tau)}{\partial R}\right\} + R^nP(\tau) \quad \text{in } 0 \le R \le 1$$

$$(8.26a)$$

$$\frac{\partial U\,(0,\tau)}{\partial R} = 0, \qquad U(1,\tau) = 0 \qquad\qquad (8.26b,c)$$

$$U(R,0) = U_0(R) \qquad\qquad (8.26d)$$

We examine the solution of problem (8.26) for a step change in the pressure gradient and a periodically varying pressure gradient as given in the following.

A Step Change in the Pressure Gradient

In this application we are concerned with transient velocities in laminar flow resulting from a sudden change in pressure gradient.

Consider a Poiseuille flow with a pressure gradient $(-dP/dz)_0 = $ constant, corresponding to an average velocity $w_{av,0}$. At the origin of the time coordinate $t = 0$, this pressure gradient is suddenly altered to $(-dP/dz)$, which is also constant, and has the corresponding steady-state average velocity $w_{av,1}$. We wish to determine the variation of the flow velocity as a function of time for times $t > 0$. To formulate the problem, we select $w_{av,1}$ as the reference velocity (i.e., $w_{av}^* \equiv w_{av,1}$) and define the normalized velocity distribution as $U(R, \tau) = w(r, t)/w_{av,1}$. Then the transient velocity field $U(R, \tau)$ is governed by the system (8.26) if the source term $P(\tau)$ and the initial velocity distribution $U_0(R)$ are defined as described.

By utilizing Equation (8.19c) and noting that for this problem the reference velocity is chosen as $w_{av,1}$ instead of $w_{av,0}$ we obtain

$$U_0(R) = \frac{w_{av,0}}{w_{av,1}} \frac{n+3}{2} (1 - R^2) \tag{8.27a}$$

From the definition (8.2b) we note $W_{av} = \mu w_{av}^*(-dP/dz)^{-1}r_1^{-2}$; and by combining this result with Equations (8.25d) and (8.19b) we write

$$P(\tau) = (n+1)(n+3) \tag{8.27b}$$

Then the problem (8.26) becomes a special case of the one-dimensional problem Equation (4.1) with various quantities defined as follows:

$$x = R, \qquad t = \tau, \qquad T(x,t) = U(R,\tau), \qquad w(x) = R^n$$

$$k(x) = R^n, \qquad d(x) = 0, \qquad x_0 = 0, \qquad x_1 = 1$$

$$\alpha_0 = 0, \qquad \beta_0 = 1, \qquad \alpha_1 = 1, \qquad \beta_1 = 0$$

$$\phi(x_k, t) = 0, \qquad f(x) = U_0(R) = \frac{w_{av,0}}{w_{av,1}} \frac{n+3}{2} (1 - R^2)$$

$$P(x,t) = R^n P(\tau) = R^n (n+1)(n+3) \tag{8.27c}$$

The solution is determined from Equation (4.8a) as now described.

By Equation (4.7c) we write $P_e(x) = 0$ and $P_0(x) = R^n(n+1)(n+3)$. Then the only contribution from Equation (4.8b) for the solution of the present problem is the functions $T_0(x)$ and $T_t(x, t)$ which are obtained from Equation (4.27a) for $j = 0$ and from Equation (4.35), respectively. The eigenfunctions, eigenvalues, and normalization integral appearing in Equation (4.35) are determined from the results in Example 3.6 for $\beta = 0$. Various integrals are evaluated by utilizing Equation (7.28a). After some manipulation and replacing n by m according to the relation $m = (1 - n)/2$, we obtain the normalized transient velocity distribution $U(R, \tau)$ as

$$U(R, \tau) \equiv \frac{w(r,t)}{w_{av,1}} = (2 - m)(1 - R^2) - 8(1 - m)(2 - m)\left(1 - \frac{w_{av,0}}{w_{av,1}}\right)$$

$$\times \sum_{i=1}^{\infty} \frac{1}{\mu_i^3} R^m \frac{J_{-m}(\mu_i R)}{J_{1-m}(\mu_i)} e^{-\mu_i^2 \tau} \tag{8.28a}$$

where the μ_i are the positive roots of the transcendental equation

$$J_{-m}(\mu) = 0 \tag{8.28b}$$

and

$$m = \begin{cases} \frac{1}{2} & \text{for parallel plate channel} \\ 0 & \text{for circular tube} \end{cases} \tag{8.28c}$$

The functions $R^m J_{-m}(R)$ and $R^m J_{1-m}(R)$ are given in Table 3.1.

The solution given by Equation (8.28a) is valid for all cases of a step change of the pressure gradient including $w_{av,0} = 0$ (i.e., the initial pressure gradient is zero) and $w_{av,1} = 0$ (i.e., the new pressure gradient is zero). However, in the case of the latter, Equation (8.28a) must be multiplied by $w_{av,1}/w_{av,0}$ and then set $w_{av,1} = 0$ to obtain

$$\frac{w(r,t)}{w_{av,0}} = 8(1-m)(2-m) \sum_{i=1}^{\infty} \frac{1}{\mu_i^3} R^m \frac{J_{-m}(\mu_i R)}{J_{1-m}(\mu_i)} e^{-\mu_i^2 \tau} \tag{8.28d}$$

which is valid for $w_{av,1} = 0$.

The transient average normalized velocity $U_{av}(\tau)$ is determined from the definition

$$U_{av}(\tau) = (n+1) \int_0^1 R^n U(R, \tau) \, dR = 2(1-m) \int_0^1 R^{1-2m} U(R, \tau) \, dR \tag{8.29a}$$

since $m = (1-n)/2$. Introducing Equation (8.28a) into (8.29a), we obtain

$$U_{av}(\tau) = 1 - 8(1-m)(2-m)\left(1 - \frac{w_{av,0}}{w_{av,1}}\right) \sum_{i=1}^{\infty} \frac{1}{\mu_i^4} e^{-\mu_i^2 \tau} \tag{8.29b}$$

where the μ_i are the roots of Equation (8.28b). A rearrangement of this result yields

$$\frac{w_{av}(\tau) - w_{av,1}}{w_{av,0} - w_{av,1}} = 8(1-m)(2-m) \sum_{i=1}^{\infty} \frac{1}{\mu_i^4} e^{-\mu_i^2 \tau} \tag{8.29c}$$

The transient wall shear stress is determined from its definition as

$$\tau_{wt} = -\mu \frac{w_{av,1}}{r_1} \frac{\partial U(1, \tau)}{\partial R} \tag{8.30a}$$

where μ is the viscosity.

The ratio of the transient wall shear stress τ_{wt} to the steady-state wall shear stress τ_{ws} for the same average velocity, determined from Equations (8.30a) and

(8.19d) is given by

$$\frac{T_{wt}}{T_{ws}} = \frac{-[\partial U(1,\tau)/\partial R]}{2(2-m)U_{av}}$$

(8.30b)

where we set $n = 1 - 2m$ and $w_{av}/w_{av,1} = U_{av}$. Introducing the values of $U(R,\tau)$ and $U_{av}(\tau)$ from Equations (8.28a) and (8.29b), respectively, we obtain

$$\frac{T_{wt}}{T_{ws}} = \frac{1 - 4(1-m)[1 - (w_{av,0}/w_{av,1})] \sum\limits_{i=1}^{\infty} e^{-\mu_i^2\tau}/\mu_i^2}{1 - 8(1-m)(2-m)[1 - (w_{av,0}/w_{av,1})] \sum\limits_{i=1}^{\infty} e^{-\mu_i^2\tau}/\mu_i^4}$$

(8.30c)

Example 8.5 Determine the transient laminar velocity distribution and the wall shear stress for a parallel plate channel and a circular tube resulting from a step change in the pressure gradient.

Solution The solution to this problem is immediately obtained from the results given by Equations (8.28) and (8.30c) by appropriately choosing the value of m.

For a parallel plate channel we set $m = \frac{1}{2}$ and obtain

$$U(R,\tau) = \frac{3}{2}(1 - R^2) - 6\left(1 - \frac{w_{av,0}}{w_{av,1}}\right) \sum_{i=1}^{\infty} \frac{(-1)^i}{\mu_i^3} \cos(\mu_i R) e^{-\mu_i^2\tau}$$

(8.31a)

$$\frac{T_{wt}}{T_{ws}} = \frac{1 - 2[1 - (w_{av,0}/w_{av,1})] \sum\limits_{i=1}^{\infty} e^{-\mu_i^2\tau}/\mu_i^2}{1 - 6[1 - (w_{av,0}/w_{av,1})] \sum\limits_{i=1}^{\infty} e^{-\mu_i^2\tau}/\mu_i^4}$$

(8.31b)

where $\mu_i = (2i-1)\pi/2$. We note that solution (8.31a) is the same as that given by Equation (2) in Reference 3.

For a circular tube we set $m = 0$ and obtain

$$U(R,\tau) = 2(1 - R^2) - 16\left(1 - \frac{w_{av,0}}{w_{av,1}}\right) \sum_{i=1}^{\infty} \frac{1}{\mu_i^3} \frac{J_0(\mu_i R)}{J_1(\mu_i)} e^{-\mu_i^2\tau}$$ (8.32a)

$$\frac{T_{wt}}{T_{ws}} = \frac{1 - 4[1 - (w_{av,0}/w_{av,1})] \sum\limits_{i=1}^{\infty} e^{-\mu_i^2\tau}/\mu_i^2}{1 - 16[1 - (w_{av,0}/w_{av,1})] \sum\limits_{i=1}^{\infty} e^{-\mu_i^2\tau}/\mu_i^4}$$

(8.32b)

where the μ_i are the roots of the transcendental equation

$$J_0(\mu) = 0$$

(8.32c)

A Periodically Varying Pressure Gradient

The velocity distribution for pulsating laminar flow in a circular tube has been treated in Reference 4. A similar problem has been solved in Reference 5 in connection with heat transfer to pulsating laminar flow between parallel plates. We now illustrate how the solution to such problems can be obtained as the special case of the one-dimensional problem (4.1).

We consider a Poiseuille flow with a pressure gradient $(-dP/dz)_0 =$ constant, corresponding to the reference average velocity w_{av}^*. Let the normalized velocity distribution at the origin of the time coordinate $t = 0$ be $U(R,0) = [(n + 3)/2](1 - R^2)$, as given by Equation (8.19c). For times $t > 0$, the flow field is pulsating as a result of varying pressure gradients given in the form

$$\left(-\frac{\partial P}{\partial z}\right) = \left(-\frac{\partial P}{\partial z}\right)_0\left(1 + \frac{\gamma}{2}\cos \omega t\right) \tag{8.33a}$$

which can be rewritten as

$$\left(-\frac{\partial P}{\partial z}\right) = \left(-\frac{\partial P}{\partial z}\right)_0\left(1 + \frac{\gamma}{4}e^{i\omega t} + \frac{\gamma}{4}e^{-i\omega t}\right) \tag{8.33b}$$

Then the transient problem (8.26) becomes

$$R^n\frac{\partial U(R,\tau)}{\partial \tau} = \frac{\partial}{\partial R}\left\{R^n\frac{\partial U(R,\tau)}{\partial R}\right\}$$

$$+ R^n(n + 1)(n + 3)\left\{1 + \frac{\gamma}{4}e^{i\Omega\tau} + \frac{\gamma}{4}e^{-i\Omega\tau}\right\} \tag{8.34a}$$

$$\frac{\partial U(0,\tau)}{\partial R} = 0, \qquad U(1,\tau) = 0 \tag{8.34b, c}$$

$$U(R,0) = \frac{n + 3}{2}(1 - R^2) \tag{8.34d}$$

where

$$\Omega = \frac{\omega r_1^2}{\nu}, \qquad \nu = \frac{\mu}{\rho} \tag{8.34e, f}$$

Problem (8.34) is now a special case of the one-dimensional problem (4.1) as follows:

$$x = R, \qquad t = \tau, \qquad T(x,t) = U(R,\tau), \qquad w(x) = R^n, \qquad k(x) = R^n,$$

$$d(x) = 0, \qquad x_0 = 0, \qquad x_1 = 1, \qquad \alpha_0 = 0, \qquad \alpha_1 = 1,$$

$$\beta_0 = 1, \qquad \beta_1 = 0, \qquad \phi_k(x_k, t) = 0, \qquad f(x) = \frac{n + 3}{2}(1 - R^2),$$

$$P(x,t) = R^n(n + 1)(n + 3)\left\{1 + \frac{\gamma}{4}e^{i\Omega\tau} + \frac{\gamma}{4}e^{-i\Omega\tau}\right\}$$

The solution is determined from Equation (4.8a) as described. By Equation (4.7c) we write

$$P_e = (n + 1)(n + 3)\frac{\gamma}{4}, \qquad d_p = (-1)^k i\Omega, \qquad k = 0 \quad \text{or} \quad 1$$

Then the contribution of Equation (4.8a) for the present solution includes the transient term $T_t(x, t)$ and the sustained portion $T_p(x)e^{-d_p t}$. For sufficiently long times, the transient portion dies out and only the sustained portion need be considered. Therefore, the solution of problem (8.34), valid for long times, is determined from Equation (4.18) as

$$U(R, \tau) = (2 - m)\{(1 - R^2) + \gamma U_s(R, \tau)\} \qquad (8.35a)$$

where

$$U_s(R, \tau) = \frac{2(1 - m)}{\Omega}$$

$$\times \left[\sin(\Omega\tau) + \cos(\Omega\tau) \frac{R_m(RM)I_m(M) - I_m(RM)R_m(M)}{R_m^2(M) + I_m^2(M)} \right.$$

$$\left. - \sin(\Omega\tau) \frac{R_m(RM)R_m(M) + I_m(RM)I_m(M)}{R_m^2(M) + I_m^2(M)} \right] \qquad (8.35b)$$

$$M = \sqrt{\Omega} = \sqrt{\frac{\omega r_1^2}{\nu}}$$

$$m = \begin{cases} \frac{1}{2} & \text{parallel plate channel} \\ 0 & \text{circular tube} \end{cases}$$

The functions $R_m(x)$ and $I_m(x)$ for $m = \frac{1}{2}$ and 0 are defined as:

$$m = \frac{1}{2}: \qquad R_{1/2}(x) = \cos\left(\frac{x}{\sqrt{2}}\right)\cosh\left(\frac{x}{\sqrt{2}}\right),$$

$$I_{1/2}(x) = \sinh\left(\frac{x}{\sqrt{2}}\right)\sin\left(\frac{x}{\sqrt{2}}\right) \qquad (8.35c, d)$$

$$m = 0: \qquad R_0(x) = ber(x), \qquad I_0(x) = bei(x) \qquad (8.35e, f)$$

Example 8.6 Determine the velocity distribution in pulsating flow between two parallel plates after transients have passed.

Solution   The solution of this problem is immediately obtained from the general solution (8.35) by setting $m = \frac{1}{2}$. We find

$$U(R, \tau) = \tfrac{3}{2}\{(1 - R^2) + \gamma U_s(R, \tau)\} \qquad (8.36a)$$

where

$$U_s(R, \tau) = \frac{\sin(\Omega\tau)}{\Omega} + \frac{1}{\Omega(A^2 + B^2)}$$

$$\times \{[A\cos(RM_0)\cosh(RM_0) - B\sin(RM_0)\sinh(RM_0)]$$

$$\times \cos(\Omega\tau) - [B\cos(RM_0)\cosh(RM_0) + A\sin(RM_0)\sinh(RM_0)]$$

$$\times \sin(\Omega\tau)\} \qquad (8.36b)$$

$$M_0 = \frac{M}{\sqrt{2}} = \sqrt{\frac{\Omega}{2}} = \sqrt{\frac{\omega r_1^2}{2\nu}} \qquad (8.36c)$$

$$A = \sinh M_0 \sin M_0 \qquad (8.36d)$$

$$B = \cosh M_0 \cos M_0 \qquad (8.36e)$$

The solution (8.36) is the same as that given by Equation(6a) in Reference 5.

In Equation (8.36a) the term $\frac{3}{2}(1 - R^2)$ represents the steady-state component of the velocity, and the term $U_s(R, \tau)$ represents the fluctuating component of the velocity. In order to illustrate the physical significance of the fluctuating component, we present in Figures 8.2a, b a plot of the function $U_s(R, \tau)$ against R for $M_0 = 0.1$ and 2.0, respectively, for different values of $(\Omega\tau)$. For small values of $M_0 = 0.1$, the oscillation frequency is so slow that the profiles are essentially quasi-steady. For the higher value of $M_0 = 2.0$, corresponding to higher frequencies, the profiles become distorted and of much smaller amplitude. The curves are drawn for $(\Omega\tau)$ less than 180° since for values greater than this there is a negative symmetry, that is, $U_s(\Omega\tau + 180) = -U_s(\Omega\tau)$.

Example 8.7   Determine the velocity distribution in pulsating flow inside a circular tube after transients have passed.

Solution   The solution of this problem is immediately obtained from the solution (8.35) by setting $m = 0$. We find

$$U(R, \tau) = 2\{(1 - R^2) + \gamma U_s(R, \tau)\} \qquad (8.37a)$$

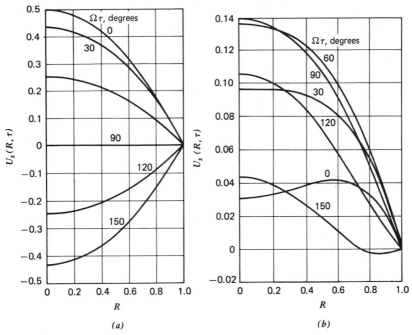

Figure 8.2 Fluctuating component $U_S(R, \tau)$ of velocity distribution (from R. Siegel and M. Perlmutter, Reference 5). (*a*) $M_0 = 0.1$. (*b*) $M_0 = 2.0$.

where

$$U_s(R, \tau) = \frac{2}{\Omega} \left[\sin(\Omega\tau) + \cos(\Omega\tau) \frac{ber(RM)bei(M) - bei(RM)ber(M)}{ber^2(M) + bei^2(M)} \right.$$

$$\left. - \sin(\Omega\tau) \frac{ber(RM)ber(M) + bei(RM)bei(M)}{ber^2(M) + bei^2(M)} \right] \quad (8.37b)$$

This solution is the same as that given by Equations (20) and (21) of Reference 4 if we consider his source term consists of the sum of a constant and a cosine term. However, the present solution by the principle of superposition is applicable to the case considered in Reference 4.

8.3 FORCED CONVECTION HEAT TRANSFER IN CONDUITS

Heat transfer by forced convection to flow inside conduits of uniform cross-section has numerous important applications in many branches of science and engineering. Such problems are generally referred to as the Graetz problem, since Graetz [6] appears to have been the first investigator to publish (in 1885) an analysis of heat transfer for laminar flow at the entrance region of a circular tube. Numerous extensions to the original Graetz problem have been reported in the literature. The solution of the problems involving *thermally developing and hydrodynamically developed* laminar or turbulent flow inside conduits of uniform cross-section can be obtained as a special case from the solutions of the one-dimensional problem (4.1) as now described.

We consider steady-state heat transfer in thermally developing, hydrodynamically developed forced flow inside a circular tube, a parallel plate channel, or an annular conduit, and make the following assumptions.

1. The fluid is incompressible with constant physical properties; consequently, the velocity problem is uncoupled from the temperature problem.
2. The velocity profile is fully developed.
3. The free convection effects are negligible; thus there are no components of velocity normal to the axis of the conduit.
4. The axial conduction of heat is negligible.
5. The energy generation rate is a function of the radial coordinate only.

The energy equation governing the temperature distribution in the flow applicable for both laminar and turbulent flow can be developed with the following considerations.

The linear diffusion law (1.24) takes the form

$$\mathbf{J} = -\left[k + \rho c_p \epsilon_h(r)\right]\frac{\partial T}{\partial r} \tag{8.38a}$$

where $\epsilon_h(r)$ is the *eddy diffusivity* of the substance and r is the distance normal to the tube wall.

The substantial derivative (1.11) reduces to

$$\frac{D}{Dt} \equiv v_z\frac{\partial}{\partial z} \equiv w(r)\frac{\partial}{\partial z} \tag{8.38b}$$

since $\partial/\partial t = 0$ for the steady flow and $v_x \equiv v_y = 0$ for the hydrodynamically developed flow.

Then the balance Equation (1.28) governing the temperature distribution $T(r, z)$ in the fluid is written as

$$w(r)\frac{\partial T(r, z)}{\partial z} = \frac{1}{r^n}\frac{\partial}{\partial r}\left[r^n(\alpha + \epsilon_h)\frac{\partial T(r, z)}{\partial r}\right] + \frac{g(r)}{\rho C_p}$$

$$\text{in } r_0 < r < r_1, \qquad z > 0 \quad (8.39a)$$

where z is the axial coordinate in the direction of flow, r is the coordinate normal to the z-axis, $w(r)$ is the flow velocity, α is the thermal diffusivity, ϵ_h is the eddy diffusivity for heat, $g(r)$ is the energy generation rate, ρ is the density, C_p is the specific heat, and

$$n = \begin{cases} 0 & \text{for parallel plate channel} \\ 1 & \text{for circular or annular tube} \end{cases}$$

The boundary conditions for Equation (8.39a) at $r = r_0$ and $r = r_1$ can be taken as any combination of the boundary conditions of the first, second, and third kind. That is, the boundary condition at $r = r_0$ can be taken as any one of the following three boundary conditions

$$T(r_0, z) = T_0(z), \qquad -k\frac{\partial T(r_0, z)}{\partial r} = q_0(z)$$

or

$$T(r_0, z) - \frac{k}{u_0}\frac{\partial T(r_0, z)}{\partial r} = T_{f0}(z) \qquad \text{(8.39b, c, d)}$$

and similarly at $r = r_1$, any one of the following three boundary conditions

$$T(r_1, z) = T_1(z), \qquad k\frac{\partial T(r_1, z)}{\partial r} = q_1(z)$$

or

$$T(r_1, z) + \frac{k}{u_1}\frac{\partial T(r_1, z)}{\partial r} = T_{f1}(z) \qquad \text{(8.39e, f, g)}$$

where $T_j(z)$, $q_j(z)$, and u_j with $j = 0, 1$ are the temperature, heat flux, and overall heat transfer coefficient at $r = r_j$, respectively; k is the thermal conductivity of fluid; $T_{fj}(z)$ with $j = 0, 1$ is the temperature of the environment.

Finally, the boundary condition at the origin of the axial coordinate $z = 0$, where the heating or cooling is started, is given as

$$T(r, 0) = f(r) \qquad \text{(8.39h)}$$

Once the temperature distribution $T(r, z)$ in the flow is determined from the solution of the system (8.39), the heat transfer coefficient $h_j(z)$ at the boundary surfaces $r = r_j$ with $j = 0, 1$ is evaluated from

$$h_j(z) = (-1)^{j+1}\frac{k[\partial T(r_j, z)/\partial r]}{T(r_j, z) - T_{av}(z)}, \qquad j = 0, 1 \qquad \text{(8.40a, b)}$$

where $T_{av}(z)$ is the average temperature of the flow, defined as

$$T_{av}(z) = \frac{\int_{r_0}^{r_1} r^n w(r) T(r, z)\, dr}{\int_{r_0}^{r_1} r^n w(r)\, dr} \tag{8.41a}$$

The average temperature $T_{av}(z)$ is also referred to as the *mixing cup temperature* or the *fluid bulk mean temperature*.

Equation (8.41a) can be written in the alternative form by utilizing the definition of the average velocity w_{av} given by Equation (8.5a) as

$$T_{av}(z) = \frac{n + 1}{r_1^{n+1} - r_0^{n+1}} \int_{r_0}^{r_1} r^n \frac{w(r)}{w_{av}} T(r, z)\, dr \tag{8.41b}$$

The Dimensionless Equations

The problem defined by Equations (8.39)–(8.41) can be expressed in the dimensionless form by introducing the following dimensionless variables

$R = \dfrac{r}{\ell_0}$, dimensionless radial coordinate

$Z = \dfrac{\alpha z}{w_{av}\ell_0^2}$, dimensionless axial coordinate where ℓ_0 is a reference length

$\theta(R, Z) = \dfrac{T(r, z) - T^*}{\Delta T}$, dimensionless temperature

$\theta_j(Z) = \dfrac{T_j(z) - T^*}{\Delta T}$, dimensionless boundary surface temperature

$\theta_{fj}(Z) = \dfrac{T_{fj}(z) - T^*}{\Delta T}$, dimensionless environment temperature

$F(R) = \dfrac{f(r) - T^*}{\Delta T}$, dimensionless temperature at $Z = 0$, where T^* and ΔT are a reference temperature and a reference temperature difference, respectively

$U(R) = \dfrac{w(r)}{w_{av}}$, dimensionless flow velocity

$E_h(R) = 1 + Pr\dfrac{\epsilon_h}{\nu}$, where Pr is the Prandtl number

$G(R, Z) = \dfrac{g(r, z)\ell_0^2}{k\,\Delta T}$, dimensionless energy generation rate

$Bi_j = \dfrac{U_j\ell_0}{k}$, $(j = 0, 1)$, Biot numbers

$Q_j = \dfrac{q_j(z)\ell_0}{k\,\Delta T}$, $(j = 0, 1)$, dimensionless heat flux at the boundary surface $R = R_k$

$Nu_j = \dfrac{h_j\ell_N}{k}$, $(j = 0, 1)$, Nusselt number, where ℓ_N is a reference length for the definition of the Nusselt number

When these dimensionless quantities are introduced into Equations (8.39) we obtain

$$R^n U(R) \frac{\partial \theta(R, Z)}{\partial Z} = \frac{\partial}{\partial R} \left[R^n E_h(R) \frac{\partial \theta(R, Z)}{\partial R} \right] + R^n G(R)$$

$$\text{in } R_0 < R < R_1 \quad (8.42\text{a})$$

$$\alpha_k \theta(R_k, Z) + (-1)^k \beta_k \frac{\partial \theta(R_k, Z)}{\partial R} = \phi_k(Z), \quad k = 0, 1$$

$$(8.42\text{b, c})$$

$$\theta(R, 0) = F(R) \qquad (8.42\text{d})$$

The heat transfer coefficients defined by Equations (8.40) and (8.41) become

$$Nu_k(Z) = (-1)^{k+1} \frac{\ell_N}{\ell_0} \frac{\partial \theta(R_k, Z)/\partial R}{\theta(R_k, Z) - \theta_{av}(Z)}, \quad (k = 0, 1)$$

$$(8.43\text{a, b})$$

where the average temperature $\theta_{av}(z)$ is

$$\theta_{av}(Z) = \frac{n+1}{R_1^{n+1} - R_0^{n+1}} \int_{R_0}^{R_1} R^n U(R) \theta(R, Z) \, dR \qquad (8.43\text{c})$$

The average value of the Nusselt number Nu_{av} over the thermal entrance region $0 \leq Z \leq Z_e$ is defined as

$$Nu_{av} = \frac{1}{Z_e} \int_0^{Z_e} Nu(Z) \, dZ \qquad (8.43\text{d})$$

Clearly, the generalized boundary conditions given by Equations (8.42b, c) can yield any of the boundary conditions of the first, second, or third kind if the coefficients α_k, β_k, and the function $\phi_k(z)$ are chosen as given in the following.

First kind: $\alpha_k = 1,$ $\beta_k = 0,$ $\phi_k(Z) = \theta_k(Z)$ (8.44a)

Second kind: $\alpha_k = 0,$ $\beta_k = 1,$ $\phi_k(Z) = Q_k(Z)$ (8.44b)

Third kind: $\alpha_k = 1,$ $\beta_k = \dfrac{1}{Bi_k},$ $\phi_k(Z) = \theta_{fk}(Z)$ (8.44c)

where $\theta_k(Z)$, $Q_k(Z)$, and $\theta_{fk}(Z)$ are as defined previously.

Solution of the Forced Convection Problem

The forced convection problem defined by Equations (8.42) is a special case of the one-dimensional problem (4.1) as follows:

$$x = R, \quad t = Z, \quad T(x, t) = \theta(R, Z), \quad w(x) = R^n U(R),$$

$$k(x) = R^n E_h(R), \quad d(x) = 0, \quad P(x, t) = R^n G(R), \quad x_0 = R_0,$$

$$x_1 = R_1, \quad \phi(x_k, t) = \phi_k(Z), \quad f(x) = F(R)$$

The general solution of the problem (4.1) is given by Equation (4.2) in which the eigenfunctions and eigenvalues are the solution of the eigenvalue problem (4.5). For simple cases, analytical solutions are possible for the eigenvalue problem, but for more complicated situations the eigenvalue problems should be solved numerically. To perform the numerical calculations, it is convenient to transform the eigenvalue problem by means of the transformation discussed in Chapter 3 into a form more suitable for numerical integration. Therefore, for generality, we present the solution of problem (8.42) in terms of the transformed eigenfunctions associated with the algorithm discussed in Chapter 3 [see Equations (3.72)–(3.76)].

$$\theta(R, Z) = \sum_{i=1}^{\infty} \frac{1}{N_i} e^{-\mu_i^2 Z} y_2(\mu_i, R) H_i(Z) \tag{8.45a}$$

where

$$N_i = \frac{1}{2\mu_i} [y_1(\mu_i, R_1) y_4(\mu_i, R_1) - y_2(\mu_i, R_1) y_3(\mu_i, R_1)] \tag{8.45b}$$

$$H_i(Z) = \int_{R_0}^{R_1} R^n U(R) y_2(\mu_i, R) F(R) \, dR + \int_0^Z e^{\mu_i^2 Z'}$$

$$\times \left[\phi_0(Z') \Omega_i(R_0) + \phi_1(Z') \Omega_i(R_1) + \int_{R_0}^{R_1} R^n G(R) y_2(\mu_i, R) \, dR \right] dZ' \tag{8.45c}$$

$$\Omega_i(R_1) = \frac{y_2(\mu_i, R_1) - y_1(\mu_i, R_1)}{\alpha_1 + \beta_1} = \frac{y_2(\mu_i, R_1)}{\beta_1} = -\frac{y_1(\mu_i, R_1)}{\alpha_1} \tag{8.45d}$$

and the function $\Omega_i(R_0)$, included in Equation (8.45c) for the sake of completeness, since for the present problem it is equal to unity, that is,

$$\Omega_i(R_0) = \frac{y_2(\mu_i, R_0) + y_1(\mu_i, R_0)}{\alpha_0 + \beta_0} = 1 \tag{8.45e}$$

This is apparent in view of Equations (3.74a, b).

If $F(R) = $ constant or $G(R) = U(R)$, then the integrals in Equation (8.45c) over the variable R can be evaluated by utilizing the result

$$\int_{R_0}^{R_1} R^n U(R) y_2(\mu_i, R) \, dR = \frac{1}{\mu_i^2} [y_1(\mu_i, R_0) - y_1(\mu_i, R_1)] \quad (8.45\text{f})$$

We note that the result given by Equation (8.45f) is obtainable from Equation (4.5d) for $f(x) = 1$, $w(x) = R^n U(R)$, $\psi_i(x) = y_2(\mu_i, R)$, and $k(x) \, d\psi(\mu_i, x)/dx = y_1(\mu_i, R)$.

The eigenvalues μ_i can be evaluated by the numerical integration of the following equations which are obtained from Equations (3.73a) and (3.73b) for $d(x) = 0$.

$$\frac{dy_1(\mu, R)}{dR} = -\mu^2 R^n U(R) y_2(\mu, R) \quad (8.46\text{a})$$

$$\frac{dy_2(\mu, R)}{dR} = \frac{1}{R^n E_h(R)} y_1(\mu, R) \quad (8.46\text{b})$$

subject to the conditions [see Equations (3.74a, b)]

$$y_1(\mu, R_0) = \alpha_0, \qquad y_2(\mu, R_0) = \beta_0 \quad (8.46\text{c, d})$$

and with the requirement that

$$\alpha_1 y_2(\mu, R_1) + \beta_1 y_1(\mu, R_1) = 0 \quad (8.46\text{e})$$

Equation (8.46e) is the eigencondition from which eigenvalues μ_i are determined. The conditions (8.46c, d) imply that the result given by Equation (8.45e) is valid.

In Equation (8.46a) if $U(R)$ is represented by $U(R) = CU^*(R)$, where C is a constant [i.e., $U(R) = (1 - R^2)(n + 3)/2 \equiv CU^*(R)$], then this constant may be combined with μ_i^2 to define a new eigenvalue λ_i^2 as $\lambda_i^2 \equiv C\mu_i^2$. Then one may apparently have a different eigenvalue λ_i^2 for the same problem.

Once the eigenvalues μ_i are available, the functions $y_1(\mu_i, R)$ and $y_2(\mu_i, R)$ are obtained from Equations (8.46a–d). The functions $y_3(\mu_i, R)$ and $y_4(\mu_i, R)$ are calculated by direct numerical integration of the following equations [see equations (3.76a, b)]

$$\frac{dy_3(\mu, R)}{dR} = -\mu R^n U(R)[\mu y_4(\mu, R) + 2 y_2(\mu, R)] \quad (8.47\text{a})$$

$$\frac{dy_4(\mu, R)}{dR} = \frac{1}{R^n E_h(R)} y_3(\mu, R) \quad (8.47\text{b})$$

subject to the conditions [see Equations (3.76c, d)]

$$y_3(\mu, R_0) = 0, \qquad y_4(\mu, R_0) = 0 \qquad (8.47c, d)$$

Then the solution for the temperature distribution $\theta(R, Z)$ can be calculated from Equation (8.45a), since N_i, $H_i(Z)$, and $\Omega_i(R_1)$ can be determined from Equations (8.45b, c, d), respectively.

Finally, by substituting $\theta(R, Z)$ given by Equation (8.45a) into Equations (8.43) and utilizing the results in Equations (8.45f) and (8.46b), we obtain the Nusselt numbers as

$$\mathrm{Nu}_k(Z) = \frac{(-1)^{k+1}\left(\dfrac{\ell_N}{\ell_0 R_k^n E_h(R)}\right) \sum\limits_{i=1}^{\infty} (1/N_i)e^{-\mu_i^2 Z}y_1(\mu_i, R_k)H_i(Z)}{K},$$

$$k = 0, 1 \qquad (8.48a, b)$$

where

$$K = \sum_{i=1}^{\infty} (1/N_i)e^{-\mu_i^2 Z}H_i(Z)$$

$$\times \left\{ y_2(\mu_i, R_k) + (n + 1)[y_1(\mu_i, R_1) - y_1(\mu_i, R_0)]/[\mu_i^2(R_1^{n+1} - R_0^{n+1})] \right\}$$

The Limiting Nusselt Number

We now consider a situation for which $\phi_0(Z) = \phi_1(Z) = G(R) = 0$. In the regions far from the inlet (i.e., $Z \to \infty$), in Equations (8.48) only the first term in the series need be considered. Hence the limiting Nusselt number is given by

$$\mathrm{Nu}_k(\infty) = \frac{(-1)^{k+1}\left(\dfrac{\ell_N}{\ell_0 R_k^n E_h(R)}\right)y_1(\mu_1, R_k)}{K}, \qquad k = 0, 1$$

$$(8.49a, b)$$

where

$$K = y_2(\mu_1, R_k) + (n + 1)[y_1(\mu_1, R_1) - y_1(\mu_1, R_0)]/[\mu_1^2(R_1^{n+1} - R_0^{n+1})]$$

where μ_1 is the first eigenvalue. This result implies that the limiting Nusselt numbers are invariant, that is, independent of Z, and the flow in regions away from the inlet is designated as *thermally fully developed*. The *thermal entry length* L_{th} is defined, somewhat arbitrarily, as the duct length required to achieve a value of $\text{Nu}(Z)$ equal to $1.05\text{Nu}(\infty)$.

Both Boundary Conditions are Second Kind

When the boundary conditions at $R = R_0$ and $R = R_1$ are both of the second kind, we have $\alpha_0 = \alpha_1 = 0$ and $\beta_0 = \beta_1 = 1$. For this special case the solution of problem (8.42) is obtained from Equations (4.6) as

$$\theta(R, Z) = \theta_{\text{av}}(Z) + \sum_{i=1}^{\infty} \frac{1}{N_i} e^{-\mu_i^2 Z} y_2(\mu_i, R) H_i(Z) \qquad (8.50\text{a})$$

where

$$\theta_{\text{av}}(Z) = \frac{n+1}{R_1^{n+1} - R_0^{n+1}} \left\{ \int_{R_0}^{R_1} R^n U(R) F(R) \, dR \right.$$

$$\left. + \int_0^Z [\phi_0(Z') + \phi_1(Z')] \, dZ' + Z \int_{R_0}^{R_1} R^n G(R) \, dR \right\}$$

$$(8.50\text{b})$$

We recall the definition $y_1(\mu_i, R) \equiv k(x) \, d\psi(\mu_i, x)/dx$ [see Equation (3.72a)] and conclude that for the boundary condition of the second kind at both boundaries we must have $y_1(\mu_i, R_0) = y_1(\mu_i, R_1) = 0$. For this special case, the functions N_i and $H_i(Z)$ are given by Equations (8.45b) and (8.45c), respectively, but the function Ω_i and the integral $\int_{R_0}^{R_1} R^n U(R) y_2(\mu_i, R) \, dR$ are determined according to Equations (8.45d) and (8.45f), respectively.

The Nusselt numbers for this case are evaluated according to Equations (8.43). Introducing the solution (8.50a) into the denominator of Equations (8.43) and noting that $\partial\theta(R_k, Z)/\partial R = \phi_k(Z)$ by the boundary conditions for the problem, the Nusselt numbers become

$$\text{Nu}_k(Z) = (-1)^{k+1} \frac{\ell_N}{\ell_0} \frac{\phi_k(Z)}{\displaystyle\sum_{i=1}^{\infty} (1/N_i) e^{-\mu_i^2 Z} y_2(\mu_i, R_k) H_i(Z)} \qquad k = 0, 1$$

$$(8.51\text{a, b})$$

Splitting Up the Solution

The expressions (8.48) and (8.51) for the Nusselt number, although representing the exact solutions for the problem, are not in a convenient form for practical purposes. When the boundary condition functions $\phi_k(Z)$ can be expressed as exponentials or q-order polynomials of Z, it is better to use solutions (4.8) for the determination of the temperature distribution. Then the Nusselt number determined from such a solution will be in a more convenient form for computational purposes. To illustrate this matter we consider the following special case defined with the parameters chosen as:

$$\alpha_0 = 0, \ \beta_0 = 1, \ \phi_0(Z) = 0, \ R_0 = 0, \ R_1 = 1, \ \ell_0 = r_1, \ \ell_N = 2r_1$$

which corresponds to a circular tube or a parallel plate channel. Then the mathematical formulation of the heat transfer problem given by Equations (8.42) reduces to

$$R^n U(R)\frac{\partial \theta(R,Z)}{\partial Z} = \frac{\partial}{\partial R}\left[R^n E_h(R)\frac{\partial \theta(R,Z)}{\partial R}\right] + R^n G(R) \quad \text{in } 0 < R < 1$$

$$(8.52a)$$

$$\frac{\partial \theta(0,Z)}{\partial R} = 0 \tag{8.52b}$$

$$\alpha \theta(1,Z) + \beta\frac{\partial \theta(1,Z)}{\partial R} = \phi(Z) \tag{8.52c}$$

$$\theta(R,0) = F(R) \tag{8.52d}$$

Here we assume that the function $\phi(Z)$ can be represented as

$$\phi(Z) = \phi_e e^{-d_\phi Z} + \sum_{j=0}^{n_\phi} \phi_j Z^j \tag{8.52e}$$

Then the solution of problem (8.52) can be split up into the solution of a set of simpler problems as given by Equation (4.8a) for $\alpha \neq 0$, and given by Equation (4.8b) for $\alpha = 0$, $\beta = 1$, that is,

$$\theta(R,Z) = \theta_\phi(R)e^{-d_\phi Z} + \sum_{j=0}^{n_\phi} \theta_{\phi j}(R)Z^j + \theta_{g0}(R) + \theta_t(R,Z), \qquad \alpha \neq 0$$

$$(8.53a)$$

and

$$\theta(R,Z) = \theta_{av}(Z) + \theta_\phi(R)e^{-d_\phi Z} + \sum_{j=0}^{n_\phi} \theta_{\phi j}(R)Z^j + \theta_{g0}(R) + \theta_t(R,Z),$$

$$\alpha = 0, \qquad \beta = 1 \quad (8.53b)$$

where

$$\theta_{\mathrm{av}}(Z) = (n + 1)\left[\int_0^1 R^n U(R) F(R)\, dR + \int_0^Z \phi(Z')\, dZ' + Z\int_0^1 R^n G(R)\, dR\right]$$

(8.53c)

Here the functions $\theta_\phi(R)$ and $\theta_{\phi j}(R)$ are associated with the nonhomogeneities ϕ_e and ϕ_j, respectively, in the boundary condition function $\phi(Z)$ given by Equation (8.52e); the function $\theta_{g0}(R)$ is associated with the generation term $R^n G(R)$ of Equation (8.52a); the function $\theta_t(R, Z)$ is for the effects of the nonhomogeneous term $F(R)$ of the condition at $Z = 0$.

The function $\theta_\phi(R)$ is the solution of the one-dimensional steady-state problem defined by Equation (4.9), the functions $\theta_{\phi j}(R)$ and $\theta_{g0}(R)$ satisfy the problem (4.23), and the function $\theta_t(R, Z)$ satisfies the one-dimensional, time dependent problem (4.34). In the following we summarize the resulting solutions for the cases of both $\alpha \neq 0$ and $\alpha = 0$, $\beta = 1$.

Solution for $\theta_\phi(R)$

The function $\theta_\phi(R)$ is determined from Equation (4.12) for $\alpha \neq 0$ and from Equation (4.14) for $\alpha = 0$, $\beta = 1$ by setting

$$\phi_e(x_0) = 0, \qquad \phi_e(x_1) = \phi_e, \qquad \alpha_0 = 0, \qquad \beta_0 = 1, \qquad \alpha_1 = \alpha, \qquad \beta_1 = \beta$$

and utilizing the functions $y_1(\mu, R)$ and $y_2(\mu, R)$, originally defined by Equations (3.72a, b). After some manipulation, Equation (4.12) yields

$$\theta_\phi(R) = \phi_e \frac{y_2\left(\sqrt{d_\phi}, R\right)}{\alpha y_2\left(\sqrt{d_\phi}, 1\right) + \beta y_1\left(\sqrt{d_\phi}, 1\right)}, \qquad \alpha \neq 0 \qquad (8.54a)$$

and Equation (4.14) gives

$$\theta_\phi(R) = \phi_e \left[\frac{n + 1}{d_\phi} + \frac{y_2\left(\sqrt{d_\phi}, R\right)}{y_1\left(\sqrt{d_\phi}, 1\right)}\right], \qquad \alpha = 0, \qquad \beta = 1 \quad (8.54b)$$

In Equation (8.54b) the term $(n + 1)$ resulted from the integration $\int_0^1 R^n U(R)\, dR = (n + 1)^{-1}$, which was performed by utilizing the results given by Equations (8.5b) and (8.7a). The functions $y_1(\sqrt{d_\phi}, R)$ and $y_2(\sqrt{d_\phi}, R)$ are the solutions of the Equations (8.46a, b) subject to the boundary conditions (8.46c, d); that is, $y_1(\mu, 1) = \alpha_0 = 0$ and $y_2(\mu, 1) = \beta_0 = 1$, in which μ is replaced by $\sqrt{d_\phi}$.

Solution for $\theta_{\phi j}(R)$

The function $\theta_{\phi j}(R)$ is determined from Equation (4.27) for $\alpha \neq 0$ and from Equation (4.33) for $\alpha = 0$, $\beta = 1$ by making the following substitutions

$$\alpha_0 = 0, \quad \beta_0 = 1, \quad \alpha_1 = \alpha, \quad \beta_1 = \beta, \quad x = R, \quad x_0 = 0,$$

$$x_1 = 1, \quad \phi_j(x_0) = 0, \quad \phi_j(x_1) = \phi_j, \quad P_j(x) = 0,$$

$$w(x) = R^n U(R), \quad k(x) = R^n E_h(R)$$

Then Equation (4.27) yields

$$\theta_{\phi j}(R) = \frac{\phi_j}{\alpha} - (1 + j)\left[\frac{\beta}{\alpha} \int_0^1 R^n U(R)\theta_{\phi, j+1}(R)\, dR \right.$$

$$\left. + \int_R^1 \frac{1}{R''^n E_h(R')} \int_0^{R'} R'''^n U(R'')\theta_{\phi, j+1}(R'')\, dR''\, dR' \right],$$

$$\alpha \neq 0 \quad (8.55\text{a})$$

and Equation (4.33) gives

$$\theta_{\phi j}(R) = \phi_j \left\{ \int_0^1 \frac{\left[(n+1)\int_0^R R'''^n U(R')\, dR'\right]^2}{R^n E_h(R)}\, dR \right.$$

$$\left. - \int_R^1 \frac{(n+1)\int_0^{R'} R'''^n U(R'')\, dR''}{R''^n E_h(R')}\, dR' \right\}$$

$$-(j+1)\int_R^1 \frac{1}{R''^n E_h(R')} \int_0^{R'} R'''^n U(R'')\theta_{\phi, j+1}(R'')\, dR''\, dR' + (j+1)$$

$$\times \int_0^1 \frac{(n+1)\int_0^R R'''^n U(R')\, dR'}{R^n E_h(R)} \int_0^R R'''^n U(R')\theta_{\phi, j+1}(R')\, dR'\, dR,$$

$$\alpha = 0, \quad \beta = 1 \quad (8.55\text{b})$$

where

$$\theta_{\phi, n_\phi+1}(R) = 0, \quad j = n_\phi, n_\phi - 1, \ldots, 2, 1, 0.$$

Here we utilized the integral $\int_0^1 R^n U(R) \, dR = (n+1)^{-1}$ to obtain the terms $(n+1)$ appearing in Equation (8.55b).

Solution for $\theta_{g0}(R)$

The function $\theta_{g0}(R)$ is also determined from Equation (4.27) for $\alpha = 0$ and from Equation (4.33) for $\alpha = 0$, $\beta = 1$ by making substitutions similar to those used in the derivation of $\theta_{\phi j}(R)$, except for this case we have

$$\phi_j(x_1) = 0, \qquad P_j(x) = R^n G(R)$$

Here the functional form of $P_j(x)$ implies that $j = 0$, hence $T_{j+1} = T_1 = 0$.
Then Equation (4.27) yields

$$\theta_{g0}(R) = \frac{\beta}{\alpha} \int_0^1 R^n G(R) \, dR + \int_R^1 \frac{1}{R'' E_h(R')} \int_0^{R'} R''' G(R'') \, dR'' \, dR',$$

$$\alpha \neq 0 \quad (8.56a)$$

and Equation (4.33) gives

$$\theta_{g0}(R) = \left[\int_0^1 R^n G(R) \, dR \right] \left\{ \int_0^1 \frac{\left[(n+1) \int_0^R R''' U(R') \, dR' \right]^2}{R^n E_h(R)} \, dR \right.$$

$$\left. - \int_R^1 \frac{(n+1) \int_0^{R'} R''' U(R'') \, dR''}{R' E_h(R')} \, dR' \right\}$$

$$+ \int_R^1 \frac{1}{R'' E_h(R')} \int_0^{R'} R''' G(R'') \, dR'' \, dR'$$

$$- \int_0^1 \frac{(n+1) \int_0^R R''' U(R') \, dR'}{R^n E_h(R)} \int_0^R R''' G(R') \, dR' \, dR,$$

$$\alpha = 0, \beta = 1 \quad (8.56b)$$

Solution for $\theta_t(R, Z)$

The function $\theta_t(R, Z)$ is determined from Equation (4.35) for $\alpha \neq 0$ by making the following substitutions:

$$\phi_e(x_0) = 0, \qquad \phi_e(x_1) = \phi_e, \qquad \phi_j(x_0) = 0, \qquad \phi_j(x_1) = \phi_j, \qquad P_e(x) = 0$$

$$P_j(x) = P_0(x) = R^n G(R), \qquad f(x) = F(R), \qquad q = n_\phi,$$

$$\psi_i(x) = y_2(\mu_i, R)$$

$$\Omega_i(x_1) = -\frac{y_1(\mu_i, R)}{\alpha_1} \qquad \text{by Equation (8.45d)}$$

$$\frac{1}{N_i} = \frac{2\mu_i}{y_4(\mu_i, 1) + (\beta/\alpha)y_3(\mu_i, 1)} \frac{1}{y_1(\mu_i, 1)}$$

by Equations (8.45b) and (8.46e)

Then Equation (4.35) yields

$$\theta_t(R, Z) = 2\sum_{i=1}^{\infty} \frac{\mu_i e^{-\mu_i^2 Z}}{y_4(\mu_i, 1) + (\beta/\alpha)y_3(\mu_i, 1)} \frac{y_2(\mu_i, R)}{y_1(\mu_i, 1)} H_i, \qquad \alpha \neq 0$$

(8.57a)

where

$$H_i = \int_0^1 R^n U(R) y_2(\mu_i, R) F(R)\, dR + \frac{\phi_e}{\alpha} \frac{y_1(\mu_i, 1)}{\mu_i^2 - d_\phi}$$

$$+ \sum_{j=0}^{n_\phi} (-1)^j \frac{\phi_j}{\alpha} \frac{j!}{\mu_i^{2(j+1)}} y_1(\mu_i, 1) - \frac{1}{\mu_i^2} \int_0^1 R^n G(R) y_2(\mu_i, R)\, dR$$

(8.57b)

For the case of $\alpha = 0$, $\beta = 1$, similar substitutions are made, but with the following exceptions

$$\Omega_i(x_1) = y_2(\mu_i, 1)$$

$$y_1(\mu_i, 1) = -k\frac{d\psi_1(1)}{dx} = 0$$

$$\frac{1}{N_i} = -\frac{2\mu_i}{y_3(\mu_i, 1)y_2(\mu_i, 1)}$$

Then Equation (4.35) yields

$$\theta_t(R, Z) = -2 \sum_{i=1}^{\infty} \frac{\mu_i e^{-\mu_i^2 Z}}{y_3(\mu_i, 1) y_2(\mu_i, 1)} y_2(\mu_i, R) H_i, \qquad \alpha = 0, \qquad \beta = 1$$

$$(8.57c)$$

where

$$H_i = \int_0^1 R^n U(R) y_2(\mu_i, R) F(R) \, dR - \phi_e \frac{y_2(\mu_i, 1)}{\mu_i^2 - d_\phi}$$

$$- \sum_{j=0}^{n_\phi} (-1)^j \phi_j \frac{j!}{\mu_i^{2(j+1)}} y_2(\mu_i, 1) - \frac{1}{\mu_i^2} \int_0^1 R^n G(R) y_2(\mu_i, R) \, dR$$

$$(8.57d)$$

Splitting Up the Nusselt Number

The Nusselt number for the heat transfer problem (8.52) can also be represented by the superposition of the results obtained from the solutions of the simpler problems defined by Equations (8.53). We consider the cases for $\alpha \neq 0$ and $\alpha = 0$, $\beta = 1$.

The Case $\alpha \neq 0$ The solutions (8.54a)–(8.57a) for $\alpha \neq 0$ are introduced into the definition of the Nusselt number given by Equations (8.43) for $0 \leq R \leq 1$, $k = 1$, and $\ell_N/\ell_0 = 2r_1/r_1 = 2$. We obtain

$$Nu(Z) = 2 \left[\frac{d\theta_\phi(1)}{dR} e^{-d_\phi Z} + \sum_{j=0}^{n_\phi} \frac{d\theta_{\phi j}(1)}{dR} Z^j + \frac{d\theta_{g0}(1)}{dR} + \frac{\partial \theta_t(1, Z)}{\partial R} \right]$$

$$\times \left\{ \left[\theta_\phi(1) - \theta_{\phi, av} \right] e^{-d_\phi Z} + \sum_{j=0}^{n_\phi} \left[\theta_{\phi j}(1) - \theta_{\phi j, av} \right] Z^j + \left[\theta_{g0}(1) - \theta_{g0, av} \right] \right.$$

$$\left. + \left[\theta_t(1, Z) - \theta_{t, av}(Z) \right] \right\}^{-1} \qquad \text{for } \alpha \neq 0 \qquad (8.58a)$$

where various terms are determined as:

$$\frac{d\theta_\phi(1)}{dR} = \phi_e \frac{y_1\left(\sqrt{d_\phi}, 1\right)}{\alpha y_2\left(\sqrt{d_\phi}, 1\right) + \beta y_1\left(\sqrt{d_\phi}, 1\right)}$$

$$(8.58b)$$

$$\theta_\phi(1) - \theta_{\phi, av} = \phi_e \frac{y_2\left(\sqrt{d_\phi}, 1\right) + \left[(n+1)/d_\phi\right] y_1\left(\sqrt{d_\phi}, 1\right)}{\alpha y_2\left(\sqrt{d_\phi}, 1\right) + \beta y_1\left(\sqrt{d_\phi}, 1\right)}$$

$$(8.58c)$$

Here we utilized the expression (8.46b) to obtain Equation (8.58b), and made use of the relations (8.43c) and (8.45f) to get the result (8.58c).

$$\frac{d\theta_{\phi j}(1)}{dR} = (j + 1)\int_0^1 R^n U(R)\theta_{\phi, j+1}(R)\, dR \qquad (8.58\text{d})$$

$$\theta_{\phi j}(1) - \theta_{\phi j, \text{av}} = (j + 1)(n + 1)$$

$$\times \int_0^1 \frac{\int_0^R R'''U(R')\, dR' \int_0^R R'''U(R')\theta_{\phi, j+1}(R')\, dR'}{R^n E_h(R)}\, dR$$

$$(8.58\text{e})$$

Here we applied the integration by parts to obtain the contribution of $\theta_{\phi j, \text{av}}$ in the result (8.58e).

$$\frac{d\theta_{g0}(1)}{dR} = -\int_0^1 R^n G(R)\, dR \qquad (8.58\text{f})$$

$$\theta_{g0}(1) - \theta_{g0, \text{av}} = -(n + 1)\int_0^1 \frac{\int_0^R R'''U(R')\, dR' \int_0^R R'''G(R')\, dR'}{R^n E_h(R)}\, dR$$

$$(8.58\text{g})$$

The derivation of Equation (8.58g) follows a similar procedure to that of Equation (8.58e).

$$\frac{\partial \theta_t(1, Z)}{\partial R} = 2\sum_{i=1}^{\infty} \frac{\mu_i e^{-\mu_i^2 Z}}{y_4(\mu_i, 1) + (\beta/\alpha)y_3(\mu_i, 1)} H_i \qquad (8.58\text{h})$$

$$\theta_t(1, Z) - \theta_{t, \text{av}}(Z) = 2\sum_{i=1}^{\infty} \frac{e^{-\mu_i^2 Z}}{y_4(\mu_i, 1) + (\beta/\alpha)y_3(\mu_i, 1)}\left(\frac{n + 1}{\mu_i} - \mu_i\frac{\beta}{\alpha}\right)H_i$$

$$(8.58\text{i})$$

where H_i is defined by Equation (8.57b). We utilized the relation (8.46b) in performing the differentiation in Equation (8.58h), and made use of the relations (8.46c, e) to present the result in the form given by Equation (8.58i).

The Case $\alpha = 0$, $\beta = 1$ The Nusselt number defined by Equation (8.43) is written for this case by setting $k = 1$, $\ell_N/\ell_0 = 2r_1/r_1 = 2$, $R_1 = 1$, and $\partial\theta(1, Z)/\partial R = \phi(Z)$ as

$$\text{Nu}(Z) = 2\frac{\phi(Z)}{\theta(1, Z) - \theta_{\text{av}}(Z)} \qquad (8.59)$$

Introducing Equations (8.52e) and (8.53b) into (8.59) we obtain

$$\text{Nu}(Z) = 2\frac{\phi_e e^{-d_\phi Z} + \sum\limits_{j=0}^{n_\phi} \phi_j Z^j}{\theta_\phi(1)e^{-d_\phi Z} + \sum\limits_{j=0}^{n_\phi} \theta_{\phi j}(1)Z^j + \theta_{g0}(1) + \theta_t(1, Z)} \tag{8.60a}$$

where various temperatures are determined from the results given by Equations (8.54)–(8.57) as

$$\theta_\phi(1) = \phi_e \left[\frac{n+1}{d_\phi} + \frac{y_2\left(\sqrt{d_\phi}, 1\right)}{y_1\left(\sqrt{d_\phi}, 1\right)} \right] \tag{8.60b}$$

$$\theta_{\phi j}(1) = \phi_j \int_0^1 \frac{\left[(n+1)\int_0^R R'^n U(R')\,dR'\right]^2}{R^n E_h(R)}\,dR$$

$$+(j+1)\int_0^1 \frac{(n+1)\int_0^R R'^n U(R')\,dR'}{R^n E_h(R)}$$

$$\times \int_0^R R'^n U(R')\theta_{\phi, j+1}(R')\,dR'\,dR \tag{8.60c}$$

$$\theta_{g0}(1) = \int_0^1 R^n G(R)\,dR \int_0^1 \frac{\left[(n+1)\int_0^R R'^n U(R')\,dR'\right]^2}{R^n E_h(R)}\,dR$$

$$+\int_R^1 \frac{1}{R'^n E_h(R')} \int_0^{R'} R^n G(R)\,dR\,dR'$$

$$-\int_0^1 \frac{(n+1)\int_0^R R'^n U(R')\,dR'}{R^n E_h(R)} \int_0^R R'^n G(R')\,dR'\,dR \tag{8.60d}$$

$$\theta_t(1, Z) = -2\sum_{i=1}^\infty \frac{\mu_i e^{-\mu_i^2 Z}}{y_3(\mu_i, 1)} H_i \tag{8.60e}$$

where H_i is defined by Equation (8.57d).

The Limiting Nusselt Number

The limiting Nusselt number $Nu(\infty)$ is determined from Equation (8.58a) for $\alpha \neq 0$ and Equation (8.60a) for $\alpha = 0$, $\beta = 1$ by letting $Z \to \infty$. The resulting expressions for some special cases are summarized in the following.

The Case $\alpha \neq 0$ From Equation (8.58a) for $n_\phi = 0$, $G(R) = 0$, and $\mu_1^2 < d_\phi$ where μ_1 is the first eigenvalue, we obtain:

$$Nu(\infty) = \frac{2}{[(n+1)/\mu_1^2] - \beta/\alpha} \tag{8.61a}$$

For $n_\phi = 0$, $G(R) = 0$, and $\mu_1^2 > d_\phi$ we find:

$$Nu(\infty) = \frac{2 y_1\left(\sqrt{d_\phi}, 1\right)}{y_2\left(\sqrt{d_\phi}, 1\right) + [(n+1)/d_\phi] y_1\left(\sqrt{d_\phi}, 1\right)} \tag{8.61b}$$

For $n_\phi = 0$, $G(R) \neq 0$, we have:

$$Nu(\infty) = \frac{2 \int_0^1 R^n G(R)\, dR}{Z}$$

where

$$Z = (n+1) \int_0^1 \left\{ \left[\int_0^R R'^n U(R')\, dR' \times \int_0^R R'^n G(R')\, dR' \right] / [R^n E_h(R)] \right\} dR \tag{8.61c}$$

The Case $\alpha = 0$, $\beta = 1$ For $n_\phi = 0$, $G(R) = 0$, Equation (8.60a) yields

$$Nu(\infty) = \frac{2}{\int_0^1 \left\{ \left[(n+1) \int_0^R R'^n U(R')\, dR' \right]^2 / [R^n E_h(R)] \right\} dR} \tag{8.61d}$$

Eigenvalues and Eigenfunctions for Slug and Laminar Flow

The solution developed in this chapter for the temperature distribution and the Nusselt number are all expressed in terms of the functions $y_r(\mu, R)$, $(r = 1, 2, 3$ or $4)$, which can be evaluated for the complicated situations by the numerical integration of the system of Equations (8.46) and (8.47). However, for the simple cases such as the slug or the laminar flow, they can be evaluated analytically by recalling their definitions given by Equations (3.72) and (3.75) and utilizing the solutions of the eigenvalue problems developed in Chapter 3. To illustrate this matter with specific examples, we now present the functions $y_r(\mu, R)$, $(r = 1, 2, 3,$ and $4)$, for the slug flow and laminar flow inside conduits.

The Slug Flow

For the slug flow we have $U(R) = 1$ and $E_h(R) = 1$. Then the eigenvalue problem appropriate for the solution of the heat transfer problem (8.52) is taken as

$$\psi''(\mu, R) + \frac{n}{R}\psi'(\mu, R) + \mu^2\psi(\mu, R) = 0 \quad \text{in } 0 < R < 1 \quad \text{(8.62a)}$$

$$\psi'(\mu, 0) = 0 \quad \text{(8.62b)}$$

$$\alpha\psi(\mu, 1) + \beta\psi'(\mu, 1) = 0 \quad \text{(8.62c)}$$

This system is identical to the eigenvalue problem (3.42) if we set $n = 1 - 2m$. Then the solution for problem (8.62) is immediately determined by utilizing the results given by Equations (3.43) and Table 3.1. Then the functions $y_r(\mu, R)$, $(r = 1, 2, 3, \text{ and } 4)$ are determined according to their definitions given by Equations (3.72) and (3.75) and noting that $k(R) = R^n$. The results are summarized as follows.
For a parallel plate channel ($n = 0$ or $m = -\frac{1}{2}$):

$$y_1(\mu_i, R) = -\mu_i\sin(\mu_i R) \quad \text{(8.63a)}$$

$$y_2(\mu_i, R) = \cos(\mu_i R) \quad \text{(8.63b)}$$

$$y_3(\mu_i, R) = -\sin(\mu_i R) - \mu_i R\cos(\mu_i R) \quad \text{(8.63c)}$$

$$y_4(\mu_i, R) = -R\sin(\mu_i R) \quad \text{(8.63d)}$$

and the μ_i are the roots of the transcendental equation

$$\alpha\cos\mu - \beta\mu\sin\mu = 0 \quad \text{(8.63e)}$$

For a circular tube ($n = 1$ or $m = 0$):

$$y_1(\mu_i, R) = -\mu_i R J_1(\mu_i R) \quad \text{(8.64a)}$$

$$y_2(\mu_i, R) = J_0(\mu_i R) \quad \text{(8.64b)}$$

$$y_3(\mu_i, R) = -\mu_i R^2 J_0(\mu_i R) \quad \text{(8.64c)}$$

$$y_4(\mu_i, R) = -R J_1(\mu_i R) \quad \text{(8.64d)}$$

and the μ_i are the roots of the transcendental equation

$$\alpha J_0(\mu) - \beta\mu J_1(\mu) = 0 \quad \text{(8.64e)}$$

The Laminar Flow

For the laminar flow we have $U(R) = 1 - R^2$ and $E_h(R) = 1$. Then the eigenvalue problem appropriate for the solution of problem (8.52) is taken as

$$\psi''(\mu, R) + \frac{n}{R}\psi'(\mu, R) + \mu^2(1 - R^2)\psi(\mu, R) = 0 \quad \text{in } 0 < R < 1$$

$$\tag{8.65a}$$

$$\psi'(\mu, 0) = 0 \tag{8.65b}$$

$$\alpha\psi(\mu, 1) + \beta\psi'(\mu, 1) = 0 \tag{8.65c}$$

This system is identical to the eigenvalue problem (3.44) considered in Example 3.7 if we set $n = 1 - 2m$. It has been shown in Example 3.7 that the function $\psi(\mu_i, R)$ has the explicit form [see Equation (3.47a)]

$$\psi(\mu_i, R) = M\left(\frac{1 + n - \mu_i}{4}, \frac{1 + n}{2}, \mu_i R^2\right)e^{-(\mu_i R^2/2)} \tag{8.66a}$$

where $M(a, b, z)$ is the confluent hypergeometric function. Then by utilizing the definitions (3.72a, b), noting that $k(R) = R^n$, and the derivative of the confluent hypergeometric function $M(a, b, z)$ is given by

$$\frac{d}{dz}M(a, b, z) = \frac{a}{b}M(a + 1, b + 1, z) \tag{8.66b}$$

the explicit expressions for the functions $y_1(\mu_i, R)$ and $y_2(\mu_i, R)$ are determined as

$$y_1(\mu_i, R) = \left[\frac{1 + n - \mu_i}{1 + n}M\left(\frac{5 + n - \mu_i}{4}, \frac{3 + n}{2}, \mu_i R^2\right)\right.$$

$$\left. - M\left(\frac{1 + n - \mu_i}{4}, \frac{1 + n}{2}, \mu_i R^2\right)\right]\mu_i R^{1+n}e^{-\mu_i R^2/2} \tag{8.67a}$$

$$y_2(\mu_i, R) = M\left(\frac{1 + n - \mu_i}{4}, \frac{1 + n}{2}, \mu_i R^2\right)e^{-\mu_i R^2/2} \tag{8.67b}$$

where the μ_i are the roots of the following transcendental equation [see Equation (3.47b)]

$$(\alpha - \beta\mu)M\left(\frac{1 + n - \mu}{4}, \frac{1 + n}{2}, \mu\right)$$

$$+ \beta\mu\frac{1 + n - \mu}{1 + n}M\left(\frac{5 + n - \mu}{4}, \frac{3 + n}{2}, \mu\right) = 0 \tag{8.67c}$$

where $n = 0$ for a parallel plate channel, $n = 1$ for a circular tube.

A survey of the literature on heat transfer to thermally developing and hydrodynamically developed flow inside conduits reveals that a considerable amount of effort has been devoted to this subject. Starting with the pioneering works of Graetz [6] and Nusselt [7], the generalization of the *Graetz–Nusselt problem* to different types of boundary conditions, inlet conditions, and velocity distributions followed. For example, in laminar flow, the extensions included the variation of the tube wall temperature axially [8–11], the heat generation in the fluid [12], the variation of the wall heat flux axially [13–15], and the convection boundary conditions [16–19]. A comprehensive literature survey of Graetz–Nusselt type heat transfer problems for laminar flow inside conduits is available in the References 20 and 21. The generalization to turbulent flow under constant uniform wall temperature and heat flux conditions are available in the References 22–25.

The analysis of forced convection in conduits presented in this chapter is sufficiently general that the solutions to the Graetz–Nusselt type problems and to their extensions available in the literature can readily be produced as special cases of the solutions developed here. One needs only the eigenvalues and eigenfunctions in order to compute the temperature distribution or the Nusselt number. We illustrate this matter with specific examples.

Example 8.8 Determine the Nusselt number for fully developed laminar flow through a parallel plate channel and a circular tube maintained at a constant and uniform temperature T_w, which is different from the uniform inlet temperature T_e of the fluid at the origin of the axial coordinate $Z = 0$.

Solution We choose $T^* = T_w$ as the reference temperature and $\Delta T = T_e - T_w$ as the reference temperature difference, and define the dimensionless temperature as $\theta(R, Z) = [T(r, z) - T_w]/[T_e - T_w]$. For fully developed laminar flow, the velocity distribution is given by Equation (8.19c) and $E_h(R) = 1$. Then the mathematical formulation of this heat transfer problem becomes

$$\frac{n + 3}{2}(1 - R^2)\frac{\partial \theta(R, Z)}{\partial Z} = \frac{1}{R^n}\frac{\partial}{\partial R}\left\{R^n\frac{\partial \theta(R, Z)}{\partial R}\right\}$$

$$\text{in } 0 < R < 1, \quad Z > 0 \quad (8.68a)$$

$$\frac{\partial \theta(0, Z)}{\partial R} = 0, \qquad \theta(1, Z) = 0 \quad \text{for } Z > 0 \qquad (8.68\text{b,c})$$

$$\theta(R, 0) = 1 \quad \text{in } 0 \le R \le 1 \qquad (8.68\text{d})$$

The problem (8.68) is a special case of the problem (8.52) as follows:

$$F(R) = 1, \qquad G(R) = 0, \qquad \phi(Z) = 0, \qquad \alpha = 1, \qquad \beta = 0$$

Then from Equations (8.58a), (8.58h), and (8.58i) we have

$$\text{Nu}(Z) = 2\frac{\displaystyle\sum_{i=1}^{\infty} \mu_i H_i e^{-\mu_i^2 Z}/y_4(\mu_i, 1)}{\displaystyle\sum_{i=1}^{\infty}(n + 1)H_i e^{-\mu_i^2 Z}/[\mu_i y_4(\mu_i, 1)]} \qquad (8.69a)$$

where H_i is determined from Equation (8.57b) by utilizing Equations (8.46a) and (8.46c). We find

$$H_i = -\frac{y_1(\mu_i, 1)}{\mu_i^2} \tag{8.69b}$$

Equation (8.69) can be rearranged in the alternative form as

$$Nu(Z^*) = \frac{4}{(n+1)(n+3)} \frac{\sum\limits_{i=1}^{\infty} G_i e^{-C_n \lambda_i^2 Z^*}}{\sum\limits_{i=1}^{\infty} (G_i/\lambda_i^2) e^{-C_n \lambda_n^2 Z^*}} \tag{8.70a}$$

where

$$C_n = \frac{32}{(n+3)(n+1)^2} \tag{8.70b}$$

$$G_i = \frac{1}{\lambda_i} \frac{y_1(\mu_i, 1)}{y_4(\mu_i, 1)} \tag{8.70c}$$

$$Nu(Z^*) = \frac{1+n}{2} \frac{hD_h}{k} \tag{8.70d}$$

$$Pe = \frac{w_{av} D_h}{\alpha} \tag{8.70e}$$

$$Z^* = \frac{z}{Pe D_h} \tag{8.70f}$$

$$\lambda_i^2 = \frac{n+3}{2} \mu_i^2 \tag{8.70g}$$

$$D_h = \begin{cases} 2r_1 & \text{for circular tube} \\ 4r_1 & \text{for parallel plates} \end{cases} \tag{8.70h}$$

For the limiting case, the fully developed Nusselt number becomes

$$Nu(\infty) = \frac{4}{(n+1)(n+3)} \lambda_1^2 \tag{8.71}$$

Circular Tubes

For a circular tube, by setting $n = 1$, Equations (8.70a) and (8.71), respectively, yield

$$\frac{h(Z^*)D}{k} = \frac{1}{2} \frac{\sum\limits_{i=1}^{\infty} G_i e^{-2\lambda_i^2 Z^*}}{\sum\limits_{i=1}^{\infty} (G_i/\lambda_i^2) e^{-2\lambda_i^2 Z^*}} \tag{8.72a}$$

Table 8.1. Values of λ_i and G_i for Laminar Flow Inside a Circular Tube[a]

i	λ_i	G_i
1	2.70436 44199	0.74877 4555
2	6.67903 14493	0.54382 7956
3	10.67337 95381	0.46286 1060
4	14.67107 84627	0.41541 8455
5	18.66987 18645	0.38291 9183
6	22.66914 33588	0.35868 5566
7	26.66866 19960	0.33962 2164
8	30.66832 33409	0.32406 2211
9	34.66807 38224	0.31101 4074
10	38.66788 33469	0.29984 4038
11	42.66773 38055	0.29012 4676

[a]Based on the results of Brown [20].

and

$$\mathrm{Nu}(\infty) = \frac{h(\infty)D}{k} = \frac{1}{2}\lambda_1^2 \tag{8.72b}$$

where D is the inside diameter of the circular tube. The eigenvalues λ_i and the constants G_i, for $i = 1$ to 11, are listed in Table 8.1. For example, the limiting Nusselt number becomes

$$\mathrm{Nu}(\infty) = \tfrac{1}{2}(2.7044)^2 \cong 3.66$$

Parallel Plates

For laminar flow between parallel plates, by setting $n = 0$, Equations (8.70a) and (8.71), respectively, yield

$$\frac{h(Z^*)D_h}{k} = \frac{8}{3} \frac{\displaystyle\sum_{i=1}^{\infty} G_i e^{-(32/3)\lambda_i^2 Z^*}}{\displaystyle\sum_{i=1}^{\infty} \left(G_i/\lambda_i^2\right) e^{-(32/3)\lambda_i^2 Z^*}} \tag{8.73a}$$

and

$$\mathrm{Nu}(\infty) = \frac{h(\infty)D_h}{k} = \frac{8}{3}\lambda_1^2 \tag{8.73b}$$

where $D_h = 4r_1$. The eigenvalues λ_i and the constants G_i, for $i = 1$ to 10, are

Table 8.2.   Values of λ_i and G_i for Laminar Flow Inside a Parallel Plate Channel[a]

i	λ_i		G_i	
1	1.68159	53222	0.85808	6674
2	5.66985	73459	0.56946	2850
3	9.66824	24625	0.47606	5463
4	13.66766	14426	0.42397	3730
5	17.66737	35653	0.38910	8706
6	21.66720	53243	0.36346	5044
7	25.66709	64863	0.34347	5506
8	29.66702	10447	0.32726	5745
9	33.66696	60687	0.31373	9318
10	37.66692	44563	0.30220	4200

[a] Based on the results of Brown [20].

listed in Table 8.2. For example, the limiting Nusselt number becomes

$$\text{Nu}(\infty) = \tfrac{8}{3}(1.6816)^2 = 7.54$$

REFERENCES

1. G. Hagen, Über die Bewegung des Wassers in Ergen Zylindrischen, *Röhren. Pogg. Ann.*, **46**, 423–442 (1839).

2. J. Poiseuille, *C. R. Acad. Sci.*, **11**, 961–967, 1041–1048 (1840); **12**, 112–115 (1841).

3. M. Perlmutter and R. Siegel, Unsteady Laminar Flow in a Duct with Unsteady Heat Addition, *J. Heat Transf.*, **83C**, 432–440 (1961).

4. S. Uchida, The Pulsating Viscous Flow Superposed on the Steady Laminar Motion of Incompressible Fluid in a Circular Pipe, *ZAMP*, **7**, 403–421 (1956).

5. R. Siegel and M. Perlmutter, Heat Transfer for Pulsating Laminar Duct Flow, *J. Heat Transf.*, **84C**, 111–123 (1962).

6. L. Graetz, Über die Wärmeleitungsfähigkeit von Flüssigkeiten, Part I. *Ann. Phys. Chem.*, **18**, 79–94 (1883); Part II. *Ann. Phys. Chem.*, **25**, 337–357 (1885).

7a. W. Nusselt, Die Abhängigkeit der Wärmeubergangs-zahl von der Rohrlänge, *VDI Z*, **54**, 1154–1158 (1910).

7b. W. Nusselt, Der Wärmeaustausch am Berieselungskühler, *VDI Z*, **67**, 206–210 (1923).

8. U. Grigull and H. Tratz, Thermischer einlauf in ausgebildeter laminarer Rohrströmung, *Int. J. Heat Mass Transf.*, **8**, 669–678 (1965).

9. J. R. Sellers, M. Tribus, and J. S. Klein, Heat Transfer to Laminar Flow in a Round Tube or Flat Conduit the Graetz Problem Extended, *Trans. ASME*, **78**, 441–448 (1956).

10. J. W. Mitchell, An Expression for Internal Flow Heat Transfer for Polynomial Wall Temperature Distributions, *J. Heat Transf.*, **91**, 175–177 (1969).

11. V. V. Shapovalov, Heat Transfer in Laminar Flow of an Incompressible Fluid in a Round Tube, *J. Eng. Phys.* (USSR) **12**, 363–364 (1967).

12. E. M. Sparrow, J. L. Novotny, and S. H. Lin, Laminar Flow of a Heat-Generating Fluid in a Parallel-Plate Channel, *A. I. Chem. Eng. J.*, **9**, 797–804 (1963).

13. R. D. Cess and E. C. Shaffer, Laminar Heat Transfer between Parallel Plates with an Unsymmetrically Prescribed Heat Flux at the Walls, *Appl. Sci. Res.*, Sect. **A9**, 64–70 (1960).

14. P. A. McCuen, W. M. Kays, and W. C. Reynolds, *Heat Transfer with Laminar and Turbulent Flow between Parallel Planes with Constant and Variable Wall Temperature and Heat Flux*, Rep. No. AHT-3, Dep. Mech. Eng., Stanford University, Stanford, California, 1962.

15. A. O. Tay and G. DeVahl Davis, Application of the Finite Element Method to Convection Heat Transfer between Parallel Planes, *Int. J. Heat Mass Transf.*, **14**, 1057–1069 (1971).

16. J. Schenk and J. M. DuMore, Heat Transfer in Laminar Flow through Cylindrical Tubes, *Appl. Sci. Res.*, **A.39**, 51 (1954).

17. S. Sideman, D. Luss, and R. E. Peck, Heat Transfer in Laminar Flow in Circular and Flat Conduits with (Constant) Surface Resistance, *Appl. Sci. Res.*, **A14**, 157–171 (1964).

18. C. J. Hsu, Laminar Flow Heat Transfer in Circular or Parallel-Plate Channels with Internal Heat Generation and the Boundary Condition of the Third Kind, *J. Chin. Inst. Chem. Eng.*, **2**, 85–100 (1971).

19. G. J. Hwang and I. Yih, Correction on the Length of Ice-Free Zone in a Convectively-Cooled Pipe, *Int. J. Heat Mass Transf.*, **16**, 681–683 (1973).

20. G. M. Brown, Heat or Mass Transfer in a Fluid in Laminar Flow in a Circular or Flat Conduit, *AIChE J.*, **6**, 179–183 (1960).

21. R. K. Shah and A. L. London, *Laminar Flow Forced Convection in Ducts*, Academic, New York, 1978.

22. C. A. Sleicher and M. Tribus, Heat Transfer in a Pipe with Turbulent Flow and Arbitrary Wall-Temperature Distribution, *Trans. Am. Soc. Mech. Eng.*, **79**, 789 (1957).

23. R. Siegel and E. M. Sparrow, Turbulent Flow in a Circular Tube with Arbitrary Internal Heat Sources and Wall Heat Transfer, *J. Heat Transf.*, **81C**, 280–290 (1959).

24. A. P. Hatton, Heat Transfer in the Thermal Entrance Region with Turbulent Flow between Parallel Plates at Unequal Temperatures, *Appl. Sci. Res.*, **A12**, 249 (1963).

25. A. Quarmby and R. K. Anand, Turbulent Heat Transfer in the Thermal Entrance Region of Concentric Annuli with Uniform Wall Heat Flux, *Int. J. Heat Mass Transf.*, **13**, 395–411 (1970).

26. W. L. Wilkinson, *Non-Newtonian Fluids*, Pergamon, New York, 1960.

CHAPTER NINE

Class II Solutions Applied to Heat Diffusion in One-Dimensional Composite Media

Wisdom denotes the pursuing of the best end by best means.

Francis Hutcheson
1694–1746

Transient temperature distribution in a composite medium consisting of several layers in contact has numerous applications in science and engineering. Various methods of analysis are available for the solution of such problems. For example, the orthogonal expansion technique and the Green's function approach have been used [1–10], the adjoint solution technique has been discussed [6, 11], the Laplace transform technique has been used [12–15], and the finite integral transform technique has been applied [16–26].

The heat conduction problem in composite media is a special case of the problem of Class II. This chapter is devoted to the application of the results of analysis of the one-dimensional Class II problem presented in Chapter 4 for the solution of one-dimensional, transient heat conduction in composite media. Therefore, we present first the mathematical formulation of a sufficiently general heat conduction problem for composite layers of slabs, cylinders, and spheres, then illustrate how the general analysis of Class II problem given in Chapter 4 can be utilized to obtain solutions for such heat conduction problems.

9.1 FORMULATION OF TRANSIENT HEAT CONDUCTION FOR ONE-DIMENSIONAL COMPOSITE MEDIA

We now consider a composite medium consisting of n parallel layers of slabs, cylinders, or spheres in contact as illustrated in Figure 9.1. For generality we

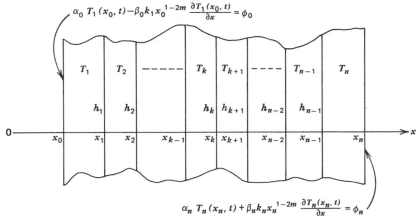

Figure 9.1 n-layer slabs ($m = \frac{1}{2}$), cylinders ($m = 0$), or spheres ($m = -\frac{1}{2}$).

assume contact resistance at the interfaces and convection from the outer boundaries. Let h_k be the film coefficient at the interfaces $x = x_k$, $k = 1, 2, \ldots, (n-1)$. Each layer is homogeneous, isotropic, and has thermal properties (i.e., ρ_k, C_k, k_k) that are constant within the layer and different from those of the adjacent layers. Initially, each layer is at specified temperature $T_k(x, 0) = f_k(x)$, in $x_{k-1} \leq x \leq x_k$, $k = 1, 2, \ldots, n$. For times $t > 0$, heat is transferred from the two outer boundaries according to boundary conditions of the first, second, or third kind. There is no energy generation in the medium. The mathematical formulation of this heat conduction problem governing the temperature distribution $T_k(x, t)$, $k = 1, 2, \ldots, n$ for times $t > 0$ is given as

$$x^{1-2m} \frac{\partial T_k(x, t)}{\partial t} = \alpha_k^* \frac{\partial}{\partial x}\left[x^{1-2m} \frac{\partial T_k(x, t)}{\partial x} \right]$$

$$\text{in } x_{k-1} < x < x_k \quad \text{for } t > 0, \qquad k = 1, 2, \ldots, n. \quad (9.1a)$$

where

$$m = \begin{cases} \frac{1}{2} & \text{slab} \\ 0 & \text{cylinder} \\ -\frac{1}{2} & \text{sphere} \end{cases}$$

$$\alpha_k^* = \frac{k_k}{\rho_k C_k} = \text{thermal diffusivity of layer-}k.$$

Subject to the boundary conditions

$$\alpha_0 T_1(x_0, t) - \beta_0 k_1 x_0^{1-2m} \frac{\partial T_1(x_0, t)}{\partial x} = \phi_0 \qquad (9.1b)$$

at the outer boundary $x = x_0$ for $t > 0$;

$$-k_k \frac{\partial T_k(x_k, t)}{\partial x} = h_k [T_k(x_k, t) - T_{k+1}(x_k, t)] \qquad (9.1c)$$

$$k_k \frac{\partial T_k(x_k, t)}{\partial x} = k_{k+1} \frac{\partial T_{k+1}(x_k, t)}{\partial x} \qquad (9.1d)$$

at the interfaces $x = x_k$, $k = 1, 2, \ldots, (n - 1)$ for $t > 0$;

$$\alpha_n T_n(x_n, t) + \beta_n k_n x_n^{1-2m} \frac{\partial T_n(x_n, t)}{\partial x} = \phi_n \qquad (9.1e)$$

at the outer boundary $x = x_n$ for $t > 0$; and the initial conditions given by

$$T_k(x, 0) = f_k(x) \quad \text{for } t = 0 \quad \text{in } x_{k-1} \le x \le x_k, \qquad k = 1, 2, \ldots, n$$

$$(9.1f)$$

By appropriate choice of the values of the parameters α_0, β_0, α_n, and β_n, various combinations of the boundary conditions of the first, second, and third kind are obtainable at the two outer boundaries. The special case of $\alpha_0 = \alpha_n = 0$, $\beta_0 = \beta_n = 1$, corresponding to the boundary conditions of the second kind at the two outer boundaries is not considered here.

The physical significance of the interface boundary conditions (9.1c) and (9.1d) is as follows.

The finite value of the film coefficient h_k, $k = 1, 2, \ldots, (n - 1)$ in the boundary condition (9.1c) implies a discontinuity of temperature at the interfaces. The boundary conditions (9.1d) state that the heat flux is continuous at the interfaces. For the special case of $h_k \to \infty$, the boundary conditions (9.1c) reduce to

$$T_k(x_k, t) = T_{k+1}(x_k, t) \quad \text{at } x = x_k, \qquad k = 1, 2, \ldots, (n - 1) \quad \text{for } t > 0$$

$$(9.1g)$$

which implies the continuity of temperature or perfect thermal contact at the interfaces.

The problem given by Equations (9.1) is a special case of the one-dimensional form of the Class II problem given by Equations (4.81). Therefore, the solution of problem (9.1) is readily obtainable as a special case from the solution of the more general problem (4.81) as described in the next section.

9.2 FORMAL SOLUTION FOR n-LAYER COMPOSITE MEDIUM

The transient heat conduction problem defined by Equations (9.1) is a special case of the Class II problem given by Equations (4.81). The correspondence between the two problems is given by

$$w_k(x) = \rho_k C_k x^{1-2m}, \qquad k_k(x) = k_k x^{1-2m}, \qquad \alpha_k^* = \frac{k_k}{C_k \rho_k}$$

$$h_{k,k+1} = h_k x_k^{1-2m}, \qquad \phi(x_0, t) = \phi_0, \qquad \phi(x_n, t) = \phi_n$$

$$P_k(x, t) = 0 \tag{9.2}$$

The general solution of problem (4.81) is given in Chapter 4 by Equations (4.82) and the alternative split-up form of the solution is given by Equation (4.84). Here we prefer to use the latter form because it is more convenient for numerical computations. The solution (4.84) consists of several different functions, each characterizing the solution of a simpler problem associated with the functional forms of the nonhomogeneous terms $\phi(x_0, t)$, $\phi(x_n, t)$, and $P_k(x, t)$ specified by Equations (4.83).

For the specific heat conduction problem given by Equations (9.1), the solution (4.84) reduces to

$$T_k(x, t) = T_k(x) + T_{kt}(x, t) \tag{9.3}$$

where $T_k(x) \equiv T_{k\phi}(x)$. The reason for this is that there is no energy generation in the present problem, and the boundary condition function $\phi(x_0, t) \equiv \phi_0$ and $\phi(x_n, t) \equiv \phi_n$ are constants.

The functions $T_k(x)$ satisfy the steady-state problem defined by Equations (4.87), which for the specific case considered here reduces to

$$\frac{d}{dx}\left\{ k_k(x) \frac{dT_k(x)}{dx} \right\} = 0 \quad \text{in } x_{k-1} < x < x_k, \qquad k = 1, 2, \ldots, n \tag{9.4a}$$

subject to the boundary conditions

$$\alpha_0 T_1(x_0) - \beta_0 k_1(x_0) T_1'(x_0) = \phi_0 \tag{9.4b}$$

$$k_k(x_k) T_k'(x_k) = k_{k+1}(x_k) T_{k+1}'(x_k) = h_k[T_{k+1}(x_k) - T_k(x_k)] \tag{9.4c}$$

$$\alpha_n T_n(x_n) + \beta_n k_n(x_n) T_n'(x_n) = \phi_n \tag{9.4d}$$

where prime denotes differentiation with respect to x, and $k_k(x)$, h_k are defined by Equations (9.2). The solution of Equations (9.4) is a straightforward matter.

The functions $T_{kt}(x, t)$ are obtained from solution (4.88) by appropriate simplifications, that is, by setting $p_{ke}(x) = 0$ and $j = 0$.

Introducing the solutions for $T_k(x)$ and $T_{kt}(x, t)$ determined in this manner into Equation (9.3), the solution for the composite medium problem (9.1) becomes

$$
T_k(x, t) = \left\{ \left[\beta_n + \alpha_n \left(\frac{1}{k_k} \int_x^{x_k} x^{2m-1} \, dx \right. \right. \right.
$$

$$
\left. \left. + \sum_{i=k+1}^{n} \frac{1}{k_i} \int_{x_{i-1}}^{x_i} x^{2m-1} \, dx + \sum_{i=k}^{n-1} \frac{x_i^{2m-1}}{h_i} \right) \right] \phi_0
$$

$$
+ \left[\beta_0 + \alpha_0 \left(\frac{1}{k_k} \int_{x_{k-1}}^{x_k} x^{2m-1} \, dx \right. \right.
$$

$$
\left. \left. + \sum_{i=1}^{k-1} \frac{1}{k_i} \int_{x_{i-1}}^{x_i} x^{2m-1} \, dx + \sum_{i=1}^{k-1} \frac{x_i^{2m-1}}{h_i} \right) \right] \phi_n \right\}
$$

$$
\times \left(\alpha_0 \beta_n + \alpha_n \beta_0 + \alpha_0 \alpha_n \sum_{k=1}^{n} \frac{1}{k_k} \int_{x_{k-1}}^{x_k} x^{2m-1} \, dx + \alpha_0 \alpha_n \sum_{k=1}^{n-1} \frac{x_k^{2m-1}}{h_k} \right)^{-1}
$$

$$
+ \sum_{i=1}^{\infty} \frac{\psi_k(\mu_i, x)}{N_i} e^{-\mu_i^2 t} \left\{ \tilde{f}_i - \frac{1}{\mu_i^2} \left[\phi_0 \Omega_i(x_0) + \phi_n \Omega_i(x_n) \right] \right\},
$$

$$
k = 1, 2, \ldots, n \quad (9.5a)
$$

where the functions N_i, $\tilde{f}_i$, $\Omega_i(x_0)$, and $\Omega_i(x_n)$ are defined by Equations (4.82b, c, e, f), respectively. That is,

$$
N_i = \sum_{k=1}^{n} \rho_k C_k \int_{x_{k-1}}^{x_k} x^{1-2m} \psi_k^2(\mu_i, x) \, dx \qquad (9.5b)
$$

$$
\tilde{f}_i = \sum_{k=1}^{n} \rho_k C_k \int_{x_{k-1}}^{x_k} x^{1-2m} \psi_k(\mu_i, x) f_k(x) \, dx \qquad (9.5c)
$$

$$
\Omega_i(x_0) = \frac{\psi_1(\mu_i, x_0) + k_1 x_0^{1-2m} \psi_1'(\mu_i, x_0)}{\alpha_0 + \beta_0} \qquad (9.5d)
$$

$$
\Omega_i(x_n) = \frac{\psi_n(\mu_i, x_n) - k_n x_n^{1-2m} \psi_n'(\mu_i, x_n)}{\alpha_n + \beta_n} \qquad (9.5e)
$$

The temperature distribution $T_k(x, t)$ in any one layer k, $(k = 1, 2, \ldots, n)$ of an n layer of *slabs* ($m = \frac{1}{2}$), *cylinder* ($m = 0$), and *spheres* ($m = -\frac{1}{2}$) can be determined from Equation (9.5) if the eigenvalues μ_i and the eigenfunctions $\psi(\mu_i, x)$ are known. Therefore, in the following section we discuss the determination of the eigenvalues and eigenfunctions.

9.3   CONSTRUCTION OF EIGENFUNCTIONS

The eigenvalue problem appropriate for the heat conduction problem defined by Equations (9.1) is obtained by proper simplification of the eigenvalue problem (3.96). By setting $k_k(x) = k_k x^{1-2m}$, $w_k(x) = \rho_k C_k x^{1-2m}$, $h_{k,k+1} = h_k$ and $d_k(x) = 0$, we obtain

$$\frac{d}{dx}\left[x^{1-2m}\frac{d\psi_k(\mu, x)}{dx} \right] + \frac{\mu^2}{\alpha_k^*}x^{1-2m}\psi_k(\mu, x) = 0$$

$$\text{in } x_{k-1} < x < x_k, \qquad k = 1, 2, \ldots, n, \qquad \alpha_k^* = \frac{k_k}{\rho_k C_k} \quad (9.6a)$$

subject to the boundary conditions

$$\alpha_0 \psi_1(\mu, x_0) - \beta_0 k_1 x_0^{1-2m}\psi_1'(\mu, x_0) = 0 \qquad (9.6b)$$

$$k_k \psi_k'(\mu, x_k) = k_{k+1}\psi_{k+1}'(\mu, x_k) = h_k\{\psi_{k+1}(\mu, x_k) - \psi_k(\mu, x_k)\},$$

$$k = 1, 2, \ldots, (n-1) \quad (9.6c, d)$$

$$\alpha_n \psi_n(\mu, x_k) + \beta_n k_n x_n^{1-2n}\psi_n'(\mu, x_n) = 0 \qquad (9.6e)$$

Let $u_k(\mu, x)$ and $v_k(\mu, x)$ be two linearly independent elementary solutions of Equation (9.6a). Table 9.1 shows these functions for the case of slab, cylinder, and sphere.

The eigenfunctions $\psi_k(\mu, x)$ of this eigenvalue problem can be constructed by taking any linear combination of these two elementary solutions. For the

Table 9.1   Linearly Independent Solutions $u_k(\mu, x)$ and $v_k(\mu, x)$ of Equation (9.6a) for Slab, Cylinder, and Sphere

Geometry	m	$u_k(\mu, x)^a$	$v_k(\mu, x)^a$
Slab	$\frac{1}{2}$	$\cos(\omega_k x)$	$\sin(\omega_k x)$
Cylinder	0	$J_0(\omega_k x)$	$Y_0(\omega_k x)$
Sphere	$-\frac{1}{2}$	$\sin(\omega_k x)/(\omega_k x)$	$\cos(\omega_k x)/(\omega_k x)$

$^a\omega_k = \mu/\sqrt{\alpha_k^*}$, $\alpha_k^* = k_k/\rho_k C_k$.

Table 9.2 Functions $U_k(\mu, x)$ and $V_k(\mu, x)$ of Equation (9.7) for Slab, Cylinder, and Sphere

Geometry	m	$U_k(\mu, x)^a$	$V_k(\mu, x)^a$
Slab	$\tfrac{1}{2}$	$\dfrac{\sin[\omega_k(x_k - x)]}{\sin[\omega_k(x_k - x_{k-1})]}$	$\dfrac{\sin[\omega_k(x - x_{k-1})]}{\sin[\omega_k(x_k - x_{k-1})]}$
Cylinder	0	$\dfrac{J_0(\omega_k x)Y_0(\omega_k x_k) - J_0(\omega_k x_k)Y_0(\omega_k x)}{J_0(\omega_k x_{k-1})Y_0(\omega_k x_k) - J_0(\omega_k x_k)Y_0(\omega_k x_k)}$	$\dfrac{J_0(\omega_k x_{k-1})Y_0(\omega_k x) - J_0(\omega_k x)Y_0(\omega_k x_{k-1})}{J_0(\omega_k x_{k-1})Y_0(\omega_k x_k) - J_0(\omega_k x_k)Y_0(\omega_k x_{k-1})}$
Sphere	$-\tfrac{1}{2}$	$\dfrac{x_{k-1}}{x}\dfrac{\sin[\omega_k(x_k - x)]}{\sin[\omega_k(x_k - x_{k-1})]}$	$\dfrac{x_k}{x}\dfrac{\sin[\omega_k(x - x_{k-1})]}{\sin[\omega_k(x_k - x_{k-1})]}$

$^a \omega_k = \mu/\sqrt{\alpha_k^*}$, $\alpha_k^* = k_k/\rho_k C_k$.

reasons discussed in Chapter 3, we prefer to construct the eigenfunctions in the form given by Equations (3.97b) and (3.98); namely,

$$\psi_k(\mu, x) = \psi_{k-1}^* U_k(\mu, x) + \psi_k^* V_k(\mu, x), \qquad k = 1, 2, \ldots, n \quad (9.7a)$$

where the constants ψ_{k-1}^* and ψ_k^* are the values of the eigenfunctions evaluated at the end points $x = x_{k-1}$ and $x = x_k$, respectively; that is,

$$\psi_k(\mu, x_{k-1}) \equiv \psi_{k-1}^* \quad \text{and} \quad \psi_k(\mu, x_k) \equiv \psi_k^* \qquad (9.7b)$$

Then the functions $U_k(\mu, x)$ and $V_k(\mu, x)$ become

$$U_k(\mu, x) = \frac{u_k(\mu, x) v_k(\mu, x_k) - u_k(\mu, x_k) v_k(\mu, x)}{u_k(\mu, x_{k-1}) v_k(\mu, x_k) - u_k(\mu, x_k) v_k(\mu, x_{k-1})} \qquad (9.8a)$$

$$V_k(\mu, x) = \frac{u_k(\mu, x_{k-1}) v_k(\mu, x) - u_k(\mu, x) v_k(\mu, x_{k-1})}{u_k(\mu, x_{k-1}) v_k(\mu, x_k) - u_k(\mu, x_k) v_k(\mu, x_{k-1})} \qquad (9.8b)$$

We note that with the choice of the functions $U_k(\mu, x)$ and $V_k(\mu, x)$ as defined previously, the following conditions are automatically satisfied.

$$U_k(\mu, x_{k-1}) = 1, \qquad U_k(\mu, x_k) = 0 \qquad (9.9a)$$

$$V_k(\mu, x_{k-1}) = 0, \qquad V_k(\mu, x_k) = 1 \qquad (9.9b)$$

Table 9.2 lists the functions $U_k(\mu, x)$ and $V_k(\mu, x)$ for the cases of slab, cylinder, and sphere.

The eigenfunctions $\psi_k(\mu, x)$ constructed in the form given by Equation (9.7) involve the eigenvalues μ_i and the constants $\psi_k^*(\mu, x)$, $k = 0, 1, 2, \ldots, n$ which can be determined as described in the following section.

9.4 DETERMINATION OF EIGENVALUES AND EIGENFUNCTIONS

For simplicity in the analysis, we first examine the case of perfect thermal contact between the layers. For such a case we have $h_k \to \infty$ for $k = 1, 2, \ldots, (n-1)$, hence the boundary conditions Equations (9.6d) reduce to

$$\psi_k(\mu, x_k) = \psi_{k+1}(\mu, x_k), \qquad k = 1, 2, \ldots, (n-1) \qquad (9.10)$$

We note that the eigenfunctions defined by Equation (9.7) automatically satisfy the boundary conditions (9.10), but they involve $(n+1)$ unknown constants, ψ_k^* $(k = 0, 1, 2, \ldots, n)$, and the unknown eigenvalues μ_i. The remaining boundary conditions, Equations (9.6b, c, e), provide $(n + 1)$ homogeneous equations for the determination of the eigenvalues and the constants as now described.

If the solution, Equation (9.7), should satisfy the boundary conditions (9.6b, c, e), we have, respectively

$$\left[\frac{\alpha_0}{\beta_0} - P_1(\mu, x_0)\right]\psi_0^* + P_1(\mu, x_1)\psi_1^* = 0 \qquad (9.11a)$$

$$P_k(\mu, x_k)\psi_{k-1}^* + \left[Q_k(\mu, x_k) - P_{k+1}(\mu, x_k)\right]\psi_k^* + P_{k+1}(\mu, x_{k+1})\psi_{k+1}^* = 0$$

$$k = 1, 2, \ldots, (n - 1) \quad (9.11b)$$

$$P_n(\mu, x_n)\psi_{n-1}^* + \left[Q_n(\mu, x_n) + \frac{\alpha_n}{\beta_n}\right]\psi_n^* = 0 \qquad (9.11c)$$

where

$$P_k(\mu, x) = k_k(x)U_k'(\mu, x) \qquad (9.11d)$$

$$Q_k(\mu, x) = k_k(x)V_k'(\mu, x) \qquad (9.11e)$$

In the derivation of Equations (9.11), the relation

$$P_k(\mu, x_k) + Q_k(\mu, x_{k-1}) = 0, \qquad k = 1, 2, \ldots, n \qquad (9.12)$$

was substantially utilized. A proof of the validity of Equation (9.12) was discussed in Chapter 3. The functions $P_k(\mu, x)$ and $Q_k(\mu, x)$ appearing in Equation (9.11) are listed in Table 9.3 for the cases of slab, cylinder, and sphere.

The system of Equations (9.11) will form the basis of our analysis for computing the eigenvalues μ_i. Once the eigenvalues are known, the constants $\psi_0^*, \psi_1^*, \ldots, \psi_n^*$ are readily evaluated. The procedure is as follows.

The eigenfunctions defined by Equation (9.7) are introduced into Equations (9.11) to yield $(n + 1)$ homogeneous equations for the determination of the $(n + 1)$ unknown constants ψ_k^* $(k = 0, 1, \ldots, n)$. This system of equations can be expressed in the matrix form as

$$[K(\mu)]\{\psi^*\} = 0 \qquad (9.13a)$$

where

$$\{\psi^*\}^T = \{\psi_0^*, \psi_1^*, \ldots, \psi_{n-1}^*, \psi_n^*\} \qquad (9.13b)$$

is the transpose of $\{\psi^*\}$. The determinant $[K(\mu)]$ is given by [see Equations

Table 9.3 Functions $P_k(\mu, x)$ and $Q_k(\mu, x)$ of Equation (9.11)

Geometry	m	$P_k(\mu, x)^a$	$Q_k(\mu, x)^a$
Slab	$\tfrac{1}{2}$	$-\omega_k k_k \dfrac{\cos[\omega_k(x_k - x)]}{\sin[\omega_k(x_k - x_{k-1})]}$	$\omega_k k_k \dfrac{\cos[\omega_k(x - x_{k-1})]}{\sin[\omega_k(x_k - x_{k-1})]}$
Cylinder	0	$-\omega_k k_k \dfrac{J_1(\omega_k x)Y_0(\omega_k x_k) - J_0(\omega_k x_k)Y_1(\omega_k x)}{J_0(\omega_k x_k)Y_0(\omega_k x_{k-1}) - J_0(\omega_k x_{k-1})Y_0(\omega_k x_k)}\, x_{k-1}$	$\omega_k k_k \dfrac{J_0(\omega_k x_{k-1})Y_1(\omega_k x) - J_1(\omega_k x)Y_0(\omega_k x_{k-1})}{J_0(\omega_k x_k)Y_0(\omega_k x_{k-1}) - J_0(\omega_k x_{k-1})Y_0(\omega_k x_k)}\, x_k$
Sphere	$-\tfrac{1}{2}$	$-k_k \dfrac{x_{k-1}}{\sin[\omega_k(x_k - x_{k-1})]}$ $\times \{x\omega_k\cos[\omega_k(x_k - x)] + \sin[\omega_k(x_k - x)]\}$	$k_k \dfrac{x_k}{\sin[\omega_k(x_k - x_{k-1})]}$ $\times \{x\omega_k\cos[\omega_k(x - x_{k-1})] - \sin[\omega_k(x - x_{k-1})]\}$

$^a \omega_k = \mu/\sqrt{\alpha_k^*}$, $\alpha_k^* = k_k/\rho_k C_k$.

(3.104)]

$$[K(\mu)] = \begin{bmatrix} a_0 & b_1 & 0 & \cdots & 0 & 0 \\ b_1 & a_1 & b_2 & & & \vdots \\ 0 & b_2 & a_2 & & & \vdots \\ \vdots & & & & & 0 \\ \vdots & & & b_{n-1} & a_{n-1} & b_n \\ 0 & 0 & \cdots & 0 & b_n & a_n \end{bmatrix}$$ (9.13c)

where

$$a_0 = \frac{\alpha_0}{\beta_0} - P_1(\mu, x_0)$$ (9.13d)

$$b_k = P_k(\mu, x_k), \qquad k = 1, 2, \ldots, n$$ (9.13e)

$$a_k = Q_k(\mu, x_k) - P_{k+1}(\mu, x_k), \qquad k = 1, 2, 3, \ldots, (n-1)$$ (9.13f)

$$a_n = Q_n(\mu, x_n) + \frac{\alpha_n}{\beta_n}$$ (9.13g)

If the system of Equations (9.13a) has a nontrivial solution, the determinant of the coefficients should vanish, that is,

$$\det[K(\mu)] = 0$$ (9.14)

The infinite number of real roots of this transcendental equation give the eigenvalues μ_i of the eigenvalue problem, Equations (9.6). Customarily, such a procedure is used to determine the eigenvalues of the system but, as discussed in Chapter 3, it has several disadvantages. Therefore, we now discuss the use of the sign-count method described in Chapter 3, for the determination of the eigenvalues for the specific case considered previously [28].

Determination of Eigenvalues by the Sign-Count Method

The general procedure as discussed in Chapter 3 involves the determination of the number of positive eigenvalues $N(\tilde{\mu})$ lying in the interval between $\mu = 0$ and some prescribed positive value $\mu = \tilde{\mu}$ of the eigenvalue parameter μ. It is given by [see Equation (3.106)]

$$N(\tilde{\mu}) = N_0(\tilde{\mu}) + s([K(\tilde{\mu})])$$ (9.15a)

where

$$N_0(\tilde{\mu}) = \sum_{k=1}^{n} N_{0,k}(\tilde{\mu}) \tag{9.15b}$$

$$s([K(\tilde{\mu})]) = \text{the sign-count of } [K(\tilde{\mu})] \tag{9.15c}$$

and $N_{0,k}(\tilde{\mu})$ is the number of eigenvalues lying in the interval $0 < \mu < \tilde{\mu}$ of the following decoupled system of simple eigenvalue problems:

$$\frac{d}{dx}\left[x^{1-2m} \frac{d\psi_k(\mu, x)}{dx} \right] + \omega_k^2 x^{1-2m} \psi_k(\mu, x) = 0$$

$$\text{in } x_{k-1} < x < x_k \quad (k = 1, 2, \ldots, n) \tag{9.16a}$$

subject to the boundary conditions

$$\psi_k(\mu, x_{k-1}) = 0 \quad \text{and} \quad \psi_k(\mu, x_k) = 0 \tag{9.16b, c}$$

where $\omega_k = \mu/\sqrt{\alpha_k^*}$. Clearly, Equations (9.16) are the eigenvalue problems for each single layer that is regarded decoupled from the composite layer. The eigencondition for each of these simple systems is given by

$$u_k(\mu, x_{k-1})v_k(\mu, x_k) - u_k(\mu, x_k)v_k(\mu, x_{k-1}) = 0 \tag{9.17}$$

where the linearly independent solutions $u_k(\mu, x)$ and $v_k(\mu, x)$ are listed in Table 9.1 for the cases of slab, cylinder, and sphere.

The number of eigenvalues $N_{0,k}(\tilde{\mu})$ lying in the interval between $\mu = 0$ and $\mu = \tilde{\mu}$ is determined for each layer from the solution of the transcendental Equation (9.17) by a standard procedure. For the cases of slab ($m = \frac{1}{2}$) and sphere ($m = -\frac{1}{2}$), an explicit relation is available for $N_{0,k}(\tilde{\mu})$ in the form

$$N_{0,k}(\tilde{\mu}) = \text{int}\left[\frac{\omega_k(x_k - x_{k-1})}{\pi} \right] \tag{9.18}$$

where the symbol "int(z)" denotes the largest integer not exceeding the value of the argument z of the function, and $\omega_k = \mu/\sqrt{\alpha_k^*}$.

In the case of a cylinder ($m = 0$), Equation (9.17) takes the form

$$J_0(\omega_k x_{k-1})Y_0(\omega_k x_k) - J_0(\omega_k x_k)Y_0(\omega_k x_{k-1}) = 0$$

$$\text{in } x_{k-1} < x < x_k \quad (k = 1, 2, \ldots, n) \tag{9.19}$$

For this particular case, no simple formula such as Equation (9.18) is available to determine the eigenvalues; however, the following procedure can be applied to determine $N_{0,k}(\tilde{\mu})$. The roots of the transcendental Equation (9.19) are computed by a standard method and stored in the memory of the computer.

Then for a given value of $\tilde{\mu}$ the number of eigenvalues lying in the interval $0 < \mu < \tilde{\mu}$ is determined for each layer, which provides the values of $N_{0,k}(\tilde{\mu})$.

Knowing $N_{0,k}(\tilde{\mu})$ for each layer, the value of $N_0(\tilde{\mu})$ is determined from Equation (9.15b).

The sign-count $s\{[K(\tilde{\mu})]\}$ is the number of negative elements along the main diagonal of the matrix $[K^{\Delta}(\tilde{\mu})]$, which is the triangulated form of the matrix $[K(\tilde{\mu})]$ defined by Equation (3.104c) or (9.13c). Therefore, by using the Gauss elimination process, the matrix $[K(\tilde{\mu})]$ is diagonalized as discussed in Chapter 3 [see Equation (3.109)] and the number of negative elements in the main diagonal are determined.

Thus knowing both $N_0(\tilde{\mu})$ and $s\{[K(\tilde{\mu})]\}$, the number of eigenvalues $N(\tilde{\mu})$ of the original eigenvalue problem, Equations (9.6), are determined from Equation (9.15a). Once such information is available, the algorithm discussed in Chapter 3 is used to determine the eigenvalues μ_i.

Contact Resistance at the Interfaces

When there is contact resistance at any of the interfaces x_k, that interface is regarded as a fictitious layer for which $P_k(\mu, x)$ and $Q_k(\mu, x)$ are specified as [see Equation (3.111)]

$$P_k(\mu) = -h_k \quad \text{and} \quad Q_k(\mu) = h_k \qquad \text{(9.20a, b)}$$

at $x = x_k$. Here h_k is the contact conductance at the interface x_k. Then the computational procedure becomes exactly the same as that for the case of perfect thermal contact at all interfaces except we are now introducing into the system fictitious layers at the interfaces where there is imperfect thermal contact.

Determination of Eigenfunctions

Once the eigenvalues μ_i are available, the eigenfunctions $\psi_k(\mu_i, x)$ for any layer k in the medium can be determined from Equation (9.7) if the constants $\psi_{k-1}^*(\mu_i)$ and $\psi_k^*(\mu_i)$ are known. These constants can be determined by utilizing Equations (9.11a–c) as now described.

From Equations (9.11a, b) we, respectively, have

$$\psi_1^* = \frac{\left[P_1(\mu, x_0) - \dfrac{\alpha_0}{\beta_0}\right]\psi_0^*}{P_1(\mu, x_1)} \qquad (9.21a)$$

$$\psi_{k+1}^* = \frac{\left[P_{k+1}(\mu, x_k) - Q_k(\mu, x_k)\right]\psi_k^* - P_k(\mu, x_k)\psi_{k-1}^*}{P_{k+1}(\mu, x_{k+1})},$$

$$k = 1, 2, \ldots, (n-1) \quad (9.21b)$$

Since the eigenfunctions are arbitrary within a multiplication constant, one can set

$$\psi_0^* = \beta_0$$

and knowing $\psi_0^* = \beta_0$, the quantity ψ_1^* is determined from Equation (9.21a).

Now ψ_0^* and ψ_1^* being available, Equation (9.21b) provides a recurrence relation for the remaining ψ_k^*, $k = 2, 3, \ldots, (n - 1)$.

In the case of boundary conditions of the first kind at $x = x_0$, *Equation* (9.21a) *is no longer applicable*; we know that for such a case $\psi_0^* = 0$. Then knowing $\psi_0^* = 0$ and choosing the value of ψ_1^* arbitrarily, say, $\psi_1^* = 1$, Equation (9.21b) provides a recurrence relation for the determination of the remaining ψ_k^*, $k = 2, 3, \ldots, (n - 1)$.

Finally, Equation (9.11c), that is,

$$P_n(\mu, x_0)\psi_{n-1}^* + \left[Q_n(\mu, x_n) + \frac{\alpha_n}{\beta_n} \right] \psi_n^* = 0 \qquad (9.21c)$$

is used to estimate the magnitude of the global error involved in the computation of ψ_k.

For the special case of the boundary condition of the first kind at $x = x_n$, *Equation* (9.21c) *is not applicable*. For such a case, Equation (9.21c) is written for $k = n - 1$ and noted that $\psi_n^* = 0$. This resulting expression is used for the estimation of the global error.

Evaluation of the Norm

Once the eigenfunctions are available, the norm N_i is determined from

$$N_i = \sum_{k=1}^{n} \int_{x_{k-1}}^{x_k} w_k(x)\psi_k^2(\mu_i, x)\, dx \qquad (9.22a)$$

and for the three specific geometries considered here, Equation (9.22a) takes the form

$$N_i = \sum_{k=1}^{n} \rho_k C_k \int_{x_{k-1}}^{x_k} x^{1-2m}\psi_k^2(\mu_i, x)\, dx \qquad (9.22b)$$

The problem of determining the norm is now reduced to the evaluation of the integral in Equation (9.22b), and the functions $\psi_k(\mu_i, x)$ appearing in this equation are, actually, the eigenfunctions for a single region problem. We recall that in Chapter 3 the general solution for the $\psi_k(\mu, x)$ functions was developed first in terms of the Bessel functions [see Equations (3.11)], and the eigenfunctions for the slab and sphere were obtained from the Bessel function solution as special cases. Therefore, regarding $\psi_k(\mu, x)$, any linear combination of the Bessel functions, the following relation holds [26, p. 89; 27, p. 135]

$$\int z W_m^2(\beta z)\, dz = \frac{z^2}{2} \left\{ \left(1 - \frac{m^2}{\beta^2 z^2} \right) W_m^2(\beta z) + \frac{1}{\beta^2} \left[\frac{dW_m(\beta z)}{dz} \right]^2 \right\} \qquad (9.23)$$

where $W_m(\beta z)$ is any linear combination of Bessel functions. The integral appearing in Equation (9.22b) is written in a more convenient form as

$$\int_{x_{k-1}}^{x_k} x^{1-2m}\psi_k^2(\mu_i, x)\, dx = \int_{x_{k-1}}^{x_k} x\left[\frac{\psi_k(\mu_i, x)}{x^m}\right]^2 dx \qquad (9.24)$$

A comparison of the left-hand side of Equation (9.23) with the right-hand side of Equation (9.24) yields

$$W_m(\beta x) = \frac{\psi_k(\mu_i, x)}{x^m} \qquad (9.25)$$

The differentiation of Equation (9.25) gives

$$\frac{dW_m(\beta x)}{dx} = \frac{1}{x^{2m}}\left[x^m \frac{d\psi_k(\mu_i, x)}{dx} - mx^{m-1}\psi_k(\mu_i, x)\right] \qquad (9.26a)$$

This relation can be rearranged as

$$\frac{dW_m(\omega_k x)}{dx} = \frac{x^{m-1}}{k_k}\left[\psi_{k-1}^* P_k(x) + \psi_k^* Q_k(x)\right] - \frac{m}{x^{m+1}}\psi_k(x)$$

$$(9.26b)$$

Introducing Equations (9.25) and (9.26b) into Equation (9.23) we obtain

$$\int_{x_{k-1}}^{x_k} x^{1-2m}\psi_k^2(\mu_i, x)\, dx$$

$$= \frac{1}{2\omega_k^2}\left\{\frac{1}{x_k^{2m}}\left[(x_k^2\omega_k^2 - m^2)\psi_k^{*2}\right.\right.$$

$$+ \left\{\frac{x_k^{2m}}{k_k}\left[\psi_{k-1}^* P_k(\mu_i, x_k) + \psi_k^* Q_k(\mu_i, x_k)\right] - m\psi_k^*\right\}^2\right]$$

$$- \frac{1}{x_{k-1}^{2m}}\left[(x_{k-1}^2\omega_k^2 - m^2)\psi_{k-1}^{*2}\right.$$

$$+ \left\{\frac{x_{k-1}^{2m}}{k_k}\left[\psi_{k-1}^* P_k(\mu_i, x_{k-1}) + \psi_k^* Q_k(\mu_i, x_{k-1})\right] - m\psi_{k-1}^*\right\}^2\right]\right\}$$

$$(9.27)$$

Equation (9.27) allows the determination of the normalization integral defined by Equation (9.22b).

To compute the function $T_k(x, t)$ from Equation (9.5a) we need not only the normalization integral N_i, but also the quantities $\Omega_i(x_0)$ and $\Omega_i(x_n)$ defined by Equations (9.5d, e), respectively. In terms of the notation used in the foregoing analysis, the functions $\Omega_i(x_0)$ and $\Omega_i(x_n)$ can be written as

$$\Omega_i(x_0) = \frac{\psi_0^*[1 + P_1(\mu_i, x_0)] - \psi_1^* P_1(\mu_i, x_1)}{\alpha_0 + \beta_0} \qquad (9.28a)$$

$$\Omega_i(x_n) = \frac{\psi_n^*[1 - Q_n(\mu_i, x_n)] - \psi_{n-1}^* P_n(\mu_i, x_n)}{\alpha_n + \beta_n} \qquad (9.28b)$$

9.5 SUMMARY OF GENERAL SOLUTIONS FOR LAYERS OF SLABS, CYLINDERS, AND SPHERES

To serve as a ready reference, we now write explicitly the general solution given by Equation (9.5a) for the cases of layers of slabs, cylinders, and spheres, by setting $m = \frac{1}{2}$, $m = 0$, and $m = -\frac{1}{2}$, respectively.
Slabs:

$$T_k(x, t) = \left\{ \left[\beta_n + \alpha_n \left(\frac{x_k - x}{k_k} + \sum_{i=k+1}^{n} \frac{x_i - x_{i-1}}{k_i} + \sum_{i=k}^{n-1} \frac{1}{h_i} \right) \right] \phi_0 \right.$$

$$+ \left[\beta_0 + \alpha_0 \left(\frac{x - x_{k-1}}{k_k} + \sum_{i=1}^{k-1} \frac{x_i - x_{i-1}}{k_i} + \sum_{i=1}^{k-1} \frac{1}{h_i} \right) \right] \phi_n \right\}$$

$$\times \left\{ \alpha_0 \beta_n + \alpha_n \beta_0 + \alpha_0 \alpha_n \sum_{k=1}^{n} \frac{x_k - x_{k-1}}{k_k} + \alpha_0 \alpha_n \sum_{k=1}^{n-1} \frac{1}{h_i} \right\}^{-1}$$

$$+ \sum_{i=1}^{\infty} \frac{\psi_k(\mu_i, x)}{N_i} e^{-\mu_i^2 t} \left\{ \tilde{f}_i - \frac{1}{\mu_i^2} [\phi_0 \Omega_i(x_0) + \phi_n \Omega_i(x_n)] \right\}$$

$$(k = 1, 2, \dots, n) \qquad (9.29)$$

Cylinders:

$$
T_k(x,t) = \left\{ \left[\beta_n + \alpha_n \left(\frac{1}{k_k} \ln \frac{x_k}{x} + \sum_{i=k+1}^{n} \frac{1}{k_i} \ln \frac{x_i}{x_{i-1}} + \sum_{i=k}^{n-1} \frac{1}{h_i x_i} \right) \right] \phi_0 \right.
$$

$$
+ \left[\beta_0 + \alpha_0 \left(\frac{1}{k_k} \ln \frac{x}{x_{k-1}} + \sum_{i=1}^{k-1} \frac{1}{k_i} \ln \frac{x_i}{x_{i-1}} + \sum_{i=1}^{k-1} \frac{1}{h_i x_i} \right) \right] \phi_n \right\}
$$

$$
\times \left\{ \alpha_0 \beta_n + \alpha_n \beta_0 + \alpha_0 \alpha_n \sum_{k=1}^{n} \frac{1}{k_k} \ln \frac{x_k}{x_{k-1}} + \alpha_0 \alpha_n \sum_{k=1}^{n-1} \frac{1}{h_k x_k} \right\}^{-1}
$$

$$
+ \sum_{i=1}^{\infty} \frac{\psi_k(\mu_i,x)}{N_i} e^{-\mu_i^2 t} \left\{ \tilde{f}_i - \frac{1}{\mu_i^2} \left[\phi_0 \Omega_i(x_0) + \phi_n \Omega_i(x_n) \right] \right\}
$$

$$
(k = 1,2,\dots,n) \quad (9.30)
$$

Spheres:

$$
T_k(x,t) = \left\{ \left[\beta_n + \alpha_n \left(\frac{1}{k_k} \left\langle \frac{1}{x} - \frac{1}{x_k} \right\rangle \right. \right. \right.
$$

$$
+ \left. \left. \sum_{i=k+1}^{n} \frac{1}{k_i} \left\langle \frac{1}{x_{i-1}} - \frac{1}{x_i} \right\rangle + \sum_{i=k}^{n-1} \frac{1}{h_i x_i^2} \right) \right] \phi_0
$$

$$
+ \left[\beta_0 + \alpha_0 \left(\frac{1}{h_k} \left\langle \frac{1}{x_{k-1}} - \frac{1}{x} \right\rangle \right. \right.
$$

$$
+ \left. \left. \sum_{i=1}^{k-1} \frac{1}{k_i} \left\langle \frac{1}{x_{i-1}} - \frac{1}{x_i} \right\rangle + \sum_{i=1}^{k-1} \frac{1}{h_i x_i^2} \right) \right] \phi_n \right\}
$$

$$
\times \left\{ \alpha_0 \beta_n + \alpha_n \beta_0 + \alpha_0 \alpha_n \sum_{k=1}^{n} \frac{1}{k_k} \left\langle \frac{1}{x_{k-1}} - \frac{1}{x_k} \right\rangle + \alpha_0 \alpha_n \sum_{k=1}^{n-1} \frac{1}{h_k x_k^2} \right\}^{-1}
$$

$$
+ \sum_{i=1}^{\infty} \frac{\psi_k(\mu_i,x)}{N_i} e^{-\mu_i^2 t} \left\{ \tilde{f}_i - \frac{1}{\mu_i^2} \left[\phi_0 \Omega_i(x_0) + \phi_n \Omega_i(x_n) \right] \right\}
$$

$$
(k = 1,2,\dots,n) \quad (9.31)
$$

Once the eigenvalues and the eigenfunctions are computed as described, and the normalization integral and the functions $\Omega_i(x_0)$ and $\Omega_i(x_n)$ are calculated, Equations (9.29)–(9.31) provide explicit relations for the determination of $T(x, t)$ at any location of a composite of slabs, cylinders, and spheres, respectively. This matter is illustrated with some examples in the following section.

Example 9.1 A two-layer slab consists of a first layer $0 \le x \le x_1$ and a second layer $x_1 \le x \le x_2$ which are in perfect thermal contact. Initially, the first and the second layers have temperature distributions specified by the functions $f_1(x)$ and $f_2(x)$, respectively. For times $t > 0$, the boundary surface at $x = 0$ is maintained at zero temperature, and that at $x = x_2$ dissipates heat by convection with a heat transfer coefficient h into an ambient at zero temperature. Develop an expression for the temperature distribution in this composite slab as a function of time and position.

Solution The mathematical formulation of this heat conduction problem is given by

$$\frac{\partial T_k(x, t)}{\partial t} = \alpha_k^* \frac{\partial^2 T_k(x, t)}{\partial x^2} \quad \text{in } x_{k-1} < x < x_k \quad \text{for } t > 0$$

$$(k = 1, 2) \quad \text{and} \quad x_0 = 0 \quad (9.32)$$

Subject to the boundary conditions

$$T_1(x, t) = 0 \qquad\qquad \text{at } x = 0, \qquad \text{for } t > 0 \qquad (9.33a)$$

$$T_1(x, t) = T_2(x, t) \qquad\qquad \text{at } x = x_1, \qquad \text{for } t > 0 \qquad (9.33b)$$

$$k_1 \frac{\partial T_1(x, t)}{\partial x} = k_2 \frac{\partial T_2(x, t)}{\partial x} \qquad \text{at } x = x_1, \qquad \text{for } t > 0 \qquad (9.33c)$$

$$T_2(x, t) + k_2^* \frac{\partial T_2(x, t)}{\partial x} = 0 \qquad \text{at } x = x_2, \qquad \text{for } t > 0 \qquad (9.33d)$$

where

$$k_2^* = \frac{k_2}{h} \qquad\qquad (9.33e)$$

and the initial conditions

$$T_k(x, t) = f_k(x) \quad \text{in } x_{k-1} \le x \le x_k \quad \text{for } t = 0 \qquad (k = 1, 2) \quad (9.34)$$

A comparison of this problem with the general problem given by Equations

(9.1) reveals the following correspondence:

$$m = \tfrac{1}{2}(\text{slab}), \qquad n = 2 \text{ (two regions)}$$

$$\alpha_0 = 1, \qquad \beta_0 = 0, \qquad \phi_0 = 0 \tag{9.35}$$

$$\alpha_2 = 1, \qquad \beta_2 = 1, \qquad \phi_2 = 0, \qquad k_2^* = k_2$$

The solution of this problem is obtainable from the general solution, Equation (9.29), as a special case by introducing the simplifications defined by Equations (9.35). We find

$$T_k(x, t) = \sum_{i=1}^{\infty} \frac{\psi_k(\mu_i, x)}{N_i} e^{-\mu_i^2 t} \tilde{f}_i \tag{9.36a}$$

where the transform of the initial condition functions $\tilde{f}_i$ is determined from Equation (9.5c) as

$$\tilde{f}_i = \rho_1 c_1 \int_0^{x_1} f_1(x) \psi_1(\mu_i, x)\, dx + \rho_2 c_2 \int_{x_1}^{x_2} f_2(x) \psi_2(\mu_i, x)\, dx \tag{9.36b}$$

The normalization integral N_i is obtained from Equation (9.5b) as

$$N_i = \rho_1 c_1 \int_0^{x_1} \psi_1^2(\mu_i, x)\, dx + \rho_2 c_2 \int_{x_1}^{x_2} \psi_2^2(\mu_i, x)\, dx \tag{9.36c}$$

where the integrals can be performed by utilizing the expression given by Equation (9.27).

The eigenfunctions $\psi_k(\mu_i, x)$ in general can be taken in the form given by Equation (9.7). In the case of boundary condition of the first kind at $x = 0$ as discussed previously, the first constant is taken as zero, $\psi_0^* = 0$, and the second constant is chosen arbitrarily as $\psi_1^* = 1$, then the eigenfunctions take the form

$$\psi_1(\mu_i, x) = V_1(\mu_i, x) \tag{9.37a}$$

$$\psi_2(\mu_i, x) = U_2(\mu_i, x) + \psi_2^* V_2(\mu_i, x) \tag{9.37b}$$

The functions $U_k(\mu_i, x)$ and $V_k(\mu_i, x)$ are obtained from Table 9.2 for $m = \tfrac{1}{2}$ as

$$V_1(\mu_i, x) = \frac{\sin(\omega_1 x)}{\sin(\omega_1 x_1)} \tag{9.38a}$$

$$V_2(\mu_i, x) = \frac{\sin[\omega_2(x - x_1)]}{\sin[\omega_2(x_2 - x_1)]} \tag{9.38b}$$

$$U_2(\mu_i, x) = \frac{\sin[\omega_2(x_2 - x)]}{\sin[\omega_2(x_2 - x_1)]} \tag{9.38c}$$

where

$$\omega_k = \mu_i / \sqrt{\alpha_k^*} . \tag{9.38d}$$

The constant ψ_2^* is determined from Equation (9.21b) by setting $k = 1$:

$$\psi_2^* = \frac{P_2(\mu_i, x_1) - Q_1(\mu_i, x_1)}{P_2(\mu_i, x_2)} \tag{9.39a}$$

When the functions P_k and Q_k are obtained from Table 9.3, Equation (9.39a) takes the form

$$\psi_2^* = \cos[\omega_2(x_2 - x_1)] + \frac{\omega_1 k_1}{\omega_2 k_2} \frac{\sin[\omega_2(x_2 - x_1)]}{\sin(\omega_1 x_1)} \cos(\omega_1 x_1) \tag{9.39b}$$

Finally, the eigenvalues μ_i are determined by the sign-count method discussed earlier.

REFERENCES

1. V. Vodicka, Wärmeleitung in Geschichteten Kugel-und Zylinderkörpern, *Schweizer Archiv*, **10**, 297–304 (1950).

2. V. Vodicka, Eindimensionale Wärmeleitung in Geschichteten Körpern, *Math. Nachrich.*, **14**, 47–55 (1955).

3. P. E. Bulavin and V. M. Kascheev, Solution of the Nonhomogeneous Heat Conduction Equation for Multilayered Bodies, *Int. Chem. Eng.*, **1**, 112–115 (1965).

4. C. W. Tittle, Boundary Value Problems in Composite Media: Quasi-orthogonal Functions, *J. Applied Physics*, **36**, 1486–1488 (1965).

5. C. W. Tittle and V. L. Robinson, *Analytical Solution of Conduction Problems in Composite Media*, ASME Paper 65-WA-HT-32, 1965.

6. M. N. Özişik, *Boundary Value Problems of Heat Conduction*, International Textbook, Scranton, PA, 1968.

7. M. H. Cobble, Heat Transfer in Composite Media Subject to Distributed Sources, and Time-Dependent Discrete Sources and Surroundings, *J. Franklin Inst.*, **290**, 5, 453–465 (1970).

8. G. P. Mulholland and M. N. Cobble, Diffusion Through Composite Media, *Int. J. Heat Mass Transf.*, **15**, 147–160 (1972).

9. C. A. Chase, D. Gidaspow, and R. E. Peck, Diffusion of Heat and Mass in Porous Medium with Bulk Flow: Part I. Solution by Green's Functions, *Chem. Eng. Progr. Symp.*, **65**, Ser. No. 92, 91–109 (1969).

10. B. S. Baker, D. Gidaspow, and D. Wasan, in *Advances in Electrochemistry and Electrochemical Engineering*, Wiley-Interscience, New York, 1971, pp. 63–156.

11. T. R. Goodman, *The Adjoint Heat-Conduction Problems for Solids*, ASTIA-AD 254-769, (AFOSR-520), April 1961.

12. J. Crank, *The Mathematics of Diffusion*, 2nd ed., Clarendon Press, London, 1975.

13. H. S. Carslaw and J. C. Jaeger, *Conduction of Heat in Solids*, Clarendon Press, London, 1959.

14. V. S. Arpaci, *Conduction Heat Transfer*, Addison-Wesley, Reading, MA, 1966.

15. A. V. Luikov, *Analytical Heat Diffusion Theory*, Academic, New York, 1968.

16. N. Y. Ölçer, Theory of Unsteady Heat Conduction in Multicomponent Finite Regions, *Ingenieur-Archiv*, **36**, 285–293 (1968).

17. N. Y. Ölçer, A General Class of Unsteady Heat Flow Problems in a Finite Composite Hollow Circular Cylinder, *Quart. Apl. Math.*, **26**, 355–371 (1968).

18. N. Y. Ölçer, A General Unsteady Heat Flow Problem in a Finite Composite Hollow Circular Cylinder Under Boundary Conditions of the Second Kind, *Nuclear Eng. Des.*, **7**, 97–112 (1968).

19. N. Y. Ölçer, Theory of Unsteady Heat Conduction in Multicomponent Finite Regions, *Ingenieur-Archiv.*, **36**, 285–293 (1968).

20. Kanae Senda, *A Family of Integral Transforms and Some Applications to Physical Problems*, Technol. Reports Osaka University, Osaka, Japan, No. 823, **18**, 261–286, 1968.

21. J. D. Lockwood and G. P. Mulholland, Diffusion Through Laminated Composite Cylinders Subjected to a Circumferentially Varying External Heat Flux, *J. Heat Transf.*, **95c**, 487–491 (1973).

22. M. D. Mikhailov, General Solutions of the Coupled Diffusion Equations, *Int. J. Eng. Sci.*, **11**, 235–241 (1973).

23. M. D. Mikhailov, General Solutions of the Diffusion Equations Coupled at Boundary Conditions, *Int. J. Heat Mass Transf.*, **16**, 2155–2164 (1973).

24. Y. Yener and M. N. Özişik, *On the Solution of Unsteady Heat Conduction in Multi-Region Media with Time Dependent Heat Transfer Coefficients, Proc. 5th Int. Heat Transfer Conference*, Tokyo, Sept. 1974.

25. J. Padovan, Generalized Sturm–Liouville Procedure for Composite Domain Anisotropic Transient Conduction Problems, *AIAA J.*, **12**, 1158–1160 (1974).

26. M. N. Özişik, *Heat Conduction*, Wiley, New York, 1980.

27. G. N. Watson, *A Treatise on the Theory of Bessel Functions*, 2nd ed., Cambridge University Press, London, 1966.

28. M. D. Mikhailov, M. N. Özişik and N. L. Vulchanov, Transient Heat Diffusion in One-Dimensional Composite Media and Automatic Solution of the Eigenvalue Problem, *Int. J. Heat Mass Transf.*, **26**, 1131–1141 (1983).

CHAPTER TEN

Class III Solutions Applied
to Heat and Mass Transfer
in Capillary Porous Body
and Heat Transfer
in Entrance Concurrent Flow

Observation is a passive science, experimentation is an active science.

Claude Bernard
1813–1878

In this chapter, we illustrate various applications of the Class III problem. We first consider the problem of temperature and moisture distribution during contact drying of a moist, porous sheet, but the method is more general, and applicable to the problem of capillary porous body of arbitrary geometry [14]. We then consider problems involving the determination of mass concentration in the flow of two immiscible fluids, that is, gas–liquid or liquid–liquid flow [15–18], temperature distributions in concurrent flow double pipe heat exchangers [19–20], and simultaneous heat and mass transfer in internal gas flow in a duct whose walls are coated with a sublimable material [21], and many others. It is shown that solutions to all these different types of problems are obtainable as special cases of the one-dimensional problem of Class III presented in Chapter 4. Numerous other applications of the problems belonging to this class can be found in References 22–26.

10.1 LUIKOV'S EQUATIONS OF HEAT AND MASS TRANSFER IN CAPILLARY-POROUS BODIES

The theory of drying is based on the phenomenon of thermodiffusion found by Luikov in 1935 [1], who proved experimentally and explained theoretically that

399

the moisture migration is caused by the temperature gradient [2, 3]. These findings enabled him, on the basis of thermodynamics of irreversible processes, to define a system of coupled differential equations [4] known as the Luikov system of equations. A survey of work on drying of solids by Luikov and his co-workers, until 1969, is given by Fulford [5].

A system of partial differential equations governing the distribution of temperature and moisture in a capillary porous body was also proposed independently by Krischer [6, 7]. However, as shown in Reference 8, Krischer's system was identical to that of Luikov. The system defined by de Vries [9] and that defined in reference 10 for the case of constant thermophysical properties were also of the Luikov type.

The interrelated differential equations of heat and mass transfer in capillary-porous bodies, which were first proposed by Luikov and discussed in References 26 and 27, have numerous applications in civil engineering, thermophysics, and migration in soils and grounds as well as the calculation of the wicks of heat pipes.

Here we present these equations for the simple case of zero gradient of total pressure inside the body. The reader should consult Reference 28 for the development of these equations on the basis of irreversible thermodynamics for more complicated system.

If u_1 and u_2 denote the *mass capacities* (or *the mass content*) of vapor and subscript 0 refers to a perfectly dry body, then by setting in Equation (1.23):

$$\rho \equiv \rho_0, \qquad \xi = u_k, \qquad \mathbf{J} \equiv \mathbf{J}_k \quad \text{and} \quad I \equiv I_k$$

we write

$$\rho_0 \frac{\partial u_k}{\partial t} + \nabla \mathbf{J}_k = I_k, \qquad k = 1, 2 \tag{10.1}$$

where ρ_0 is the density of a perfectly dry body, $\mathbf{J}_k$ is the mass flux vector, and I_k is the mass source or sink due to phase transition of the kth component. In transient processes the source of vapor mass I_1 or the sink of liquid mass I_2 are related by

$$I_1 = -I_2 = \epsilon \rho_0 \frac{\partial u}{\partial t} \tag{10.2}$$

where ϵ is the *phase conversion factor* of liquid into vapor [26].

If the temperature T and the mass content u are taken as the potentials of diffusional mass transfer, then the nonisothermal moisture diffusion law is taken in the form [26]

$$\mathbf{J}_m = \mathbf{J}_1 + \mathbf{J}_2 = -\rho_0 a_m (\nabla u + \delta \nabla T) \tag{10.3}$$

where $u = u_1 + u_2$, δ is the *thermogradient coefficient*, and a_m is the *moisture diffusivity*. Here for ordinary conditions of heat and mass transfer the mass of vapor is extremely small compared with the liquid mass. Therefore, mass content of the liquid can be taken, with a high degree of accuracy, equal to the total mass, namely, $u = u_2$.

Summing up Equations (10.1) over $k = 1$ and 2, and utilizing Equations (10.2) and (10.3) we obtain

$$\frac{\partial u}{\partial t} = \nabla(a_m \nabla u) + \nabla(a_m \delta \nabla T) \tag{10.4}$$

If the mass flow rate in the body is slow, then as an approximation the body skeleton temperature may be considered equal to that of any phase of the body material. It follows that vapor in body capillaries is at thermodynamic equilibrium with the liquid.

The energy balance Equation (1.32) can then be taken in the form

$$\rho_0 c \frac{\partial T}{\partial t} = \nabla(k \nabla T) + r\epsilon\rho_0 \frac{\partial u}{\partial t} \tag{10.5}$$

where r is *the latent heat of evaporation*, c is the *reduced specific heat* of the body taken as [see Equation (1.31)]

$$c \equiv \left(\frac{\partial \xi}{\partial T}\right)_\rho = c_0 + \sum_{k=1}^{2} c_k u_k \tag{10.6}$$

c_k is the specific heat of the kth component and k is the thermal conductivity. If the thermodynamic properties (i.e., specific heat and thermogradient coefficient) and the transfer coefficients (i.e., thermal diffusivity and moisture diffusivity) are assumed to be constant over the body, then Equations (10.4) and (10.5) lead to the linear *Luikov's system of equations* of heat and mass transfer in capillary porous bodies given as

$$\frac{\partial T}{\partial t} = a \nabla^2 T + \frac{\epsilon r}{c} \frac{\partial u}{\partial t} \tag{10.7a}$$

$$\frac{\partial u}{\partial t} = a_m \nabla^2 u + a_m \delta \nabla^2 T \tag{10.7b}$$

where $a = k/\rho_0 c$ is the *thermal diffusivity* and a_m the *moisture diffusivity*.

In the foregoing analysis we used Luikov's nomenclature, since it is widely used.

10.2 TEMPERATURE AND MOISTURE DISTRIBUTIONS DURING CONTACT DRYING OF A SHEET OF POROUS MOIST MATERIAL

The coupled system of partial differential equations of the Luikov type can be reduced to two decoupled partial differential equations of the diffusion type coupled only at the boundaries. The decoupling procedure, in general, is similar to that followed by Henry [11] and later by Crank [12] and Smirnov [13], who applied the method for the case of monotype boundary conditions. In the case of its application to a more general system, the resulting transformed equations belong to the problem of Class III.

In this section we illustrate the application of the one-dimensional Class III problem given in Chapter 4 to obtain solutions for the temperature and moisture distributions during contact drying of a moist porous sheet on a hot plate, considered in reference 23. Heat flows through the sheet from bottom to top as illustrated in Figure 10.1. Depending on whether the temperature at the top surface is higher or lower than that of surrounding air, the sheet loses or gains heat by free or forced convection.

The distributions of temperature $T(x, t)$ and mass $u(x, t)$ in the sheet are described by the Luikov system of Equations (10.7), that is,

$$\frac{\partial T(x, t)}{\partial t} = a\frac{\partial^2 T(x, t)}{\partial x^2} + \epsilon\frac{r}{c}\frac{\partial u(x, t)}{\partial t} \quad \text{in } 0 < x < \ell, \quad t > 0 \qquad (10.8a)$$

$$\frac{\partial u(x, t)}{\partial t} = a_m\frac{\partial^2 u(x, t)}{\partial x^2} + a_m\delta\frac{\partial^2 T(x, t)}{\partial x^2} \quad \text{in } 0 < x < \ell, \quad t > 0 \qquad (10.8b)$$

where a is the thermal diffusivity of the porous medium, a_m is the diffusion coefficient of moisture in the porous medium, ϵ is the phase change criterion (i.e., $\epsilon = 1$ all vapor, $\epsilon = 0$ all liquid), r is the latent heat of evaporation, c is the specific heat of the porous medium, δ is the thermogradient coefficient (i.e., $°C^{-1}$) for the transfer of mass, and $u(x, t)$ is the concentration of the diffusing matter given as mass of the diffusing matter per unit mass of the porous material.

Equations (10.8) are to be solved subject to the initial condition

$$T(x, 0) = T_0, \qquad u(x, 0) = u_0 \qquad (10.9a, b)$$

and the boundary conditions given in the following.

A prescribed constant heat flux q supplied by the hot plate at the lower surface $x = 0$ of the sheet:

$$k\frac{\partial T(0, t)}{\partial x} = -q \qquad (10.9c)$$

where k is the thermal conductivity of the sheet.

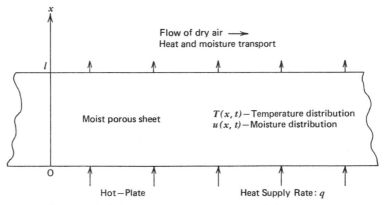

Figure 10.1   Schematic picture of the contact drying process.

The mass balance at the lower surface $x = 0$ of the sheet is written as

$$\frac{\partial u(0, t)}{\partial x} + \delta \frac{\partial T(0, t)}{\partial x} = 0 \qquad (10.9d)$$

The heat balance on the free surface $x = \ell$ of the sheet is

$$k \frac{\partial T(\ell, t)}{\partial x} + h[T(\ell, t) - T_s] + (1 - \epsilon) r h_m [u(\ell, t) - u^*] = 0 \qquad (10.9e)$$

where h is heat transfer coefficient, h_m is the mass transfer coefficient, T_s is the temperature of surrounding air, and u^* is the moisture in equilibrium with surrounding air.

The mass balance on the free surface $x = \ell$ of the sheet gives

$$k_m \left[\frac{\partial u(\ell, t)}{\partial x} + \delta \frac{\partial T(\ell, t)}{\partial x} \right] + h_m [u(\ell, t) - u^*] = 0 \qquad (10.9f)$$

where h_m is the mass transfer coefficient and k_m is the moisture conductivity.

To nondimensionalize the problem given by Equations (10.8) and (10.9) we define the following dimensionless variables

$\text{Bi}_m = \dfrac{h_m \ell}{k_m}$, dimensionless mass transfer coefficient

$\text{Bi}_q = \dfrac{h\ell}{k}$, dimensionless heat transfer coefficient

$\text{Ko} = \dfrac{r}{c} \dfrac{u_0 - u^*}{T_s - T_0}$, Kossovitch number

$\text{Lu} = \dfrac{a_m}{a}$, Luikov number or (Lewis number)$^{-1}$

$\text{Pn} = \delta \dfrac{T_s - T_0}{u_0 - u^*}$, Possnov number

$Q = \dfrac{q\ell}{k(T_s - T_0)}$, dimensionless heat flux

$X = \dfrac{x}{\ell}$, dimensionless coordinate

$\theta_1(X, \tau) = \dfrac{T(x, t) - T_0}{T_s - T_0}$, dimensionless temperature

$\theta_2(X, \tau) = \dfrac{u_0 - u(x, t)}{u_0 - u^*}$, dimensionless moisture

$\tau = \dfrac{at}{\ell^2}$, dimensionless time $\qquad (10.10)$

Introducing these dimensionless quantities into the problem given by Equations (10.8) and (10.9) we obtain the system in the dimensionless form as

$$\frac{\partial \theta_1(X, \tau)}{\partial \tau} = \frac{\partial^2 \theta_1(X, \tau)}{\partial X^2} - \epsilon \, \text{Ko} \frac{\partial \theta_2(X, \tau)}{\partial \tau} \quad \text{in } 0 < X < 1, \quad \tau > 0,$$

$$(10.11\text{a})$$

$$\frac{\partial \theta_2(X, \tau)}{\partial \tau} = \text{Lu} \frac{\partial^2 \theta_2(X, \tau)}{\partial X^2} - \text{Lu Pn} \frac{\partial^2 \theta_1(X, \tau)}{\partial X^2} \quad \text{in } 0 < X < 1, \quad \tau > 0$$

$$(10.11\text{b})$$

Subject to the following initial conditions

$$\theta_1(X, 0) = 0, \qquad \theta_2(X, 0) = 0 \qquad\qquad (10.11\text{c, d})$$

and the boundary conditions given by

$$\frac{\partial \theta_1(0, \tau)}{\partial X} = -Q \qquad\qquad (10.11\text{e})$$

$$\frac{\partial \theta_2(0, \tau)}{\partial X} - \text{Pn} \frac{\partial \theta_1(0, \tau)}{\partial X} = 0 \qquad\qquad (10.11\text{f})$$

$$\frac{\partial \theta_1(1, \tau)}{\partial X} - \text{Bi}_q\left[1 - \theta_1(1, \tau)\right] + (1 - \epsilon)\text{Ko Lu Bi}_m\left[1 - \theta_2(1, \tau)\right] = 0$$

$$(10.11\text{g})$$

$$-\frac{\partial \theta_2(1, \tau)}{\partial X} + \text{Pn} \frac{\partial \theta_1(1, \tau)}{\partial X} + \text{Bi}_m\left[1 - \theta_2(1, \tau)\right] = 0 \quad (10.11\text{h})$$

In the foregoing system the governing partial differential Equations (10.11a) and (10.11b) are coupled. However, these partial differential equations can be decoupled by a transformation similar to that used in References 11–13. Then the resulting system becomes a special case of the one-dimensional Class III problem given in Chapter 4.

The transformation from the functions $\theta_k(X, \tau)$ to new functions $Z_k(X, \tau)$, $k = 1, 2$, is taken as (see Note 1 for the derivation)

$$Z_k(X, \tau) = \theta_1(X, \tau) + \frac{\nu_k^2 - 1}{\text{Pn}} \theta_2(X, \tau), \qquad k = 1, 2 \quad (10.12\text{a})$$

where

$$\nu_k^2 = \tfrac{1}{2}\left[\left(1 + \epsilon\,\mathrm{Ko}\,\mathrm{Pn} + \frac{1}{\mathrm{Lu}}\right) + (-1)^k\sqrt{\left(1 + \epsilon\,\mathrm{Ko}\,\mathrm{Pn} + \frac{1}{\mathrm{Lu}}\right)^2 - \frac{4}{\mathrm{Lu}}}\,\right],$$

$$k = 1,2 \quad (10.12\mathrm{b})$$

Thus from Equations (10.12) it follows that

$$\theta_1(X,\tau) = \frac{1}{\nu_2^2 - \nu_1^2}\left[(\nu_2^2 - 1)Z_1(X,\tau) - (\nu_1^2 - 1)Z_2(X,\tau)\right]$$

$$(10.13\mathrm{a})$$

$$\theta_2(X,\tau) = \frac{\mathrm{Pn}}{\nu_2^2 - \nu_1^2}\left[Z_2(X,\tau) - Z_1(X,\tau)\right] \qquad (10.13\mathrm{b})$$

and from Equations (10.12b) we also write

$$\nu_1^2\nu_2^2 = \frac{1}{\mathrm{Lu}} \qquad (10.14\mathrm{a})$$

$$\left(\nu_1^2 - 1\right)\left(\nu_2^2 - 1\right) = -\epsilon\,\mathrm{Ko}\,\mathrm{Pn} \qquad (10.14\mathrm{b})$$

Now introducing Equations (10.13) into problem (10.11) and utilizing the relations (10.14), the system of equations for the determination of the functions $Z_k(X,\tau)$ becomes

$$\nu_k^2\frac{\partial Z_k(X,\tau)}{\partial\tau} = \frac{\partial^2 Z_k(X,\tau)}{\partial X^2} \quad \text{in } 0 < X < 1, \qquad \tau > 0, \qquad k = 1,2$$

$$(10.15\mathrm{a,b})$$

subject to the initial conditions

$$Z_k(X,0) = 0, \qquad k = 1,2 \qquad (10.15\mathrm{c,d})$$

and the boundary conditions given by

$$-\frac{\partial Z_k(0,\tau)}{\partial X} = \nu_k^2 Q, \qquad k = 1,2 \qquad (10.15\mathrm{e,f})$$

$$\sum_{k=1}^{2}(-1)^k\left(1 - \nu_{3-k}^2\right)\left\{\left[1 + \left(1 - \nu_k^2\right)\frac{1-\epsilon}{\epsilon}\mathrm{Lu}\frac{\mathrm{Bi}_m}{\mathrm{Bi}_q}\right]Z_k(1,\tau) + \frac{1}{\mathrm{Bi}_q}\frac{\partial Z_k(1,\tau)}{\partial X}\right\}$$

$$= \left(\nu_2^2 - \nu_1^2\right)\left[1 - (1-\epsilon)\mathrm{Ko}\,\mathrm{Lu}\frac{\mathrm{Bi}_m}{\mathrm{Bi}_q}\right] \qquad (10.15\mathrm{g})$$

$$\mathrm{Pn}\sum_{k=1}^{2}(-1)^k\left[Z_k(1,\tau) + \frac{\nu_{3-k}^2}{\mathrm{Bi}_m}\frac{\partial Z_k(1,\tau)}{\partial X}\right] = \left(\nu_2^2 - \nu_1^2\right) \quad (10.15\mathrm{h})$$

Clearly, problem (10.15) for the functions $Z_k(X, \tau)$, $k = 1, 2$ is a special case of the Class III problem given by Equations (4.94) and its solution can be readily obtained from the solution of Class III problems given in Chapter 4. Once the functions $Z_k(X, \tau)$ are known, the functions $\theta_1(X, \tau)$ and $\theta_2(X, \tau)$ for the temperature and moisture distribution in the medium are determined according to the relations (10.13). We now describe the solution for the functions $Z_k(X, \tau)$.

The Solution for $Z_k(X, \tau)$

Problem (10.15) for the functions $Z_k(X, \tau)$, $k = 1, 2$ is a special case of the Class III problem defined by Equations (4.94) as follows:

$$x = X, \qquad t = \tau, \qquad T_k(x, t) = Z_k(X, \tau), \qquad w_k(x) = \nu_k^2$$

$$k_k(x) = 1, \qquad d_k(x) = 0, \qquad P_k(x, t) = 0, \qquad x_0 = 0, \qquad x_1 = 1$$

$$\alpha_k = 0, \qquad \beta_k = 1, \qquad \phi_k(x_0, t) = \nu_k^2 Q$$

$$\alpha_{k1} = (-1)^k \left(1 - \nu_{3-k}^2\right) \left[1 + \left(1 - \nu_k^2\right) \frac{1 - \epsilon}{\epsilon} \operatorname{Lu} \frac{\operatorname{Bi}_m}{\operatorname{Bi}_q}\right]$$

$$= (-1)^k \left[1 - \nu_{3-k}^2 - (1 - \epsilon)\operatorname{Ko}\operatorname{Lu}\frac{\operatorname{Bi}_m}{\operatorname{Bi}_q}\right] \quad \left[\text{by utilizing Equation (10.14b)}\right]$$

$$\beta_{k1} = (-1)^k \left(1 - \nu_{3-k}^2\right) \frac{1}{\operatorname{Bi}_q}$$

$$\alpha_{k2} = (-1)^k \operatorname{Pn}, \qquad \beta_{k2} = (-1)^k \nu_{3-k}^2 \frac{\operatorname{Pn}}{\operatorname{Bi}_m}$$

$$\phi_1(x_1, t) = \left(\nu_2^2 - \nu_1^2\right)\left[1 - (1 - \epsilon)\operatorname{Ko}\operatorname{Lu}\frac{\operatorname{Bi}_m}{\operatorname{Bi}_q}\right]$$

$$\phi_2(x_2, t) = \nu_2^2 - \nu_1^2, \qquad f_k(x) = 0 \quad \text{and} \quad k = 1, 2 \qquad (10.16)$$

The solution of problem (4.94) is given by Equations (4.95). However, we prefer to use the solution that was split up into the solution of simpler problems as given by Equation (4.97). A comparison of the nonhomogeneous terms for the present problem with those given by Equations (4.96) for the general problem (4.94) reveals that we only need the simpler solutions $T_{k0}(x) \equiv Z_{k0}(x)$ and $T_{kt}(x, t) \equiv Z_{kt}(X, \tau)$. Then the solution (4.97) reduces to

$$Z_k(X, \tau) = Z_{k0}(X) + Z_{kt}(X, \tau), \qquad k = 1, 2 \qquad (10.17)$$

The function $Z_{k0}(X)$ satisfies the steady-state problem (4.100) for $j = 0$, with various coefficients specified by Equations (10.16); the resulting solution for $Z_{k0}(X)$ is given by

$$Z_{k0}(X) = 1 + Q\left(1 + \frac{1}{\mathrm{Bi}_q}\right) - \left(1 - \nu_k^2\right)\left(\frac{1}{\mathrm{Pn}} + Q\right) - Q\nu_k^2 X \quad (10.18)$$

The function $Z_{kt}(X, \tau)$ is obtained from the transient solution (4.101), and the eigenvalues and the eigenfunctions for the problem are determined from the one-dimensional eigenvalue problem for Class III. For the simpler situation considered here, the eigenvalue problem (3.115) reduces to

$$\psi_k''(X) + \mu^2 \nu_k^2 \psi_k(X) = 0 \quad \text{in } 0 < X < 1, \qquad k = 1,2 \quad (10.19\mathrm{a,b})$$

subject to the boundary conditions

$$\psi_k'(0) = 0, \qquad k = 1,2 \qquad\qquad (10.19\mathrm{c,d})$$

$$\sum_{k=1}^{2} (-1)^k \left(1 - \nu_{3-k}^2\right)\left\{\left[1 + \left(1 - \nu_k^2\right)\frac{1 - \epsilon}{\epsilon} \mathrm{Lu}\frac{\mathrm{Bi}_m}{\mathrm{Bi}_q}\right]\psi_k(1)\right.$$

$$\left. + \frac{1}{\mathrm{Bi}_q}\frac{d\psi_k(1)}{dX}\right\} = 0 \qquad\qquad (10.19\mathrm{e})$$

$$\sum_{k=1}^{2} (-1)^k \left[\psi_k(1) + \frac{\nu_{3-k}^2}{\mathrm{Bi}_m}\frac{d\psi_k(1)}{dX}\right] = 0 \qquad\qquad (10.19\mathrm{f})$$

The solution of Equations (10.19a, b) is taken in the form

$$\psi_k(\mu, X) = C_k\cos(\nu_k\mu X) + D_k\sin(\nu_k\mu X) \qquad (10.20\mathrm{a,b})$$

where C_k and D_k are constants to be determined by utilizing the boundary conditions for the problem.

From the boundary condition (10.19c, d) it follows that

$$D_k = 0 \qquad\qquad (10.20\mathrm{c})$$

The substitution of solution (10.20) into the boundary conditions (10.19e, f) and recalling that $\nu_k^2\nu_{3-k}^2 = 1/\mathrm{Lu}$ gives

$$\sum_{k=1}^{2} (-1)^k \left(1 - \nu_{3-k}^2\right)a_k(\mu)C_k = 0 \qquad\qquad (10.21\mathrm{a})$$

$$\sum_{k=1}^{2} (-1)^k b_k(\mu)C_k = 0 \qquad\qquad (10.21\mathrm{b})$$

where

$$a_k(\mu) = \left[1 + \left(1 - \nu_k^2\right)\frac{1-\epsilon}{\epsilon}\mathrm{Lu}\frac{\mathrm{Bi}_m}{\mathrm{Bi}_q}\right]\cos(\nu_k\mu) - \frac{\nu_k\mu}{\mathrm{Bi}_q}\sin(\nu_k\mu)$$

(10.21c)

$$b_k(\mu) = \cos(\nu_k\mu) - \frac{\mu}{\mathrm{Lu}\,\mathrm{Bi}_m\nu_k}\sin(\nu_k\mu)$$

(10.21d)

The eigencondition for the determination of the eigenvalues μ_i is obtained from Equations (10.21a, b) as

$$\sum_{k=1}^{2}(-1)^k\left(1 - \nu_k^2\right)a_{3-k}(\mu)b_k(\mu) = 0$$

(10.22)

The normalization integral N_i is determined according to Equation (4.95b) which contains the coefficients σ_k defined by Equation (4.95c). For the present problem, the coefficients σ_k are determined from Equation (4.95c), by utilizing the relations in Equations (10.16) as

$$\sigma_{3-k} = (-1)^{k+1}\frac{\mathrm{Pn}}{\nu_{3-k}^2}e_k$$

(10.23a)

where

$$e_k = \left(\nu_k^2 - 1\right)\left(\frac{1}{\mathrm{Bi}_q} - \frac{1}{\mathrm{Lu}\,\mathrm{Bi}_m}\right) - \frac{\mathrm{Ko}\,\mathrm{Pn}}{\mathrm{Bi}_q}$$

(10.23b)

Then introducing Equations (10.20) and (10.23) into Equation (4.95b), the normalization integral N_i is determined as

$$N_i = -\frac{C_1^2\,\mathrm{Pn}}{2b_2^2(\mu_i)}\left[e_2b_2^2(\mu_i)d_1(\mu_i) - e_1b_1^2(\mu_i)d_2(\mu_i)\right]$$

(10.23c)

where

$$d_k(\mu_i) = 1 + \cos(\nu_k\mu_i)\frac{\sin(\nu_k\mu_i)}{\nu_k\mu_i}$$

(10.23d)

Finally, the solution for $Z_{kt}(X, \tau)$ is determined from Equation (4.101) as

$$Z_{kt}(X, \tau) = -\left(\nu_2^2 - \nu_1^2\right)\sum_{i=1}^{\infty}A_ib_{3-k}(\mu_i)\cos(\mu_i\nu_k X)e^{-\mu_i^2\tau}$$

(10.24a)

where

$$A_i = \frac{2}{\mu_i} \left\{ \frac{Q}{\nu_2^2 - \nu_1^2} \left[b_2(\mu_i)e_2 - b_1(\mu_i)e_1 \right] \right.$$

$$+ \left[\left\langle 1 - (1 - \epsilon) \text{Ko Lu} \frac{\text{Bi}_m}{\text{Bi}_q} \right\rangle b_1(\mu_i) + \frac{\nu_2^2 - 1}{\text{Pn}} a_1(\mu_i) \right] b_2(\mu_i) \right\}$$

$$\times \left[b_2^2(\mu_i)e_2 d_1(\mu_i) - b_1^2(\mu_i)e_1 d_2(\mu_i) \right]^{-1} \tag{10.24b}$$

Introducing the functions $Z_{kt}(X, \tau)$ and $Z_{k0}(X)$, given by Equations (10.24a) and (10.18), respectively, into Equation (10.17), the functions $Z_k(X, \tau)$, $k = 1, 2$ are obtained. Knowing the functions $Z_k(X, \tau)$, the temperature and moisture distributions $\theta_1(X, \tau)$ and $\theta_2(X, \tau)$ are determined from Equations (10.13) as

$$\theta_1(X, \tau) = 1 + Q \left(1 + \frac{1}{\text{Bi}_q} - X \right)$$

$$- \sum_{i=1}^{\infty} A_i e^{-\mu_i^2 \tau} \sum_{k=1}^{2} (-1)^k (\nu_k^2 - 1) b_k(\mu_i) \cos(\nu_{3-k} \mu_i X)$$

$$\tag{10.25a}$$

$$\theta_2(X, \tau) = 1 + \text{Pn } Q(1 - X)$$

$$+ \text{Pn} \sum_{i=1}^{\infty} A_i e^{-\mu_i^2 \tau} \sum_{k=1}^{2} (-1)^k b_k(\mu_i) \cos(\nu_{3-k} \mu_i X) \tag{10.25b}$$

where ν_k, A_i, and $b_k(\mu_i)$ are defined by Equations (10.12b), (10.24b), and (10.21d), respectively, and the μ_i are the roots of the eigencondition (10.22). The utility of presenting comprehensive tables of roots is doubted, because it is more convenient to write a computer subroutine for the calculation of the roots whenever they are needed. However, to illustrate some typical results that may serve as reference values for root calculations, we present in Table 10.1 the first 25 roots of the transcendental Equation (10.22) for representative values of the parameters Lu, Pn, Ko, Bi$_q$, Bi$_m$, and ϵ.

We now illustrate with a specific example, the application of the foregoing results to solve the Luikov system of equations associated with a problem of contact drying.

Table 10.1 Roots of the Transcendental Equation (10.22) for Lu = 0.4, Pn = 0.6, Ko = 5, $Bi_q = 5$[a]

	$\epsilon = 0.2$					$\epsilon = 0.8$			
i	$Bi_m = 1$	$Bi_m = 2.5$	$Bi_m = 5$	$Bi_m = 10$	i	$Bi_m = 1$	$Bi_m = 2.5$	$Bi_m = 5$	$Bi_m = 10$
1	0.51519	0.63831	0.69360	0.72335	1	0.47217	0.56011	0.59744	0.61754
2	1.34300	3.61562	3.70757	3.78244	2	1.34923	2.89373	2.97822	3.07664
3	1.79173	4.36375	4.73523	6.98836	3	1.56302	4.20726	4.29142	4.37686
4	3.52453	5.11397	4.91463	10.32595	4	2.82690	5.29285	6.85339	6.92607
5	4.22759	6.91995	6.94944	13.72017	5	4.13418	5.38607	8.19809	8.26442
6	5.16807	7.71052	7.87753	17.13578	6	5.11142	6.81454	10.84592	10.89357
7	6.89770	8.55753	8.48721	20.56010	7	5.45586	8.15659	12.19444	12.24075
8	7.62573	10.31443	10.31890	23.98821	8	6.79104	10.82354	13.54706	13.59208
9	8.58915	11.22749	11.34959	27.41802	9	8.12897	12.16937	14.15944	14.26027
10	10.31135	11.99226	11.94316	30.84856	10	9.47411	13.51774	14.86293	14.89580
11	11.16115	13.73204	13.72768	32.60510	11	9.52488	14.10957	16.21477	16.24946
12	12.01553	14.79911	14.90019	32.93866	12	10.81071	14.84915	17.56219	17.59594
13	13.73489	15.42423	15.38302	34.27936	13	12.15381	16.19714	18.77126	20.27114
14	14.74381	17.15704	17.14951	36.01884	14	13.49775	17.54347	18.86232	21.61777
15	15.44358	18.39636	18.48794	36.57647	15	14.08092	18.70734	20.24370	22.97085
16	17.16181	18.85436	18.81354	37.71014	16	14.84166	18.87321	21.59084	23.38183
17	18.34788	20.58487	20.57632	39.44282	17	16.18652	20.23031	22.94318	24.29856
18	18.87225	22.00971	22.11613	40.20748	18	17.53165	21.57676	23.32264	25.64652
19	20.59016	22.28135	22.21891	41.14069	19	18.67239	22.92579	24.27626	26.99378
20	21.96365	24.01393	24.00522	42.87024	20	18.87654	23.29531	25.62412	27.99579
21	22.30093	27.44357	27.43504	43.83796	21	20.22241	24.26587	26.97161	28.32543
22	24.01923	30.87351	30.86529	44.57075	22	21.56814	25.61279	27.94214	39.67876
23	25.58828	32.59780	32.60072	46.29918	23	22.91410	26.95961	28.30859	31.02550
24	25.72771	32.85158	32.87995	47.46984	24	23.27993	27.91393	29.65963	32.38121
25	27.44870	34.30361	34.29571	47.99976	25	24.25987	28.30183	31.00663	32.58617

[a] From Reference 23.

Example 10.1 Calculate the temperature and moisture distribution during contact drying of a moist porous sheet, illustrated in Figure 10.1 for the values of various parameters taken as

$$Lu = 0.4, \quad \epsilon = 0.2, \quad Bi_m = Bi_q = 2.5, \quad Ko = 5, \quad Pn = 0.6, \quad Q = 0.9$$

at dimensionless times

$$\tau = 0.05, 0.1, 0.2, 0.4, 0.8, 1.6, 3.2, 6.4$$

Solution For the values of parameters as specified, the dimensionless temperature distribution $\theta_1(X, \tau)$ and the concentration distribution $\theta_2(X, \tau)$ can be calculated from Equations (10.25a) and (10.25b), respectively. Such calcula-

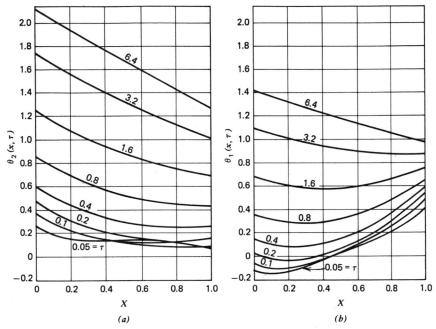

Figure 10.2 Dimensionless distributions for (*a*) temperature and (*b*) moisture during contact drying for Lu = 0.4, ϵ = 0.2, Bi_m = Bi_q = 2.5, Ko = 5, Pn = 0.6, and Q = 0.9 (from Reference 23).

tions have already been performed in Reference 23 and typical results for the temperature and moisture distribution in the conduit are presented in Figures (10.2a, b).

10.3 ENTRANCE REGION HEAT OR MASS TRANSFER IN CONCURRENT FLOW

The usual double pipe, concurrent flow heat exchanger consists of two concentric circular tubes with fluids at different temperatures entering the annular space and the central tube at the same side of the heat exchanger. As the fluids flow through their respective channels, heat is transferred from the hot to the cold fluid. The traditional methods of predicting heat transfer in such situations are based on the assignment of heat transfer coefficients for each flow independent of the actual coupling of the boundary conditions. The influence of coupling of the boundary conditions can be important in the thermal entry regions, especially with laminar flow. Therefore, the problem of heat transfer at the entrance region in concurrent flow double pipe heat exchangers has been analytically investigated [19, 20, 22].

The problems of simultaneous heat and mass transfer in internal flows with strong mutual interaction between the heat transfer processes, such as the one studied in Reference 21, are also governed by differential equations which are coupled at the boundary similar to those encountered in concurrent flow double pipe heat exchangers.

In this section we first present the mathematical formulation of such problems coupled at the boundary and then illustrate how their solutions are obtained as a special case of the one-dimensional Class III problem.

The differential equations governing the heat or mass transfer in the entrance region for concurrent flow inside two concentric pipes or two parallel plate channels can be written in general form as

$$\omega^{2(k-1)}R^{1-2m_k}U_k(R)\frac{\partial\theta_k(R,Z)}{\partial Z} = \frac{\partial}{\partial R}\left\{R^{1-2m_k}E_k(R)\frac{\partial\theta_k(R,Z)}{\partial R}\right\}$$

$$\text{in } 0 < R < 1, \quad Z > 0 \quad (10.26a)$$

where $k = 1, 2$, $E_k(1) = 1$ for laminar flow, $m_k = \frac{1}{2}$ for parallel plate channel, and $m_k = 0$ for circular tube.

The inlet conditions for the problem are taken as

$$\theta_k(R,0) = f_k \qquad (10.26b)$$

The boundary conditions, assuming symmetry at $R = 0$, are taken as

$$\frac{\partial\theta_k(0,Z)}{\partial R} = 0 \qquad (10.26c)$$

$$\alpha_{k1}\theta_1(1,Z) + \alpha_{k2}\theta_2(1,Z) + \beta_{k1}\frac{\partial\theta_1(1,Z)}{\partial R} + \beta_{k2}\frac{\partial\theta_2(1,Z)}{\partial R} = 0$$

$$(10.26d)$$

where $\alpha_{11} = 0$, $\alpha_{12} = 0$, $\alpha_{21}\alpha_{22}\beta_{11}\beta_{12} < 0$ (i.e., $\beta_{11}\alpha_{22}$ and $\beta_{12}\alpha_{21}$ are the opposite sign as discussed in Chapter 3), and for an insulated boundary at the outer surface we set $\alpha_{11}\alpha_{22} - \alpha_{12}\alpha_{21} = 0$.

Figure 10.3 illustrates the physical significance of the dimensionless transverse coordinate R for each of the regions $k = 1$ and $k = 2$ in the case of concurrent flow inside two concentric tubes. Clearly, the inner tube $R \equiv r/a_1$ is measured from the center of the tube and the tube radius a_1 is used to nondimensionalize the coordinate. In the case of the outer tube $R = r/a_2$ is measured from the outer wall of the outer tube and the spacing a_2 between the tube walls is used to nondimensionalize the coordinates. As a result, the range of R for both regions lies in $0 \le R \le 1$.

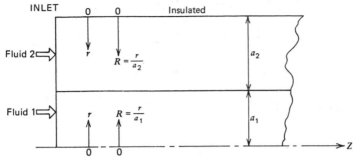

Figure 10.3   Geometry and coordinates for concurrent flow double pipe heat exchangers.

In Equation (10.26a), the $U_k(R)$ are the normalized velocities [see Equations (8.5b) and (8.7a) for the definition]; therefore, we have

$$\int_0^1 R^{1-2m_k} U_k(R) \, dR = \frac{1}{2(1-m_k)} \tag{10.27}$$

The problem (10.26) is a special case of the Class III problem defined by Equations (4.94) as follows:

$$x = R, \qquad t = Z, \qquad T_k(x, t) = \theta_k(R, Z),$$

$$w_k(x) = \omega^{2(k-1)} R^{1-m_k} U_k(R), \qquad k_k(x) = R^{1-2m_k} E_k(R), \qquad d_k(x) = 0,$$

$$P_k(x, t) = 0, \qquad x_0 = 0, \qquad x_1 = 1 \tag{10.28}$$

$$\alpha_k = 0, \qquad \beta_k = 1, \qquad \phi_k(x_0, t) = 0, \qquad \phi_k(x_1, t) = 0$$

Since $\alpha_k = 0$, $\beta_k = 1$, and $\alpha_{11}\alpha_{22} - \alpha_{12}\alpha_{21} = 0$, the boundary conditions at $R = 0$ are of the second kind for both regions. Then the solution of problem (10.26) is obtained from Equation (4.102a) as

$$\theta_k(R, Z) = \psi_{k0} \frac{\displaystyle\sum_{k=1}^{2} \sigma_k \psi_{k0} \omega^{2(k-1)} f_k \int_0^1 R^{1-2m_k} U_k(R) \, dR}{\displaystyle\sum_{k=1}^{2} \sigma_k \psi_{k0}^2 \omega^{2(k-1)} \int_0^1 R^{1-2m_k} U_k(R) \, dR}$$

$$+ \sum_{i=1}^{\infty} \frac{1}{N_i} e^{-\mu_i^2 Z} \psi_k(\mu_i, R) \tilde{f}_i \tag{10.29}$$

In this result the integrals appearing in the first term are evaluated according to Equation (10.27) and the first term is simplified further by making use of the relations given by Equations (4.102b) and (4.95c); that is,

$$\left(\frac{\psi_{k0}}{\psi_{10}}\right) = -\left(\frac{\alpha_{21}}{\alpha_{22}}\right)^{k-1} \quad \text{and} \quad \sigma_k = (-1)^{k-1}\alpha_{2k}\beta_{1k}$$

After some manipulation, the solution (10.28) takes the form

$$\theta_k(R,Z) = (-1)^{k-1}\alpha_{2,3-k}\frac{(1-m_2)\beta_{11}f_1 + (1-m_1)\beta_{12}f_2\omega^2}{(1-m_2)\beta_{11}\alpha_{22} - (1-m_1)\beta_{12}\alpha_{21}\omega^2}$$

$$+ \sum_{i=1}^{\infty}\frac{1}{N_i}e^{-\mu_i^2 Z}\psi_k(\mu_i,R)\tilde{f}_i, \qquad k=1,2 \qquad (10.30a)$$

where the normalization integral N_i is determined according to Equation (4.95b) as

$$N_i = \frac{1}{2\mu_i}\sum_{k=1}^{2}(-1)^{k-1}\alpha_{2k}\beta_{1k}\left\{\left[\frac{\partial\psi_k(\mu,1)}{\partial\mu}\right]_{\mu=\mu_i}\frac{\partial\psi_k(\mu_i,1)}{\partial R}\right.$$

$$\left. - \left[\frac{\partial^2\psi_k(\mu,1)}{\partial R\,\partial\mu}\right]_{\mu=\mu_i}\psi_k(\mu_i,1)\right\} \qquad (10.30b)$$

To obtain Equation (10.30b) from (4.95b) we made the substitution $\sigma_k = (-1)^{k-1}\alpha_{2k}\beta_{1k}$ and utilized the fact that

$$\frac{\partial\psi_k(\mu_i,0)}{\partial\mu} = 0 \quad \text{and} \quad \frac{\partial^2\psi_k(\mu_i,0)}{\partial R\,\partial\mu} = 0 \qquad (10.30c,d)$$

The integral transform $\tilde{f}_i$ appearing in Equation (10.30a) is defined by Equation (4.95d). The integral associated with the definition of $\tilde{f}_i$ is evaluated by the integration of the differential Equation (10.19a) for the eigenvalue problem. After some manipulation $\tilde{f}_i$ is determined as

$$\tilde{f}_i = -\frac{1}{\mu_i^2}\sum_{k=1}^{2}(-1)^{k-1}\alpha_{2k}\beta_{1k}f_k\frac{\partial\psi_k(\mu_i,1)}{\partial R} \qquad (10.30e)$$

To evaluate the functions $\theta_k(R,Z)$ from solution (10.30), the eigenvalues μ_i and the eigenfunctions $\psi_k(\mu_i,R)$ are needed. For the specific problem given by Equations (10.26), the Class III eigenvalue problem defined by Equations (3.115) simplifies to

$$\frac{d}{dR}\left\{R^{1-2m_k}E_k(R)\frac{d\psi_k(\mu,R)}{dR}\right\} + \mu^2\omega^{2(k-1)}R^{1-2m_k}U_k(R)\psi_k(\mu,R) = 0$$

$$\text{in } 0 < R < 1 \qquad (10.31a)$$

$$\psi_k'(\mu,0) = 0 \qquad (10.31b)$$

$$\alpha_{k1}\psi_1(\mu,1) + \alpha_{k2}\psi_2(\mu,1) + \beta_{k1}\psi_1'(\mu,1) + \beta_{k2}\psi_2'(\mu,1) = 0 \qquad (10.31c)$$

Thus the solution of problem (10.26) is given by Equation (10.30), and the eigenfunctions $\psi_k(\mu_i, R)$ eigenvalues μ_i are determined from the solution of the eigenvalue problem (10.31).

An Alternative Rearrangement of the Solution

In Chapter 3 we discussed an algorithm for the numerical solution of the eigenvalue problem of Class III. We now describe this algorithm for the solution of the eigenvalue problem given by Equations (10.31), and then present an alternative rearrangement of solution (10.30) in terms of the new variables associated with this algorithm.

We introduce new variables $y_{4k-2}(\mu, R)$ and $y_{4k-3}(\mu, R)$ by following the definitions given by Equations (3.123) as

$$\psi_k(\mu, R) = C_k y_{4k-2}(\mu, R) \tag{10.32a}$$

$$R^{1-2m_k} E_k(R) \frac{\partial \psi_k(\mu, R)}{\partial R} = C_k y_{4k-3}(\mu, R) \tag{10.32b}$$

and the derivatives of these new functions with respect to μ are denoted by

$$y_{4k}(\mu, R) = \frac{\partial y_{4k-2}(\mu, R)}{\partial \mu} \tag{10.32c}$$

$$y_{4k-1}(\mu, R) = \frac{\partial y_{4k-3}(\mu, R)}{\partial \mu} \tag{10.32d}$$

By utilizing the new functions defined by Equations (10.32), solution (10.30) is rearranged as (see Note 2 for the derivation)

$$\theta_k(R, Z) = (-1)^{k-1} \alpha_{2,3-k} \frac{(1 - m_2)\beta_{11} f_1 + (1 - m_1)\beta_{22} f_2 \omega^2}{(1 - m_2)\beta_{11}\alpha_{22} - (1 - m_1)\beta_{12}\alpha_{21}\omega^2}$$

$$- (-1)^{k-1} \beta_{1,3-k}(\alpha_{21} f_1 + \alpha_{22} f_2)$$

$$\times \sum_{i=1}^{\infty} D_i y_{9-4k}(\mu_i, 1) y_{4k-2}(\mu_i, R) e^{-\mu_i^2 Z} \tag{10.33a}$$

where

$$D_i = \frac{2}{\mu_i} y_1(\mu_i, 1) y_5(\mu_i, 1) \left\{ \alpha_{21}\beta_{12} y_5^2(\mu_i, 1) \begin{vmatrix} y_4(\mu_i, 1) & y_3(\mu_i, 1) \\ y_2(\mu_i, 1) & y_1(\mu_i, 1) \end{vmatrix} \right.$$

$$\left. - \alpha_{22}\beta_{11} y_1^2(\mu_i, 1) \begin{vmatrix} y_8(\mu_i, 1) & y_7(\mu_i, 1) \\ y_6(\mu_i, 1) & y_5(\mu_i, 1) \end{vmatrix} \right\}^{-1}$$

$$\tag{10.33b}$$

The eigenvalue problem (10.31) is now rearranged in terms of the new variables defined by Equations (10.32) as [see Equations (3.124)]

$$y'_{4k-3}(\mu, R) = -\mu^2 \omega^{2(k-1)} R^{1-2m_k} U_k(R) y_{4k-2}(\mu, R) \qquad (10.34a)$$

$$y'_{4k-2}(\mu, R) = \frac{y_{4k-3}(\mu, R)}{R^{1-2m_k} E_k(R)} \qquad (10.34b)$$

subject to the boundary conditions [see Equations (3.125)]

$$y_{4k-3}(\mu, 0) = 0, \qquad y_{4k-2}(\mu, 0) = 1 \qquad (10.34c, d)$$

The eigencondition is given by [see Equation (3.126b)]

$$-\alpha_{21}\beta_{12}\frac{y_2(\mu, 1)}{y_1(\mu, 1)} + \alpha_{22}\beta_{11}\frac{y_6(\mu, 1)}{y_5(\mu, 1)} + \beta_{11}\beta_{22} - \beta_{12}\beta_{21} = 0 \quad (10.34e)$$

Thus the eigenvalues μ_i are determined from the solution of the eigencondition (10.34e); knowing the μ_i, the solution of Equations (10.34a, b) subject to the boundary conditions (10.34c, d) yields the functions $y_{4k-2}(\mu, R)$ and $y_{4k-3}(\mu, R)$.

Now Equations (10.34a, b) are differentiated with respect to μ and the definitions (10.34c, d) are utilized. We find

$$y'_{4k-1}(\mu, R) = -\mu \omega^{2(k-1)} R^{1-2m_k} U_k(R) [\mu y_{4k}(\mu, R) + 2 y_{4k-2}(\mu, R)]$$

$$(10.35a)$$

$$y'_{4k}(\mu, R) = \frac{y_{4k-1}(\mu, R)}{R^{1-2m_k} E_k(R)} \qquad (10.35b)$$

and the differentiation of the boundary conditions (10.34c, d) with respect to μ yields

$$y_{4k-1}(\mu, 0) = 0, \qquad y_{4k}(\mu, 0) = 0 \qquad (10.35c, d)$$

Thus once the functions $y_{4k-2}(\mu, R)$ and the eigenvalues μ_i are determined from the solution of system (10.34), the functions $y_{4k}(\mu, R)$ and $y_{4k-1}(\mu, R)$ are obtained from the solution of system (10.35). A discussion of the algorithm for the numerical solution of the eigenvalue problem (10.34) is given in Chapter 3.

The Slug Flow

If the velocity fields in the ducts can be approximated as slug flow, we have $U_k(R) = 1$, $E_k(R) = 1$ for $k = 1, 2$. For this particular case, system (10.34) has exact analytical solution. This is apparent if Equations (10.34a, b) are combined into a single equation for the function $y_{4k-2}(\mu, R)$ as

$$\frac{d}{dR}\left(R^{1-2m_k}\frac{dy_{4k-2}}{dR}\right) + \left(\mu\omega^{k-1}\right)^2 R^{1-2m_k} y_{4k-2}(\mu, R) = 0 \quad \text{in } 0 < R < 1$$

(10.36a)

The solution of this equation which satisfies condition (10.34d), that is,

$$y_{4k-2}(\mu, 0) = 1 \tag{10.36b}$$

gives the function $y_{4k-2}(\mu, R)$ as

$$y_{4k-2}(\mu_i, R) = \left(\mu_i\omega^{k-1}R\right)^{m_k} J_{-m_k}\left(\mu_i\omega^{k-1}R\right) \tag{10.36c}$$

We note that this result is similar to solution (3.122a) of Equation (3.121a). Then the function $y_{4k-3}(\mu, R)$ is determined by Equation (10.34b) as

$$y_{4k-3}(\mu_i, R) = -\mu_i\omega^{k-1}R^{1-2m_k}\left(\mu_i\omega^{k-1}\right)^{m_k} J_{1-m_k}\left(\mu_i\omega^{k-1}R\right) \tag{10.36d}$$

and the functions $y_{4k}(\mu, R)$ and $y_{4k-1}(\mu, R)$ are obtained by utilizing their definitions given by Equations (10.32c) and (10.32d), respectively.

The eigencondition for this problem is determined by introducing the functions (10.36c, d) into Equation (10.34e). We find

$$\alpha_{21}\beta_{12}\frac{J_{-m_1}(\mu)}{\mu J_{1-m_1}(\mu)} - \alpha_{22}\beta_{11}\frac{J_{-m_2}(\mu\omega)}{\mu\omega J_{1-m_2}(\mu\omega)} + \beta_{11}\beta_{22} - \beta_{12}\beta_{21} = 0$$

(10.37)

The positive nonzero roots of Equation (10.37) give the eigenvalues μ_i. We note that Equation (10.37) is identical to Equation (3.126c).

We now introduce the functions defined by Equations (10.36c, d) into the solution (10.33) and note that the function y_{9-4k} is obtainable from y_{4k-3} by replacing in Equation (10.36d) the subscript k by $(3 - k)$. After some manipulation, the solution $\theta_k(R, Z)$ for the slug flow is determined as

$$\theta_k(R, Z) = (-1)^{k-1}\alpha_{2,3-k}\frac{(1 - m_2)\beta_{11}f_1 + (1 - m_1)\beta_{22}f_2\omega^2}{(1 - m_2)\beta_{11}\alpha_{22} - (1 - m_1)\beta_{12}\alpha_{21}\omega^2}$$

$$+ (-1)^{k-1}\beta_{1,3-k}(\alpha_{21}f_1 + \alpha_{22}f_2)$$

$$\times \sum_{i=1}^{\infty} D_i\mu_i\omega^{2-k}\left(\mu_i\omega^{2-k}\right)^{m_{3-k}} J_{1-m_{3-k}}\left(\mu_i\omega^{2-k}\right)$$

$$\times \left(\mu_i\omega^{k-1}R\right)^{m_k} J_{-m_k}\left(\mu_i\omega^{k-1}R\right)e^{-\mu_i^2 Z} \tag{10.38a}$$

where

$$D_i = 2\mu_i^{m_1-1} J_{1-m_1}(\mu_i)(\mu_i\omega)^{m_2-1} J_{1-m_2}(\mu_i\omega) \times \{\alpha_{21}\beta_{12}(\mu_i\omega)^{2m_2} J_{1-m_2}^2(\mu_i\omega)$$

$$\times \left[\mu_i^{2m_1} J_{-m_1}^2(\mu_i) + \mu_i^{2m_1} J_{1-m_1}^2(\mu_i) + 2m_1\mu_i^{m_1} J_{-m_1}(\mu_i)\mu_i^{m_1-1} J_{1-m_1}(\mu_i)\right]$$

$$- \alpha_{22}\beta_{11}\mu_i^{2m_1} J_{1-m_1}^2(\mu_i)\left[(\mu_i\omega)^{2m_2} J_{-m_2}^2(\mu_i\omega) + (\mu_i\omega)^{2m_2} J_{1-m_2}^2(\mu_i\omega)\right]$$

$$+ 2m_2(\mu_i\omega)^{m_2} J_{-m_2}(\mu_i\omega)(\mu_i\omega)^{m_2-1} J_{1-m_2}(\mu_i\omega)\right]\}^{-1} \tag{10.38b}$$

Applications

We now illustrate the application of the foregoing results to obtain solutions for heat and mass transfer problems at the entrance region in concurrent flow in parallel plate channels and circular concentric tubes, or for problems coupled at the boundary conditions. For this purpose, we have chosen some representative examples from the literature and shown that such solutions are indeed obtainable as special cases from the present results.

Example 10.2 This example is concerned with the problem of simultaneous heat and mass transfer in an internal flow in which there is interaction between the transfer processes at the wall surface. We consider the hydrodynamically developed flow of a gas inside a parallel plate channel whose walls are coated with a sublimated material and thermally insulated from the external environment. Sublimation occurs at the duct wall provided that the entering gas stream is not saturated with the vapor of the sublimated material. As the walls are thermally insulated from the external environment, the latent heat of sublimation is supplied to the wall by the gas itself, thus causing the coupling of these two transfer processes. The mathematical formulation of this problem leads to the following set of dimensionless equations given in Reference 21 in the form

Energy: $\quad U(R)\dfrac{\partial\theta(R,Z)}{\partial Z} = \dfrac{\partial^2\theta(R,Z)}{\partial R^2} \quad$ in $0 < R < 1, \quad Z > 0$

$$\tag{10.39a}$$

Mass: $\quad LU(R)\dfrac{\partial\Phi(R,Z)}{\partial Z} = \dfrac{\partial^2\Phi(R,Z)}{\partial R^2} \quad$ in $0 < R < 1, \quad Z > 0$

$$\tag{10.39b}$$

subject to the inlet conditions

$$\theta(R,0) = 1, \quad \Phi(R,0) = 1 \tag{10.39c, d}$$

and the boundary conditions

$$\frac{\partial \theta(0, Z)}{\partial R} = 0, \qquad \frac{\partial \Phi(0, Z)}{\partial R} = 0 \qquad \text{(10.39e, f)}$$

$$L\frac{\partial \theta(1, Z)}{\partial R} - \frac{\partial \Phi(1, Z)}{\partial R} = 0 \qquad \text{(10.39g)}$$

$$K\theta(1, Z) + \Phi(1, Z) = 0 \qquad \text{(10.39h)}$$

where θ and Φ are the dimensionless temperature and concentration distributions in the gas stream, $U(R)$ is the dimensionless laminar velocity profile, L is the ratio of the thermal diffusivity to the binary diffusion coefficient, K is a dimensionless parameter, and Z and R are, respectively, the dimensionless axial and normal coordinates. The dimensionless temperature and mass distributions in the gas stream are needed.

Solution   In this particular problem, the functions $\theta(R, Z)$ and $\Phi(R, Z)$ both lie in the region $0 \le R \le 1$, $Z \ge 0$, and problem (10.39) is a special case of problem (10.26) as follows.

$$\omega^2 = L, \qquad m_1 = m_2 = \tfrac{1}{2}, \qquad U_1(R) = U_2(R) = U(R)$$

$$E_1(R) = E_2(R) = 1, \qquad f_1 = f_2 = 1, \qquad \theta_1 = \theta$$

$$\theta_2 = \Phi, \qquad \beta_{11} = L, \qquad \beta_{12} = -1, \qquad \alpha_{21} = K$$

$$\alpha_{22} = 1, \qquad \beta_{21} = \beta_{22} = 0 \qquad \text{(10.40)}$$

and for fully developed laminar flow inside a circular tube we have

$$U(R) = \tfrac{3}{2}(1 - R^2)$$

Then the solution for $\theta(R, Z)$ and $\phi(R, Z)$ can be readily calculated from the results given by Equations (10.33) provided that the eigenfunctions and the eigenvalues for the problem are available. Such calculations have already been performed in Reference 22 using 10 eigenvalues. We present in Figure 10.4 the dimensionless temperature $\theta(R, Z)$ and mass $\Phi(R, Z)$ profiles at different axial locations Z along the channel for the case $L = 0.81$ and $K = 0.1$.

Example 10.3   In this example we consider mass transfer between two immiscible fluids, a liquid and a gas flowing separately in laminar, concurrent flow inside a parallel plate channel. The mathematical formulation of this problem is defined in References 17 and 18 through the following set of dimensionless

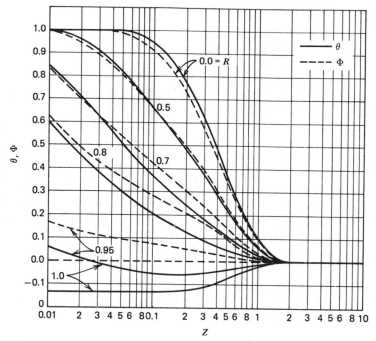

Figure 10.4 Dimensionless temperature $\theta(R, Z)$ and mass $\Phi(R, Z)$ distribution in the gas stream for $L = 0.81$ and $K = 0.1$ (from Reference 22).

equations

$$\beta^2 \frac{3}{2}(1 - R^2)\frac{\partial C_1(R, Z)}{\partial Z} = \frac{\partial^2 C_1(R, Z)}{\partial R^2} \quad \text{in } 0 < R < 1, \quad Z > 0$$

$$(10.41a)$$

$$\frac{3}{2}R(1 - R^2)\frac{\partial C_2(R, Z)}{\partial Z} = \frac{\partial^2 C_2(R, Z)}{\partial R^2} \quad \text{in } 0 < R < 1, \quad Z > 0$$

$$(10.41b)$$

subject to the inlet conditions

$$C_1(R, 0) = 1, \qquad C_2(R, 0) = 0 \qquad (10.41c, d)$$

and the boundary conditions

$$\frac{\partial C_1(0, Z)}{\partial R} = 0, \qquad \frac{\partial C_2(0, Z)}{\partial R} = 0 \qquad (10.41e, f)$$

$$\frac{\partial C_1(1, Z)}{\partial R} + \beta^2 \epsilon \frac{\partial C_2(1, Z)}{\partial R} = 0 \qquad (10.41g)$$

$$C_1(1, Z) - C_2(1, Z) = 0 \qquad (10.41h)$$

where $C_1(R, Z)$ and $C_2(R, Z)$ are the dimensionless mass concentrations in the gas and liquid streams, respectively, and β^2 and ϵ are dimensionless parameters defined in References 17 and 18. The dimensionless mass concentrations $C_1(R, Z)$ and $C_2(R, Z)$ are to be determined.

Solution   In this particular problem we are concerned with concurrent flow of two immiscible fluids, a liquid and a gas in a parallel plate channel, each stream lying in a separate region $0 < R < 1$ as illustrated in Figure 10.3, but with no separating wall between the fluids. The problem defined by Equations (10.41) is a special case of problem (10.26) as follows:

$$\omega^2 = \beta^2, \qquad m_1 = m_2 = \tfrac{1}{2}, \qquad U_1(R) = \tfrac{3}{2}(1 - R^2)$$

$$U_2(R) = \tfrac{3}{2}R(1 - R^2), \qquad E_1(R) = E_2(R) = 1, \qquad \theta_1 = C_2$$

$$\theta_2 = C_1, \qquad f_1 = 0, \qquad f_2 = 1, \qquad \beta_{11} = \beta^2 \epsilon$$

$$\beta_{12} = 1, \qquad \alpha_{21} = -1, \qquad \alpha_{22} = 1, \qquad \beta_{21} = \beta_{22} = 0 \qquad (10.42)$$

Therefore, the concentration distributions $C_1(R, Z)$ and $C_2(R, Z)$ are readily determined from solution (10.33) once the eigenvalues and the eigenfunctions for the problem are available. Such calculations are performed in Reference 22 and we present in Figure 10.5 the dimensionless mass concentra-

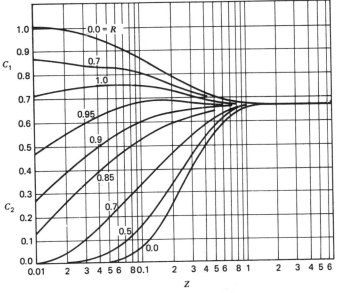

Figure 10.5   Dimensionless mass concentrations $C_1(R, Z)$ and $C_2(R, Z)$ in the gas liquid streams, respectively, for $\epsilon = 0.5$, $\beta^2 = 0.2$ (from Reference 22).

tions $C_1(R, Z)$ and $C_2(R, Z)$ plotted as a function of the axial position Z for the values of the parameters $\epsilon = 0.5$ and $\beta^2 = 0.2$.

Example 10.4 The usual double pipe heat exchanger consists of two channels (i.e., parallel plane channels or concentric circular pipes), similar to those illustrated in Figure 10.3. In the case of concurrent flow, two fluids at different temperatures enter through the same inlet. Such heat transfer problems have been studied in References 19, 20, and 25 and the mathematical formulation is given in the dimensionless form as

$$R^{1-2m}U_1(R)\frac{\partial\theta_1(R, Z)}{\partial Z} = \frac{\partial}{\partial R}\left[R^{1-2m}\frac{\partial\theta_1(R, Z)}{\partial R}\right] \qquad (10.43\text{a})$$

$$\frac{KH}{2(1-m)}U_2(R)\frac{\partial\theta_2(R, Z)}{\partial Z} = \frac{\partial^2\theta_2(R, Z)}{\partial R^2} \qquad (10.43\text{b})$$

subject to the inlet conditions

$$\theta_1(R, 0) = 0, \qquad \theta_2(R, 0) = 0 \qquad (10.43\text{c, d})$$

and the boundary conditions

$$\frac{\partial\theta_1(0, Z)}{\partial R} = 0, \qquad \frac{\partial\theta_2(0, Z)}{\partial R} = 0 \qquad (10.43\text{e, f})$$

$$K\frac{\partial\theta_1(1, Z)}{\partial R} + \frac{\partial\theta_2(1, Z)}{\partial R} = 0 \qquad (10.43\text{g})$$

$$K_w\frac{\partial\theta_1(1, Z)}{\partial R} + \theta_1(1, Z) - \theta_2(1, Z) = 0 \qquad (10.43\text{h})$$

Here Equation (10.43a) is for the inner region, which can be a circular tube $m = 0$ or a parallel plate channel $m = \frac{1}{2}$. Equation (10.43b) for the outer region assumes that the annular space is approximated by a parallel plate geometry. The functions θ_1 and θ_2 refer to temperatures of the fluids at the inner and outer regions, respectively. The normalized velocities $U_1(R)$ and $U_2(R)$ for the laminar flow are taken as

$$U_1(R) = 2(1 - R^2) \qquad (10.43\text{i})$$

$$U_2(R) = 6R(1 - R^2) \qquad (10.43\text{j})$$

or, if slug flow is assumed, they simplify to

$$U_1(R) = U_2(R) = 1 \qquad (10.43\text{k})$$

The dimensionless parameter H is the heat capacity flow rate ratio, K is the relative thermal resistance of fluids, and K_w is the relative thermal resistance of the wall.

The temperature distributions in the channels are needed.

Solution Problem (10.43) is a special case of the general problem (10.26) as follows:

$$m_1 = m, \qquad m_2 = 0, \qquad E_1 = E_2 = 1, \qquad f_1 = 0, \qquad f_2 = 1$$

$$\omega^2 = \frac{KH}{2(1 - m)}, \qquad \beta_{10} = K, \qquad \beta_{12} = 1 \qquad (10.44)$$

$$\alpha_{21} = 1, \qquad \alpha_{22} = -1, \qquad \beta_{21} = K_w, \qquad \beta_{22} = 0$$

The calculations are performed in Reference 22 for this particular case using the laminar velocity profile for both streams. We present in Figure 10.6 the

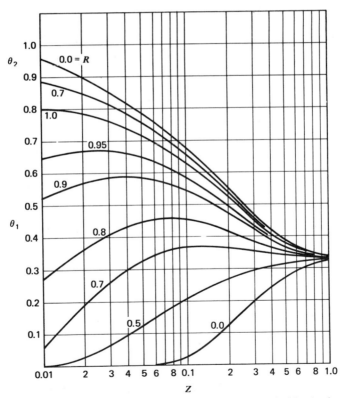

Figure 10.6 Dimensionless temperature distributions in a concurrent double pipe heat exchanger (from Reference 22).

resulting dimensionless temperatures θ_1 and θ_2 for $K = 0.1$, $H = 0.5$, $K_w = 0$, and $m = \frac{1}{2}$.

REFERENCES

1. A. V. Luikov, *J. Thermotechn. Inst.* (Russ.), **3**, 43 (1935); **6**, 20 (1935).

2. A. V. Luikov, *J. Appl. Chem.* (Russ.), **8**, 1354 (1935).

3. A. Luikov, Moisture Content Gradients in the Drying of Clay, *Trans. Ceramic Soc.*, **35**, 123–129 (1936).

4. A. V. Luikov and Yu. A. Mikhailov, *Theory of Energy and Mass Transfer*, Pergamon, Oxford, 1965.

5. G. D. Fulford, A Survey of Recent Soviet Research on the Drying of Solids, *Can. J. Chem. Eng.*, **47**, 378–391 (1969).

6. O. Kirscher, Der Wärme-und Stoffaustausch im Trocknungsgut, *VDI-Forschungsheft*, 415 (1942).

7. O. Krischer, *Die Wissenschaftlichen Grundlagen der Trocknungstechnik*, Springer, 1965.

8. Zw. Iwantschewa and M. D. Mikhailov, Über Differentialgleichungen für Wärme-und Stoffaustausch von Hygroskopischen Stoffen, *Wärme-u. Stoffübertragung*, **1**, 117–120 (1968).

9. A. De Vries, Simultaneous Transfer of Heat and Moisture in Porous Media, *Trans. Am. Geophys. Union*, **39**, 909–916 (1968).

10. D. A. De Vries and A. J. Peck, On the Cylindrical Probe Method of Measuring Thermal Conductivity with Special Reference to Soils, II. Analysis of Moisture Effect, *Aust. J. Phys.*, **11**, 409–423 (1958).

11. P. S. Henry, Diffusion in Absorbing Media, *Proc. R. Soc.*, **171A**, 215–241 (1939).

12. J. Crank, *The Mathematics of Diffusion*, Clarendon Press 2nd ed., London, 1975.

13. M. S. Smirnov, A. System of Differential Equations of the Drying Process, *J. Eng. Phys.*, **4**, 40–44 (1961).

14. M. D. Mikhailov, General Solutions of the Diffusion Equations Coupled at Boundary Conditions, *Int. J. Heat Mass Transf.*, **16**, 2155–2167 (1973).

15. Y. P. Tang and D. M. Himmelblau, Interphase Mass Transfer for Laminar Concurrent Flow of Carbon Dioxide and Water Between Parallel Plates, *A.I. Chem. Eng. J.*, **9**, 630 (1963).

16. A. Apelblat and A. Katchalsky, Mass Transfer with a Moving Interface, *Int. J. Heat Mass Transf.*, **11**, 1053–1067 (1968).

17. V. N. Baback, L. I. Cholpanov, V. A. Moliussov, and V. N. Javoronkov, *Theor. Osn. Chim. Tech.* (Russ.) **5**, 179 (1971).

18. V. N. Baback, T. B. Baback, L. I. Cholpanov, V. A. Moliussov, and V. N. Javoronkov, Numerical Solution of the Problem for Stationary Two-Phase Mass Transfer for Laminar Flow of a Liquid or Gaseous Films in Co- and Countercurrent Phases (Russ.) *Toplomassoobmen*, **4**, 238 (1972).

19. R. P. Stein, Heat Transfer Coefficients in Liquid Metal Concurrent Flow Double Pipe Heat Exchangers, *Chem. Eng. Prog. Symp. Ser.*, **59**, 64–75 (1965).

20. R. P. Stein, The Graetz Problem in Concurrent Flow Double Pipe Heat Exchangers, *Chem. Eng. Prog. Symp. Ser.*, **59**, 76–87, 1965.

21. E. M. Sparrow and E. C. Spalding, Coupled Laminar Heat Transfer and Sublimation Mass Transfer in a Duct, *J. Heat Transf.*, **90C**, 115–124 (1968).

22. M. D. Mikhailov and B. K. Shishedjiev, Coupled at Boundary Mass or 1   . Transfer in Entrance Concurrent Flow, *Int. J. Heat Mass Transf.*, **19**, 553–557 (1976).

23. M. D. Mikhailov and B. K. Shishedjiev, Temperature and Moisture Distributions During Contact Drying of a Moist Porous Sheet, *Int. J. Heat Mass Transf.*, **18**, 15–24 (1975).

24. S. Bruin, Calculation of Temperature and Moisture Distributions During Contact Drying of a Sheet of Moist Material, *Int. J. Heat Mass Transf.*, **12**, 45–49 (1969).

25. M. D. Mikhailov and B. K. Shishedjiev, Heat Transfer in Concurrent Flow Double Pipe Heat Exchangers, in *Proc. Heat Exchangers Conf.*, Istanbul, August 1980, Hemisphere Publishing Corp., Washington, D. C.

26. A. V. Luikov, Systems of Differential Equations of Heat and Mass Transfer in Capillary-Porous Bodies (Review), *Int. J. Heat Mass Transf.*, **18**, 1–14 (1975).

NOTES

1. Derivation of the transformations given by Equations (10.12)

We write Equation (10.11a) as

$$\frac{\partial \theta_1}{\partial \tau} + \epsilon \, \mathrm{Ko} \frac{\partial \theta_2}{\partial \tau} = \frac{\partial^2 \theta_1}{\partial X^2} \tag{1}$$

We eliminate $\partial^2 \theta_1 / \partial X^2$ from Equation (10.11b) by means of Equation (10.11a); then Equation (10.11b) becomes

$$\mathrm{Pn} \frac{\partial \theta_1}{\partial \tau} + \left(\mathrm{Pn} \, \epsilon \, \mathrm{Ko} + \frac{1}{\mathrm{Lu}} \right) \frac{\partial \theta_2}{\partial \tau} = \frac{\partial^2 \theta_2}{\partial X^2} \tag{2}$$

We multiply Equation (2) by d and add the resulting expression to Equation (1).

$$(1 + d\,\mathrm{Pn}) \frac{\partial}{\partial \tau} \left[\theta_1 + \frac{\epsilon \, \mathrm{Ko} + d(\mathrm{Pn} \, \epsilon \, \mathrm{Ko} + 1/\mathrm{Lu})}{1 + d\,\mathrm{Pn}} \theta_2 \right] = \frac{\partial^2}{\partial X^2} (\theta_1 + d\theta_2) \tag{3}$$

We now define a new function Z as

$$Z = \theta_1 + \frac{\epsilon \, \mathrm{Ko} + d(\mathrm{Pn} \, \epsilon \, \mathrm{Ko} + 1/\mathrm{Lu})}{1 + d\,\mathrm{Pn}} \theta_2 = \theta_1 + d\theta_2 \tag{4}$$

Then we find

$$\frac{\epsilon \, \mathrm{Ko} + d(\mathrm{Pn} \, \epsilon \, \mathrm{Ko} + 1/\mathrm{Lu})}{1 + d\,\mathrm{Pn}} = d \tag{5}$$

We let

$$1 + d\,\mathrm{Pn} = \nu^2 \quad \text{or} \quad d = \frac{\nu^2 - 1}{\mathrm{Pn}} \tag{6a, b}$$

Then Equation (5) becomes

$$\epsilon \, \text{Ko} + \frac{\nu^2 - 1}{\text{Pn}}\left(\text{Pn}\,\epsilon\,\text{Ko} + \frac{1}{\text{Lu}}\right) = \frac{\nu^2 - 1}{\text{Pn}}\nu^2$$

or

$$\nu^4 - \nu^2\left(1 + \text{Pn}\,\epsilon\,\text{Ko} + \frac{1}{\text{Lu}}\right) + \frac{1}{\text{Lu}} = 0 \tag{7}$$

The solution for ν^2 yields

$$\nu_k^2 = \frac{1}{2}\left[\left(1 + \epsilon\,\text{Pn}\,\text{Ko} + \frac{1}{\text{Lu}}\right) + (-1)^k\sqrt{\left(1 + \epsilon\,\text{Pn}\,\text{Ko} + \frac{1}{\text{Lu}}\right)^2 - \frac{4}{\text{Lu}}}\,\right],$$

$$k = 1,2 \tag{8}$$

and from definition (4) we write

$$Z_k = \theta_1 + d_k\theta_2 = \theta_1 + \frac{\nu_k^2 - 1}{\text{Pn}}\theta_2 \tag{9}$$

Equations (8) and (9) are the transformation given by Equations (10.12).

2. Derivation of Equation (10.33)

To rearrange Equation (10.30) in the form given by Equation (10.33) we first simplify the terms N_i and $\tilde{f}_i$ as now described.

Introducing Equations (10.32a, b) into equation (10.30b), N_i becomes

$$N_i = \frac{1}{2\mu_i}\sum_{k=1}^{2}(-1)^{k-1}\alpha_{2k}\beta_{1k}C_k^2\left|\begin{array}{cc}\left[\dfrac{\partial}{\partial\mu}y_{4k-2}(\mu,1)\right]_{\mu=\mu_i} & \left[\dfrac{\partial}{\partial\mu}y_{4k-3}(\mu,1)\right]_{\mu=\mu_i} \\ y_{4k-2}(\mu,1) & y_{4k-3}(\mu,1)\end{array}\right|$$

$$\tag{1}$$

By utilizing the definitions (10.32c, d), Equation (1) becomes

$$N_i = \frac{1}{2\mu_i}\sum_{k=1}^{2}(-1)^{k-1}\alpha_{2k}\beta_{1k}C_k^2\left|\begin{array}{cc}y_{4k}(\mu_i,1) & y_{4k-1}(\mu_i,1) \\ y_{4k-2}(\mu_i,1) & y_{4k-3}(\mu_i,1)\end{array}\right| \tag{2}$$

From the boundary condition (10.31c) for $k = 1$, and noting that $\alpha_{11} = 0$, $\alpha_{12} = 0$, we obtain

$$\beta_{11}\psi_1'(\mu,1) + \beta_{12}\psi_2'(\mu,1) = 0 \tag{3}$$

From Equation (10.32b) for $k = 1$ and 2 we have, respectively,

$$\psi_1'(\mu, 1) = C_1 y_1(\mu, 1) \tag{4a}$$

$$\psi_2'(\mu, 1) = C_2 y_5(\mu, 1) \tag{4b}$$

Then Equation (3) becomes

$$C_1 \beta_{11} y_1(\mu_i, 1) + C_2 \beta_{12} y_5(\mu_i, 1) = 0 \tag{5a}$$

or

$$\frac{C_2}{C_1} = -\frac{\beta_{11}}{\beta_{12}} \frac{y_1(\mu_i, 1)}{y_5(\mu_i, 1)} \tag{5b}$$

or

$$\left(\frac{C_k}{C_1}\right)^{k-1} = \left(-\frac{\beta_{11}}{\beta_{12}} \frac{y_1(\mu_i, 1)}{y_5(\mu_i, 1)}\right)^{k-1} \tag{5c}$$

Introducing Equation (5c) into Equation (2), N_i becomes

$$N_i = \frac{C_1^2}{2\mu_i} \frac{\beta_{11}}{\beta_{12} y_5^2} \left\{ \alpha_{21} \beta_{12} y_5^2 \begin{vmatrix} y_4 & y_3 \\ y_2 & y_1 \end{vmatrix} - \alpha_{22} \beta_{11} y_1^2 \begin{vmatrix} y_8 & y_7 \\ y_6 & y_5 \end{vmatrix} \right\} \tag{6}$$

The integral transform $\tilde{f}_i$ is defined by Equation (4.95d) as

$$\tilde{f}_i = \sum_{k=1}^{2} \sigma_k f_k \int_0^1 w_k(R) \psi_{ki}(R) \, dR \tag{7}$$

By noting the definition of $w_k(R)$ given by Equation (10.28), the integral term in Equation (7) is evaluated by the integration of Equation (10.31a). Then Equation (7) takes the form

$$\tilde{f}_i = -\frac{1}{\mu_i^2} \sum_{k=1}^{2} (-1)^{k-1} \alpha_{2k} \beta_{1k} f_k \frac{\partial \psi_k(\mu_i, 1)}{\partial R} \tag{8}$$

By utilizing the relation (10.32b), Equation (8) becomes

$$\tilde{f}_i = -\frac{1}{\mu_i^2} C_1 y_1 \sum_{k=1}^{2} (-1)^{k-1} \alpha_{2k} \beta_{1k} f_k \left(\frac{C_k}{C_1} \frac{y_{4k-3}(\mu_i, 1)}{y_1}\right)^{k-1} \tag{9}$$

we write Equation (5a) as

$$\left[\frac{C_k}{C_1} \frac{y_{4k-3}(\mu_i, 1)}{y_1(\mu_i, 1)}\right]^{k-1} = \left(-\frac{\beta_{11}}{\beta_{12}}\right)^{k-1} \tag{10}$$

Then Equation (9) is written as

$$\tilde{f}_i = -\frac{1}{\mu_i^2} C_1 y_1 \sum_{k=1}^{2} \alpha_{2k} \beta_{1k} f_k \left(\frac{\beta_{11}}{\beta_{12}}\right)^{k-1} \tag{11}$$

or

$$\tilde{f}_i = -C_1 \frac{\beta_{11}}{\mu_i^2} y_1(\mu_i, 1)(\alpha_{21} f_1 + \alpha_{22} f_2) \tag{12}$$

Introducing the results given by Equations (6) and (12) into Equation (10.30a) we obtain the solution given by Equations (10.33).

CHAPTER ELEVEN

Class IV Solutions Applied to Heat or Mass Diffusion in One-Dimensional Heterogeneous Media

*Man can learn nothing unless he proceeds
from the known to the unknown.*

**Claude Bernard
1813–1878**

The boundary value problems of Class IV as discussed in Chapter 1 are characterized by a set of diffusion equations in which the potentials (i.e., the temperature or the mass concentration) $T_k(\mathbf{x}, t)$, $k = 1, 2, 3, \ldots, n$ in every point in space $\mathbf{x} \in V$ are coupled through source or sink terms. Such problems usually appear in mass diffusion processes involving several chemically reacting components [1] in which the presence of chemical reactions are coupled through source-sink terms.

A similar mathematical modeling has been introduced [2] to describe the transient temperature distribution in the components of a heterogeneous medium (i.e., a medium consisting of concrete, soil, water and/or oil-saturated sand layers, etc.). In such modeling, the coupling of heat transfer between the components, say, the components m and p, is realized through a term $\sigma_{mp}(T_m - T_p)$, where $\sigma_{mp} = \sigma_{pm}$ is the specific conductance or heat exchange coefficient between the components, and T_p and T_m are the temperatures.

The heat or mass diffusion problems in heterogeneous media as discussed is a special case of the Class IV problem. This chapter is devoted to the application of the analysis of one-dimensional Class IV problems presented in Chapter 4 for the solution of one-dimensional transient heat or mass diffusion in heterogeneous media. Specific solutions and numerical results are given for the distribution of dimensionless potential (i.e., temperature or concentration) for each component in a heterogeneous slab, cylinder, and sphere subjected to the boundary condition of the first kind at the outer boundary.

11.1 FORMULATION OF TRANSIENT HEAT OR MASS DIFFUSION IN ONE-DIMENSIONAL HETEROGENEOUS MEDIA

We consider a heterogeneous slab, cylinder, or sphere consisting of n different components. The potentials $T_k(x, t)$, $(k = 1, 2, \ldots, n)$ for each of the n different components are continuous functions of space and time, and in every point of the region there are n such potentials corresponding to each of the components. The physical significance of potentials defined in this manner is quite apparent in the case of multicomponent, chemically reacting mass diffusion problems; that is, when there are n different species at every point of space, then there are n different concentration distributions that are continuous functions of space and time over the region under consideration.

In the case of heat transfer problems, however, the physical significance of n different temperature distributions over an heterogeneous region may not be so obvious; some clarification of this matter is in order. Let us consider a medium consisting of uniformly mixed particles (or species) of n different types of materials, each having different thermal properties. The size of the particles is sufficiently large so that the medium cannot be regarded as homogeneous; then it is called an *heterogeneous* medium consisting of n different materials. Suppose such an heterogeneous medium has a geometry in the form of a long solid cylinder of finite radius and subject to one-dimensional radial heat transfer as now described. Initially, the entire medium is at a prescribed uniform temperature; that is, all the particles (or species) in the medium are at the same temperature. Suddenly, the temperature of the cylindrical boundary surface is altered and kept at a constant temperature different from the initial temperature. During the transients, each component in this heterogeneous medium will obey a temperature distribution which will be different from those for the other components. Therefore, one can envision the presence of n different temperatures $T_k(x, t)$, $(k = 1, 2, \ldots, n)$ in the medium associated with each of the n components in the system. Such a heat transfer problem can be modelled with a coupling between the components, say, m and p, represented in the form $\sigma_{mp}(T_m - T_p)$, where $\sigma_{mp} = \sigma_{pm}$ is the heat exchange coefficient between the components m and p.

We now consider a one-dimensional transient diffusion problem in an heterogeneous medium consisting of n different components. Each component is homogeneous, isotropic, and has constant properties; the coupling coefficients between the components are constant. Initially, each component is at the same constant potential (i.e., temperature or mass concentration) T_0. For times $t > 0$, the potential at the outer boundary of the region is kept at a constant potential T_w which is different from T_0. The mathematical formulation of this transient diffusion problem for each of the n components is given by

$$C_k \rho_k x^{1-2m} \frac{\partial T_k(x,t)}{\partial t} = k_k \frac{\partial}{\partial x} \left[x^{1-2m} \frac{\partial T_k(x,t)}{\partial x} \right]$$

$$+ x^{1-2m} \sum_{p=1}^{n} \alpha_{kp} \left[T_p(x,t) - T_k(x,t) \right],$$

$$\text{in } 0 < x < \ell, \quad \text{for } t > 0, \qquad k = 1,2,\ldots,n \quad (11.1a)$$

Subject to the boundary conditions

$$\frac{\partial T_k(0,t)}{\partial x} = 0, \qquad k = 1,2,\ldots,n \qquad (11.1b)$$

$$T_k(\ell,t) = T_w, \qquad k = 1,2,\ldots,n \qquad (11.1c)$$

and the initial conditions

$$T_k(x,0) = T_0, \qquad k = 1,2,\ldots,n \qquad (11.1d)$$

where

$$m = \begin{cases} \frac{1}{2} & \text{slab} \\ 0 & \text{cylinder} \\ -\frac{1}{2} & \text{sphere} \end{cases} \qquad (11.1e)$$

We now define the following dimensionless variables

$$R = \frac{x}{\ell}, \qquad \tau = \frac{k^* t}{\rho^* C^* \ell^2}, \qquad \theta_k(R,\tau) = \frac{T_k(x,t) - T_w}{T_0 - T_w}$$

$$\omega_k = \frac{C_k \rho_k}{C^* \rho^*}, \qquad \kappa_k = \frac{k_k}{k^*}, \qquad \sigma_{kp} = \frac{\alpha_{kp} \ell^2}{k^*} \qquad (11.2)$$

where C^*, ρ^*, and k^* are the reference properties.

Then the problem defined by Equations (11.1) is expressed in the dimensionless form as

$$\omega_k R^{1-2m} \frac{\partial \theta_k(R,\tau)}{\partial \tau} = \kappa_k \frac{\partial}{\partial R} \left[R^{1-2m} \frac{\partial \theta_k(R,\tau)}{\partial R} \right]$$

$$+ R^{1-2m} \sum_{p=1}^{n} \sigma_{kp} \left[\theta_p(R,\tau) - \theta_k(R,\tau) \right]$$

$$\text{in } 0 < R < 1, \quad \text{for } \tau > 0, \qquad k = 1,2,\ldots,n \quad (11.3a)$$

Subject to the boundary conditions

$$\frac{\partial \theta_k(0, \tau)}{\partial R} = 0, \qquad \theta_k(1, \tau) = 0, \qquad k = 1, 2, \ldots, n \qquad (11.3\text{b,c})$$

and the initial conditions

$$\theta_k(R, 0) = 1, \qquad k = 1, 2, \ldots, n \qquad (11.3\text{d})$$

The dimensionless potentials $\theta_k(R, \tau)$ are defined in such a way that the resulting boundary conditions (11.3c) are homogeneous. Therefore, there is no need to split up the problem into steady-state and transient parts.

The problem defined by Equations (11.3) is a special case of the one-dimensional Class IV problem given by Equations (4.103). The solution to diffusion problem (11.3) can be obtained readily as a special case from solution (4.104) of problem (4.103) as described in the following section.

11.2   SOLUTIONS FOR COMPONENTS' POTENTIALS (TEMPERATURE OR MASS CONCENTRATION)

The one-dimensional, transient heat, or mass diffusion problem for an n-component heterogeneous medium defined by Equations (11.3) implies that during the transients there are n different potentials (i.e., temperature or mass concentration) $\theta_k(R, \tau)$, $(k = 1, 2, \ldots, n)$ at each point of the space for each of the n different components in the medium. We are concerned with the solution of the heat or mass diffusion problem defined by Equations (11.3) in order to determine the distribution of potentials $\theta_k(R, \tau)$ for each of the n components (i.e., $k = 1, 2, \ldots, n$) in the medium. The problem defined by Equations (11.3) is a special case of the one-dimensional Class IV problem given by Equations (4.103). The correspondence between these two problems is as follows.

$$x = R, \qquad t = \tau, \qquad T_k(x, t) = \theta_k(R, \tau), \qquad w_k(x) = \omega_k R^{1-2m}$$

$$k_k(x) = \kappa_k R^{1-2m}, \qquad d_k(x) = 0, \qquad P_k(x, t) = 0, \qquad b(x) = R^{1-2m}$$

$$\alpha_{kp} = \sigma_{kp}, \qquad x_0 = 0, \qquad x_1 = 1, \qquad \alpha_{0k} = 0 \qquad (11.4)$$

$$\beta_{0k} = 1, \qquad \phi_k(x_0, t) = 0, \qquad \alpha_{1k} = 1, \qquad \beta_{1k} = 0$$

$$\phi_k(x_1, t) = 0, \qquad f_k(x) = 1$$

Then the solution of problem (11.3) is immediately obtained from the solution (4.104) as

$$\theta_k(R, \tau) = \sum_{i=1}^{\infty} \frac{\displaystyle\sum_{k=1}^{n} \omega_k \int_0^1 R^{1-2m} \psi_k(\mu_i, R)\, dR}{\displaystyle\sum_{k=1}^{n} \omega_k \int_0^1 R^{1-2m} \psi_k^2(\mu_i, R)\, dR} \psi_k(\mu_i, R) e^{-\mu_i^2 \tau} \qquad (11.5)$$

The eigenvalue problem for the determination of the eigenfunctions $\psi_k(\mu_i, R)$ and the eigenvalues μ_i is obtained by the simplification of the eigenvalue problem (3.129) for the special case defined by Equation (11.4). We find

$$\kappa_k \frac{d}{dR}\left\{ R^{1-2m}\frac{d\psi_k(\mu, R)}{dR}\right\} + \mu^2\omega_k R^{1-2m}\psi_k(\mu, R) + R^{1-2m}$$

$$\sum_{p=1}^{n}\sigma_{kp}\{\psi_p(\mu, R) - \psi_k(\mu, R)\} = 0 \quad \text{in } 0 < R < 1$$

$$(k = 1, 2, \ldots, n) \quad (11.6a)$$

Subject to the boundary conditions

$$\psi_k'(\mu, 0) = 0, \qquad \psi_k(\mu, 1) = 0 \qquad (k = 1, 2, \ldots, n) \quad (11.6b, c)$$

The eigenvalue problem defined by Equations (11.6) is a special case of the more general eigenvalue problem (3.129). The determination of the eigenfunctions and eigenvalues of the eigenvalue problem (3.129) was discussed in Section 3.4. Therefore, we follow the general procedure described in Section 3.4 in order to solve the eigenvalue problem (11.6).

In the present problem, the boundary conditions being of the same kind for all components, the solution for the eigenfunctions $\psi_k(\mu_i, R)$ can be assumed in the form [see Equation (3.130a)]

$$\psi_k(\mu_i, R) = E_k\psi(\lambda_i, R) \tag{11.7}$$

where the E_k are constants that are yet to be determined and the function $\psi(\lambda_i, R)$ is taken as the solution of the following eigenvalue problem [see Equation (3.130b)]

$$\frac{d}{dR}\left[R^{1-2m}\frac{d\psi(\lambda, R)}{dR}\right] + \lambda^2 R^{1-2m}\psi(\lambda, R) = 0 \quad \text{in } 0 < R < 1 \quad (11.8a)$$

subject to the boundary conditions

$$\psi'(\lambda, 0) = 0, \qquad \psi(\lambda, 1) = 0 \tag{11.8b, c}$$

Equations for the determination of the constants E_k are obtained by introducing solution (11.7) into Equation (11.6a) and utilizing Equation (11.8a). Then the constants E_k satisfy the following n linear homogeneous equations [see Equation (3.131)]

$$\left(\mu^2\omega_k - \kappa_k\lambda_i^2\right)E_k + \sum_{p=1}^{n}\sigma_{kp}\left(E_p - E_k\right) = 0 \qquad (k = 1, 2, \ldots, n) \quad (11.9)$$

Equations (11.9) can be expressed in the matrix form as

$$[A]\{E\} = \mu^2[\omega]\{E\} \tag{11.10a}$$

where $[A]$ is a real, symmetric, square matrix whose elements are all known and given by

$[A]$

$$= \begin{bmatrix} \kappa_1\lambda_i^2 + \sum_{p=1}^{n}\sigma_{1p} - \sigma_{11} & -\sigma_{12} & \cdots & -\sigma_{1n} \\ -\sigma_{21} & \kappa_2\lambda_i^2 + \sum_{p=1}^{n}\sigma_{2p} - \sigma_{22} & & -\sigma_{2n} \\ \vdots & \vdots & & \vdots \\ -\sigma_{n1} & -\sigma_{n2} & \cdots & \kappa_n\lambda_i^2 + \sum_{p=1}^{n}\sigma_{np} - \sigma_{nn} \end{bmatrix}$$

$$\tag{11.10b}$$

$[\omega]$ is a diagonal matrix with known elements given by

$$[\omega] = \begin{bmatrix} \omega_1 & & & & \\ & \omega_2 & & 0 & \\ & & \ddots & & \\ & 0 & & & \\ & & & & \omega_n \end{bmatrix} \tag{11.10c}$$

and $\{E\}$ is column vector for the unknown constants E_k given by

$$\{E\} = \begin{bmatrix} E_1 \\ E_2 \\ \vdots \\ E_n \end{bmatrix} \tag{11.10d}$$

Once the constants E_k, $(k = 1, 2, \ldots, n)$ are determined from the solution of the homogeneous system (11.10), the general solution $\psi_k(\mu_i, R)$ of the eigenvalue problem (11.6), assumed in the form given by Equation (11.7), is constructed by a linear combination of these E_k in the form

$$\psi_k(\mu_i, R) = \sum_{j=1}^{n} E_{kij}\psi(\lambda_i, R) \tag{11.11}$$

Introducing Equation (11.11) into Equation (11.5), the solution for $\theta_k(R, \tau)$ takes the form

$$\theta_k(R, \tau) = \sum_{i=1}^{\infty} \frac{\int_0^1 R^{1-2m} \psi(\lambda_i, R) \, dR}{\int_0^1 R^{1-2m} \psi^2(\lambda_i, R) \, dR} \psi(\lambda_i, R) F_k(\lambda_i, \tau) \quad (11.12a)$$

where

$$F_k(\lambda_i, \tau) = \sum_{j=1}^{n} E_{kij} \frac{\sum\limits_{k=1}^{n} \omega_k E_{kij}}{\sum\limits_{k=1}^{n} \omega_k E_{kij}^2} e^{-\mu_{ij}^2 \tau} \quad (11.12b)$$

and the $\psi(\lambda_i, R)$ are the eigenfunctions of the eigenvalue problem (11.8). This eigenvalue problem is a special case of the eigenvalue problem (3.42) for $\beta_1 = 0$. Therefore, the solution for $\psi(\lambda_i, R)$ is immediately obtained from Equation (3.43a) as

$$\psi(\lambda_i, R) = (\lambda_i R)^m J_{-m}(\lambda_i R) \quad (11.13a)$$

which satisfies Equation (11.8a) and the boundary condition (11.8b). The boundary condition (11.8c) is satisfied if the λ_i are taken as the roots of the transcendental equation

$$J_{-m}(\lambda) = 0 \quad (11.13b)$$

When Equation (11.13a) is introduced into Equation (11.12) and the integrals are evaluated, the solution $\theta_k(R, \tau)$ of the problem (11.3) becomes (see note 1 for the evaluation of integrals)

$$\theta_k(R, \tau) = 2 \sum_{i=1}^{\infty} R^m \frac{J_{-m}(\lambda_i R)}{\lambda_i J_{1-m}(\lambda_i)} F_k(\lambda_i, \tau) \quad (k = 1, 2, \ldots, n)$$

$$(11.14a)$$

where

$$F_k(\lambda_i, \tau) = \sum_{j=1}^{n} E_{kij} \frac{\sum\limits_{k=1}^{n} \omega_k E_{kij}}{\sum\limits_{k=1}^{n} \omega_k E_{kij}^2} e^{-\mu_{ij}^2 \tau} \quad (11.14b)$$

and the λ_i are the roots of the transcendental Equation (11.13b).

We recall that if the matrix [A] in Equations (11.10) is real and symmetric, the parameter $\omega_k > 0$, $(k = 1, 2, \ldots, n)$, then the eigenvalues μ_{ij} are real and $\mu_{ij}^2 > 0$. Various methods are available for the computation of all eigenvalues μ_{ij} and the corresponding eigenvectors E_{kij} from Equation (11.10) [3].

Once μ_{ij} and E_{kij} are determined, the distribution of the potentials $\theta_k(R, \tau)$, $(k = 1, 2, \ldots, n)$ for each of n components in the one-dimensional region $0 \le R \le 1$ is computed from the solution given by Equations (11.14). This solution is valid for a slab, long solid cylinder, and solid sphere, depending on the choice of the value of m. In the following section we present the resulting solutions for each of these three different geometries, and give some numerical results.

11.3   DISTRIBUTION OF POTENTIALS IN AN HETEROGENEOUS SLAB, LONG SOLID CYLINDER, AND SPHERE

Specific solutions for the distribution of potentials $\theta_k(R, \tau)$, $(k = 1, 2, \ldots, n)$ for each of the n components in an heterogeneous slab, long solid cylinder, and sphere are obtained from Equation (11.14) by setting $m = \frac{1}{2}$, 0, and $-\frac{1}{2}$, respectively, and making use of the relations given in Table 3.1 as now described.

Slab ($m = \frac{1}{2}$)

By setting $m = \frac{1}{2}$ in Equation (11.14a) and replacing the Bessel functions by the equivalent trigonometric functions given in Table 3.1, the solution for a slab becomes

$$\theta_k(R, \tau) = 2 \sum_{i=1}^{\infty} (-1)^{i+1} \frac{\cos(\lambda_i R)}{\lambda_i} F_k(\lambda_i, \tau) \qquad (k = 1, 2, 3, \ldots, n)$$

(11.15a)

where the function $F_k(\lambda_i, \tau)$ is defined by Equation (11.14b) and the eigenvalues λ_i are the roots of $\cos \lambda_i = 0$, that is,

$$\lambda_i = (2i - 1)\frac{\pi}{2} \qquad (i = 1, 2, 3, \ldots) \qquad (11.15b)$$

Long Solid Cylinder ($m = 0$)

By setting $m = 0$ in Equation (11.14a), the solution for a long solid cylinder becomes

$$\theta_k(R, \tau) = 2 \sum_{i=1}^{\infty} \frac{J_0(\lambda_i R)}{\lambda_i J_1(\lambda_i)} F_k(\lambda_i, \tau) \qquad (k = 1, 2, \ldots, n) \quad (11.16a)$$

where the function $F_k(\lambda_i, \tau)$ is defined by Equation (11.14b) and the λ_i are the roots of

$$J_0(\lambda) = 0 \qquad (11.16b)$$

Solid Sphere ($m = -\frac{1}{2}$)

In the case of a solid sphere, we set $m = -\frac{1}{2}$ in Equation (11.14a), utilize the results in Table 3.1 together with the definition of spherical Bessel functions given by Equations (3.36d, e). Then the solution for a solid sphere becomes

$$\theta_k(R,\tau) = 2 \sum_{i=1}^{\infty} (-1)^{i+1} \frac{\sin(\lambda_i R)}{\lambda_i R} F_k(\lambda_i, R) \qquad (k = 1, 2, \ldots, n)$$

$$(11.17a)$$

where $F_k(\lambda_i, R)$ is defined by Equation (11.14b) and the λ_i are the roots of $\sin \lambda_i / \lambda_i = 0$, that is,

$$\lambda_i = i\pi \qquad (i = 1, 2, 3, \ldots) \qquad (11.17b)$$

Example 11.1   Calculate the distribution of potentials (i.e., dimensionless temperature or mass concentration) $\theta_k(0, \tau)$ in the center of a slab, long solid cylinder, and sphere as a function of the dimensionless time τ from the solutions given by Equations (11.15)–(11.17), respectively, for a three-component (i.e., $n = 3$) heterogeneous medium. The system parameters for the components are $\omega_1 = \omega_2 = \omega_3 = 1$ and $\kappa_1 = 0.1$, $\kappa_2 = 1$, $\kappa_3 = 10$. The calculations are to be performed for the values of $\sigma_{kp} \equiv \sigma = 0$, 1, and 10.

Solution   This particular problem was considered in Reference 3. The dimensionless potentials $\theta_k(0, \tau)$ at the center, plotted as a function of the dimensionless time τ, are shown in Figures 11.1–11.3 for a slab, long solid cylinder, and sphere, respectively. To illustrate the effects of various system parameters on the variation of the potentials $\theta_k(0, \tau)$ at the center of the region as a function of time we examine Figure 11.1 for the slab geometry.

All three components initially (i.e., $\tau = 0$) are at the same potential; that is, $\theta_k(0, \tau) = 1$. During the transients, at any instant τ they are at different potentials; as the steady-state (i.e., $\tau \to \infty$) is reached, they all attain the same steady-state potential $\theta_k(0, \infty) \to 0$.

The parameter σ_k is the measure of the relative magnitude of the coupling between the components; the larger is the value of σ_k, the stronger is the coupling between the components. For $\sigma = 10$, the curves for the three components are much closer to each other than for $\sigma = 0$. The case $\sigma = 0$ implies that the three components are decoupled. For the case $\sigma \to \infty$, the curves for the components would merge into a single curve.

Figures 11.2 and 11.3 are, respectively, for a long solid cylinder and a solid sphere. The curves in these figures follow the same general trend as those in Figure 11.1 for a slab, except they are slightly shrunk in the time scale. The crossover of the curves for $\sigma = 0$ and $\sigma = 1$ for $\theta_2(0, \tau)$ are apparent in Figures 11.2 and 11.3.

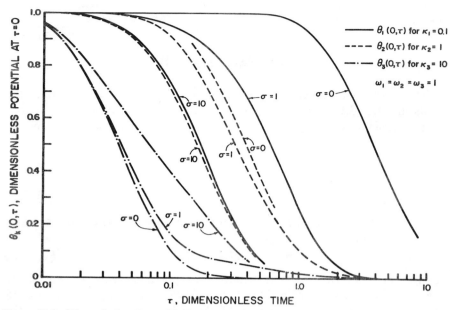

Figure 11.1 Dimensionless potentials $\theta_k(0, \tau)$ at the surface $R = 0$ of a slab as a function of dimensionless time τ for each of the three components.

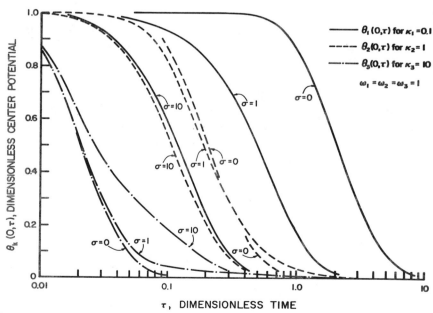

Figure 11.2 Dimensionless potentials $\theta_k(0, \tau)$ at the center of a long solid cylinder as a function of dimensionless time τ for each of the three components.

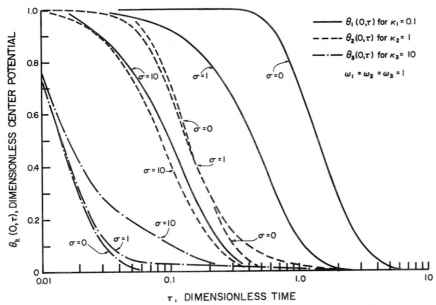

Figure 11.3 Dimensionless potentials $\theta_k(0, \tau)$ at the center of a solid sphere as a function of dimensionless time τ for each of the three components.

REFERENCES

1. G. Nicolis and I. Prigogine, *Self-Organization in Nonequilibrium Systems*, Wiley, New York, 1977.

2. L. I. Rubinstein, *Temperature Fields in Oil Saturated Layers* (Russ.), Nedra Press, Moscow, 1972.

3. M. D. Mikhailov, M. N. Özişik, and B. K. Shishedjiev, Diffusion in Heterogeneous Media, *J. Heat Transf.*, **104**, 781–787 (1982).

NOTES

1. The Derivation of Equation (11.14)

We consider Equation (11.12) given by

$$\theta_k(R, \tau) = \sum_{i=1}^{\infty} \frac{\int_0^1 R^{1-2m} \psi(\lambda_i, R)\, dR}{\int_0^1 R^{1-2m} \psi^2(\lambda_i, R)\, dR} \psi(\lambda_i, R) F_i(\lambda_i, \tau) \qquad (1)$$

where

$$\psi(\lambda_i, R) = (\lambda_i R)^m J_{-m}(\lambda_i R) \qquad (2)$$

Then the integrals in Equation (1) are evaluated as

$$\int_0^1 R^{1-2m}\psi(\lambda_i, R)\, dR = \lambda_i^m \int_0^1 R^{1-m} J_{-m}(\lambda_i R)\, dR = \lambda_i^{m-1} J_{1-m}(\lambda_i) \tag{3}$$

$$\int_0^1 R^{1-2m}\psi^2(\lambda_i, R)\, dR = \lambda_i^{2m} \int_0^1 R J_{-m}^2(\lambda_i R)\, dR$$

$$= \lambda_i^{2m}\left\{\frac{1}{2}R^2\left[J'^2_{-m}(\lambda_i R) + \left(1 - \frac{m^2}{\lambda_i^2 R^2}\right) J_{-m}^2(\lambda_i R)\right]\right\}_0^1$$

$$= \frac{1}{2}\lambda_i^{2m} J'^2_{-m}(\lambda_i)$$

$$= \frac{1}{2}\lambda_i^{2m} J_{1-m}^2(\lambda_i) \tag{4}$$

where we made use of the result

$$J_{-m}(\lambda_i) = 0 \tag{5}$$

When Equations (3) and (4) are introduced into Equation (1), we obtain

$$\theta_k(R, \tau) = 2\sum_{i=1}^{\infty} R^m \frac{J_{-m}(\lambda_i R)}{\lambda_i J_{1-m}(\lambda_i)} F_k(\lambda_i, \tau) \tag{6}$$

which is the result given by Equation (11.14a).

CHAPTER TWELVE

Class V Solutions Applied to Mass Diffusion in Chemically Reacting Systems

Learn what is true in order to do what is right.

Thomas Henry Huxley
1825–1895

The boundary value problems belonging to Class V are characterized by two diffusion equations in which the potentials $T_k(x, t)$, $(k = 1, 2)$ in every point of the space are coupled through the source or sink terms, but the coupling in this case is more general than the coupling considered for the Class IV problems. Namely, the coupling coefficients σ_{kp} for the Class V problems are not symmetrical; $\sigma_{kp} \neq \sigma_{pk}$. Such problems are encountered in chemically reacting diffusing systems as discussed later.

Consider a chemically reacting n component system. If the medium is motionless (i.e., $\mathbf{v} = 0$) and the components diffuse according to the Fick law, Equation (1.19) reduces to

$$\frac{\partial \rho_k}{\partial t} = D_k \nabla^2 \rho_k + I_k \qquad (k = 1, 2, \dots, n) \qquad (12.1a)$$

where ρ_k is the mass per unit volume of the kth component and the production rate per unit volume I_k of the kth component is, in general, a *nonlinear* function of masses per unit volume of all components, namely,

$$I_k \equiv I_k(\rho_1, \rho_2, \dots, \rho_n) \qquad (12.1b)$$

then the system (12.1a) is nonlinear. To linearize the system we define a new quantity ρ_k^* as

$$\rho_k^* = \rho_k - \rho_{k\infty} \qquad (12.1c)$$

441

where $\rho_{k\infty}$ is the steady-state solution of Equations (12.1a) subject to specified boundary conditions with constant source terms. Introducing Equation (12.1c) into (12.1a) and linearizing the source terms I_k we obtain

$$\frac{\partial \rho_k^*}{\partial t} = D_k \nabla^2 \rho_k^* + \sum_{p=1}^{n} \left(\frac{\partial I_k}{\partial \rho_p^*} \right) \rho_p^* \quad (k = 1, 2, \ldots, n) \quad (12.1d)$$

subject to the same but *homogeneous* boundary conditions.

The problems governed by the system given by Equations (12.1d) are important in order to understand self-organization in nonequilibrium systems [1].

Clearly, the system of Equations (12.1d) for $n = 2$ leads to the Class V problem described in the Chapters 2 and 4. The mathematical modeling of some diffusion problems such as the heat transfer in a fluid–solid mixture flowing in turbulent flow inside a pipe [2, 3] or diffusion in a tubular reactor leads to a system of equations similar to that of the Class V problem.

In this chapter we first examine the solution of a one-dimensional Class V problem for a two-component system having a monotype *homogeneous* boundary condition (i.e., the boundary condition is the same for both components). In the following sections we illustrate how the general analysis of the one-dimensional Class V problem developed in Chapter 4 can be utilized for the solution of heat and mass diffusion problems in a chemically reacting plug flow when the boundary conditions are all of the second kind.

12.1 SOLUTION FOR POTENTIALS FOR MONOTYPE HOMOGENEOUS BOUNDARY CONDITIONS

We consider the one-dimensional transient diffusion of a two-component system modeled as a Class V problem. Initially, each component has a uniform distribution f_k, $(k = 1, 2)$, throughout the medium. For times $t > 0$, the outer surface of the region is subject to the same homogeneous boundary condition of the third kind for each of the components, whereas the other boundary satisfies the symmetry condition. The mathematical formulation of this transient diffusion problem for each of the two components is given by

$$\omega_k x^{1-2m} \frac{\partial T_k(x, t)}{\partial t} = \kappa_k \frac{\partial}{\partial x} \left[x^{1-2m} \frac{\partial T_k(x, t)}{\partial x} \right]$$

$$+ (-1)^k x^{1-2m} \left[\sigma_{k-1} T_1(x, t) - \sigma_{k+1} T_2(x, t) \right]$$

$$\text{in } 0 \leq x \leq 1, \quad t > 0, \quad k = 1, 2, \ldots \quad (12.2a)$$

subject to the boundary conditions

$$\frac{\partial T_k(0, t)}{\partial x} = 0, \quad \alpha T_k(1, t) + \beta \frac{\partial T_k(1, t)}{\partial x} = 0 \quad (12.2b, c)$$

and the initial conditions

$$T_k(x,0) = f_k, \qquad k = 1,2 \qquad (12.2\text{d})$$

where $m = \frac{1}{2}, 0$, and $-\frac{1}{2}$ for slab, cylinder, and sphere, respectively.

This problem is a special case of problem (4.106) as follows:

$$w_k(x) = \omega_k x^{1-2m}, \qquad k_k(x) = \kappa_k x^{1-2m}, \qquad d_k(x) = 0, \qquad P_k(x,t) = 0$$

$$b(x) = x^{1-2m}, \qquad x_0 = 0, \qquad x_1 = 1, \qquad \alpha_{0k} = 0, \qquad \beta_{0k} = 1 \qquad (12.3)$$

$$\alpha_{1k} = \alpha, \qquad \beta_{1k} = \beta, \qquad \phi_k(x_0, t) = 0, \qquad \phi_k(x_1, t) = 0, \qquad f_k(x) = f_k$$

Then the solution for the potentials $T_k(x, t)$ is immediately obtained from solution (4.107) as

$$T_k(x,t) = \sum_{i=1}^{\infty} \frac{\displaystyle\sum_{k=1}^{2} \sigma_k \omega_k f_k \int_0^1 x^{1-2m} \psi_k(\mu_i, x)\, dx}{\displaystyle\sum_{k=1}^{2} \sigma_k \omega_k \int_0^1 x^{1-2m} \psi_k^2(\mu_i, x)\, dx} \psi_k(\mu_i, x) e^{-\mu_i^2 t} \qquad (12.4)$$

where $\psi_k(\mu_i, x)$ and μ_i are the eigenfunctions and eigenvalues, respectively, of the eigenvalue problem (3.134) which, for the special case defined by Equations (12.3), reduces to

$$\kappa_k \frac{d}{dx}\left[x^{1-2m} \frac{d\psi_k(\mu, x)}{dx} \right] + \mu^2 \omega_k x^{1-2m} \psi_k(\mu, x) + (-1)^k x^{1-2m}$$

$$\times \left[\sigma_{\ell-1} \psi_1(\mu, x) - \sigma_{k+1} \psi_2(\mu, x) \right] = 0 \quad \text{in } 0 < x < 1, \qquad k = 1,2$$

$$(12.5\text{a})$$

subject to the boundary conditions

$$\psi_k'(\mu, 0) = 0, \qquad \alpha \psi_k(\mu, 1) + \beta \psi_k'(\mu, 1) = 0 \qquad (12.5\text{b},\text{c})$$

The boundary conditions being the same for each of the two components, the solution of this eigenvalue problem may be taken in the form [see Equation (3.135a)]

$$\psi_k(\mu, x) = E_k \psi(\lambda, x), \qquad k = 1,2 \qquad (12.6\text{a})$$

where E_k ($k = 1, 2$) are constants and $\psi(\lambda, x)$ is taken as the solution of the following eigenvalue problem [see Equation (3.135b)]

$$\frac{d}{dx}\left[x^{1-2m} \frac{d\psi(\lambda, x)}{dx} \right] + \lambda^2 x^{1-2m} \psi(\lambda, x) = 0 \quad \text{in } 0 < x < 1 \qquad (12.6\text{b})$$

subject to the boundary conditions

$$\psi'(\lambda, 0) = 0, \qquad \alpha\psi(\lambda, 1) + \beta\psi'(\lambda, 1) = 0 \qquad (12.6c, d)$$

The eigenvalue problem is the same as that given by Equations (3.42) and its solution is immediately obtained from Equations (3.43a) as

$$\psi(\lambda_i, x) = (\lambda_i x)^m J_{-m}(\lambda_i, x) \qquad (12.7a)$$

and the λ_i are the roots of the transcendental equation

$$\alpha J_{-m}(\lambda) - \beta\lambda J_{1-m}(\lambda) = 0 \qquad (12.7b)$$

To determine the constants E_k ($k = 1, 2$) Equation (12.6a) is introduced into Equation (12.5a) and Equation (12.6b) is utilized. We find that E_k ($k = 1, 2$) must satisfy the following two homogeneous equations

$$\left(\omega_k\mu^2 - \kappa_k\lambda_i^2\right)E_k + (-1)^k\left(\sigma_{k-1}E_1 - \sigma_{k+1}E_2\right) = 0, \qquad k = 1, 2 \quad (12.8)$$

This relation is written more explicitly by setting $k = 1$ and 2 to yield, respectively,

$$\left(\omega_1\mu^2 - \kappa_1\lambda_i^2 - \sigma_0\right)E_1 + \sigma_2 E_2 = 0 \qquad (12.9a)$$

$$\sigma_1 E_1 + \left(\omega_2\mu^2 - \kappa_2\lambda_i^2 - \sigma_3\right)E_2 = 0 \qquad (12.9b)$$

We solve Equations (12.9) for the ratio E_2/E_1

$$\frac{E_2}{E_1} = \frac{\sigma_0 + \kappa_1\lambda_i^2 - \omega_1\mu^2}{\sigma_2} = \frac{\sigma_1}{\sigma_3 + \kappa_2\lambda_i^2 - \omega_2\mu^2} \qquad (12.10)$$

Equation (12.10) implies that the ratio E_2/E_1 is a function of the parameter μ. That is, E_1 and E_2 are not independent, but they are related to each other according to Equation (12.10).

The right-hand side equality in Equation (12.10) yields the following relation for the determination of the eigenvalues μ_{ij}^2.

$$2\mu_{ij}^2 = \left[\frac{\sigma_o + \kappa_1\lambda_i^2}{\omega_1} + \frac{\sigma_3 + \kappa_2\lambda_i^2}{\omega_2}\right]$$

$$+ (-1)^j\sqrt{\left[\frac{\sigma_0 + \kappa_1\lambda_i^2}{\omega_1} - \frac{\sigma_3 + \kappa_2\lambda_i^2}{\omega_2}\right]^2 + 4\frac{\sigma_1\sigma_2}{\omega_1\omega_2}} \qquad (12.11)$$

where $j = 1$ and 2, and the λ_i are the roots of the transcendental Equation (12.7b).

Once the μ_{ij} are determined from Equation (12.11) and the ratio E_{2ij}/E_{1ij} from Equation (12.10), then the eigenfunctions $\psi_k(\mu, x)$ are established according to Equation (12.6a) as

$$\psi_k(\mu_{ij}, x) = C_{ij}\left(\frac{\sigma_0 + \kappa_1\lambda_i^2 - \omega_1\mu_{ij}^2}{\sigma_2}\right)^{k-1}\psi(\lambda_i, x) \qquad (12.12a)$$

where

$$\psi(\lambda_i, x) = (\lambda_i x)^m J_{-m}(\lambda_i x) \qquad k = 1,2 \quad \text{and} \quad j = 1,2 \qquad (12.12b)$$

The eigenfunctions $\psi_k(\mu_{ij}, x)$ given by Equation (12.12) are introduced into Equation (12.4); the constant C_{ij} cancels out and the resulting expressions are rearranged to yield

$$T_k(x, t) = \sum_{i=1}^{\infty}\left[\frac{\displaystyle\int_0^1 x^{1-2m}\psi(\lambda_i, x)\,dx}{\displaystyle\int_0^1 x^{1-2m}\psi^2(\lambda_i, x)\,dx}\right]\psi(\lambda_i, x)F_k(\lambda_i, t) \qquad (12.13a)$$

where

$$F_k(\lambda_i, t) = \sum_{j=1}^{2} E_{ij}^{k-1}\frac{\displaystyle\sum_{k=1}^{2}\sigma_k\omega_k f_k E_{ij}^{k-1}}{\displaystyle\sum_{k=1}^{2}\sigma_k\omega_k E_{ij}^{2(k-1)}}e^{-\mu_{ij}^2 t} \qquad (12.13b)$$

$$E_{ij} = \frac{\sigma_0 + \kappa_1\lambda_i^2 - \omega_1\mu_{ij}^2}{\sigma_2} \qquad (12.13c)$$

$$\psi(\lambda_i, x) = (\lambda_i x)^m J_{-m}(\lambda_i x) \qquad (12.13d)$$

The integrals in Equation (12.13a) can be evaluated and the resulting expression for the potentials $T_k(x, t)$, $(k = 1, 2)$ for the case $\alpha \neq 0$ takes the form

$$T_k(x, t) = 2\sum_{i=1}^{\infty}\frac{1}{\lambda_i}\left[1 + 2m\frac{\beta}{\alpha} + \left(\lambda_i\frac{\beta}{\alpha}\right)^2\right]^{-1}x^m\frac{J_{-m}(\lambda_i x)}{J_{1-m}(\lambda_i)}F_k(\lambda_i, t)$$

$$(12.14)$$

where $F_k(\lambda_i, t)$ is defined by Equation (12.13b).

The case $\alpha = 0$, $\beta = 1$, representing the situation when both boundary conditions are of the second kind, is considered in the following sections.

12.2   SOLUTION FOR BOTH BOUNDARY CONDITIONS OF THE SECOND KIND

In this section we consider the solution of a nonhomogeneous Class V problem with both the boundary conditions of the second kind. The problem is given in the form

$$\omega_k x^{1-2m}\frac{\partial T_k(x,t)}{\partial t} = \kappa_k \frac{\partial}{\partial x}\left\{ x^{1-2m}\frac{\partial T_k(x,t)}{\partial x}\right\}$$

$$+(-1)^k x^{1-2m}\left\{\sigma_{k-1}T_1(x,t) - \sigma_{k+1}T_2(x,t)\right\}$$

$$\text{in } 0 < x < 1, \qquad t > 0, \qquad k = 1,2 \quad (12.15a)$$

Subject to the boundary and initial conditions

$$\frac{\partial T_k(0,t)}{\partial x} = 0, \qquad \kappa_k \frac{\partial T_k(1,t)}{\partial x} = \phi_k, \qquad T_k(x,0) = f_k \quad (12.15b,c,d)$$

where ϕ_k and f_k are constants.

To solve this nonhomogeneous problem we first split it up into a set of simpler problems and then seek solutions to each of the simpler problems.

The solution $T_k(x,t)$ of problem (12.15) can be split up into three potentials defined as

$$T_k(x,t) = T_{av,k}(t) + T_{0k}(x) + T_{tk}(x,t), \qquad k = 1,2 \quad (12.16)$$

where $T_{av,k}(t)$ are the *average potentials*, $T_{0k}(x)$ the *steady-state potentials*, and $T_{tk}(x,t)$ the *transient homogeneous potentials*. The simpler problems satisfied by each of these potentials are established as now described.

Problem Defining $T_{av,k}(t)$

The average potentials $T_{av,k}(t)$ are defined by

$$T_{av,k}(t) = \frac{\int_0^1 x^{1-2m}T_k(x,t)\,dx}{\int_0^1 x^{1-2m}\,dx} = 2(1-m)\int_0^1 x^{1-2m}T_k(x,t)\,dx \quad (12.17)$$

The differential equation satisfied by $T_{av,k}(t)$ is determined by integrating Equation (12.15a) from $x = 0$ to 1, an utilizing the definition (12.17) and the boundary conditions (12.15b,c). We obtain

$$\omega_k \frac{dT_{av,k}(t)}{dt} = 2(1-m)\phi_k +(-1)^k\left\{\sigma_{k-1}T_{av,1}(t) - \sigma_{k+1}T_{av,2}(t)\right\},$$

$$k = 1,2 \quad (12.18a)$$

The application of definition (12.17) to Equation (12.15d) gives the initial condition for this equation as

$$T_{\text{av},k}(0) = f_k, \qquad k = 1,2 \tag{12.18b}$$

The solution of Equation (12.18) for $T_{\text{av},k}(t)$ will be discussed later.

Problems Defining $T_{0k}(x)$ and $T_{tk}(x,t)$

To establish the simpler problems defining the functions $T_{0k}(x)$ and $T_{tk}(x,t)$, we introduce Equation (12.16) into the original problem (12.15), take into account the Equations (12.18), and choose the steady-state problem defining the functions $T_{0k}(x)$ as

$$\kappa_k \frac{d}{dx}\left\{ x^{1-2m}\frac{dT_{0k}(x)}{dx} \right\} + (-1)^k x^{1-2m}\{ \sigma_{k-1}T_{01}(x) - \sigma_{k+1}T_{02}(x) \}$$

$$= 2(1-m)x^{1-2m}\phi_k \quad \text{in } 0 < x < 1, \qquad k = 1,2 \tag{12.19a}$$

Subject to the boundary conditions

$$\frac{dT_{0k}(0)}{dx} = 0, \qquad \kappa_k \frac{dT_{0k}(1)}{dx} = \phi_k \tag{12.19b, c}$$

and the condition

$$\int_0^1 x^{1-2m}T_{0k}(x)\, dx = 0 \tag{12.19d}$$

Then the functions $T_{tk}(x,t)$ satisfy the following transient homogeneous problem

$$\omega_k x^{1-2m}\frac{\partial T_{tk}(x,t)}{\partial t} = \kappa_k \frac{\partial}{\partial x}\left[x^{1-2m}\frac{\partial T_{tk}(x,t)}{\partial x} \right]$$

$$+ (-1)^k x^{1-2m}\{ \sigma_{k-1}T_{t1} - \sigma_{k+1}T_{t2} \}$$

$$\text{in } 0 < x < 1, \qquad t > 0, \qquad k = 1,2 \tag{12.20a}$$

Subject to the boundary and initial conditions

$$\frac{\partial T_{tk}(0,t)}{\partial x} = 0, \qquad \frac{\partial T_{tk}(1,t)}{\partial x} = 0, \qquad T_{tk}(x,0) = -T_{0k}(x)$$

$$\tag{12.20b, c, d}$$

and the condition

$$\int_0^1 x^{1-2m} T_{tk}(x, t)\, dx = 0 \qquad (12.20e)$$

The constraints specified by Equations (12.19d) and (12.20e) are the out-come of the substitution of Equation (12.16) into Equation (12.17).

For the case of boundary conditions both of the second kind considered here, $\lambda_0 = 0$ is also an eigenvalue of the eigenvalue problem (12.6b) with $\psi(\lambda_0) =$ constant being the corresponding eigenfunction. If the eigenvalue Equation (12.5a) should also satisfy the requirement that $\psi_k =$ constant is also an eigenfunction with $\mu = 0$ being the corresponding eigenvalue, then the relation $\sigma_{k-1}\psi_1 - \sigma_{k+1}\psi_2 = 0$ should hold for $k = 1, 2$. By writing this con-straint for $k = 1$ and 2, it can be shown that the following relation should hold for the case of boundary conditions both of the second kind

$$\sigma_0\sigma_3 - \sigma_1\sigma_2 = 0 \qquad (12.21)$$

We now examine the solutions for the functions $T_{av,k}(t)$, $T_{0k}(x)$, and $T_{tk}(x, t)$ defined by the simpler problems given, respectively, by Equations (12.18)–(12.20).

12.3   SOLUTION FOR THE AVERAGE POTENTIALS $T_{av,k}(t)$

The average potentials $T_{av,k}(t)$ are satisfied by the two coupled ordinary differential Equations (12.18). To solve these equations, we first decouple them as described in the following.

We multiply both sides of Equation (12.18a) by σ_{2-k}, write the resulting expression for $k = 1$ and $k = 2$, add the results, and utilize the identity given by Equation (12.21) to obtain

$$\sigma_1\omega_1 \frac{dT_{av,1}(t)}{dt} + \sigma_0\omega_2 \frac{dT_{av,2}(t)}{dt} = 2(1 - m)(\sigma_1\phi_1 + \sigma_0\phi_2)$$

which can be rewritten more compactly in the form

$$\sum_{k=1}^2 \sigma_{2-k}\omega_k \frac{dT_{av,k}(t)}{dt} = 2(1 - m) \sum_{k=1}^2 \sigma_{2-k}\phi_k \qquad (12.22a)$$

The initial conditions are obtained from Equation (12.18b) as

$$T_{av,k}(0) = f_k, \qquad k = 1, 2 \qquad (12.22b)$$

The integration of Equation (12.22a) from $t = 0$ to t by taking into account

(12.22b) yields

$$\sum_{k=1}^{2} \sigma_{2-k}\omega_k T_{av,k}(t) = \sum_{k=1}^{2} \sigma_{2-k}\omega_k f_k + \left[2(1-m) \sum_{k=1}^{2} \sigma_{2-k}\phi_k \right] t \quad (12.23)$$

which can be rewritten in the form

$$\sigma_1\omega_1 T_{av,1}(t) + \sigma_0\omega_2 T_{av,2}(t) = a \quad (12.24a)$$

where

$$a = a_0 + a_1 t \quad (12.24b)$$

$$a_0 = \sum_{k=1}^{2} \sigma_{2-k}\omega_k f_k \quad (12.24c)$$

$$a_1 = 2(1-m) \sum_{k=1}^{2} \sigma_{2-k}\phi_k \quad (12.24d)$$

The relations (12.24) can be utilized to decouple Equation (12.18a) as now described.

We rewrite Equation (12.18a) in the form

$$\omega_k \frac{dT_{av,k}(t)}{dt} = 2(1-m)\phi_k + D_k \quad (12.25a)$$

where

$$D_k = (-1)^k \left[\sigma_{k-1} T_{av,1}(t) - \sigma_{k+1} T_{av,2}(t) \right], \quad k = 1,2 \quad (12.25b)$$

Equation (12.25b) is first written for $k = 1$ and $T_{av,2}(t)$ is replaced by its value given by Equation (12.24a) to obtain

$$D_1 = \frac{1}{\omega_2} \left[\frac{\sigma_2}{\sigma_0} a - \left(\sigma_0\omega_2 + \frac{\sigma_2\sigma_1}{\sigma_0}\omega_1 \right) T_{av,1}(t) \right]$$

$$= \frac{1}{\omega_2} \left[\frac{\sigma_2}{\sigma_0} a - (\sigma_0\omega_2 + \sigma_3\omega_1) T_{av,1}(t) \right] \quad (12.26a)$$

since $\sigma_3 = \sigma_1\sigma_2/\sigma_0$ by equation (12.21).

Then Equation (12.25b) is written for $k = 2$ and $T_{av,1}(t)$ is replaced by its value given by Equation (12.24a) to yield

$$D_2 = \frac{1}{\omega_1} \left[a - (\sigma_0\omega_2 + \sigma_3\omega_1) T_{av,2}(t) \right] \quad (12.26b)$$

The results given by Equations (12.26a, b) can be written more compactly as

$$D_k = \frac{1}{\omega_{3-k}}\left[a\frac{\sigma_{3-k}}{\sigma_{k-1}} - (\sigma_0\omega_2 + \sigma_3\omega_1)T_{\text{av},\,k}(t)\right], \qquad k = 1, 2 \quad (12.26c)$$

When Equation (12.26c) is introduced into Equation (12.25a) we note that the resulting system for $k = 1$ and 2 becomes decoupled and is written in the form

$$\frac{dT_{\text{av},\,k}(t)}{dt} + \omega T_{\text{av},\,k}(t) = g_{0k} + g_{1k}t, \qquad k = 1, 2 \quad (12.27a)$$

where

$$\omega = \frac{\sigma_0\omega_2 + \sigma_3\omega_1}{\omega_1\omega_2} \quad (12.27b)$$

$$g_{0k} = 2(1 - m)\frac{\phi_k}{\omega_k} + \frac{\sigma_{3-k}}{\sigma_{k-1}}\frac{a_0}{\omega_1\omega_2} \quad (12.27c)$$

$$g_{1k} = \frac{\sigma_{3-k}}{\sigma_{k-1}}\frac{a_1}{\omega_1\omega_2} \quad (12.27d)$$

The solution of the first order linear differential Equation (12.27a), subject to the initial conditions (12.18b), gives the average potentials $T_{\text{av},\,k}(t)$ as

$$T_{\text{av},\,k}(t) = \left(\frac{g_{0k}}{\omega} - \frac{g_{1k}}{\omega^2}\right)(1 - e^{-\omega t}) + \frac{g_{1k}}{\omega}t + f_k e^{-\omega t}, \qquad k = 1, 2 \quad (12.28a)$$

where

$$\omega = \frac{\sigma_0\omega_2 + \sigma_3\omega_1}{\omega_1\omega_2} \quad (12.28b)$$

$$g_{0k} = 2(1 - m)\frac{\phi_k}{\omega_k} + \frac{\sigma_{3-k}}{\sigma_{k-1}}\frac{a_0}{\omega_1\omega_2} \quad (12.28c)$$

$$g_{1k} = \frac{\sigma_{3-k}}{\sigma_{k-1}}\frac{a_1}{\omega_1\omega_2} \quad (12.28d)$$

$$a_0 = \sum_{k=1}^{2} \sigma_{2-k}\omega_k f_k \quad (12.28e)$$

$$a_1 = 2(1 - m)\sum_{k=1}^{2} \sigma_{2-k}\phi_k \quad (12.28f)$$

12.4 SOLUTION FOR THE STEADY-STATE POTENTIALS $T_{0k}(x)$

The steady-state potentials $T_{0k}(x)$ are satisfied by the two, coupled, second order, linear ordinary differential Equations (12.19). To solve these equations, we first decouple them by the procedure described.

We multiply both sides of Equation (12.19a) by σ_{2-k}, write the resulting expression for $k = 1$ and 2, add these two results, and utilize the identity $\sigma_0\sigma_3 = \sigma_1\sigma_2$ given by Equation (12.21) to obtain

$$\sum_{k=1}^{2} \sigma_{2-k}\kappa_k \frac{d}{dx} \left\{ x^{1-2m} \frac{dT_{0k}(x)}{dx} \right\} = 2(1-m)x^{1-2m} \sum_{k=1}^{2} \sigma_{2-k}\phi_k$$

$$(12.29a)$$

We note that the coupling of the functions $T_{0k}(x)$ in Equation (12.29a) is similar to that of function $T_{av,k}(t)$ in Equation (12.22a). Therefore, we follow a similar procedure in the decoupling process.

Equation (12.29a) is integrated from $x = 0$ to x, and the boundary condition (12.19b) is taken into account.

$$\sum_{k=1}^{2} \sigma_{2-k}\kappa_k \frac{dT_{0k}(x)}{dx} = x \sum_{k=1}^{2} \sigma_{2-k}\phi_k \qquad (12.29b)$$

This result is integrated once more from $x = 0$ to x to yield

$$\sum_{k=1}^{2} \sigma_{2-k}\kappa_k T_{0k}(x) = \sum_{k=1}^{2} \sigma_{2-k}\kappa_k T_{0k}(0) + \tfrac{1}{2}x^2 \sum_{k=1}^{2} \sigma_{2-k}\phi_k \quad (12.29c)$$

Equation (12.29c) is multiplied by $2(1-m)x^{1-2m}$ and the result is integrated from $x = 0$ to $x = 1$; then the left-hand side vanishes because of the constraint given by Equation (12.19d) and we obtain

$$\sum_{k=1}^{2} \sigma_{2-k}\kappa_k T_{0k}(0) = -\frac{1-m}{2(2-m)} \sum_{k=1}^{2} \sigma_{\sigma-k}\phi_k \qquad (12.29d)$$

When the summation term involving $T_{0k}(0)$ is eliminated between Equations (12.29c) and (12.29d) we find

$$\sum_{k=1}^{2} \sigma_{2-k}\kappa_k T_{0k}(x) = \frac{1}{2} \left(x^2 - \frac{1-m}{2-m} \right) \sum_{k=1}^{2} \sigma_{2-k}\phi_k \qquad (12.29e)$$

This result can be rewritten in the form

$$\sigma_1\kappa_1 T_{01}(x) + \sigma_0\kappa_2 T_{02}(x) = b \qquad (12.30a)$$

where

$$b = \frac{1}{2}\left(x^2 - \frac{1-m}{2-m}\right) \sum_{k=1}^{2} \sigma_{2-k}\phi_k \qquad (12.30b)$$

Now we rewrite Equations (12.19a) as

$$\kappa_k \frac{d}{dx}\left[x^{1-2m}\frac{dT_{0k}(x)}{dx}\right] + x^{1-2m}D_k^* = 2(1-m)x^{1-2m}\phi_k \qquad (12.31a)$$

where

$$D_k^* = (-1)^k\left[\sigma_{k-1}T_{01}(x) - \sigma_{k+1}T_{02}(x)\right], \qquad k = 1,2 \qquad (12.31b)$$

We note that Equation (12.30a) is of the same form as Equation (12.24a), and Equation (12.31b) is of the same form as Equation (12.25b). Therefore, Equation (12.31b) can be manipulated by utilizing Equation (12.30a) in exactly the same way we manipulated Equation (12.25b) by utilizing Equation (12.24a). The correspondence between the two systems is as follows.

$$T_{av,k}(t) \equiv T_{0k}(x), \qquad \omega_k \equiv \kappa_k, \qquad a \equiv b = \frac{1}{2}\left(x^2 - \frac{1-m}{2-m}\right) \sum_{k=1}^{2} \sigma_{2-k}\phi_k$$

$$(12.32)$$

Then the alternative form of D_k^* is immediately determined from Equation (12.26c) by making the substitutions defined by Equation (12.32); we find

$$D_k^* = \frac{1}{\kappa_{3-k}}\left[\frac{\sigma_{3-k}}{\sigma_{k-1}}b - (\sigma_0\kappa_2 + \sigma_3\kappa_1)T_{0k}(x)\right], \qquad k = 1,2 \qquad (12.33)$$

When D_k^* as given by Equation (12.33) is introduced into Equation (12.31a), the system for $k = 1$ and 2 becomes uncoupled. The resulting equations, after some rearrangement, are written in the form

$$\frac{d^2T_{0k}(x)}{dx^2} + \frac{1-2m}{x}\frac{dT_{0k}(x)}{dx} - \kappa T_{0k}(x) = 2(1-m)\frac{\phi_k}{\kappa_k}$$

$$-\frac{\sigma_{3-k}}{\sigma_{k-1}}\frac{1}{2\kappa_1\kappa_2}\left(x^2 - \frac{1-m}{2-m}\right) \sum_{k=1}^{2} \sigma_{2-k}\phi_k \quad \text{in } 0 \le x \le 1, \qquad k = 1,2$$

$$(12.34a)$$

where

$$\kappa = \frac{\sigma_0\kappa_2 + \sigma_3\kappa_1}{\kappa_1\kappa_2} \qquad (12.34b)$$

The boundary conditions for Equations (12.34) are obtained from Equations (12.19b, c) as

$$\frac{dT_{0k}(0)}{dx} = 0, \qquad \kappa_k \frac{dT_{0k}(1)}{dx} = \phi_k, \qquad k = 1, 2 \qquad (12.34c, d)$$

Thus we transformed the coupled problem (12.19) into an uncoupled problem (12.34) for the steady-state potentials $T_{0k}(x)$, $(k = 1, 2)$.

The solution of Equation (12.34a) is taken in the form

$$T_{0k}(x) = A_k + B_k x^2 + H_k(x), \qquad k = 1, 2 \qquad (12.35)$$

where A_k and B_k are constants.

When solution (12.35) is introduced into the differential Equation (12.34a) and the boundary conditions (12.34c, d) are utilized, the coefficients A_k and B_k are determined as

$$A_k = \frac{2(1 - m)}{\kappa}\left[2 - \frac{\kappa}{2(2 - m)} B_k - \frac{\phi_k}{\kappa_k}\right] \qquad (12.36a)$$

$$B_k = \frac{1}{2} \frac{\sigma_{3-k}}{\sigma_{k-1}} \frac{\displaystyle\sum_{k=1}^{2} \sigma_{2-k}\phi_k}{\sigma_0 \kappa_2 + \sigma_3 \kappa_1} \qquad (12.36b)$$

and the functions $H_k(x)$ become the solution of the following problem

$$H_k''(x) + \frac{1 - 2m}{x} H_k'(x) - \kappa H_k(x) = 0 \quad \text{in } 0 \le x \le 1, \qquad k = 1, 2$$

$$(12.37a)$$

$$H_k'(0) = 0, \qquad H_k'(1) = \frac{\phi_k}{\kappa_k} - 2B_k \qquad (12.37b, c)$$

Problem (12.37) is similar to the one considered in Example 3.6. A comparison of Equations (12.37) with Equations (3.42) reveals that $\mu^2 = -\kappa$. Then the solution of Equation (12.37a) satisfying the boundary condition (12.37b) is immediately obtained from Equation (3.43a) as

$$H_k(x) = C_k(i\sqrt{\kappa}\, x)^m J_{-m}(i\sqrt{\kappa}\, x) \qquad (12.38a)$$

Now by making use of the relation (7.60a), that is,

$$I_{-m}(z) = (i)^m J_{-m}(iz) \qquad (12.38b)$$

Solution (12.38a) is written as

$$H_k(x) = C_k(\sqrt{\kappa}\, x)^m I_{-m}(\sqrt{\kappa}\, x) \tag{12.39}$$

The constant C_k is determined from the requirement that solution (12.39) satisfy the boundary condition (12.37c). Then the solution for $H_k(x)$ becomes

$$H_k(x) = \frac{1}{\sqrt{\kappa}}\left(\frac{\phi_k}{\kappa_k} - 2B_k\right)x^m \frac{I_{-m}(x\sqrt{\kappa})}{I_{1-m}(\sqrt{\kappa})}, \qquad k = 1, 2 \tag{12.40}$$

Finally, substituting Equations (12.36) and (12.40) into Equation (12.35), we obtain the solution for the steady-state potentials $T_{0k}(x)$ as

$$T_{0k}(x) = B_k\left(x^2 - \frac{1-m}{2-m}\right) + \left(2B_k - \frac{\phi_k}{\kappa_k}\right)\left[\frac{2(1-m)}{\kappa} - \frac{x^m}{\sqrt{\kappa}}\frac{I_{-m}(x\sqrt{\kappa})}{I_{1-m}(\sqrt{\kappa})}\right] \tag{12.41a}$$

where B_k and κ are defined by Equations (12.36b) and (12.34b), that is,

$$B_k = \frac{1}{2}\frac{\sigma_{3-k}}{\sigma_{k-1}}\frac{\sigma_1\phi_1 + \sigma_0\phi_2}{\sigma_0\kappa_2 + \sigma_3\kappa_1} \tag{12.41b}$$

$$\kappa = \frac{\sigma_0\kappa_2 + \sigma_3\kappa_1}{\kappa_1\kappa_2} \tag{12.41c}$$

and $k = 1$ and 2.

12.5 SOLUTION FOR THE TRANSIENT HOMOGENEOUS POTENTIALS $T_{tk}(x, t)$

The transient homogeneous potentials $T_k(x, t)$, $(k = 1, 2)$ are satisfied by the two coupled transient homogeneous system defined by Equations (12.20).

We note that the transient homogeneous problem defined by Equations (12.20) is a special case of the one-dimensional more general problem given by Equations (4.106). The correspondence between the two problems is as follows.

$$T_k = T_{tk}, \qquad w_k(x) = \omega_k x^{1-2m}, \qquad k_k(x) = \kappa_k x^{1-2m}, \qquad d_k(x) = 0$$

$$P_k(x) = 0, \qquad b(x) = x^{1-2m}, \qquad x_o = 0, \qquad x_1 = 1$$

$$\alpha_{0k} = \alpha_{1k} = 0, \qquad \beta_{0k} = \beta_{1k} = 1, \qquad \phi_k(x_0, t) = \phi_k(x_1, t) = 0$$

$$f_k(x) = -T_{0k}(x) \tag{12.42}$$

The solution of problem (4.106) subject to the boundary conditions of the second kind for both boundaries was given previously by Equation (4.108). Therefore, for the present problem (12.20), the solution is immediately obtained from Equation (4.108) by applying the simplifications given by Equations (12.42). We find

$$T_{tk}(x, t) = \sum_{i=1}^{\infty} \frac{\psi_k(\mu_i, x) e^{-\mu_i^2 t}}{\sum_{k=1}^{2} \sigma_k \omega_k \int_0^1 x^{1-2m} \psi_k^2(\mu_i, x)\, dx} \tilde{f}_i \qquad (12.43a)$$

where the integral transform of the initial conditions $\tilde{f}_i$ is defined as

$$\tilde{f}_i = -\sum_{k=1}^{2} \sigma_k \omega_k \int_0^1 x^{1-2m} \psi_k(\mu_i, x) T_{0k}(x)\, dx \qquad (12.43b)$$

Here $\psi_k(\mu_i, x)$ and μ_i are, respectively, the eigenfunctions and eigenvalues of the eigenvalue problem (12.5) for $\alpha = 0$, $\beta = 1$.

To evaluate the integral in Equation (12.43b) we multiply Equation (12.19a) by $-\sigma_k \psi_k(\mu_i, x)$ for $k = 1, 2$, Equation (12.5a) by $\sigma_k T_{0k}(x)$ for $k = 1, 2$, add the resulting expressions, and integrate from $x = 0$ to $x = 1$. After some manipulation we find

$$\tilde{f}_i = \frac{1}{\mu_i^2} \sum_{k=1}^{2} \sigma_k \phi_k \psi_k(\mu_i, 1) \qquad (12.44)$$

We introduce Equation (12.44) into Equation (12.43a), replace the function $\psi_k(\mu_i, x)$ by the expressions given by Equations (12.12), and finally obtain the solution for the transient homogeneous potential $T_{tk}(x, t)$ for the special case of $\alpha = 0$ and $\beta = 1$ as

$$T_{tk}(x, t) = -2 \sum_{i=1}^{\infty} \frac{(\lambda_i x)^m J_{-m}(\lambda_i x)}{\lambda_i^m J_{-m}(\lambda_i)} \sum_{j=1}^{2} \frac{E_{ij}^{k-1}}{\mu_{ij}^2} \frac{\sigma_1 \phi_1 + \sigma_2 \phi_2 E_{ij}}{\sigma_1 \omega_1 + \sigma_2 \omega_2 E_{ij}^2} e^{-\mu_{ij}^2 t}$$

$$(12.45a)$$

where E_{ij} was defined by Equation (12.13c) as

$$E_{ij} = \frac{\sigma_0 + \kappa_1 \lambda_i^2 - \omega_1 \mu_{ij}^2}{\sigma_2} \qquad (12.45b)$$

We now present one worked out example to illustrate the application of the foregoing analysis to develop solutions to problems belonging to Class V.

Example 12.1   Determine the concentration and temperature distribution in a chemically reacting fluid in plug flow inside a parallel plate channel, when the problem is defined by the following system of equations

$$\rho C_p u \frac{\partial T(x, z)}{\partial z} = k \frac{\partial^2 T(x, z)}{\partial x^2} + I_T(T, C) \qquad (12.46a)$$

$$u \frac{\partial C(x, z)}{\partial z} = D \frac{\partial^2 C(x, z)}{\partial x^2} + I_C(T, C) \qquad (12.46b)$$

where ρ, C_p, k, and D are, respectively, density, specific heat, thermal conductivity, and diffusion coefficient, u is the average velocity, $I_C(T, C)$ is the mass production rate due to chemical reaction, and $I_T(T, C)$ is a heat sink. The boundary conditions for this problem are taken as

$$\frac{\partial T(0, z)}{\partial x} = 0, \qquad \frac{\partial C(0, z)}{\partial x} = 0 \qquad (12.46c, d)$$

and

$$k \frac{\partial T(\ell, z)}{\partial x} = q, \qquad \frac{\partial C(\ell, z)}{\partial x} = 0 \qquad (12.46e, f)$$

The gas enters the channel in thermodynamic equilibrium, that is, $I_C(T_e, C_e) = 0$, with a uniform temperature T_e and concentration C_e; then at the inlet $z = 0$ we have

$$T(x, 0) = T_e, \qquad C(x, 0) = C_e \qquad (12.46g, h)$$

Solution   This problem is a special case of the Class V problem with both boundary conditions of the second kind. To solve this problem by utilizing the solutions presented in this chapter, we should first express the problem in the dimensionless form in order to bring it into the form comparable with that given previously by Equations (12.15). Therefore, the following dimensionless variables are defined

$$X = \frac{x}{\ell}, \qquad Z = \frac{kz}{\rho C_p u \ell^2} = \frac{\alpha z}{u \ell^2}, \qquad \theta = \frac{T - T_e}{\Delta T}$$

$$\phi = \frac{C - C_e}{\Delta C}, \qquad Lu = \frac{D \rho C_p}{k} = \frac{D}{\alpha} \qquad (12.47a, b, c, d, e)$$

where ΔC and ΔT are, respectively, reference concentration difference and temperature difference which are specified later on; ℓ is the spacing between the plates and α is the thermal diffusivity.

With the dimensionless variables as defined by Equations (12.47), the system of equations (12.46) is expressed in the dimensionless form as

$$\frac{\partial \theta(X, Z)}{\partial Z} = \frac{\partial^2 \theta(X, Z)}{\partial X^2} + \frac{\ell^2}{k \Delta T} I_T \qquad (12.48a)$$

$$\frac{1}{Lu} \frac{\partial \phi(X, Z)}{\partial Z} = \frac{\partial^2 \phi(X, Z)}{\partial X^2} + \frac{\ell^2}{D \Delta C} I_C \qquad (12.48b)$$

subject to the boundary conditions

$$\frac{\partial \theta(0, Z)}{\partial X} = 0, \qquad \frac{\partial \phi(0, Z)}{\partial X} = 0 \qquad (12.48c, d)$$

$$\frac{\partial \theta(1, Z)}{\partial X} = \frac{q\ell}{k \Delta T}, \qquad \frac{\partial \phi(1, Z)}{\partial X} = 0 \qquad (12.48e, f)$$

and the inlet conditions

$$\theta(X, 0) = 0, \qquad \phi(X, 0) = 0 \qquad (12.48g, h)$$

The heat sink is related to the production rate resulting from the chemical reaction by

$$I_T(T, C) = -r I_C(T, C) \qquad (12.49a)$$

where r is a constant.

The term $I_C(T, C)$ can be linearized and represented by

$$I_C(T, C) = \left(\frac{\partial I_C}{\partial C}\right)_e (C - C_e) + \left(\frac{\partial I_C}{\partial T}\right)_e (T - T_e) \qquad (12.49b)$$

where from physical considerations we have

$$\left(\frac{\partial I_C}{\partial C}\right)_e > 0, \qquad \left(\frac{\partial I_C}{\partial T}\right)_e > 0 \qquad (12.49c, d)$$

Now the dimensionless temperature θ and concentration ϕ defined by Equations (12.47c, d) are introduced into Equation (12.49b) and the result is expressed in the form

$$I_C = \left(\frac{\partial I_C}{\partial C}\right)_e \Delta C (\phi - \kappa \theta) \qquad (12.49e)$$

where κ is a positive dimensionless quantity defined by

$$\kappa = -\frac{(\partial I_C / \partial C)_e \Delta T}{(\partial I_C / \partial T)_e \Delta C} \qquad (12.49f)$$

By utilizing Equations (12.49), we can rewrite Equations (12.48a, b) in the form

$$\frac{\partial \theta}{\partial Z} = \frac{\partial^2 \theta}{\partial X^2} + \beta_T^2 (\phi - \kappa \theta) \qquad (12.50a)$$

$$\frac{1}{Lu} \frac{\partial \phi}{\partial Z} = \frac{\partial^2 \phi}{\partial X^2} - \beta_C^2 (\phi - \kappa \theta) \qquad (12.50b)$$

where

$$\beta_T^2 = -\frac{r \ell^2}{k \, \Delta T} \left(\frac{\partial I_C}{\partial C} \right)_e \Delta C \qquad (12.50c)$$

$$\beta_C^2 = -\frac{\ell^2}{D \, \Delta C} \left(\frac{\partial I_C}{\partial C} \right)_e \Delta C \qquad (12.50d)$$

now the reference temperature difference ΔT is specified in such a way as to make unity the right-hand side of the boundary condition Equation (12.48e); that is, ΔT is chosen as

$$\Delta T = \frac{q \ell}{k} \qquad (12.51a)$$

The reference concentration difference ΔC is determined so that $\beta_T^2 = \beta_C^2$, that is,

$$\Delta C = \frac{q \ell}{rD} \qquad (12.51b)$$

Therefore, the final dimensionless form of the system (12.46) takes the form

$$\frac{\partial \theta(X, Z)}{\partial Z} = \frac{\partial^2 \theta(X, Z)}{\partial X^2} + \beta^2 [\phi(X, Z) - \kappa \theta(X, Z)] \qquad (12.52a)$$

$$\frac{1}{Lu} \frac{\partial \phi(X, Z)}{\partial Z} = \frac{\partial^2 \phi(X, Z)}{\partial X^2} - \beta^2 [\phi(X, Z) - \kappa \theta(X, Z)] \qquad (12.52b)$$

Subject to the following boundary conditions

$$\frac{\partial \theta(0, Z)}{\partial X} = 0, \qquad \frac{\partial \phi(0, Z)}{\partial X} = 0 \qquad (12.52c, d)$$

$$\frac{\partial \theta(1, Z)}{\partial X} = 1, \qquad \frac{\partial \phi(1, Z)}{\partial X} = 0 \qquad (12.52e, f)$$

and the inlet conditions

$$\theta(X, 0) = 0, \qquad \phi(X, 0) = 0 \qquad (12.52g, h)$$

where

$$\theta = k\frac{T - T_e}{q\ell}, \qquad \phi = Dr\frac{C - C_e}{q\ell}, \qquad \beta^2 = -\frac{\ell^2}{D}\left(\frac{\partial I_C}{\partial C}\right)_e \qquad (12.53\text{a, b, c})$$

The problem defined by Equations (12.52) is a special case of the problem given by Equations (12.15). By comparing these two problems we write

$$m = \frac{1}{2}, \qquad \omega_1 = \frac{1}{Lu}, \qquad \omega_2 = 1, \qquad \kappa_1 = \kappa_2 = 1$$

$$\sigma_0 = \sigma_1 = \beta^2, \qquad \sigma_2 = \sigma_3 = \beta^2\kappa, \qquad \phi_1 = 0, \qquad \phi_2 = 1$$

$$f_k = 0, \qquad x = X, \qquad t = Z, \qquad T_1 = \phi, \qquad T_2 = \theta \qquad (12.54)$$

Then the solutions (12.16) become

$$\theta(X, Z) = \theta_{av}(Z) + \theta_0(X) + \theta_t(X, Z) \qquad (12.55\text{a})$$

$$\phi(X, Z) = \phi_{av}(Z) + \phi_0(X) + \phi_t(X, Z) \qquad (12.55\text{b})$$

The functions on the right-hand side of Equations (12.55) are readily obtained from the general solutions given in this chapter as now described.

The average potentials $\theta_{av}(Z)$ and $\phi_{av}(Z)$ are determined from the solution (12.28) by utilizing the results given by Equations (12.54). We find

$$\theta_{av}(Z) = \frac{Lu}{\kappa + Lu}Z + \frac{\kappa}{\beta^2(\kappa + Lu)^2}\left[1 - e^{-\beta^2(\kappa + Lu)Z}\right] \qquad (12.56\text{a})$$

$$\phi_{av}(Z) = \kappa\frac{Lu}{\kappa + Lu}Z - \frac{\kappa}{\beta^2(\kappa + Lu)^2}\left[1 - e^{-\beta^2(\kappa + Lu)Z}\right] \qquad (12.56\text{b})$$

The steady-state potentials $\theta_0(X)$ and $\phi_0(X)$ are determined from solution (12.41) by utilizing the results given by Equations (12.54) and noting that the following relations hold:

$$\sqrt{\frac{\pi/2}{X}}\,I_{-1/2}(X) = \frac{\cosh X}{X} \quad \text{and} \quad \sqrt{\frac{\pi/2}{X}}\,I_{1/2}(X) = \frac{\sinh X}{X}$$

We obtain

$$\theta_0(x) = \frac{X^2 - (1/3)}{2(1 + \kappa)} - \frac{1}{1 + \kappa}\left[1 - \sqrt{\kappa}\,\frac{\cosh(X\sqrt{\kappa})}{\sinh(\sqrt{\kappa})}\right] \qquad (12.57\text{a})$$

$$\phi_0(x) = \kappa\frac{X^2 - (1/3)}{2(1 + \kappa)} + \frac{1}{1 + \kappa}\left[1 - \sqrt{\kappa}\,\frac{\cosh(X\sqrt{\kappa})}{\sinh(\sqrt{\kappa})}\right] \qquad (12.57\text{b})$$

Finally, the potentials $\phi(X, Z)$ and $\theta(X, Z)$ are obtained from the solution Equation (12.45) by setting, respectively, $k = 1$ and 2, by utilizing the results given by Equations (12.54), and noting that

$$\sqrt{\frac{\pi/2}{X}} J_{-1/2}(X) = \frac{\cos X}{X}$$

We obtain

$$\theta(X, Z) = 2\kappa \sum_{i=1}^{\infty} (-1)^{i+1} \cos(\lambda_i X) \sum_{j=1}^{2} \frac{E_{ij}^2}{(1/Lu) + \kappa E_{ij}^2} \frac{1}{\mu_{ij}^2} e^{-\mu_{ij}^2 Z} \quad (12.58a)$$

$$\phi(X, Z) = 2\kappa \sum_{i=1}^{\infty} (-1)^{i+1} \cos(\lambda_i X) \sum_{j=1}^{2} \frac{E_{ij}}{(1/Lu) + \kappa E_{ij}^2} \frac{1}{\mu_{ij}^2} e^{-\mu_{ij}^2 Z} \quad (12.58b)$$

where the λ_i are the roots of

$$J_{1/2}(\lambda_i) = 0 \quad \text{or} \quad \sin \lambda_i = 0$$

hence they are given by

$$\lambda_i = i\pi \quad (i = 1, 2, 3, \ldots) \quad (12.58c)$$

the μ_{ij} are determined from Equation (12.11) as

$$\mu_{ij}^2 = \tfrac{1}{2}\left\{ (1 + Lu)\lambda_i^2 + (\kappa + Lu)\beta^2 \right.$$

$$\left. + (-1)^j \sqrt{[(Lu - 1)\lambda_i^2 - (\kappa - Lu)\beta^2]^2 + 4 Lu \beta^4 \kappa} \right\} \quad (12.58d)$$

and the E_{ij} are obtained from Equation (12.10) or (12.13c) as

$$E_{ij} = \frac{\beta^2 + \lambda_i^2 - (\mu_{ij}^2/Lu)}{\beta^2\kappa} = \frac{\beta^2}{\beta^2\kappa + \lambda_i^2 - \mu_{ij}^2} \quad (12.58e)$$

REFERENCES

1. G. Nicolis and I. Prigogine, *Self-Organization in Nonequilibrium Systems*, Wiley, New York, 1977.
2. C. L. Tien, Heat Transfer by a Turbulently Flowing Fluid–Solids Mixture in a Pipe, *J. Heat Transf.*, **83C**, 183–188 (1961).
3. S. L. Soo, *Fluid Dynamics of Multiphase Systems*, Blaisdel Publishing Co., Waltham, MA, 1967.

CHAPTER THIRTEEN

Class VI Solution Applied to Heat and Mass Diffusion and Developing Duct Flow

The man of science has learned to believe in justification, not by faith, but by verification.

Thomas Henry Huxley
1825–1895

In this chapter we illustrate the application of the one-dimensional Class VI problem to the analysis of: (1) heating of bodies in a limited volume of well-stirred fluid studied in References 1 and 2, (2) mass diffusion into a body from a limited volume of well-stirred fluid [3], and (3) flow development in the hydrodynamic entrance region of a circular tube and a parallel plate channel [5]. Solutions to all these problems are obtained by appropriate simplification of the solution of the one-dimensional Class VI problem developed in Chapter 4.

13.1 HEATING OF BODIES IN A LIMITED VOLUME OF WELL-STIRRED FLUID

We consider n different, one-dimensional bodies (i.e., slabs of thickness ℓ_k with the boundary at $x = 0$ insulated, or long solid cylinders and solid spheres of radius ℓ_k, $k = 1, 2, \ldots, n$ in number) placed inside a limited volume of well-stirred fluid. The temperature of the fluid is always uniform but varies with time as a result of heat transfer to (or from) the bodies. Initially, the fluid is at temperature T_{f0} and the bodies are at temperatures $T_k(x)$. For times $t > 0$, there is heat transfer by convection between the well-stirred fluid and the bodies through the surfaces at $x = \ell_k$ of the bodies. The mathematical formu-

461

lation of this heat diffusion problem is given by

$$x^{1-2m_k}\frac{\partial T_k(x,t)}{\partial t} = \alpha_k \frac{\partial}{\partial x}\left\{ x^{1-2m_k}\frac{\partial T_k(x,t)}{\partial x}\right\}$$

$$\text{in } 0 < x < \ell_k, \quad t > 0 \quad (k = 1,2,\ldots,n) \quad (13.1\text{a})$$

with $m_k = \frac{1}{2}$ for slab, $m_k = 0$ for cylinder, and $m_k = -\frac{1}{2}$ for sphere; and subject to the boundary conditions

$$\frac{\partial T_k(0,t)}{\partial x} = 0 \tag{13.1b}$$

$$k_k \frac{\partial T_k(\ell_k,t)}{\partial x} + h_k\left[T_k(\ell_k,t) - T_f(t)\right] = 0 \tag{13.1c}$$

and the initial condition

$$T_k(x,0) = T_k(x) \tag{13.1d}$$

Since the fluid is finite in volume, the temperature of the fluid $T_f(t)$ as a function of time depends on the heat exchange between the fluid and the bodies, hence on the temperatures $T_k(x, t)$ of the bodies. Thus the temperatures $T_f(t)$ and $T_k(x, t)$ are coupled. The differential equation governing this coupling is obtained by writing a heat balance between the heat gained (or lost) by the fluid and heat transferred across the boundaries of the bodies as

$$c_f \rho_f V_f \frac{dT_f(t)}{dt} + \sum_{k=1}^{n} k_k \frac{\partial T_k(\ell_k,t)}{\partial x} A_k = 0 \tag{13.1e}$$

with the initial condition

$$T_f(0) = T_{f0} \tag{13.1f}$$

Here, c_f, ρ_f, and V_f are, respectively, the specific heat, density, and volume of the well-stirred fluid; A_k and k_k are, respectively, the surface area and the thermal conductivity of the body k with $k = 1, 2, \ldots, n$.

If the volume V_k of the well-stirred fluid were infinite or its temperature $T_f(t)$ were a given function of time, then the heat diffusion problem defined by Equations (13.1a,d) would be uncoupled from the constraint given by Equations (13.1e, f). For such a case, the uncoupled heat diffusion problem (13.1a, d) coincides with the one-dimensional Class I problem given by Equations (7.12) and its solution is obtained from the results given in Chapter 7.

The coupled problem given by Equations (13.1) is a special case of the one-dimensional Class VI problem as now demonstrated.

For convenience in the analysis, we introduce the following dimensionless variables

$$X = \frac{x}{\ell_k}, \qquad \tau = \frac{\alpha_0 t}{\ell_0^2}, \qquad \theta_k(X, \tau) = \frac{T_k(x, t) - T_0}{\Delta T}$$

$$\theta_f(\tau) = \frac{T_f(t) - T_0}{\Delta T}, \qquad \theta_0 = \frac{T_{f0} - T_0}{\Delta T}, \qquad F_k(X) = \frac{T_k(x) - T_0}{\Delta T}$$

$$\mathrm{Bi}_k = \frac{h_k \ell_k}{k_k}, \qquad \omega_k^2 = \frac{\alpha_0}{\alpha_k} \frac{\ell_k^2}{\ell_0^2}, \qquad K_k = \frac{2(1 - m_k)}{\omega_k^2} \frac{c_k \rho_k V_k}{c_f \rho_f V_f} \qquad (13.2)$$

where T_0 is a reference temperature, ΔT a reference temperature difference, and ℓ_0 a reference length.

With various dimensionless quantities as defined, the mathematical formulation of the preceding heat diffusion problem takes the form

$$\omega_k^2 X^{1-2m_k} \frac{\partial \theta_k(X, \tau)}{\partial \tau} = \frac{\partial}{\partial X} \left\{ X^{1-2m_k} \frac{\partial \theta_k(X, \tau)}{\partial X} \right\}$$

$$\text{in } 0 < X < 1, \qquad \tau > 0 \qquad (k = 1, 2, \ldots, n) \quad (13.3a)$$

subject to the boundary conditions

$$\frac{\partial \theta_k(0, \tau)}{\partial X} = 0 \qquad (13.3b)$$

$$\theta_k(1, \tau) + \frac{1}{\mathrm{Bi}_k} \frac{\partial \theta_k(1, \tau)}{\partial X} = \theta_f(\tau) \qquad (13.3c)$$

the coupling condition given by

$$\frac{d\theta_f(\tau)}{d\tau} + \sum_{k=1}^{n} K_k \frac{\partial \theta_k(1, \tau)}{\partial X} = 0 \qquad (13.3d)$$

and the initial conditions

$$\theta_k(X, 0) = F_k(X) \qquad (13.3e)$$

$$\theta_f(0) = \theta_0 \qquad (13.3f)$$

The problem given by Equations (13.3) is a special case of the one-dimensional Class VI problem (4.109). The correspondence between the two problems is given by

$$x = X, \qquad x_0 = 0, \qquad x_1 = 1, \qquad t = \tau, \qquad w_k(x) = \omega_k^2 X^{1-2m_k}$$

$$k_k(x) = X^{1-2m_k}, \qquad \alpha_k = 1, \qquad \beta_k = \frac{1}{\text{Bi}_k}$$

$$T_k(x, t) = \theta_k(X, \tau), \qquad \phi(t) = \theta_f(\tau), \qquad \gamma_k = K_k$$

$$Q(t) = 0, \qquad \phi_0 = \theta_0, \qquad f_k(x) = F_k(X), \qquad P_k(x, t) = 0 \quad (13.4)$$

Then the solution of problem (13.3) is immediately obtained from the general solution (4.110) by appropriate simplifications. The eigenvalues and the eigenfunctions needed for the solution are determined from the eigenvalue problem (3.147). The resulting expression for the body temperatures $\theta_k(X, \tau)$ after some manipulations is determined in the form (see Note 1 for details of the derivation of this result)

$$\theta_k(X, \tau) = \frac{\theta_0 + \sum_{k=1}^{n} \omega_k^2 K_k \int_0^1 X^{1-2m_k} F_k(X) \, dX}{1 + \sum_{k=1}^{n} \omega_k^2 K_k / [2(1 - m_k)]}$$

$$+ \sum_{i=1}^{\infty} \frac{\theta_0 + \sum_{k=1}^{n} \omega_k^2 K_k \int_0^1 X^{1-2m_k} W_{m_k}(\mu_i \omega_k X) F(X) \, dX}{1 + \sum_{k=1}^{n} \omega_k^2 K_k A_{m_k}(\mu_i \omega_k)}$$

$$\times W_{m_k}(\mu_i \omega_k X) e^{-\mu_i^2 \tau} \qquad (13.5a)$$

where

$$W_{m_k}(\mu_i \omega_k X) = \frac{X^{m_k} J_{-m_k}(\mu_i \omega_k X)}{J_{-m_k}(\mu_i \omega_k) - (\mu_i \omega_k / \text{Bi}_k) J_{1-m_k}(\mu_i \omega_k)} \qquad (13.5b)$$

$$A_{m_k}(\mu_i \omega_k)$$

$$= \frac{J_{-m_k}^2(\mu_i \omega_k) + J_{1-m_k}^2(\mu_i \omega_k) + (2m_k / \mu_i \omega_k) J_{1-m_k}(\mu_i \omega_k) J_{-m_k}(\mu_i \omega_k)}{2 \left[J_{-m_k}(\mu_i \omega_k) - (\mu_i \omega_k / \text{Bi}_k) J_{1-m_k}(\mu_i \omega_k) \right]^2}$$

$$(13.5c)$$

and the μ_i are the roots of the transcendental equation [see Equation (3.148b) subject to the conditions (13.4)]

$$\mu^2 + \sum_{k=1}^{n} \frac{K_k \omega_k J_{1-m_k}(\mu \omega_k)}{J_{-m_k}(\mu \omega_k) - (\mu \omega_k / Bi_k) J_{1-m_k}(\mu \omega_k)} = 0 \qquad (13.5d)$$

A Special Case

If all the bodies are of the same type ($m_k = m$) (i.e., they are all slabs, cylinders, or spheres), of the same kind and size ($\omega_k = 1$), and the heat transfer coefficients are the same ($Bi_k = Bi$), the mathematical formulation of problem (13.3) reduces to

$$X^{1-2m} \frac{\partial \theta(X, \tau)}{\partial \tau} = \frac{\partial}{\partial X} \left\{ X^{1-2m} \frac{\partial \theta(X, \tau)}{\partial X} \right\} \quad \text{in } 0 < X < 1, \qquad \tau > 0$$

$$(13.6a)$$

with $m = \frac{1}{2}$ for slab, $m = 0$ for cylinder, and $m = -\frac{1}{2}$ for sphere; subject to the conditions

$$\frac{\partial \theta(0, \tau)}{\partial X} = 0 \qquad (13.6b)$$

$$\theta(1, \tau) + \frac{1}{Bi} \frac{\partial \theta(1, \tau)}{\partial X} = \theta_f(\tau) \qquad (13.6c)$$

$$\frac{d\theta_f(\tau)}{d\tau} + K \frac{\partial \theta(1, \tau)}{\partial X} = 0 \qquad (13.6d)$$

$$\theta(X, 0) = F(X), \qquad \theta_f(0) = \theta_0 \qquad (13.6e, f)$$

where

$$K = \sum_{k=1}^{n} K_k = 2(1 - m) \frac{n C_0 \rho_0 V_0}{C_f \rho_f V_f} \qquad (13.6g)$$

Here subscript 0 refers to the bodies and n is the number of bodies. Then the solution of problem (13.6) is obtained from solution (13.5) by appropriate simplifications. We find (see Note 2 for details of the derivation)

$$\theta(X, \tau) = \frac{\theta_0 + K \int_0^1 X^{1-2m} F(X) \, dX}{1 + K/[2(1 - m)]}$$

$$- \sum_{i=1}^{\infty} E_i \left[\theta_0 - \mu_i \int_0^1 X^{1-m} \frac{J_{-m}(\mu_i X)}{J_{1-m}(\mu_i)} F(X) \, dX \right] \frac{X^m J_{-m}(\mu_i X)}{J_{1-m}(\mu_i)} e^{-\mu_i^2 \tau}$$

$$(13.7a)$$

where

$$E_i = \frac{2}{\mu_i} \left[1 + \frac{2m}{\mu_i} \left(\frac{\mu_i}{\text{Bi}} - \frac{K}{\mu_i} \right) + \left(\frac{\mu_i}{\text{Bi}} - \frac{K}{\mu_i} \right)^2 + \frac{2K}{\mu_i^2} \right]^{-1} \qquad (13.7b)$$

and the μ_i are the roots of the following transcendental equation which is obtained from Equation (13.5d)

$$J_{-m}(\mu) + \left(\frac{K}{\mu} - \frac{\mu}{\text{Bi}} \right) J_{1-m}(\mu) = 0 \qquad (13.7c)$$

If all the bodies are initially at the same constant temperature, that is, $F(X) = F_0$, the solution (13.7) simplifies to

$$\theta(X, \tau) = \frac{\theta_0 + F_0 K/[2(1-m)]}{1 + K/[2(1-m)]} - (\theta_0 - F_0) \sum_{i=1}^{\infty} E_i X^m \frac{J_{-m}(\mu_i X)}{J_{1-m}(\mu_i)} e^{-\mu_i^2 \tau}$$

$$(13.8)$$

where E_i and μ_i are defined, respectively, by Equations (13.7b) and (13.7c).
The mean temperature $\theta_{av}(\tau)$ is defined by Equation (7.17) as

$$\theta_{av}(\tau) = 2(1-m) \int_0^1 X^{1-2m} \theta(X, \tau) \, dX \qquad (13.9a)$$

Now by introducing Equation (13.8) into Equation (13.9a), the mean temperature $\theta_{av}(\tau)$ for bodies having constant initial temperature is determined as

$$\theta_{av}(\tau) = \frac{\theta_0 + F_0 K/[2(1-m)]}{1 + K/[2(1-m)]} - (\theta_0 - F_0) 2(1-m) \sum_{i=1}^{\infty} \frac{1}{\mu_i} E_i e^{-\mu_i^2 \tau}$$

$$(13.9b)$$

The relation between $\theta_f(\tau)$ and $\theta_{av}(\tau)$ is now determined by integrating Equation (13.6a) from $X = 0$ to $X = 1$ and utilizing the boundary condition (13.6d) and the relation given by Equation (13.9b). We find

$$\theta_f(\tau) = \theta_0 + \frac{K}{2(1-m)} [F_0 - \theta_{av}(\tau)] \qquad (13.10)$$

In the definition of the dimensionless quantities given by Equations (13.2) we can choose the reference temperature T_0 and the reference temperature difference ΔT such that the initial temperatures for the fluid become unity (i.e.,

$\theta_0 = 1$) and for the bodies become zero (i.e., $F_0 = 0$). Then for such a case, by setting $\theta_0 = 1$ and $F_0 = 0$, Equations (13.8), (13.9b), and (13.10), respectively, become

$$\theta(X, \tau) = \frac{1}{1 + K/[2(1 - m)]} - \sum_{i=1}^{\infty} E_i X^m \frac{J_{-m}(\mu_i X)}{J_{1-m}(\mu_i)} e^{-\mu_i^2 \tau} \quad (13.11a)$$

$$\theta_{av}(\tau) = \frac{1}{1 + K/[2(1 - m)]} - 2(1 - m) \sum_{i=1}^{\infty} \frac{1}{\mu_i} E_i e^{-\mu_i^2 \tau} \quad (13.11b)$$

$$\theta_f(\tau) = 1 - \frac{K}{2(1 - m)} \theta_{av}(\tau)$$

$$= \frac{1}{1 + K/[2(1 - m)]} + K \sum_{i=1}^{\infty} \frac{1}{\mu_i} E_i e^{-\mu_i^2 \tau} \quad (13.11c)$$

where E_i and μ_i are defined by Equations (13.7b) and (13.7c), respectively.

The functions $X^m J_{-m}(X)$ and $X^m J_{1-m}(X)$, appearing in the foregoing solutions are listed in Table 3.1 for slab ($m = \frac{1}{2}$), cylinder ($m = 0$), and sphere ($m = -\frac{1}{2}$). We illustrate the application of these solutions for slabs, cylinders, and spheres with the following examples.

Example 13.1 n identical bodies in the form of slabs, initially at dimensionless temperature $F(X) = 0$ are placed in a limited volume of well-stirred fluid which is initially at a dimensionless temperature $\theta_f(0) = 1$. The mathematical formulation of this heat transfer problem is given by Equations (13.6) with $m = \frac{1}{2}$ (for slab), $F(X) = 0$, and $\theta_0 = 1$. Determine the dimensionless temperature distribution $\theta(X, \tau)$, the average temperature $\theta_{av}(\tau)$ in the slab, and the temperature $\theta_f(\tau)$ of the fluid.

Solution The solution of this problem is immediately obtained from Equations (13.11) by setting $m = \frac{1}{2}$ and utilizing the relations given in Table 3.1. We find

$$\theta(X, \tau) = \frac{1}{1 + K} - \sum_{i=1}^{\infty} E_i \frac{\cos(\mu_i X)}{\sin \mu_i} e^{-\mu_i^2 \tau} \quad (13.12a)$$

$$\theta_{av}(\tau) = \frac{1}{1 + K} - \sum_{i=1}^{\infty} \frac{1}{\mu_i} E_i e^{-\mu_i^2 \tau} \quad (13.12b)$$

$$\theta_f(\tau) = 1 - K\theta_{av}(\tau) \quad (13.12c)$$

The eigenvalues μ_i are the roots of the transcendental Equation (13.7c), which for this special case reduces to

$$\cot \mu = \frac{\mu}{\text{Bi}} - \frac{K}{\mu} \qquad (13.12\text{d})$$

E_i and K are defined, respectively, by Equations (13.7b) and (13.6g).

Figure 13.1 shows the dimensionless temperature $\theta(0, \tau)$ at $X = 0$ as a function of the dimensionless time τ for several different values of the parameters Bi and K as determined from Equation (13.12a) by setting $X = 0$. Figure 13.2 shows the dimensionless average temperature $\theta_{av}(\tau)$ of the body as a function of τ for the same values of the parameters Bi and K considered in Figure 13.1. Knowing $\theta_{av}(\tau)$, the fluid temperature $\theta_f(\tau)$ is determined by Equation (13.12c).

Example 13.2   Repeat the heat transfer problem considered in Example 13.1 for the case of n identical cylinders placed in a limited volume of well-stirred fluid.

Solution   The solution of this problem is obtained from Equations (13.11) by setting $m = 0$. We find

$$\theta(X, \tau) = \frac{1}{1 + (K/2)} - \sum_{i=1}^{\infty} E_i \frac{J_0(\mu_i X)}{J_1(\mu_i)} e^{-\mu_i^2 \tau} \qquad (13.13\text{a})$$

$$\theta_{av}(\tau) = \frac{1}{1 + (K/2)} - 2 \sum_{i=1}^{\infty} \frac{1}{\mu_i} E_i e^{-\mu_i^2 \tau} \qquad (13.13\text{b})$$

$$\theta_f(\tau) = 1 - \frac{K}{2} \theta_{av}(\tau) \qquad (13.13\text{c})$$

The eigenvalues μ_i are the roots of the transcendental Equation (13.7c) with $m = 0$,

$$J_0(\mu) + \left(\frac{K}{\mu} - \frac{\mu}{\text{Bi}} \right) J_1(\mu) = 0 \qquad (13.13\text{d})$$

E_i and K are defined, respectively, by Equations (13.7b) and (13.6g).

Figure 13.3 shows the dimensionless center temperature $\theta(0, \tau)$ as a function of the dimensionless time τ for several different values of the parameters Bi and K, as determined from Equation (13.13a) with $X = 0$. Figure 13.4 shows the dimensionless average temperature $\theta_{av}(\tau)$ of the body as a function of τ for the same values of the parameters Bi and τ considered in Equation 13.13. Knowing $\theta_{av}(\tau)$, the fluid temperature $\theta_f(\tau)$ is determined by Equation (13.13c).

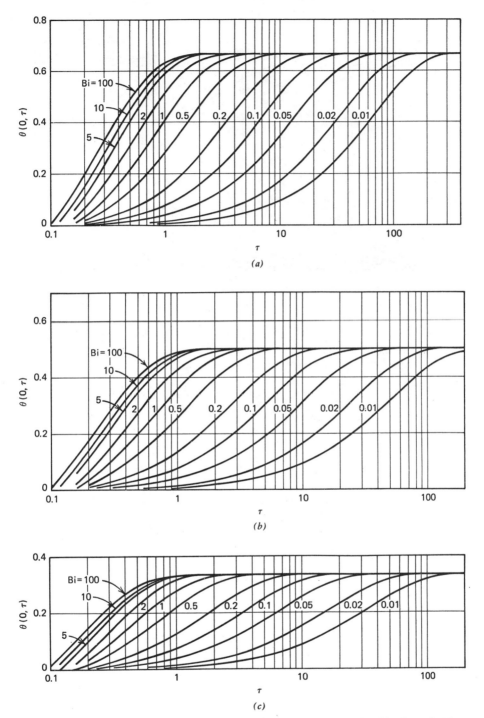

Figure 13.1 Dimensionless temperature at $X = 0$, $\theta(0, \tau)$ for slab considered in Example 13.1: (a) for $K = 0.5$; (b) for $K = 1.0$; (c) for $K = 2.0$.

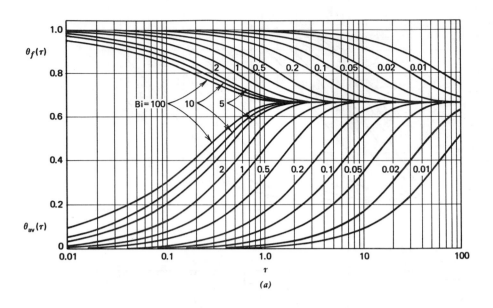

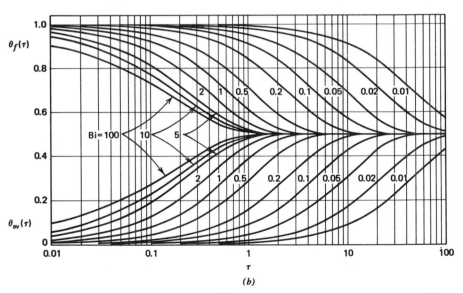

Figure 13.2 Dimensionless fluid temperature $\theta_f(\tau)$ and dimensionless average temperature $\theta_{av}(\tau)$ for slab considered in Example 13.1: (*a*) for $K = 0.5$; (*b*) for $K = 1.0$; (*c*) for $K = 2.0$.

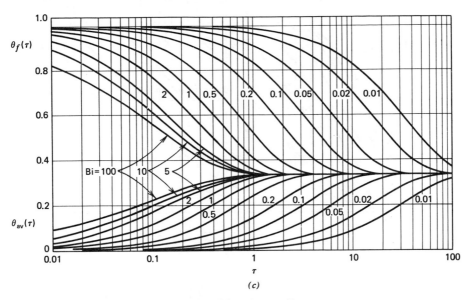

Figure 13.2   (*Continued*).

Example 13.3 Repeat the heat transfer problem considered in Example 13.1 for the case of n identical spheres placed in a limited volume of well-stirred fluid.

Solution The solution for this problem is again obtained from Equations (13.11) by setting $m = -\frac{1}{2}$ and utilizing the results in Table 3.1. We find

$$\theta(X, \tau) = \frac{1}{1+(K/3)} - \sum_{i=1}^{\infty} E_i \frac{j_0(\mu_i X)}{j_1(\mu_i X)} e^{-\mu_i^2 \tau} \tag{13.14a}$$

$$\theta_{av}(\tau) = \frac{1}{1+(K/3)} - 3 \sum_{i=1}^{\infty} \frac{1}{\mu_i} E_i e^{-\mu_i^2 \tau} \tag{13.14b}$$

$$\theta_f(\tau) = 1 - \frac{K}{3} \theta_{av}(\tau) \tag{13.14c}$$

The eigenvalues of the μ_i are the roots of the transcendental Equation (13.7c) with $m = 0$,

$$j_0(\mu) + \left(\frac{K}{\mu} - \frac{\mu}{Bi} \right) j_1(\mu) = 0 \tag{13.14d}$$

where $j_0(\mu)$ and $j_1(\mu)$ are the spherical Bessel functions related to trigonomet-

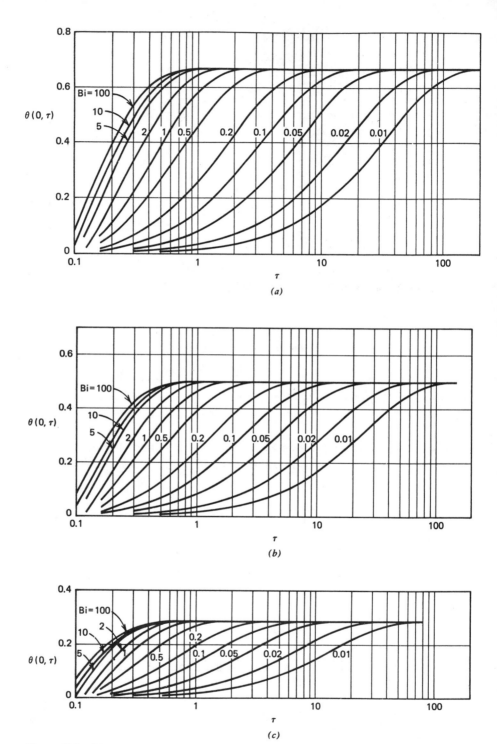

Figure 13.3 Dimensionless center temperature $\theta(0, \tau)$ for cylinder considered in Example 13.2: (*a*) for $K = 1$; (*b*) for $K = 2$; (*c*) for $K = 5$.

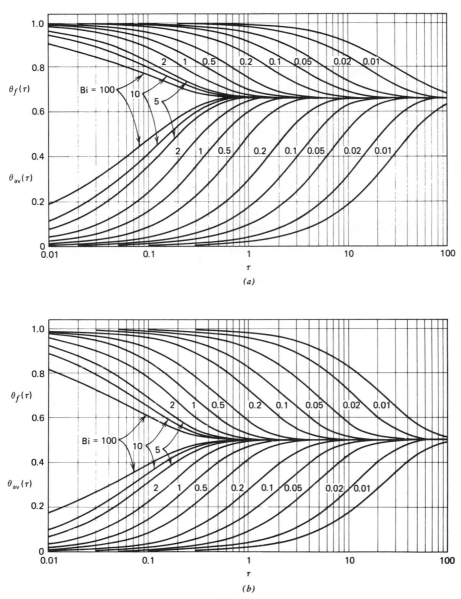

Figure 13.4 Dimensionless fluid temperature $\theta_f(\tau)$ and dimensionless average temperature $\theta_{av}(\tau)$ for cylinder considered in Example 13.2. (*a*) for $K = 1$; (*b*) for $K = 2$; (*c*) for $K = 5$.

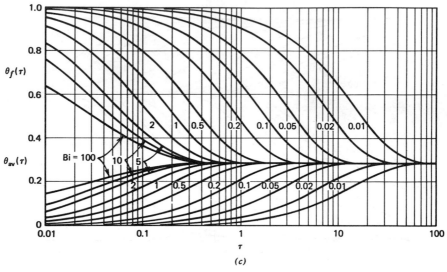

Figure 13.4 (*Continued*).

ric functions according to the relation given by Equations (3.36d) and (3.36e), respectively. E_i and K are as defined previously.

Figure 13.5 shows the dimensionless center temperature $\theta(0, \tau)$ as a function of the dimensionless time τ for several different values of the parameters Bi and K, as determined from Equation (13.14a) with $X = 0$. Figure 13.6 shows the dimensionless average temperature $\theta_{av}(\tau)$ of the body as a function of τ for the same values of the parameters Bi and τ considered in Figure 13.5. Knowing $\theta_{av}(\tau)$, the fluid temperature $\theta_f(\tau)$ is determined by Equation (13.14c).

3.2 MASS DIFFUSION INTO A BODY FROM A LIMITED VOLUME OF WELL-STIRRED FLUID

In this section we consider one-dimensional, transient mass diffusion into a body such as a plate, a long solid cylinder, or a solid sphere suspended into a limited volume of well-stirred fluid. The concentration of the solute in the solution falls as solute enters the body. The solution being well stirred, the concentration of the solute $C_s(t)$ in the solution depends only on time and is governed essentially by the condition that the total amount of solute in the solution and in the body remain constant as diffusion proceeds. The analysis of such problems is of interest in the experimental study of diffusion because the uptake of solute by the body can be deduced from observations of the uniform concentration in the solution since only a limited amount of solution is available. It is often simpler to observe the concentration of solute in the solution than the amount of solute in the body.

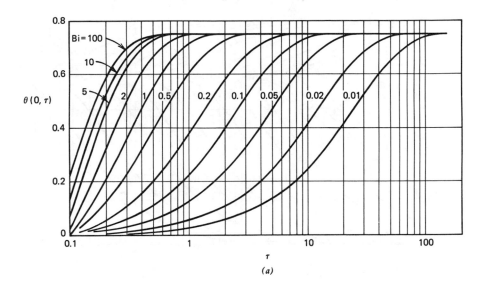

(a)

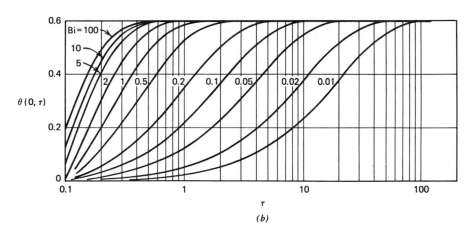

(b)

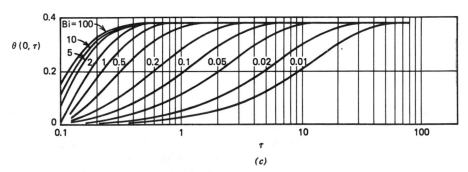

(c)

Figure 13.5 Dimensionless center temperature $\theta(0, \tau)$ for sphere considered in Example 13.3: (a) for $K = 1$; (b) for $K = 2$; (c) for $k = 5$.

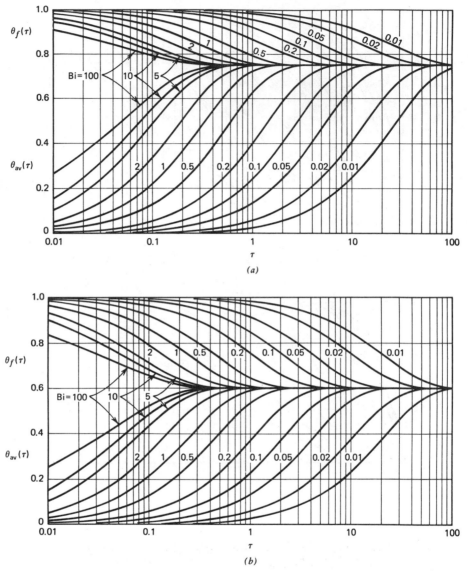

Figure 13.6 Dimensionless fluid temperature $\theta_f(\tau)$ and dimensionless average temperature $\theta_{av}(\tau)$ for sphere considered in Example 13.3: (a) for $K = 1$; (b) for $K = 2$; (c) for $K = 5$.

To present the mathematical formulation of this problem, we consider a body such as a slab of thickness r_1 with no mass diffusion from the boundary at $r = 0$ (or a long solid cylinder or a solid sphere of radius r_1) suspended in a finite volume of well-stirred fluid. The solute diffuses into the body from the boundary surface at $r = r_1$. We assume no sources or sinks within the body. The diffusion within the body is governed by the following diffusion equation

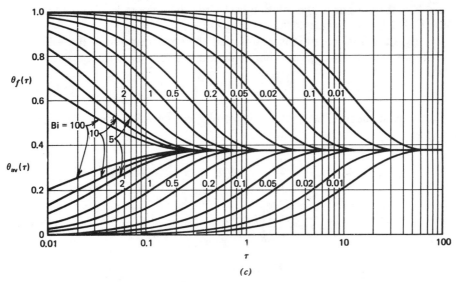

(c)

Figure 13.6 (*Continued*).

[see the one-dimensional version of Equation (7.4b) with no sources]

$$\frac{\partial C(r, t)}{\partial t} = D\frac{1}{r^{1-2m}}\frac{\partial}{\partial r}\left[r^{1-2m}\frac{\partial C(r, t)}{\partial r}\right] \quad \text{in } 0 < r < r_1, \qquad t > 0$$

(13.15a)

where $C(r, t)$ is the concentration of the solute within the body, D is the binary diffusion coefficient, and $m = \frac{1}{2}$ for slab, $m = 0$ for cylinder, and $m = -\frac{1}{2}$ for sphere.

The boundary conditions at $r = 0$ and $r = r_1$ are taken as

$$\frac{\partial C(0, t)}{\partial r} = 0$$

(13.15b)

$$C(r_1, t) = pC_s(t)$$

(13.15c)

where p is a *partition factor*, so that the concentration within the body at $r = r_1$ is p times the concentration in the solution $C_s(t)$.

A mass balance at the boundary $r = r_1$ of the body, expressing the fact that the rate at which the solute leaves the solution is always equal to that which enters the body over the boundary surface at $r = r_1$, is given by

$$V_s\frac{dC_s(t)}{dt} + D\frac{\partial C(r_1, t)}{\partial r}A = 0$$

(13.15d)

where V_s is the volume of the solution excluding the volume V_b occupied by the body, and A is the surface of the body at $r = r_1$.

Finally, the initial conditions stating that the concentration of the solute in the solution is initially C_{s0}, and that in the body is C_0, are written as

$$C(r,0) = C_0, \qquad C_s(0) = C_{s0} \qquad (13.15\text{e, f})$$

For convenience in the analysis we introduce the following dimensionless variables

$$R = \frac{r}{r_1}, \qquad \tau = \frac{Dt}{r_1^2}, \qquad \theta(R, \tau) = \frac{C(r, t) - C_0}{pC_{s0} - C_0}$$

$$\theta_s(\tau) = \frac{pC_s(t) - C_0}{pC_{s0} - C_0}, \qquad \alpha = \frac{V_s}{pV_b} \qquad (13.16)$$

where V_b is the space occupied by the body.

With various dimensionless quantities as defined by Equations (13.16), problem (13.15) takes the form

$$R^{1-2m} \frac{\partial \theta(R, \tau)}{\partial \tau} = \frac{\partial}{\partial R} \left\{ R^{1-2m} \frac{\partial \theta(R, \tau)}{\partial R} \right\} \qquad 0 < R < 1, \qquad \tau > 0$$

$$(13.17\text{a})$$

$$\frac{\partial \theta(0, \tau)}{\partial R} = 0 \qquad (13.17\text{b})$$

$$\theta(1, \tau) = \theta_s(\tau) \qquad (13.17\text{c})$$

$$\frac{d\theta_s(\tau)}{d\tau} + \frac{2(1 - m)}{\alpha} \frac{\partial \theta(1, \tau)}{\partial R} = 0 \qquad (13.17\text{d})$$

$$\theta_s(0) = 1, \qquad \theta(R, 0) = 0 \qquad (13.17\text{e, f})$$

The problem defined by Equations (13.17) is a special case of problem (13.6) with the following correspondence between the two problems:

$$x = R, \qquad \frac{1}{\text{Bi}} = 0, \qquad \theta_f(\tau) = \theta_s(\tau), \qquad K = \frac{2(1 - m)}{\alpha}$$

$$F(X) = 0, \qquad \theta_0 = 1 \qquad (13.18)$$

In the present mass diffusion problem the interest is in the determination of the ratio $M_t(\tau)/M_\infty$, where $M_t(\tau)$ is the total amount of solute in the body at time τ and M_∞ is the corresponding quantity after an infinite time. This ratio is identical to the ratio $\theta_{av}(\tau)/\theta_{av}(\infty)$ obtainable from solution (13.9b) of problem (13.6) by making simplifications according to the correspondence given by Equations (13.18). Thus we obtain

$$\frac{M_t(\tau)}{M_\infty} = 1 - 2(1-m) \sum_{i=1}^{\infty} \frac{2\alpha(1+\alpha)}{4(1-m)^2(1+\alpha)+(\mu_i\alpha)^2} e^{-\mu_i^2\tau}$$

(13.19a)

where the μ_i are the roots of the transcendental equation [see Equation (13.7c)]

$$\mu\alpha J_{-m}(\mu) + 2(1-m)J_{1-m}(\mu) = 0 \qquad (13.19b)$$

The roots μ_i of this transcendental equation depend on the value of α, which is the ratio of V_s the volume of the solution, and pV_b the volume of the body multiplied by the partition factor p. It is sometimes convenient to relate α to the fraction of the total solute finally taken up by the body; such a relationship is determined as now described.

Let V_b and V_s be the volumes of the body and the solution, respectively, C_{s0} be the initial concentration of the solute in the solution, and initially the body be free of solute, that is, $C_0 = 0$. We assume a partition factor p. In the final equilibrium state, the total amount of solute in the solution and in the body should be equal to that originally contained in the solution. Hence we write

$$V_s\frac{C_\infty}{p} + V_bC_\infty = V_sC_{s0} \qquad (13.20a)$$

where C_∞ is the uniform concentration in the body finally and C_∞/p represents the uniform concentration of the solute in the solution finally. The total content of the solute M_∞ in the body finally is determined from Equation (13.20a) as

$$M_\infty = V_bC_\infty = \frac{V_sC_{s0}}{1+[V_s/(pV_b)]} = \frac{V_sC_{s0}}{1+\alpha} \qquad (13.20b)$$

since, by definition, $\alpha = V_s/(pV_b)$.

Then the expression relating α to the *final fractional uptake of the solute by the body* $M_\infty/(V_sC_{s0})$ is obtained from Equation (13.20b) as

$$\frac{M_\infty}{V_sC_{s0}} = \frac{1}{1+\alpha} \qquad (13.20c)$$

For example, $\alpha = 1$ corresponds to 50 percent of the total solute initially in the solute is finally taken up by the body.

We now illustrate the application of the foregoing results for the cases of a plane sheet ($m = \frac{1}{2}$), a long solid cylinder ($m = 0$), and a solid sphere ($m = -\frac{1}{2}$) with the following examples.

Example 13.4 A plane sheet occupying the space $-\ell \le r \le \ell$ is inserted in a limited volume of solution occupying the spaces $(-\ell - a) \le r \le -\ell$ and $\ell \le r \le (\ell + a)$. Initially, the sheet is free from solute, whereas the initial concentration of the solute in the solution is C_{s0}. Determine the ratio $M_t(\tau)/M_\infty$, where $M_t(\tau)$ is the total amount of solute in the sheet at time τ, and M_∞ is the corresponding quantity after an infinite time.

Solution The solution of this problem is immediately obtained from Equations (13.19) by setting $m = \frac{1}{2}$ and utilizing the relations given in Table 3.1. We find

$$\frac{M_t(\tau)}{M_\infty} = 1 - \sum_{i=1}^{\infty} \frac{2\alpha(1 + \alpha)}{1 + \alpha + (\mu_i \alpha)^2} e^{-\mu_i^2 \tau} \qquad (13.21a)$$

where $\alpha = a/(p\ell)$, and the μ_i are the roots of the transcendental equation

$$\tan \mu = -\mu\alpha \qquad (13.21b)$$

The relations given by Equations (13.21) are identical to those given in Reference 3, p. 57, Equations 4.37 and 4.38.

For the slab problem considered here, Equation (13.20b) takes the form

$$M_\infty = 2\ell C_\infty = \frac{2aC_{s0}}{1 + [a/(p\ell)]} = \frac{2aC_{s0}}{1 + \alpha} \qquad (13.21c)$$

Hence the final fractional uptake of the solute by the sheet is given by

$$\frac{M_\infty}{2aC_{s0}} = \frac{1}{1 + \alpha} \qquad (13.21d)$$

Table 13.1 gives the first six roots of the transcendental Equation (13.21b) for values of α corresponding to different values of the final fractional uptake of the solute by the sheet.

Figure 13.7 shows a plot of $M_t(\tau)/M_\infty$ against $\sqrt{\tau}$, determined from Equation (13.21a), for five different values of the final fractional uptake. This figure shows that the rate of removal of the solute from the solution increases with increasing final fractional uptake of the solute by the sheet.

Example 13.5 Determine the ratio $M_t(\tau)/M_\infty$ for a solid cylinder of radius $r = a$, immersed in a well-stirred solution whose cross-section area (excluding the space occupied by the cylinder) is A. Initially, the cylinder is free from solute, whereas the concentration of the solute in the solution is initially C_{s0}.

Table 13.1 First Six Roots μ_i of tan $\mu = -\mu\alpha$

Final Fractional Uptake	α	μ_1	μ_2	μ_3	μ_4	μ_5	μ_6
0	∞	1.5708	4.7124	7.8540	10.9956	14.1372	17.2788
0.1	9.0000	1.6385	4.7359	7.8681	11.0057	14.1451	17.2852
0.2	4.0000	1.7155	4.7648	7.8857	11.0183	14.1549	17.2933
0.3	2.3333	1.8040	4.8014	7.9081	11.0344	14.1674	17.3036
0.4	1.5000	1.9071	4.8490	7.9378	11.0558	14.1841	17.3173
0.5	1.0000	2.0288	4.9132	7.9787	11.0856	14.2075	17.3364
0.6	0.6667	2.1746	5.0037	8.0385	11.1296	14.2421	17.3649
0.7	0.4286	2.3521	5.1386	8.1334	11.2010	14.2990	17.4119
0.8	0.2500	2.5704	5.3540	8.3029	11.3349	14.4080	17.5034
0.9	0.1111	2.8363	5.7172	8.6587	11.6532	14.6870	17.7481
1.0	0	3.1416	6.2832	9.4248	12.5664	15.7080	18.8496

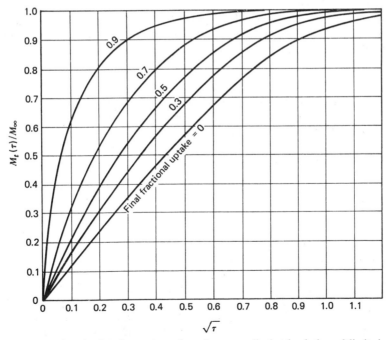

Figure 13.7 Uptake of solute by a plane sheet from a well-stirred solution of limited volume (from Reference 3).

481

Solution   The solution of this problem is immediately obtained from Equations (13.19) by setting $m = 0$. We find

$$\frac{M_t(\tau)}{M_\infty} = 1 - \sum_{i=1}^{\infty} \frac{4\alpha(1 + \alpha)}{4(1 + \alpha) + (\mu_i\alpha)^2} e^{-\mu_i^2\tau} \qquad (13.22\text{a})$$

where $\alpha = A/(p\pi a^2)$, and the μ_i are the roots of the transcendental equation

$$\mu\alpha J_0(\mu) + 2J_1(\mu) = 0 \qquad (13.22\text{b})$$

The relations given by Equations (13.22a, b) are identical to those given in Reference 3, p. 77, Equations 5.33 and 5.34. The final fractional uptake of the solute by the cylinder is given by

$$\frac{M_\infty}{AC_{s0}} = \frac{1}{1 + \alpha} \qquad (13.22\text{c})$$

Table 13.2 gives the first six roots of the transcendental Equation (13.22b) for values of α corresponding to different values of the final fractional uptake of the solute by the cylinder.

Figure 13.8 shows a plot of $M_t(\tau)/M_\infty$ against $\sqrt{\tau}$, determined from Equation (13.22a), for five different values of the final fractional uptake of the solute by the cylinder.

Table 13.2   First Six Roots of $\alpha\mu J_0(\mu) + 2J_1(\mu) = 0$

Final Fractional Uptake	α	μ_1	μ_2	μ_3	μ_4	μ_5	μ_6
0	∞	2.4048	5.5201	8.6537	11.7915	14.9309	18.0711
0.1	9.0000	2.4922	5.5599	8.6793	11.8103	14.9458	18.0833
0.2	4.0000	2.5888	5.6083	8.7109	11.8337	14.9643	18.0986
0.3	2.3333	2.6962	5.6682	8.7508	11.8634	14.9879	18.1183
0.4	1.5000	2.8159	5.7438	8.8028	11.9026	15.0192	18.1443
0.5	1.0000	2.9496	5.8411	8.8727	11.9561	15.0623	18.1803
0.6	0.6667	3.0989	5.9692	8.9709	12.0334	15.1255	18.2334
0.7	0.4286	3.2645	6.1407	9.1156	12.1529	15.2255	18.3188
0.8	0.2500	3.4455	6.3710	9.3397	12.3543	15.4031	18.4754
0.9	0.1111	3.6374	6.6694	9.6907	12.7210	15.7646	18.8215
1.0	0	3.8317	7.0156	10.1735	13.3237	16.4706	19.6159

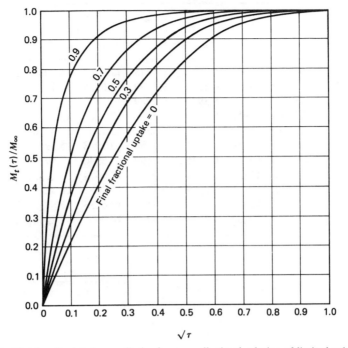

Figure 13.8   Uptake of solute by a cylinder from a well-stirred solution of limited volume (from Reference 3).

Example 13.6   Determine the ratio $M_t(\tau)/M_\infty$ for a solid sphere of radius $r = a$, immersed in a well-stirred solution of volume V_s (excluding the space occupied by the sphere). Initially, the sphere is free from solute, whereas the concentration of the solute in the solution is initially C_{s0}.

Solution   The solution of this problem is immediately obtained from Equations (13.19) by setting $m = -\tfrac{1}{2}$ and utilizing the relations given in Table 3.1. We find

$$\frac{M(\tau)}{M_\infty} = 1 - \sum_{i=1}^{\infty} \frac{6\alpha(1 + \alpha)}{9(1 + \alpha) + (\mu_i\alpha)^2} e^{-\mu_i^2\tau} \tag{13.23a}$$

where $\alpha = 3V_s/(4\pi a^3 p)$ and the μ_i are the roots of the transcendental equation

$$\tan \mu = \frac{3\mu}{3 + \alpha\mu^2} \tag{13.23b}$$

Table 13.3   First Six Roots of tan $\mu = 3\mu / (3 + \alpha\mu^2)$

Final Fractional Uptake	α	μ_1	μ_2	μ_3	μ_4	μ_5	μ_6
0	∞	3.1416	6.2832	9.4248	12.5664	15.7080	18.8496
0.1	9.0000	3.2410	6.3353	9.4599	12.5928	15.7292	18.8671
0.2	4.0000	3.3485	6.3979	9.5029	12.6254	15.7554	18.8891
0.3	2.3333	3.4650	6.4736	9.5567	12.6668	15.7888	18.9172
0.4	1.5000	3.5909	6.5665	9.6255	12.7205	15.8326	18.9541
0.5	1.0000	3.7264	6.6814	9.7156	12.7928	15.8924	19.0048
0.6	0.6667	3.8711	6.8246	9.8369	12.8940	15.9779	19.0784
0.7	0.4286	4.0236	7.0019	10.0039	13.0424	16.1082	19.1932
0.8	0.2500	4.1811	7.2169	10.2355	13.2689	16.3211	19.3898
0.9	0.1111	4.3395	7.4645	10.5437	13.6133	16.6831	19.7564
1.0	0	4.4934	7.7253	10.9041	14.0662	17.2208	20.3713

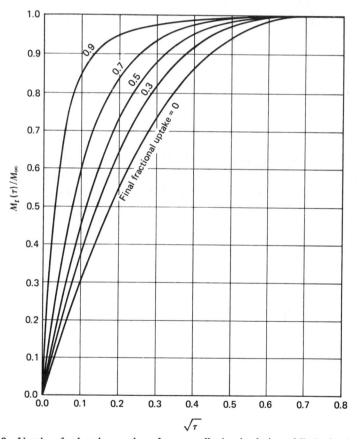

Figure 13.9   Uptake of solute by a sphere from a well-stirred solution of limited volume (from Reference 3).

The final fractional uptake of the solute by the sphere is given by

$$\frac{M_\infty}{V_s C_{s0}} = \frac{1}{1 + \alpha} \tag{13.23c}$$

The relations given by Equations (13.23) are identical to those given in Reference 3, p. 94, Equations 6.30–6.32.

Table 13.3 gives the first six roots of the transcendental Equation (13.23b) for values of α corresponding to different values of the final fractional uptake of the solute by the sphere.

Figure 13.9 shows a plot of $M_t(\tau)/M_\infty$ against $\sqrt{\tau}$, determined from Equation (13.23a), for five different values of the final fractional uptake of the solute by the sphere.

13.3 FLOW DEVELOPMENT IN THE HYDRODYNAMIC ENTRANCE REGION OF DUCTS

The velocity distribution for flow through a duct or tube undergoes a development from its initial profile at the inlet to a fully developed profile at locations far downstream. The length of duct or tube in which a velocity development takes place is called the hydrodynamic entrance length or entrance region. The pressure gradient in this region is also different from that in the fully developed flow region. The determination of the developing flow velocity and the corresponding pressure drop in the entrance region of ducts or tubes is of interest. The exact analysis of such problems, however, even for laminar flow conditions is difficult because of the nonlinearity of the inertia terms appearing in the momentum equations. Various approximate analytical approaches have been devised for the solution of such problems. One method of analysis is achieved by linearizing the inertia terms and then solving the linearized problem as has been done by Targ [4] and by Sparrow and Lin [5].

In this section it is shown that the velocity problem solved in References 4 and 5 for laminar flow inside a parallel plate channel and circular tube are special cases of the solution of the Class VI problem given in Chapter 4.

We consider laminar flow of an incompressible fluid inside a parallel plate channel and circular tube. The duct axis is taken along the positive z direction and the transverse coordinate r is measured from the z axis. It is assumed that: (1) the static pressure is uniform across each section, and (2) the component $\mu \partial^2 u / \partial z^2$ of the axial shear is negligible relative to the component $\mu(1/r^{1-2m})\partial/\partial r[r^{1-2m}(\partial u/\partial r)]$, where u is the axial velocity component, $m = \frac{1}{2}$ for parallel plate channel, and $m = 0$ for circular tube. These are the assumptions that have been generally applied to all entrance region analyses. Then the continuity and the z-momentum equations governing the flow development in the entrance region of a parallel plate channel and circular tube are, respectively, written as

$$\frac{\partial u}{\partial z} + \frac{1}{r^{1-2m}} \frac{\partial(r^{1-2m}v)}{\partial r} = 0 \tag{13.24a}$$

$$u\frac{\partial u}{\partial z} + v\frac{\partial u}{\partial r} = -\frac{1}{\rho}\frac{dP}{dz} + \frac{v}{r^{1-2m}}\frac{\partial}{\partial r}\left(r^{1-2m}\frac{\partial u}{\partial r}\right) \tag{13.24b}$$

in $0 < r < r_1$ and $z > 0$. Here r_1 is the half-spacing between the parallel plates ($m = \frac{1}{2}$) or the tube radius ($m = 0$), r is the transverse coordinate measured from the z axis, and $u \equiv u(r, z)$ and $v \equiv v(r, z)$ are the velocity components in the z and r directions, respectively. In addition, P, ρ, and v are the pressure, density, and kinematic viscosity, respectively.

In the momentum Equation (13.24b), the left-hand side represents the inertia terms. Instead of trying to solve this nonlinear equation directly, by following Reference 5 we propose to seek the solution of the following linearized z-momentum equation

$$\epsilon(z)u_{av}\frac{\partial u}{\partial z} = \Lambda(z) + \frac{v}{r^{1-2m}}\frac{\partial}{\partial r}\left(r^{1-2m}\frac{\partial u}{\partial r}\right) \quad \text{in } 0 < r < r_1, \qquad z > 0$$

$$\tag{13.25a}$$

where $\epsilon(z)$ is a yet-undetermined function of z that weights the mean axial velocity u_{av}. The second unknown function $\Lambda(z)$ includes the pressure gradient term as well as the residual of the inertia terms.

Now both sides of Equation (13.25a) are operated on with the operator

$$\int_0^{r_1} r^{1-2m}\, dr$$

and it is noted that

$$\frac{\partial}{\partial z}\left(\int_0^{r_1} r^{1-2m}u\, dr\right) = 0$$

to satisfy the continuity Equation (13.24a). Then we obtain

$$-\Lambda(z) = v\frac{2(1-m)}{r_1}\frac{\partial u(r_1, z)}{\partial r} \tag{13.25b}$$

Next, a stretched axial coordinate z^* is defined as

$$dz = \epsilon(z^*)\, dz^* \tag{13.25c}$$

In view of Equations (13.25c, b), the linearized momentum Equation (13.25a) becomes

$$u_{\text{av}}\frac{\partial u(r, z^*)}{\partial z^*} + v\frac{2(1 - m)}{r_1}\frac{\partial u(r_1, z^*)}{\partial r} = \frac{v}{r^{1-2m}}\frac{\partial}{\partial r}\left[r^{1-2m}\frac{\partial u(r, z^*)}{\partial r}\right]$$

$$\text{in } 0 < r < r_1, \qquad z^* > 0 \quad (13.26a)$$

This equation is to be solved subject to the symmetry condition at $r = 0$ and the no-slip condition at the wall; that is,

$$\frac{\partial u(0, z^*)}{\partial r} = 0, \qquad u(r_1, z^*) = 0 \qquad (13.26b, c)$$

In addition, the velocity profile at the duct inlet is assumed to be uniform across the section, that is,

$$u(r, 0) = u_{\text{av}} \qquad (13.26d)$$

Before proceeding with the solution of the velocity problem defined by Equations (13.26), we establish the relationship between the actual axial coordinate z and the stretched coordinate z^* as now described.

The relationship between z^* and z is determined in Reference 5 by imposing an additional physical constraint on the pressure gradient such that the local pressure gradient dP/dz determined from momentum considerations should be the same as that determined from mechanical-energy considerations. If the solution of the velocity $u(r, z)$ were exact, a unique value of the pressure gradient would result whether one employs momentum or mechanical-energy considerations. Since the entrance region analysis is approximate, we impose this requirement to establish the relationship between z^* and z.

To determine the local pressure gradient dP/dz from momentum considerations, we operate on both sides of the momentum Equation (13.24b) with the operator

$$\int_0^{r_1} r^{1-2m} \, dr$$

integrate the term involving the velocity component v by parts, and utilize the continuity Equation (13.24a). We find

$$-\frac{d}{dz}\left(\frac{p}{\rho}\right) = \frac{2(1 - m)}{r_1^{2(1-m)}}\frac{d}{dz}\left\{\int_0^{r_1} r^{1-2m} u^2 \, dr\right\} - v\frac{2(1 - m)}{r_1}\frac{\partial u(r_1, z)}{\partial r}$$

$$(13.27a)$$

To determine dp/dz from a mechanical-energy consideration we operate on both sides of the momentum Equation (13.24b) with the operator

$$\int_0^{r_1} u r^{1-2m} \, dr$$

and utilize the continuity Equation (13.24a). We obtain

$$
-\frac{d}{dz}\left(\frac{p}{\rho}\right) = \frac{2(1-m)}{r_1^{2(1-m)}}\frac{d}{dz}\left(\int_0^{r_1} r^{1-2m}\frac{u^3}{2u_{av}}\,dr\right)
$$

$$
+\nu\frac{2(1-m)}{r_1^{2(1-m)}}\frac{1}{u_{av}}\int_0^{r_1} r^{1-2m}\left(\frac{\partial u}{\partial r}\right)^2 dr \qquad (13.27b)
$$

The pressure gradients given by Equations (13.27a) and (13.27b) are equated, and the relation (13.25c) is utilized. After rearrangement, $\epsilon(z^*)$ is determined as

$$
\epsilon(z^*) = \frac{(d/dz^*)\left\{\int_0^{r_1} r^{1-2m}\left[u^2 - u^3/(2u_{av})\right]\,dr\right\}}{\nu r_1^{1-2m}\left[\partial u(r_1, z^*)/\partial r\right] + (\nu/u_{av})\int_0^{r_1} r^{1-2m}\left(\frac{\partial u}{\partial r}\right)^2 dr}
$$

$$
(13.28a)
$$

Clearly, when $u(r, z^*)$ is a known function of r and z^*, the right-hand side of Equation (13.28a) can be evaluated, hence the variation of $\epsilon(z^*)$ with z^* is determined. Knowing $\epsilon(z^*)$, the relationship between the actual coordinate z and the stretched coordinate z^* follows directly from Equation (13.25c) as

$$
z = \int_0^{z^*}\epsilon(z^*)\,dz^* \qquad (13.28b)
$$

We now proceed to the determination of $u(r, z^*)$ from the solution of problem (13.26).

The Solution of Problem (13.26)

For convenience in the analysis, we introduce the following dimensionless variables

$$
R = \frac{r}{r_1}, \qquad Z^* = \frac{\nu z^*}{u_{av}r_1^2}, \qquad Z = \frac{\nu z}{u_{av}r_1^2}, \qquad U(R, Z^*) = \frac{u(r, z^*)}{u_{av}}
$$

$$
(13.29)
$$

Then the problem (13.25) takes the form

$$
\frac{\partial U(R, Z^*)}{\partial Z^*} + 2(1-m)\frac{\partial U(1, Z^*)}{\partial R} = \frac{1}{R^{1-2m}}\frac{\partial}{\partial R}\left[R^{1-2m}\frac{\partial U(R, Z^*)}{\partial R}\right]
$$

$$
\text{in } 0 < R < 1, \qquad Z^* > 0 \quad (13.30a)
$$

subject to the boundary conditions

$$\frac{\partial U(0, Z^*)}{\partial R} = 0, \qquad U(1, Z^*) = 0, \qquad U(R, 0) = 1$$

$$(13.30b, c, d)$$

Equations (13.28), giving the relationship between z^* and z, become

$$Z = \int_0^{Z^*} \epsilon(Z^*) \, dZ^* \qquad (13.31a)$$

where

$$\epsilon(Z^*) = \frac{(d/dZ^*)\left\{ \int_0^1 R^{1-2m} \left[U^2 - U^3/2 \right] dR \right\}}{\left[\partial U(1, Z^*)/\partial R \right] + \int_0^1 R^{1-2m} (\partial U/\partial R)^2 \, dR} \qquad (13.31b)$$

As indicated in Reference 6, the denominator of Equation (13.31b) can be simplified by integrating the term $\int_0^1 R^{1-2m} (\partial U/\partial R)^2 \, dR$ by parts. That is,

$$\int_0^1 R^{1-2m} \left(\frac{\partial U}{\partial R} \right)^2 dR = - \int_0^1 U \frac{\partial}{\partial R} \left\{ R^{1-2m} \frac{\partial U}{\partial R} \right\} dR \qquad (13.32a)$$

Also from Equation (13.30a) we write

$$\frac{\partial}{\partial R} \left(R^{1-2m} \frac{\partial U}{\partial R} \right) = R^{1-2m} \left[\frac{\partial U}{\partial Z^*} + 2(1 - m) \frac{\partial U(1, Z^*)}{\partial R} \right] \qquad (13.32b)$$

Introducing Equation (13.32b) into (13.32a) we obtain

$$\int_0^1 R^{1-2m} \left(\frac{\partial U}{\partial R} \right)^2 dR = - \frac{\partial U(1, Z^*)}{\partial R} - \int_0^1 R^{1-2m} U \frac{\partial U}{\partial Z^*} dR \qquad (13.32c)$$

Introducing Equation (13.32c) into (13.31b), we find the alternative form of the expression for $\epsilon(Z^*)$ as

$$\epsilon(Z^*) = -2 + \frac{3}{2} \frac{\int_0^1 R^{1-2m} U^2 (\partial U/\partial Z^*) \, dR}{\int_0^1 R^{1-2m} U (\partial U/\partial Z^*) \, dR} \qquad (13.33)$$

To facilitate the solving of problem (13.30), it is convenient to express the dimensionless velocity $U(R, Z^*)$ as the sum of the following three functions

$$U(R, Z^*) = U_0(R) + U_1(Z^*) - U_2(R, Z^*) \qquad (13.34)$$

By introducing Equation (13.34) into problem (13.30) it can be shown that: the function $U_0(R)$ is the solution of the following problem

$$\frac{1}{R^{1-2m}}\frac{d}{dR}\left\{R^{1-2m}\frac{dU_0(R)}{dR}\right\} = 2(1-m)\frac{dU_0(1)}{dR} \quad \text{in } 0 < R < 1$$

(13.35a)

$$\frac{dU_0(0)}{dR} = 0, \qquad U_0(1) = 0 \qquad\qquad \text{(13.35b, c)}$$

The functions $U_1(Z^*)$ and $U_2(R, Z^*)$ satisfy the problem

$$R^{1-2m}\frac{\partial U_2(R, Z^*)}{\partial Z^*} = \frac{\partial}{\partial R}\left[R^{1-2m}\frac{\partial U_2(R, Z^*)}{\partial R}\right] \quad \text{in } 0 < R < 1, \qquad Z^* > 0$$

(13.36a)

$$\frac{\partial U_2(0, Z^*)}{\partial R} = 0 \qquad\qquad\qquad \text{(13.36b)}$$

$$U_2(1, Z^*) = U_1(Z^*) \qquad\qquad\qquad \text{(13.36c)}$$

$$\frac{dU_1(Z^*)}{dZ} - 2(1-m)\frac{\partial U_2(1, Z^*)}{\partial R} = 0 \qquad\qquad \text{(13.36d)}$$

$$U_2(R, 0) = U_0(R), \qquad U_1(0) = 1 \qquad\qquad \text{(13.36e, f)}$$

Problem (13.35) defines the fully developed velocity distribution and its solution is immediately obtained as

$$U_0(R) = (2-m)(1-R^2) \qquad\qquad\qquad \text{(13.37)}$$

The problem given by Equations (13.36) is a special case of the Class VI problem given by Equations (13.6), hence its solution is determined from solutions (13.7) as now described.

The correspondence between problems (13.36) and (13.6) is given by

$$X = R, \qquad \tau = Z^*, \qquad \theta(X, \tau) = U_2(R, Z^*)$$

$$\theta_f(\tau) = U_1(Z^*), \qquad \frac{1}{Bi} = 0, \qquad K = -2(1-m) \qquad \text{(13.38)}$$

$$F(X) = (2-m)(1-R^2), \qquad \theta_0 = 1$$

Then the solution of problem (13.36) is immediately obtained from solutions

(13.7) as (see Note 3 for the derivation)

$$U_2(R, Z^*) = 1 - 4(1 - m) \sum_{i=1}^{\infty} \frac{1}{\mu_i^2} \frac{R^m J_{-m}(\mu_i R)}{J_{-m}(\mu_i)} e^{-\mu_i^2 Z^*} \quad (13.39a)$$

where the μ_i are the roots of the transcendental equation

$$\mu J_{-m}(\mu) = 2(1 - m) J_{1-m}(\mu) \quad (13.39b)$$

The function $U_1(Z^*)$ is determined from Equations (13.36c) and (13.39a) as

$$U_1(Z^*) = 1 - 4(1 - m) \sum_{i=1}^{\infty} \frac{1}{\mu_i^2} e^{-\mu_i^2 Z^*} \quad (13.39c)$$

Introducing the solutions (13.37), (13.39a), and (13.39c) into Equation (13.34), the velocity distribution $U(R, Z^*)$ is determined as

$$U(R, Z^*) = (2 - m)(1 - R^2) + 4(1 - m) \sum_{i=1}^{\infty} \frac{1}{\mu_i^2} \left[\frac{R^m J_{-m}(\mu_i R)}{J_{-m}(\mu_i)} - 1 \right] e^{-\mu_i^2 Z^*}$$

$$(13.40)$$

where the μ_i are the roots of Equation (13.39b).

We now illustrate the application of the foregoing result for the determination of the velocity profile in the hydrodynamic entrance region for laminar flow inside a parallel plate channel and a circular tube.

Example 13.7   Determine the velocity profile in the hydrodynamic entrance region for laminar flow between a parallel plate channel.

Solution   The velocity profile $U(R, Z^*)$ is obtained from the solution given by Equation (13.40) by setting $m = \frac{1}{2}$ and utilizing the relations given in Table 3.1. We find

$$U(R, Z^*) = \frac{3}{2}(1 - R^2) + 2 \sum_{i=1}^{\infty} \frac{1}{\mu_i^2} \left[\frac{\cos(\mu_i R)}{\cos \mu_i} - 1 \right] e^{-\mu_i^2 Z^*} \quad (13.41a)$$

where the μ_i are the roots of Equation (13.39b), which takes the form

$$\tan \mu = \mu \quad (13.41b)$$

This solution is the same as that give in Reference 5, p. 345, and the first 25 values of μ_i are also listed in this reference. Figure 13.10 shows the development of the velocity profile in the hydrodynamic entrance region for laminar flow between parallel plates.

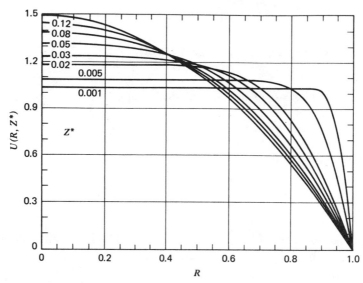

Figure 13.10 Development of velocity profile in the hydrodynamic entrance region for laminar flow inside a parallel plate channel (from Reference 5).

The velocity profile given by Equation (13.41a) is in terms of the stretched axial coordinate Z^*. Therefore, the relationship between the physical coordinate Z and the stretched coordinate Z^* is needed. This relationship is determined by the integration of the stretching factor $\epsilon(Z^*)$ according to Equation (13.31a), where $\epsilon(Z^*)$ is obtained from Equation (13.31b) by setting $m = \frac{1}{2}$. The resulting expression for $\epsilon(Z^*)$ is written in the form

$$\epsilon(Z^*) = \frac{\int_0^1 [2U - (3/2)U^2](\partial U/\partial Z^*)\, dR}{[\partial U(1, Z^*)/\partial R] + \int_0^1 (\partial U/\partial R)^2\, dR} \qquad (13.42)$$

This result is the same as that given in Reference 5, p. 345.

Figure 13.11 shows the variation of the stretching factor $\epsilon(Z^*)$ with the stretched coordinate Z^* given in the right-hand coordinate. Included in this figure is the relationship between Z and Z^* as shown in the left-hand coordinate.

Knowing the relationship between Z and Z^*, the velocity $U(R, Z^*)$ at any location R, Z in the channel can be evaluated from Equations (13.41).

Example 13.8 Repeat the velocity problem in Example 13.7 for laminar flow inside a circular tube.

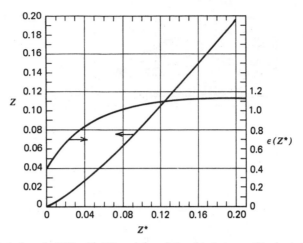

Figure 13.11 Variation of $\epsilon(Z^*)$ with Z^*, and the relationship between the physical coordinate Z and the stretched coordinate Z^* for a parallel plate channel (from Reference 5).

Solution The velocity $U(R, Z^*)$ is now determined from Equation (13.40) by setting $m = 0$.

$$U(R, Z^*) = 2(1 - R^2) + 4\sum_{i=1}^{\infty} \frac{1}{\mu_i^2}\left[\frac{J_0(\mu_i R)}{J_0(\mu_i)} - 1\right]e^{-\mu_i^2 \tau} \quad (13.43a)$$

and the μ_i are the roots of Equation (13.39b) for $m = 0$,

$$\frac{2J_1(\mu)}{J_0(\mu)} = \mu \quad (13.43b)$$

These results are the same as those given in Reference 5, p. 342; the first 25 roots of the transcendental Equation (13.43b) are listed in this reference.

Figure 13.12 shows the development of velocity profile in the hydrodynamic entrance region for laminar flow inside a circular tube. The relationship between the physical coordinate Z and the stretched coordinate Z^* is determined by the integration of the stretching factor $\epsilon(Z^*)$ according to Equation (13.31a). The stretching factor $\epsilon(Z^*)$ is computed from Equation (13.31b) for $m = 0$ as

$$\epsilon(Z^*) = \frac{\int_0^1 R[2U - (3/2)U^2](\partial U/\partial Z^*)\, dR}{[\partial U(1, Z^*)/\partial R] + \int_0^1 R(\partial U/\partial R)^2\, dR} \quad (13.44)$$

Figure 13.13 shows the variation of $\epsilon(Z^*)$ with Z^*, given in the right-hand side

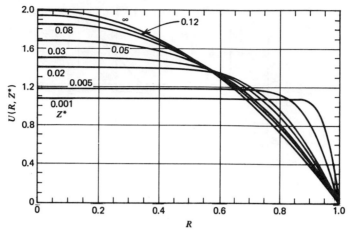

Figure 13.12   Development of velocity profile in a circular tube (from Reference 5).

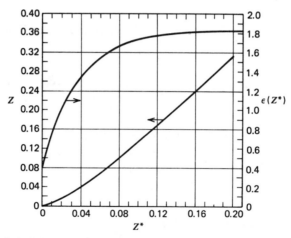

Figure 13.13   Variation of $\epsilon(Z^*)$ with Z^*, and the relationship between the physical coordinate Z and the stretched coordinate Z^* for a circular tube (from Reference 5).

coordinate for a circular tube. Included in this figure is the relationship between Z and Z^* as shown in the left-hand coordinate.

REFERENCES

1. M. D. Mikhailov, Wärme-Oder Stroffübertragung an Einfachen Körpern in Einen Begrenzten Flüssigkeitsraum, *Forschung im Ingenieurwesen*, **32**, 101–110, 147–150 (1966).
2. M. D. Mikhailov, Non-Stationary Heat and Mass Transfer in One-Dimensional Bodies (Russ.), *AN BSSR* (1969).

3. J. Crank, *The Mathematics of Diffusion*, 2nd ed., Oxford University Press, London, 1975.
4. N. A. Slezkin, *Dynamics of Viscous Incompressible Fluids* (Russ.), Gostekhizdat, Moscow, 1955.
5. E. M. Sparrow and S. H. Lin, Flow Development in the Hydrodynamic Entrance Region of Tubes and Ducts, *Phys. Fluids*, **7**, 338–347 (1964).
6. C. L. Wiginton and R. L. Wendt, Flow in the Entrance Region of Ducts, *Phys. Fluids*, **12**, 465–466 (1969).

NOTES

1. The Derivation of Equation (13.5)

Equation (4.110) with various quantities as defined by Equation (13.4) takes the form

$$\theta_k(X,\tau) = \left[1 + \sum_{k=1}^{n}\frac{\omega_k^2 K_k}{2(1-m_k)}\right]^{-1}\left\{\theta_0 + \sum_{k=1}^{n}K_k\omega_k^2\int_0^1 X^{1-2m_k}F_k(X)\,dX\right\}$$

$$+ \sum_{i=1}^{\infty}\frac{e^{-\mu_i^2\tau}}{N_i}\psi_{ki}(X)$$

$$\times\left[\sum_{k=1}^{n}K_k\omega_k^2\int_0^1 X^{1-2m_k}\psi_{ki}(x)F(x)\,dx + \theta_0\sum_{k=1}^{n}K_k\frac{\psi_k'(\mu_i,1)}{\mu_i^2}\right] \tag{1}$$

where the eigenvalues μ_i and the eigenfunctions $\psi_{ki}(X)$ are the solution of the eigenvalue problem (3.147). That is, the eigenvalues are the roots of the transcendental equation [see Equation (3.148b)]

$$\mu^2 + \sum_{k=1}^{n}\frac{\mu K_k\omega_k J_{1-m_k}(\mu\omega_k)}{J_{-m_k}(\mu\omega_k)-(\mu\omega_k/\mathrm{Bi}_k)J_{1-m_k}(\mu\omega_k)} = 0 \tag{2}$$

The eigenfunctions $\psi_{ki}(X)$ are given by [see Equation (3.148c)]

$$\psi_k(\mu_i,X) = C_1\frac{U_1(\mu_i)}{U_k(\mu_i)}(\mu_i\omega_k X)^{m_k}J_{-m_k}(\mu_i\omega_k X) \tag{3}$$

where $U_k(\mu_i)$ is defined as [see Equation (3.148d)]

$$U_k(\mu_i) = (\mu_i\omega_k)^{m_k}\left[J_{-m_k}(\mu_i\omega_k) - \frac{\mu_i\omega_k}{\mathrm{Bi}_k}J_{1-m_k}(\mu_i\omega_k)\right] \tag{4}$$

The normalization integral N_i is given by [see Equation (2.153b)]

$$N_i = \sum_{i=1}^{n} K_k \omega_k^2 \int_0^1 X_k^{1-2m_k} \psi_k^2(\mu_i, X) \, dX + \left[\sum_{k=1}^{n} K_k \omega_k^2 \int_0^1 X^{1-2m_k} \psi_k(\mu_i, X) \, dX \right]^2 \tag{5}$$

Various integrals in this normalization integral are evaluated as the following. From Equation (3.141d) we have

$$\omega_k^2 \int_0^1 X^{1-2m_k} \psi_k(\mu_i, X) \, dX = -\frac{1}{\mu_i^2} \psi_k'(\mu_i, 1) \tag{6}$$

From Equation (3)

$$\psi_k'(\mu_i, X) = -C_1 \frac{U_1(\mu_i)}{U_k(\mu_i)} \mu_i \omega_k (\mu_i \omega_k X)^{m_k} J_{1-m_k}(\mu_i \omega_k X) \tag{7}$$

Introducing Equation (7) into (6) we obtain

$$\omega_k^2 \int_0^1 X^{1-2m_k} \psi_k(\mu_i, X) \, dX = C_1 \frac{1}{\mu_i^2} \frac{U_1(\mu_i)}{U_k(\mu_i)} (\mu_i \omega_k)^{1+m_k} J_{1-m_k}(\mu_i \omega_k) \tag{8}$$

or

$$\sum_{i=1}^{n} K_k \omega_k^2 \int_0^1 X^{1-2m_k} \psi_k^2(\mu_i, X) \, dX$$

$$= C_1 U_1(\mu_i) \frac{1}{\mu_i^2} \sum_{k=1}^{n} \frac{K_n \mu_i \omega_k J_{1-m_k}(\mu_i \omega_k)}{J_{-m_k}(\mu_i \omega_k) - (\mu_i \omega_k / \mathrm{Bi}_k) J_{1-m_k}(\mu_i \omega_k)}$$

$$= -C_1 U_1(\mu_i) \tag{9}$$

where we utilized the relation given by Equation (2).

The remaining term in the normalization integral (5) is evaluated as

$$\int_0^1 X^{1-2m_k} \psi_k^2(\mu_i, X) \, dX = \left[\frac{C_1 U_1(\mu_i)}{J_{-m_k}(\mu_i \omega_k) - (\mu_i \omega_k / \mathrm{Bi}_k) J_{1-m_k}(\mu_i \omega_k)} \right]^2$$

$$\times \int_0^1 X J_{-m_k}^2(\mu_i \omega_k X) \, dX$$

$$= \left[\frac{C_1 U_1(\mu_i)}{J_{-m_k}(\mu_i \omega_k) - (\mu_i \omega_k / \mathrm{Bi}_k) J_{1-m_k}(\mu_i \omega_k)} \right]^2$$

$$\times \frac{1}{2} \left[J_{-m_k}^2(\mu_i \omega_k) + J_{1-m_k}(\mu_i \omega_k) + \frac{2m_k}{\mu_i \omega_k} \right.$$

$$\left. \times J_{1-m_k}(\mu_i \omega_k) J_{-m_k}(\mu_i \omega_k) \right] \tag{10}$$

When Equations (9) and (10) are introduced into Equation (5), the normalization integral becomes

$$N_i = C_1^2 U_1^2(\mu_i)\left[1 + \sum_{i=1}^{n} K_k \omega_k^2 A_{m_k}(\mu_i \omega_k)\right] \tag{11a}$$

where

$$A_{m_k}(\mu_i \omega_k)$$

$$= \frac{J_{-m_k}^2(\mu_i \omega_k) + J_{1-m_k}^2(\mu_i \omega_k) + (2m_k/\mu_i \omega_k)J_{1-m_k}(\mu_i \omega_k)J_{-m_k}(\mu_i \omega_k)}{2\left[J_{-m_k}(\mu_i \omega_k) - (\mu_i \omega_k/\mathrm{Bi}_k)J_{1-m_k}(\mu_i \omega_k)\right]^2}$$

$$\tag{11b}$$

Introducing Equations (3), (4), (7), and (11) into Equation (1) and utilizing Equation (2) to simplify the result, we obtain the solution for $\theta_k(X, \tau)$ as given by Equations (13.5).

2. The Derivation of Equation (13.7)

For the special case considered here, Equation (13.5b) reduces to

$$W_m(\mu_i X) = \frac{X^m J_{-m}(\mu_i X)}{J_{-m}(\mu_i) - (\mu_i/\mathrm{Bi})J_{1-m}(\mu_i)} \tag{1}$$

and Equation (13.5d) becomes

$$J_{-m}(\mu) + \left(\frac{K}{\mu} - \frac{\mu}{\mathrm{Bi}}\right)J_{1-m}(\mu) = 0 \tag{2}$$

In view of Equation (2), Equation (1) is written as

$$W_m(\mu_i X) = -\frac{\mu_i}{K} X^m \frac{J_{-m}(\mu_i X)}{J_{1-m}(\mu_i)} \tag{3}$$

By utilizing Equation (2), Equation (13.5c) is simplified as

$$A_m(\mu_i) = \frac{\mu_i^2}{2K^2}\left[1 + \left(\frac{\mu_i}{\mathrm{Bi}} - \frac{K}{\mu_i}\right)^2 + \frac{2m}{\mu_i}\left(\frac{\mu_i}{\mathrm{Bi}} - \frac{K}{\mu_i}\right)\right] \tag{4}$$

Introducing Equations (3) and (4) into solution (13.5a) and simplifying, we obtain the result given by Equations (13.7).

3. The Derivation of Equation (13.39)

By utilizing the results given by Equations (13.38), various terms in Equation (13.7a) are simplified as described.

$$F(X) = (2 - m)(1 - R^2) = U_0(R) \tag{1a}$$

$$\int_0^1 X^{1-2m} F(X)\, dX = \int_0^1 R^{1-2m} U_0(R)\, dR = \frac{1}{2(1 - m)} \tag{1b}$$

$$\theta_0 = 1 \tag{1c}$$

$$\frac{\theta_0 + K\int_0^1 X^{1-2m} F(X)\, dX}{1 + K/[2(1 - m)]} = \frac{1 + K/[2(1 - m)]}{1 + K/[2(1 - m)]} = 1 \tag{2}$$

We note the integrals

$$\mu_i \int_0^1 R^{1-m} \frac{J_{-m}(\mu_i R)}{J_{1-m}(\mu_i)}\, dR = 1 \tag{3a}$$

$$\mu_i \int_0^1 R^{1-m+2} \frac{J_{-m}(\mu_i R)}{J_{1-m}(\mu_i)}\, dR = 1 + 2\frac{J_{-m}(\mu_i)}{\mu_i J_{1-m}(\mu_i)} - \frac{4(1 - m)}{\mu_i^2} \tag{3b}$$

Then the following integral is evaluated as

$$\mu_i \int_0^1 x^{1-m} \frac{J_{-m}(\mu_i X)}{J_{1-m}(\mu_i)} F(X)\, dX = (2 - m)\mu_i \int_0^1 R^{1-m} \frac{J_{-m}(\mu_i R)}{J_{1-m}(\mu_i)}(1 - R^2)\, dR$$

$$= (2 - m)\left[\frac{4(1 - m)}{\mu_i^2} - \frac{2}{\mu_i}\frac{J_{-m}(\mu_i)}{J_{1-m}(\mu_i)} \right]$$

$$= 0 \tag{4}$$

since the term inside the bracket vanishes in view of the transcendental Equation (13.7c); that is, for $1/\text{Bi} = 0$ and $K = -2(1 - m)$, Equation (13.7c) reduces to Equation (13.39b), that is

$$J_{-m}(\mu) - \frac{2(1 - m)}{\mu} J_{1-m}(\mu) = 0 \tag{5a}$$

or this result is rearranged as

$$\frac{4(1 - m)}{\mu_i^2} - \frac{2}{\mu_i}\frac{J_{-m}(\mu_i)}{J_{1-m}(\mu_i)} = 0 \tag{5b}$$

which proves that the term in the bracket in Equation (4) vanishes. The term

E_i becomes

$$E_i = \frac{2}{\mu_i}\left[1 + \frac{2m}{\mu_i}\frac{2(1-m)}{\mu_i} + \frac{2^2(1-m)^2}{\mu_i^2} - \frac{2\times 2(1-m)}{\mu_i^2}\right] = \frac{2}{\mu_i} \quad (6)$$

Introducing the results given by Equations (2, 4, and 6) into Equation (13.7a), we obtain

$$U_2(R, Z^*) = 1 - 2\sum_{i=1}^{\infty}\frac{1}{\mu_i}R^m\frac{J_{-m}(\mu_i R)}{J_{1-m}(\mu_i)}e^{-\mu_i^2 Z^*} \quad (7)$$

By utilizing the transcendental Equation (5a), this result is written as

$$U_2(R, Z^*) = 1 - 4(1-m)\sum_{i=1}^{\infty}\frac{1}{\mu_i^2}R^m\frac{J_{-m}(\mu_i R)}{J_{-m}(\mu_i)}e^{-\mu_i^2 Z^*} \quad (8)$$

which is the same as Equation (13.39a).

═══ CHAPTER FOURTEEN ═══

Class VII Solutions Applied
to Diffusion
With Chemical Reaction

Science moves, but slowly and slowly,
creeping on from point to point.

Alfred Lord Tennyson
1809–1892

In this chapter we illustrate the application of the one-dimensional Class VII solutions developed in Chapter 4 to the analysis of diffusion problems in which the amount of diffusing substance is finite and, while diffusing, some of the diffusing substance is absorbed by another with which it reacts chemically. The problem can be regarded as one in which some of the diffusing substance becomes immobilized as diffusion proceeds, or as one encountered in chemical kinetics in which the reaction rate depends on the rate of supply of one of the reactants by diffusion [1].

Simultaneous diffusion with chemical reaction has numerous practical applications. For example, in diffusion within the pores of a solid body some of the diffusing substance may be absorbed in the pores, hence considered immobilized; or in diffusion through a gel, some of the diffusing molecules are attracted to fixed sites within the medium, hence considered immobilized. Similar processes occur in biology and biochemistry associated with diffusion into living cells and microorganisms. For example, in the problem of diffusion into a textile fiber in which there are a number of active groups, the diffusing molecules can become attached to them and thus become immobilized [2–4]. The diffusion of oxygen and carbon monoxide through the outer membrane of red blood corpuscles is accompanied by diffusion and chemical reaction in the case of concentrated hemoglobin [5].

In this chapter, as an illustration, we consider a situation in which diffusion is accompanied by a first order, reversible chemical reaction. In such problems, depending on the relative rates of diffusion and chemical reaction, certain extreme cases can be envisioned. The first is the one in which the reaction rate is very rapid so that the immobilized component can be assumed to be always in equilibrium with the component free-to-diffuse; the diffusion is called the *rate-controlling* process. The second is the one in which diffusion is so rapid compared with the reaction rate that the concentrations of the diffusing substance and the immobilized product are considered uniform throughout the medium, and the process is said to be controlled solely by the reversible reaction. The solution for these two extreme cases is relatively easy. However, the analysis of the problem becomes rather involved for the more general case, in which the diffusion and reaction rates are comparable. The solution for this general case has been developed in References 6 and 7 and the results are reproduced in Reference 1. More recently, these solutions are obtained as special cases of the Class VII solutions in Reference 8.

In the following analysis, we first present the solution of the diffusion problem with first-order, reversible chemical reaction when the diffusion and reaction rates are comparable. Then as special cases we obtain solutions for problems of instantaneous reaction and irreversible reaction. Finally, we present some numerical results for the solutions developed in this chapter.

14.1 REVERSIBLE REACTION

Consider a material region in the form of an infinite plate of thickness $2a$ (or a long solid cylinder or a solid sphere of diameter $2a$) immersed into a well-stirred solution of finite volume and the solute allowed to diffuse into the material region. The concentration of the solute in the solution is always uniform because the solution is continuously stirred. Let C_0 be the initial concentration of the solute in the solution and the material volume be free from the solute initially. For times $t > 0$, the diffusion of the solute into the material volume proceeds and a first order, reversible reaction occurs inside the material volume. As a result, a nondiffusing product is formed (i.e., some of the solute is immobilized). Because of symmetry, we consider the material volume confined to the region $0 \le r \le a$ and the solution confined to the space $a \le r \le a + \ell$. We wish to determine the concentrations of the solute free-to-diffuse and the nondiffusing product as a function of time and space in the material volume.

Let $C(r, t)$ be the concentration of the solute *free-to-diffuse* within the body and $S(r, t)$ be that of *immobilized solute*, each being expressed as an amount of mass per unit volume of the body. The mathematical formulation of this mass diffusion problem is given by

$$\frac{\partial C(r, t)}{\partial t} = D \frac{1}{r^{1-2m}} \frac{\partial}{\partial r} \left[r^{1-2m} \frac{\partial C(r, t)}{\partial r} \right] - \frac{\partial S(r, t)}{\partial t}$$

$$\text{in } 0 < r < a, \qquad t > 0 \quad (14.1a)$$

$$\frac{\partial S(r, t)}{\partial t} = \delta C(r, t) - \eta S(r, t) \quad \text{in } 0 < r < a, \qquad t > 0 \quad (14.1b)$$

where

$$m = \begin{cases} \frac{1}{2} & \text{for slab} \\ 0 & \text{for cylinder} \\ -\frac{1}{2} & \text{for sphere} \end{cases}$$

D is the diffusion coefficient, and δ and η are the forward and backward reaction rate constants, respectively. Thus the immobilized solute $S(r, t)$ is formed at a rate proportional to the concentration of the solute free-to-diffuse $C(r, t)$ and disappears at a rate proportional to its own concentration as apparent from Equation (14.1b).

The boundary condition for this problem at $r = 0$ is written by the symmetry consideration as

$$\frac{\partial C(0, t)}{\partial r} = 0 \quad \text{for } t > 0 \tag{14.1c}$$

and the boundary condition at $r = a$ is taken as

$$V\frac{\partial C(a, t)}{\partial t} + AD\frac{\partial C(a, t)}{\partial r} = 0 \quad \text{for } t > 0 \tag{14.1d}$$

which is determined by the fact that the rate at which the solute leaves the solution of volume V should be equal to that at which the solute enters the material volume over the surface A. Here we assume that the concentration of solute free-to-diffuse just within the surface A of the material region is the same as that in the solution. It is also possible that the concentration just within the surface of the material is K times that in the solution ($K \neq 1$). Such a situation can also be allowed in the analysis by modifying the volume V of the solution as V/K.

Usually, the initial condition for the solute free-to-diffuse and immobilized are taken as equal to zero and the initial concentration of the solute within the solution is taken as, say, C_s. Then the problem becomes one of diffusion of the solute from the solution into the material volume. There is also the complementary problem in which all the solute is initially uniformly distributed throughout the material region such that subsequently the solute diffuses out into the solution. Then the problem describes *desorption*.

For generality in the analysis, we assume that the initial concentrations of the solute free-to-diffuse and immobilized are not zero, and taken as

$$C(r, 0) = C_0, \qquad S(r, 0) = S_0 \quad \text{in } 0 \leq r \leq a \tag{14.1e, f}$$

and equilibrium condition exist initially everywhere, that is,

$$\delta C_0 - \eta S_0 = 0 \quad \text{for } t = 0 \quad \text{in } 0 \leq r \leq a \tag{14.1g}$$

which is obtained from Equation (14.1b) for $\partial S/\partial t = 0$.

Finally, the initial concentration of the solute in the solution is taken as

$$C_s \quad \text{for } t = 0 \text{ in the solution} \tag{14.1h}$$

In order to bring problem (14.1) into a form readily comparable with the general problem (4.111), we define the following dimensionless variables

$$\theta_1(\xi, \tau) = \frac{C(r, t) - C_0}{C_s - C_0}, \qquad \theta_2(\xi, \tau) = \frac{S(r, t) - S_0}{C_s - C_0}$$

$$\xi = \frac{r}{a}, \qquad \tau = \frac{Dt}{a^2}, \qquad K_\delta = \frac{\delta a^2}{D}, \qquad K_\eta = \frac{\eta a^2}{D}, \qquad K_v = \frac{Aa}{V} \tag{14.2}$$

Then the mass diffusion problem defined by Equations (14.1) takes the form

$$\xi^{1-2m} \left[\frac{\partial \theta_1(\xi, \tau)}{\partial \tau} + \frac{\partial \theta_2(\xi, \tau)}{\partial \tau} \right] = \frac{\partial}{\partial \xi} \left[\xi^{1-2m} \frac{\partial \theta_1(\xi, \tau)}{\partial \xi} \right]$$

$$\text{in } 0 < \xi < 1, \qquad \tau > 0 \quad (14.3\text{a})$$

$$\frac{\partial \theta_2(\xi, \tau)}{\partial \tau} = K_\delta \theta_1(\xi, \tau) - K_\eta \theta_2(\xi, \tau) \quad \text{in } 0 < \xi < 1, \qquad \tau > 0 \quad (14.3\text{b})$$

$$\frac{\partial \theta_1(0, \tau)}{\partial \xi} = 0 \quad \text{for } \tau > 0 \tag{14.3c}$$

$$K_v \frac{\partial \theta_1(1, \tau)}{\partial \xi} + \frac{\partial \theta_1(1, \tau)}{\partial \tau} = 0 \quad \text{for } \tau > 0 \tag{14.3d}$$

$$\theta_1(\xi, 0) = 0, \qquad \theta_2(\xi, 0) = 0 \quad \text{in } 0 \le \xi \le 1 \tag{14.3e}$$

and the concentration of the solute in the solution is initially C_s.

Clearly, problem (14.3) is a special case of the general problem (4.111). By comparison of these two problems we write

$$x = \xi, \qquad t = \tau, \qquad T_1(x, t) = \theta_1(\xi, \tau), \qquad T_2(x, t) = \theta_2(\xi, \tau)$$

$$w(x) = k(x) = \xi^{1-2m}, \qquad P(x, t) = 0, \qquad x_0 = 0, \qquad x_1 = 1$$

$$\sigma_1 = K_\delta, \qquad \sigma_2 = K_\eta, \qquad \alpha = 1, \qquad \beta = 0, \qquad \gamma = K_v$$

$$\phi(t) = \theta_1(1, \tau), \qquad f_1(x) = f_2(x) = 0 \tag{14.4}$$

Then the solution of problem (14.3) is obtainable from the general solution (4.112), and the eigenfunctions and eigenvalues needed for this solution from the results in Example (3.10). Hence Equations (3.154a) and (3.154d), respectively, yield

$$\mu^2 J_{-m}(\lambda) + K_v \lambda J_{1-m}(\lambda) = 0 \tag{14.5}$$

and

$$\lambda = \mu \sqrt{1 + \frac{K_\delta}{K_\eta - \mu^2}} \tag{14.6}$$

where the functions $x^m J_{-m}(x)$ and $x^m J_{1-m}(x)$, for $m = \frac{1}{2}$, 0, and $-\frac{1}{2}$ are listed in Table 3.1. The eigenvalues λ_i are the roots of the transcendental Equations (14.5) and (14.6). These equations can be expressed in alternative form if we put

$$x = \lambda^2, \qquad y = -\mu^2 \tag{14.7a, b}$$

Then Equation (14.5) takes the form

$$y = K_v \sqrt{x} \, \frac{J_{1-m}(\sqrt{x})}{J_{-m}(\sqrt{x})} \qquad \text{for } x > 0 \tag{14.8a}$$

or

$$y = -K_v \sqrt{|x|} \, \frac{I_{1-m}(\sqrt{|x|})}{I_{-m}(\sqrt{|x|})} \qquad \text{for } x < 0 \tag{14.8b}$$

since $\sqrt{-|x|} = i\sqrt{|x|}$ and $J_v(i\sqrt{|x|}) = i^v I_v(\sqrt{|x|})$. Equation (14.6) becomes

$$x = -\frac{y(y + K_\eta + K_\delta)}{y + K_\eta} \tag{14.9a}$$

or this result can be solved for y to give

$$y = -\frac{1}{2}\left[(K_\eta + K_\delta + x) \mp \sqrt{(K_\eta + K_\delta + x)^2 - 4K_\eta x} \right] \tag{14.9b}$$

Here alternative signs of the square root term give rise to two different values of y for each value of x; thus there are two branches of Equation (14.9a) when it is plotted. Equations (14.8) and (14.9) are to be solved for the roots.

Physical Significance of the Roots

To illustrate the physical significance of the roots of the preceding transcendental equations we consider the case $m = \frac{1}{2}$, characterizing the slab. Equation (14.8a) for $m = \frac{1}{2}$, in view of the results in Table 3.1, takes the form

$$y = K_v \sqrt{x} \tan(\sqrt{x}) \qquad \text{for } x > 0 \tag{14.10}$$

and Equation (14.9a) remains unaltered, that is,

$$x = -\frac{y(y + K_\eta + K_\delta)}{y + K_\eta}$$

(14.11)

The two values of y in Equation (14.11) are given by Equation (14.9b).

Figure 14.1 shows a plot of Equations (14.10) and (14.11), and illustrates the general location of the roots, which are the points where the two curves intersect. As Equation (14.10) does not depend on the parameters K_η and K_δ, the resulting graphs of Equation (14.10) are valid for all values of K_η and K_δ; the graphs of Equation (14.11) depend on K_η and K_δ because Equation (14.11) contains these parameters. The roots are located at the intersection of the two branches of Equation (14.11) with the successive branches of Equation (14.10). It is easy to envision from the graphs shown in Figure 14.1 how the roots vary with K_η and K_δ.

Figure 14.1 is constructed for a slab geometry. In the case of a long solid cylinder and a solid sphere, Equation (14.8a) takes a form similar to that of Equation (14.10), except the term $\tan(\sqrt{x})$ is replaced, respectively, by $J_0(\sqrt{x})/J_1(\sqrt{x})$ and $j_0(\sqrt{x})/j_1(\sqrt{x})$, as apparent from the results in Table 3.1. As the behavior of these functions is similar to that of $\tan(\sqrt{x})$, then Figure 14.1 can be considered to represent qualitatively the location of the roots for a solid cylinder and a solid sphere.

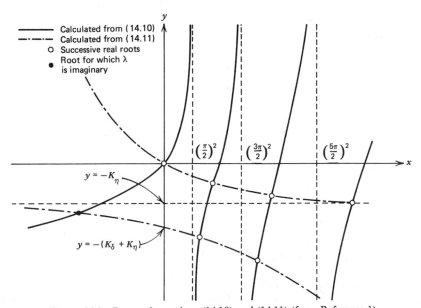

Figure 14.1   Roots of equations (14.10) and (14.11) (from Reference 1).

Once x and y values at the roots are determined, the eigenvalues λ and μ are calculated according to the relations (14.7a, b), that is,

$$\lambda = \sqrt{x}, \qquad \mu = \sqrt{-y} \qquad (14.12a, b)$$

An examination of the location of the roots illustrated in Figure 14.1 reveals the following in relation to the values of λ and μ.

1. For all roots contained in the region $x > 0$, the value of y is negative (i.e., $y < 0$), hence by Equations (14.12a, b) we conclude that *all the λ and μ are real when $x > 0$*. Also, there are two infinite sets of μ_i (or λ_i) corresponding to the intersections of two branches of Equation (14.11) with the successive branches of Equation (14.10).

2. In the region $x < 0$, there is a root for which, by Equation (14.12a), λ is imaginary. We represent the imaginary λ as

$$\lambda = \sqrt{-|x|} \equiv i\lambda^*, \qquad \lambda^* = \sqrt{|x|} \qquad (14.13a)$$

The value of μ, by Equation (14.12b), is real because y is negative; we denote the corresponding μ as

$$\mu^* = \sqrt{-y} \quad \text{for } x < 0 \qquad (14.13b)$$

The Solution of Problem (14.3)

It is apparent from the foregoing discussion of the roots that the solution of the problem should comprise two infinite series resulting from the two sets of real values of μ, and a term corresponding to the root for which λ is imaginary in addition to the steady-state solution for the problem. Therefore, the solution of the problem (14.3), obtained from the solution (4.112) by introducing Equations (14.4), is written as

$$\theta_k(\xi, \tau) = \frac{\left(K_\delta/K_\eta\right)^{k-1}}{1 + \left[1 + \left(K_\delta/K_\eta\right)\right] K_v/[2(1-m)]}$$

$$- \frac{2\lambda^*\mu^{*2}\left[K_\delta/\left(K_\eta - \mu^{*2}\right)\right]^{k-1} e^{-\mu^{*2}\tau}}{\left\{1 + \left[K_\eta K_\delta/\left(K_\eta - \mu^{*2}\right)^2\right]\right\}\left(\mu^{*4} - 2m\mu^{*2}K_v - \lambda^{*2}K_v^2\right) - 2\lambda^{*2}K_v}$$

$$\times \frac{(\lambda^*\xi)^m I_{-m}(\lambda^*\xi)}{\lambda^{*m} I_{1-m}(\lambda^*)}$$

$$- 2 \sum_{i=1}^{\infty} \frac{\lambda_i \mu_i^2 \left[K_\delta/\left(K_\eta - \mu_i^2\right)\right]^{k-1} e^{-\mu_i^2\tau}}{\left\{1 + \left[K_\eta K_\delta/\left(K_\eta - \mu_i^2\right)^2\right]\right\}\left(\mu_i^4 - 2m\mu_i^2 + \lambda_i^2 K_v^2\right) + 2\lambda_i^2 K_v}$$

$$\times \frac{(\lambda_i \xi)^m J_{-m}(\lambda_i \xi)}{\lambda_i^m J_{1-m}(\lambda_i)} \qquad (k = 1, 2) \qquad (14.14)$$

Here the summation term characterizes two infinite series resulting from the two sets of values of μ; the terms containing the parameters λ^* and μ^* result from the root for which λ is imaginary; the first term represents the steady-state value of the function.

The concentration of the solute free-to-diffuse $C(\xi, \tau)$ and the immobilized solute $S(\xi, \tau)$ are obtained from Equation (14.14) by setting $k = 1$ and 2, respectively, for a slab ($m = \frac{1}{2}$), a long solid cylinder ($m = 0$), and a solid sphere ($m = -\frac{1}{2}$).

Total Amount of Solute for $C_0 = S_0 = 0$

The total amount of solute per unit volume of the body, including that of solute free-to-diffuse and immobilized is determined from

$$\frac{M(\tau)}{V_b} = 2(1 - m) \int_0^1 \xi^{1-2m} [C(\xi, \tau) + S(\xi, \tau)] \, d\xi \qquad (14.15a)$$

where

$$V_b = \begin{cases} a & \text{slab} \\ \pi a^2 & \text{solid cylinder} \\ \frac{4}{3}\pi a^3 & \text{solid sphere} \end{cases}$$

We now consider the special situation in which $C_0 = S_0 = 0$. Introducing Equation (14.14) into Equation (14.15a) for $k = 1$ and 2, setting $C_0 = S_0 = 0$, and utilizing the result of the integral given by Equation (3.149e), we obtain

$$\frac{M(\tau)}{C_s V_b} = \frac{1 + (K_\delta / K_\eta)}{1 + [1 + (K_\delta / K_\eta)] K_v / [2(1 - m)]}$$

$$+ \frac{4(1 - m)\lambda^{*2} e^{-\mu^{*2}\tau}}{\left\{1 + \left[K_\eta K_\delta / (K_\eta - \mu^{*2})^2\right]\right\}(\mu^{*4} - 2m\mu^{*2}K_v - \lambda^{*2}K_v^2) - 2\lambda^{*2}K_v}$$

$$- \sum_{i=1}^{\infty} \frac{4(1 - m)\lambda_i^2 e^{-\mu_i^2\tau}}{\left\{1 + K_\eta K_\delta / (K_\eta - \mu_i^2)^2\right\}\{\mu_i^4 - 2m\mu_i^2 K_v + \lambda_i^2 K_v^2\} + 2\lambda_i^2 K_v}$$

$$(14.15b)$$

The steady-state value of $M(\infty)$ is obtained from Equation (14.15b) by setting $\tau \to \infty$. We find

$$\frac{M(\infty)}{C_s V_b} = \frac{1 + (K_\delta / K_\eta)}{1 + [1 + (K_\delta / K_\eta)] K_v / [2(1 - m)]} \qquad (14.15c)$$

By utilizing the results (14.15b) and (14.15c), the total amount of solute $M(\tau)$ at any time τ is expressed as a fraction of $M(\infty)$ as

$$
\frac{M(\tau)}{M(\infty)} = 1 + \frac{2\lambda^{*2}\left\{ \dfrac{2(1-m)}{1+\dfrac{K_\delta}{K_\eta}} + K_v \right\} e^{-\mu^{*2}\tau}}{\left\{ 1 + \dfrac{K_\eta K_\delta}{\left(K_\eta - \mu^{*2}\right)^2} \right\}\left\{ \mu^{*4} - 2m\mu^{*2}K_v - \lambda^{*2}K_v^2 \right\} - 2\lambda^{*2}K_v}
$$

$$
- \sum_{i=1}^{\infty} \frac{2\lambda_i^2\left\{ \dfrac{2(1-m)}{1+\dfrac{K_\delta}{K_\eta}} + K_v \right\} e^{-\mu_i^2\tau}}{\left\{ 1 + \dfrac{K_\eta K_\delta}{\left(K_\eta - \mu_i^2\right)^2} \right\}\left\{ \mu_i^4 - 2m\mu_i^2 K_v + \lambda_i^2 K_v^2 \right\} + 2\lambda_i^2 K_v}
$$

$$(14.16)$$

Once the roots μ_i, λ_i, and μ^*, λ^* are available, the solutions given by Equations (14.14)–(14.16) can readily be evaluated for a slab ($m = \frac{1}{2}$), a long solid cylinder ($m = 0$), and a solid sphere ($m = -\frac{1}{2}$), provided that $K_v \neq 0$ (i.e., V is finite).

In the case of infinite amount of solute, that is, $V \rightarrow \infty$ or $K_v = 0$, the solution (14.16) needs further simplifications because of the convergence considerations. This matter is discussed in the following.

The Special Case, $K_v = 0$

It follows from Equation (14.8) that, when $K_v = 0$, there is no root for which λ is imaginary. Then for $K_v = 0$ we omit the term resulting from the imaginary value of λ, and Equation (14.16) reduces to

$$
\frac{M(\tau)}{M(\infty)} = 1 - \frac{4(1-m)}{1+R} \sum_{i=1}^{\infty} \frac{\left[R + 1 - \left(\mu_i^2/K_\eta\right)\right]^2}{\lambda_i^2\left\{R + \left[1 - \left(\mu_i^2/K_\eta\right)\right]^2\right\}} e^{-\mu_i^2\tau} \quad (14.17a)
$$

where the λ_i are the roots of the following transcendental equation [see Equation (14.5) for $K_v = 0$]

$$
J_{-m}(\lambda) = 0 \qquad\qquad (14.17b)
$$

the μ_i are determined from [see Equation (14.9b) for $y = -\mu_i^2$ and $x = \lambda_i^2$]

$$\mu_i^2 = \tfrac{1}{2}\left[\left(K_\eta + K_\delta + \lambda_i^2\right) \mp \sqrt{\left(K_\eta + K_\delta + \lambda_i^2\right)^2 - 4K_\eta\lambda_i^2}\right] \quad (14.17c)$$

and R is defined as

$$R = \frac{K_\delta}{K_\eta} \qquad\qquad (14.17d)$$

Clearly, for each value of λ_i there are two values of μ_i resulting from the alternate sign for the square root term in Equation (14.17c). Therefore, the summation in solution (14.17a) implies two infinite series resulting from the two sets of values of μ_i.

The numerical evaluation of Equation (14.17a) can cause some computational difficulties because of the convergence of the terms for which μ_i^2 approaches K_η, especially for small values of K_η. This difficulty can be circumvented as now described.

We focus our attention on the series associated with the positive square root in Equation (14.17c), since these are the terms for which μ_i^2 approaches K_η when λ_i is large.

Let μ_i and μ_j be the roots associated, respectively, with the positive and negative sign in Equation (14.17c). If $\mu_i^2 \approx K_\eta$ to the order of accuracy required, after the first p nonzero roots arising from the positive sign, the solution (14.17a) is written as

$$\frac{M(\tau)}{M(\infty)} = 1 - \frac{4(1-m)}{1+R}\sum_{i=1}^{p}\frac{\left[R + 1 - \left(\mu_i^2/K_\eta\right)\right]^2}{\lambda_i^2\left\{R + \left[1 - \left(\mu_i^2/K_\eta\right)\right]^2\right\}}e^{-\mu_i^2\tau}$$

$$- \frac{4(1-m)}{1+R}R\sum_{i=p+1}^{\infty}\frac{e^{-\mu_i^2\tau}}{\lambda_i^2}$$

$$- \frac{4(1-m)}{1+R}\sum_{j=1}^{\infty}\frac{\left[R + 1 - \left(\mu_j^2/K_\eta\right)\right]^2}{\lambda_j^2\left\{R + \left[1 - \left(\mu_j^2/K_\eta\right)\right]^2\right\}}e^{-\mu_j^2\tau} \quad (14.18a)$$

Here the third term on the right-hand side of Equation (14.18a) can be represented approximately as

$$4(1-m)\sum_{i=p+1}^{\infty}\frac{e^{-\mu_i^2\tau}}{\lambda_i^2} = \left[1 - 4(1-m)\sum_{i=1}^{p}\frac{1}{\lambda_i^2}\right]e^{-K_\eta\tau} \quad (14.18b)$$

since

$$4(1 - m) \sum_{i=1}^{\infty} \frac{1}{\lambda_i^2} = 1 \tag{14.18c}$$

The relationship (14.18c) follows, for example, from Equation (14.17a) for $R = 0$, since $M(\tau)/M(\infty) = 0$ when $\tau = 0$.

Introducing Equation (14.18b) into Equation (14.18a), we obtain

$$\frac{M(\tau)}{M(\infty)} = 1 - \frac{4(1 - m)}{1 + R} \sum_{i=1}^{p} \frac{\left[R + 1 - \left(\mu_i^2/K_\eta\right)\right]^2}{\lambda_i^2 \left\{R + \left[1 - \left(\mu_i^2/K_\eta\right)\right]^2\right\}} e^{-\mu_i^2 \tau}$$

$$- \frac{R}{1 + R} \left\{1 - 4(1 - m) \sum_{i=1}^{p} \frac{1}{\lambda_i^2}\right\} e^{-K_\eta \tau}$$

$$- \frac{4(1 - m)}{1 + R} \sum_{j=1}^{\infty} \frac{\left[R + 1 - \left(\mu_j^2/K_\eta\right)\right]^2}{\lambda_j^2 \left\{R + \left[1 - \left(\mu_j^2/K_\eta\right)\right]^2\right\}} e^{-\mu_j^2 \tau} \tag{14.19}$$

Clearly, the error involved in the approximate form (14.19) can be made as small as desired by the choice of p.

14.2 INSTANTANEOUS REACTION

If the reaction rate for the formation of immobilized product is very rapid compared with the diffusion process, local equilibrium can be assumed between the immobilized and free-to-diffuse components. In this case we have

$$K_\eta \to \infty, \qquad \frac{K_\delta}{K_\eta} = R \neq 0 \tag{14.20a, b}$$

and Equation (14.3b) simplifies to

$$S(\xi, \tau) = RC(\xi, \tau) \tag{14.20c}$$

Since $(1/K_\eta)(\partial S(\xi, \tau)/\partial \tau) \to 0$. The result (14.20b) implies that the concentration $S(\xi, \tau)$ of the immobilized substance is directly proportional to the concentration $C(\xi, \tau)$ of the substance free-to-diffuse.

This problem is an extreme case of the more general problem considered previously; therefore, its solution can be obtained as a special case from the general problem as now described.

Equation (14.9b) for $K_\eta \to \infty$ implies that the terms that arise by taking the positive sign in Equation (14.9b) and the term from the imaginary root are no longer needed. In view of the restrictions (14.20a, b), Equation (14.6) leads to

$$\mu^2 = \frac{\lambda^2}{1 + R} \qquad (14.20d)$$

Introducing the expressions (14.20a, b, d) into Equation (14.16) and omitting the term resulting from the imaginary root, we obtain the solution as

$$\frac{M(\tau)}{M(\infty)} = 1 - 2 \sum_{i=1}^{\infty} \frac{2(1 - m) + K_v(1 + R)}{\lambda_i^2 + [K_v(1 + R) + (1 - m)]^2 - (1 - m)^2} e^{-\lambda_i^2[\tau/(1 + R)]}$$

$$(14.21a)$$

where the λ_i approach the roots of the following transcendental equation [see Equations (14.5) and (14.20d)]

$$\lambda J_{-m}(\lambda) + K_v(1 + R)J_{1-m}(\lambda) = 0 \qquad (14.21b)$$

The Special Case of $K_v = 0$

For the special case of an infinite amount of solute, we have $V \to \infty$, hence $K_v = 0$, and the solution (14.21) simplifies to

$$\frac{M(\tau)}{M(\infty)} = 1 - 4(1 - m) \sum_{i=1}^{\infty} \frac{1}{\lambda_i^2} e^{-\lambda_i^2 \tau/(1 + R)} \qquad (14.22a)$$

where the λ_i are the roots of

$$J_{-m}(\lambda) = 0 \qquad (14.22b)$$

Equation (14.22a) is a special case of Equation (7.39) for $p \to \infty$. Recalling the definition of τ given by Equation (14.2), the factor $(1 + R)$ in Equation (14.22a) implies that the diffusion is governed by an effective diffusion coefficient $D/(1 + R)$. That is, the effect of instantaneous reaction is to slow down the diffusion process as pointed out by Crank [1]. For example, if $R + 1 = 100$, the overall diffusion process with instantaneous reaction is slower than the simple diffusion process by a hundredfold. This matter is better envisioned if the linear relationship (14.20c) is introduced into the differential Equation (14.3a). We obtain

$$\xi^{1-2m} \frac{\partial C(\xi, \tau)}{\partial(\tau/(1 + R))} = \frac{\partial}{\partial \xi} \left\{ \xi^{1-2m} \frac{\partial C(\xi, \tau)}{\partial \xi} \right\} \qquad (14.23)$$

which is the same as the diffusion Equation (7.12a) with $G(x, \tau) = 0$ and the modified Fourier number $\tau/(1 + R)$. Therefore, the method of solution discussed in connection with the problem (7.12) is applicable for the solution of a problem with instantaneous reaction.

14.3   IRREVERSIBLE REACTION

Another special case of the general solution (14.16) is that of an irreversible, first order chemical reaction in which the rate of formation of immobilized solute is directly proportional to the concentration of the solute free-to-diffuse. That is, we have

$$K_\eta = 0 \qquad (14.24a)$$

Then Equation (14.3b) of the problem (14.3) reduces to

$$\frac{\partial S(\xi, \tau)}{\partial \tau} = K_\delta C(\xi, \tau) \qquad (14.24b)$$

We consider the following two situations.

Case 1   We assume

$$K_\eta = 0, \qquad K_v \neq 0, \qquad K_\delta \neq 0 \qquad (14.25)$$

Then Equations (14.6) and (14.15c), respectively, yield

$$\mu^2 = K_\delta + \lambda^2 \qquad (14.26a)$$

and

$$M(\infty) = C_s V_b \frac{2(1 - m)}{K_v} \qquad (14.26b)$$

Introducing Equations (14.25) and (14.26) into Equation (14.15b), the solution for this case is obtained as

$$\frac{M(\tau)}{C_s V_b} = \frac{2(1 - m)}{K_v}$$

$$+ \frac{4(1 - m)\lambda^{*2}e^{-(K_\delta - \lambda^{*2})\tau}}{\left(K_\delta - \lambda^{*2}\right)^2 - 2m\left(K_\delta - \lambda^{*2}\right)K_v - \lambda^{*2}K_v^2 - 2\lambda^{*2}K_v}$$

$$- \sum_{i=1}^{\infty} \frac{4(1 - m)\lambda_i^2 e^{-(K_\delta + \lambda_i^2)\tau}}{\left(K_\delta + \lambda_i^2\right)^2 - 2m\left(K_\delta + \lambda_i^2\right)K_v + \lambda_i^2 K_v^2 + 2\lambda_i^2 K_v}$$

$$(14.27a)$$

where the λ_i are the roots of [see Equation (14.5)]

$$\left(K_\delta + \lambda^2\right) J_{-m}(\lambda) + K_v \lambda J_{1-m}(\lambda) = 0 \qquad (14.27b)$$

and λ^* is the root of

$$\left(K_\delta - \lambda^{*2}\right) I_{-m}(\lambda^*) - K_v \lambda^* I_{1-m}(\lambda^*) = 0 \qquad (14.27c)$$

which is obtained from Equation (14.27b) by setting $\lambda = i\lambda^*$ and noting that

$$J_v(i\lambda^*) = i^v I_v(\lambda^*).$$

Case 2 We assume

$$K_\eta = 0, \qquad K_v = 0, \qquad K_\delta \neq 0 \qquad (14.28)$$

The resulting solution for this case is less obvious because $K_v = 0$. Some manipulations are needed as now described.

By setting $\lambda = i\lambda^*$, $\mu = \mu^*$, and $J_v(i\lambda^*) = i^v I_v(\lambda^*)$, Equation (14.5) yields

$$\frac{K_v}{\mu^{*2}} = \frac{1}{\lambda^*} \frac{I_{-m}(\lambda^*)}{I_{1-m}(\lambda^*)} \qquad (14.29a)$$

By setting $K_\eta = 0$, $\lambda = i\lambda^*$, and $\mu = \mu^*$ in Equation (14.6) and in the resulting expression by letting $\mu^* = 0$, we obtain

$$\lambda^* = \sqrt{K_\delta} \qquad (14.29b)$$

From Equations (14.29a) and (14.29b) we write

$$\frac{K_v}{\mu^{*2}} = \frac{1}{\sqrt{K_\delta}} \frac{I_{-m}\left(\sqrt{K_\delta}\right)}{I_{1-m}\left(\sqrt{K_\delta}\right)} \qquad (14.29c)$$

This expression will be utilized in the simplification of Equation (14.15b).

First we set in Equation (14.15b), $K_\eta = 0$ and let $e^{-\mu_i^{*2}\tau} \cong 1 - \mu^{*2}\tau$ for $\mu^{*2} \to 0$, to obtain

$$\frac{M(\tau)}{C_s V_b} = \left[\frac{2(1-m)}{K_v} + \frac{4(1-m)\lambda^{*2}}{\mu^{*4} - 2m\mu^{*2}K_v - \lambda^{*2}K_v^2 - 2\lambda^{*2}K_v}\right]$$

$$- \left[\frac{4(1-m)\lambda^{*2}\mu^{*2}\tau}{\mu^{*4} - 2m\mu^{*2}K_v - \lambda^{*2}K_v^2 - 2\lambda^{*2}K_v}\right]$$

$$- \left[\sum_{i=1}^\infty \frac{4(1-m)\lambda_i^2 e^{-\mu_i^2\tau}}{\mu_i^4 - 2m\mu_i^2 K_v + \lambda_i^2 K_v^2 + 2\lambda_i^2 K_v}\right]$$

$$\equiv A_1 - A_2 - A_3 \qquad (14.30)$$

The three groups appearing in Equation (14.30) are simplified by letting $K_v \to 0$, $\mu^* \to 0$, but noting that K_v/μ^{*2} remains finite according to Equation (14.29c); we find

$$A_1 = (1 - m)\left\{1 + \frac{2m}{\sqrt{K_\delta}} \frac{I_{1-m}\left(\sqrt{K_\delta}\right)}{I_{-m}\left(\sqrt{K_\delta}\right)} - \left[\frac{I_{1-m}\left(\sqrt{K_\delta}\right)}{I_{-m}\left(\sqrt{K_\delta}\right)}\right]^2\right\} \qquad (14.31a)$$

$$A_2 = -2(1 - m)\sqrt{K_\delta}\, \frac{I_{1-m}\left(\sqrt{K_\delta}\right)}{I_{-m}\left(\sqrt{K_\delta}\right)} \tau \qquad (14.31b)$$

$$A_3 = 4(1 - m)\sum_{i=1}^{\infty} \frac{\lambda_i^2}{\mu_i^4} e^{-\mu_i^2 \tau} \qquad (14.31c)$$

When Equations (14.31) are introduced into Equation (14.30), the solution for $K_\eta = 0$, $K_v = 0$, $K_\delta \neq 0$ becomes

$$\frac{M(\tau)}{C_s V_b} = (1 - m)\left[1 + \frac{2m}{\sqrt{K_\delta}} \frac{I_{1-m}\left(\sqrt{K_\delta}\right)}{I_{-m}\left(\sqrt{K_\delta}\right)} - \frac{I_{1-m}^2\left(\sqrt{K_\delta}\right)}{I_{-m}^2\left(\sqrt{K_\delta}\right)}\right]$$

$$+ 2(1 - m)\sqrt{K_\delta}\, \frac{I_{1-m}\left(\sqrt{K_\delta}\right)}{I_{-m}\left(\sqrt{K_\delta}\right)} \tau$$

$$- 4(1 - m)\sum_{i=1}^{\infty} \frac{\lambda_i^2}{\mu_i^4} e^{-\mu_i^2 \tau} \qquad (14.32a)$$

where

$$\mu_i^2 = K_\delta + \lambda_i^2 \qquad (14.32b)$$

by Equation (14.6) for $K_\eta = 0$, and the λ_i are the roots of the transcendental equation [see Equation (14.5) for $K_v = 0$]

$$J_{-m}(\lambda) = 0 \qquad (14.32c)$$

14.4 NUMERICAL RESULTS

To illustrate the physical significance of the solutions presented in this chapter, we now present some calculated results. The computation of the solution for the case of irreversible reaction is relatively easy, but for the case of reversible reaction the computations are rather involved. We focus our attention on the calculation of the quantity $M(\tau)/M(\infty)$ for a slab geometry (i.e., $m = \frac{1}{2}$) for the case of infinite amount of solute (i.e., $K_v = 0$). The desired solution is obtained from Equation (3.17) by setting $m = \frac{1}{2}$; this solution coincides with that given by Crank [1, Equation 14.97]. Figures 14.2, 14.3, and 14.4 show a plot of $M(\tau)/M(\infty)$ for some representative cases.

Figure 14.2 shows the graphs of $M(\tau)/M(\infty)$ as a function of $[\tau/(R + 1)]^{1/2}$, for $R = 10$ and several different values of the parameter K_η. The extreme case $\eta \to \infty$ represents the infinitely rapid reaction rate, and the other extreme case $\eta \to 0$ represents simple diffusion with no reaction.

Figure 14.3 shows a plot of $M(\tau)/M(\infty)$ against $(\tau K_\eta)^{1/2}$ for $R = 10$ and several different values of K_η. The curve marked $D = \infty$ represents a purely reaction controlled situation. The purpose of plotting against $(\tau K_\eta)^{1/2}$ is to illustrate how the curves approach the $D = \infty$ curve as K_η tends to zero when D becomes very large. The $D = \infty$ curve has a discontinuity in the gradient; for small values of K_η, as in the case of $K_\eta = 0.01$, this discontinuity appears as a *shoulder*. As K_η is increased, the shoulder appears as a point of inflection and at still higher values of K_η the inflection becomes less noticeable.

To illustrate the effects of the parameter R on the shape of the curves, we present in Figure 14.4 a plot similar to that Figure 14.3, but for $R = 1$. Clearly, decreasing the value of R increases both the heights of the discontinuity and of the shoulder. If the curves were plotted for a value of $R = 100$, the shoulder would not be detected for any value of K_η.

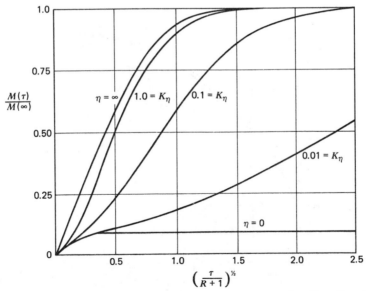

Figure 14.2 Slab geometry for the case of infinite amount of solute (i.e., $K_v = 0$) and $R = 10$ (from Reference 1).

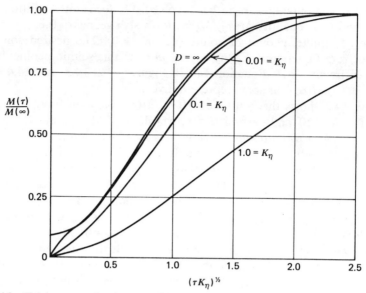

Figure 14.3 Slab geometry for the case of infinite amount of solute (i.e., $K_v = 0$) and $R = 10$ (from Reference 1).

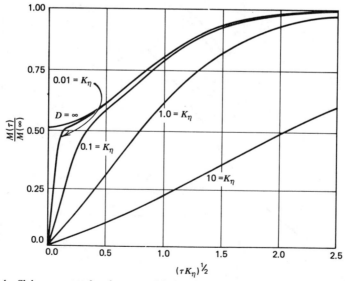

Figure 14.4 Slab geometry for the case of infinite amount of solute (i.e., $K_v = 0$) and $R = 1$ (from Reference 1).

516

REFERENCES

1. J. Crank, *The Mathematics of Diffusion*, 2nd ed., Clarendon Press, London, 1975.

2. S. M. Katz, E. T. Kubu, and J. H. Wakelin, The Chemical Attack on Polymeric Materials as Modified by Diffusion, *Text Res. J.*, **20**, 754–760 (1950).

3. C. E. Reese and H. Eyring, Mechanical Properties and Structure of Hair, *Text Res. J.*, **20**, 743–753 (1950).

4. A. Petropoulos and P. P. Roussis, *Organic Solid State Chemistry* (G. Adler, Ed.), Chapter 19, Gordon and Breach, London, 1969.

5. P. Nicolson and F. J. W. Roughton, A Theoretical Study of Influence of Diffusion and Chemical Reaction Velocity on the Rate of Exchange of Carbon Monoxide and Oxygen between the Red Blood Corpuscle and the Surrounding Fluid, *Proc. R. Soc. (B)*, **138**, 241–264 (1951).

6. A. H. Wilson, A Diffusion Problem in which the Amount of Diffusion Substance is Finite-I, *Phil. Mag.*, **39**, 45–58 (1948).

7. J. Crank, A Diffusion Problem in which the Amount of Diffusing Substance is Finite-IV, *Phil. Mag.*, **39**, 362–376 (1948).

8. M. D. Mikhailov and M. N. Özişik, A General Solution of Solute Diffusion with Reversible Reaction, *Int. J. Heat and Mass Transf.*, **24**, 81–87 (1981).

INDEX

A CATALOG OF SELECTED

DOVER BOOKS
IN SCIENCE AND MATHEMATICS

A CATALOG OF SELECTED
DOVER BOOKS
IN SCIENCE AND MATHEMATICS

QUALITATIVE THEORY OF DIFFERENTIAL EQUATIONS, V.V. Nemytskii and V.V. Stepanov. Classic graduate-level text by two prominent Soviet mathematicians covers classical differential equations as well as topological dynamics and ergodic theory. Bibliographies. 523pp. 5⅜ × 8½.   65954-2 Pa. $10.95

MATRICES AND LINEAR ALGEBRA, Hans Schneider and George Phillip Barker. Basic textbook covers theory of matrices and its applications to systems of linear equations and related topics such as determinants, eigenvalues and differential equations. Numerous exercises. 432pp. 5⅜ × 8½.   66014-1 Pa. $9.95

QUANTUM THEORY, David Bohm. This advanced undergraduate-level text presents the quantum theory in terms of qualitative and imaginative concepts, followed by specific applications worked out in mathematical detail. Preface. Index. 655pp. 5⅜ × 8½.   65969-0 Pa. $13.95

ATOMIC PHYSICS (8th edition), Max Born. Nobel laureate's lucid treatment of kinetic theory of gases, elementary particles, nuclear atom, wave-corpuscles, atomic structure and spectral lines, much more. Over 40 appendices, bibliography. 495pp. 5⅜ × 8½.   65984-4 Pa. $12.95

ELECTRONIC STRUCTURE AND THE PROPERTIES OF SOLIDS: The Physics of the Chemical Bond, Walter A. Harrison. Innovative text offers basic understanding of the electronic structure of covalent and ionic solids, simple metals, transition metals and their compounds. Problems. 1980 edition. 582pp. 6⅛ × 9¼.   66021-4 Pa. $15.95

BOUNDARY VALUE PROBLEMS OF HEAT CONDUCTION, M. Necati Özisik. Systematic, comprehensive treatment of modern mathematical methods of solving problems in heat conduction and diffusion. Numerous examples and problems. Selected references. Appendices. 505pp. 5⅜ × 8½.   65990-9 Pa. $11.95

A SHORT HISTORY OF CHEMISTRY (3rd edition), J.R. Partington. Classic exposition explores origins of chemistry, alchemy, early medical chemistry, nature of atmosphere, theory of valency, laws and structure of atomic theory, much more. 428pp. 5⅜ × 8½. (Available in U.S. only)   65977-1 Pa. $10.95

A HISTORY OF ASTRONOMY, A. Pannekoek. Well-balanced, carefully reasoned study covers such topics as Ptolemaic theory, work of Copernicus, Kepler, Newton, Eddington's work on stars, much more. Illustrated. References. 521pp. 5⅜ × 8½.   65994-1 Pa. $12.95

PRINCIPLES OF METEOROLOGICAL ANALYSIS, Walter J. Saucier. Highly respected, abundantly illustrated classic reviews atmospheric variables, hydrostatics, static stability, various analyses (scalar, cross-section, isobaric, isentropic, more). For intermediate meteorology students. 454pp. 6⅛ × 9¼. 65979-8 Pa. $14.95

CATALOG OF DOVER BOOKS

HANDBOOK OF MATHEMATICAL FUNCTIONS WITH FORMULAS, GRAPHS, AND MATHEMATICAL TABLES, edited by Milton Abramowitz and Irene A. Stegun. Vast compendium: 29 sets of tables, some to as high as 20 places. 1,046pp. 8 × 10½. 61272-4 Pa. $24.95

MATHEMATICAL METHODS IN PHYSICS AND ENGINEERING, John W. Dettman. Algebraically based approach to vectors, mapping, diffraction, other topics in applied math. Also generalized functions, analytic function theory, more. Exercises. 448pp. 5⅜ × 8¼. 65649-7 Pa. $9.95

A SURVEY OF NUMERICAL MATHEMATICS, David M. Young and Robert Todd Gregory. Broad self-contained coverage of computer-oriented numerical algorithms for solving various types of mathematical problems in linear algebra, ordinary and partial, differential equations, much more. Exercises. Total of 1,248pp. 5⅜ × 8½. Two volumes. Vol. I 65691-8 Pa. $14.95
Vol. II 65692-6 Pa. $14.95

TENSOR ANALYSIS FOR PHYSICISTS, J.A. Schouten. Concise exposition of the mathematical basis of tensor analysis, integrated with well-chosen physical examples of the theory. Exercises. Index. Bibliography. 289pp. 5⅜ × 8½. 65582-2 Pa. $8.95

INTRODUCTION TO NUMERICAL ANALYSIS (2nd Edition), F.B. Hildebrand. Classic, fundamental treatment covers computation, approximation, interpolation, numerical differentiation and integration, other topics. 150 new problems. 669pp. 5⅜ × 8½. 65363-3 Pa. $14.95

INVESTIGATIONS ON THE THEORY OF THE BROWNIAN MOVEMENT, Albert Einstein. Five papers (1905–8) investigating dynamics of Brownian motion and evolving elementary theory. Notes by R. Fürth. 122pp. 5⅜ × 8½. 60304-0 Pa. $4.95

CATASTROPHE THEORY FOR SCIENTISTS AND ENGINEERS, Robert Gilmore. Advanced-level treatment describes mathematics of theory grounded in the work of Poincaré, R. Thom, other mathematicians. Also important applications to problems in mathematics, physics, chemistry and engineering. 1981 edition. References. 28 tables. 397 black-and-white illustrations. xvii + 666pp. 6⅛ × 9¼. 67539-4 Pa. $16.95

AN INTRODUCTION TO STATISTICAL THERMODYNAMICS, Terrell L. Hill. Excellent basic text offers wide-ranging coverage of quantum statistical mechanics, systems of interacting molecules, quantum statistics, more. 523pp. 5⅜ × 8½. 65242-4 Pa. $12.95

ELEMENTARY DIFFERENTIAL EQUATIONS, William Ted Martin and Eric Reissner. Exceptionally clear, comprehensive introduction at undergraduate level. Nature and origin of differential equations, differential equations of first, second and higher orders. Picard's Theorem, much more. Problems with solutions. 331pp. 5⅜ × 8½. 65024-3 Pa. $8.95

STATISTICAL PHYSICS, Gregory H. Wannier. Classic text combines thermodynamics, statistical mechanics and kinetic theory in one unified presentation of thermal physics. Problems with solutions. Bibliography. 532pp. 5⅜ × 8½. 65401-X Pa. $11.95

GEOMETRY OF COMPLEX NUMBERS, Hans Schwerdtfeger. Illuminating, widely praised book on analytic geometry of circles, the Moebius transformation, and two-dimensional non-Euclidean geometries. 200pp. 5⅜ × 8¼.
63830-8 Pa. $8.95

MECHANICS, J.P. Den Hartog. A classic introductory text or refresher. Hundreds of applications and design problems illuminate fundamentals of trusses, loaded beams and cables, etc. 334 answered problems. 462pp. 5⅜ × 8½.   60754-2 Pa. $9.95

TOPOLOGY, John G. Hocking and Gail S. Young. Superb one-year course in classical topology. Topological spaces and functions, point-set topology, much more. Examples and problems. Bibliography. Index. 384pp. 5⅜ × 8¼.
65676-4 Pa. $9.95

STRENGTH OF MATERIALS, J.P. Den Hartog. Full, clear treatment of basic material (tension, torsion, bending, etc.) plus advanced material on engineering methods, applications. 350 answered problems. 323pp. 5⅜ × 8½.   60755-0 Pa. $8.95

ELEMENTARY CONCEPTS OF TOPOLOGY, Paul Alexandroff. Elegant, intuitive approach to topology from set-theoretic topology to Betti groups; how concepts of topology are useful in math and physics. 25 figures. 57pp. 5⅜ × 8½.
60747-X Pa. $3.50

ADVANCED STRENGTH OF MATERIALS, J.P. Den Hartog. Superbly written advanced text covers torsion, rotating disks, membrane stresses in shells, much more. Many problems and answers. 388pp. 5⅜ × 8½.   65407-9 Pa. $9.95

COMPUTABILITY AND UNSOLVABILITY, Martin Davis. Classic graduate-level introduction to theory of computability, usually referred to as theory of recurrent functions. New preface and appendix. 288pp. 5⅜ × 8½. 61471-9 Pa. $7.95

GENERAL CHEMISTRY, Linus Pauling. Revised 3rd edition of classic first-year text by Nobel laureate. Atomic and molecular structure, quantum mechanics, statistical mechanics, thermodynamics correlated with descriptive chemistry. Problems. 992pp. 5⅜ × 8½.   65622-5 Pa. $19.95

AN INTRODUCTION TO MATRICES, SETS AND GROUPS FOR SCIENCE STUDENTS, G. Stephenson. Concise, readable text introduces sets, groups, and most importantly, matrices to undergraduate students of physics, chemistry, and engineering. Problems. 164pp. 5⅜ × 8½.   65077-4 Pa. $6.95

THE HISTORICAL BACKGROUND OF CHEMISTRY, Henry M. Leicester. Evolution of ideas, not individual biography. Concentrates on formulation of a coherent set of chemical laws. 260pp. 5⅜ × 8½.   61053-5 Pa. $6.95

THE PHILOSOPHY OF MATHEMATICS: An Introductory Essay, Stephan Körner. Surveys the views of Plato, Aristotle, Leibniz & Kant concerning propositions and theories of applied and pure mathematics. Introduction. Two appendices. Index. 198pp. 5⅜ × 8½.   25048-2 Pa. $7.95

THE DEVELOPMENT OF MODERN CHEMISTRY, Aaron J. Ihde. Authoritative history of chemistry from ancient Greek theory to 20th-century innovation. Covers major chemists and their discoveries. 209 illustrations. 14 tables. Bibliographies. Indices. Appendices. 851pp. 5⅜ × 8½.   64235-6 Pa. $18.95

CHALLENGING MATHEMATICAL PROBLEMS WITH ELEMENTARY SOLUTIONS, A.M. Yaglom and I.M. Yaglom. Over 170 challenging problems on probability theory, combinatorial analysis, points and lines, topology, convex polygons, many other topics. Solutions. Total of 445pp. 5⅜ × 8½. Two-vol. set.
Vol. I 65536-9 Pa. $7.95
Vol. II 65537-7 Pa. $6.95

FIFTY CHALLENGING PROBLEMS IN PROBABILITY WITH SOLUTIONS, Frederick Mosteller. Remarkable puzzlers, graded in difficulty, illustrate elementary and advanced aspects of probability. Detailed solutions. 88pp. 5⅜ × 8½.
65355-2 Pa. $4.95

EXPERIMENTS IN TOPOLOGY, Stephen Barr. Classic, lively explanation of one of the byways of mathematics. Klein bottles, Moebius strips, projective planes, map coloring, problem of the Koenigsberg bridges, much more, described with clarity and wit. 43 figures. 210pp. 5⅜ × 8½.
25933-1 Pa. $5.95

RELATIVITY IN ILLUSTRATIONS, Jacob T. Schwartz. Clear nontechnical treatment makes relativity more accessible than ever before. Over 60 drawings illustrate concepts more clearly than text alone. Only high school geometry needed. Bibliography. 128pp. 6⅛ × 9¼.
25965-X Pa. $6.95

AN INTRODUCTION TO ORDINARY DIFFERENTIAL EQUATIONS, Earl A. Coddington. A thorough and systematic first course in elementary differential equations for undergraduates in mathematics and science, with many exercises and problems (with answers). Index. 304pp. 5⅜ × 8½.
65942-9 Pa. $8.95

FOURIER SERIES AND ORTHOGONAL FUNCTIONS, Harry F. Davis. An incisive text combining theory and practical example to introduce Fourier series, orthogonal functions and applications of the Fourier method to boundary-value problems. 570 exercises. Answers and notes. 416pp. 5⅜ × 8½.
65973-9 Pa. $9.95

THE THEORY OF BRANCHING PROCESSES, Theodore E. Harris. First systematic, comprehensive treatment of branching (i.e. multiplicative) processes and their applications. Galton-Watson model, Markov branching processes, electron-photon cascade, many other topics. Rigorous proofs. Bibliography. 240pp. 5⅜ × 8½.
65952-6 Pa. $6.95

AN INTRODUCTION TO ALGEBRAIC STRUCTURES, Joseph Landin. Superb self-contained text covers "abstract algebra": sets and numbers, theory of groups, theory of rings, much more. Numerous well-chosen examples, exercises. 247pp. 5⅜ × 8½.
65940-2 Pa. $7.95

Prices subject to change without notice.
Available at your book dealer or write for free Mathematics and Science Catalog to Dept. GI, Dover Publications, Inc., 31 East 2nd St., Mineola, N.Y. 11501. Dover publishes more than 175 books each year on science, elementary and advanced mathematics, biology, music, art, literature, history, social sciences and other areas.